W0256271

Berichte des German Chapter of the ACM

Band 4: Schneider, Portable Software
Tagung I/1980 am 18. 1. 1980 in Erlangen. 176 Seiten, DM 36,–/ÖS 263,–/SFr. 32,–

Band 6: Hauer/Seeger, Hardware für Software
Tagung III/1980 am 10./11. 10.1980 in Konstanz. 303 Seiten, DM 54,–/ÖS 394,–/SFr. 49,–

Band 7: Nehmer, Implementierungssprachen für nichtsequentielle Programmsysteme
Tagung I/1981 am 20. 2. 1981 in Kaiserslautern. 208 Seiten, DM 38,–/ÖS 277,–/SFr. 34,–

Band 8: Schlier, Personal Computing
Tagung II/1981 am 12. 10. 1981 in Freiburg i. Br. 195 Seiten, DM 40,–/ÖS 292,–/SFr. 36,–

Band 10: Kulisch/Ullrich, Wissenschaftliches Rechnen und Programmiersprachen
Fachseminar am 2./3. 4. 1982 in Karlsruhe. 231 Seiten, DM 52,–/ÖS 380,–/SFr. 47,–

Band 11: Langmaack/Schlender/Schmidt, Implementierung PASCAL-artiger Programmiersprachen
Tagung II/1982 am 12. 7. 1982 in Kiel. 221 Seiten, DM 46,–/ÖS 336,–/SFr. 41,–

Band 13: Schneider, Proceedings of the International Computing Symposium 1983 on Application Systems Development
March 22 – 24, 1983 Nürnberg. 528 Seiten, DM 90,–/ÖS 657,–/SFr. 81,–

Band 18: Morgenbrod/Sammer, Programmierumgebungen und Compiler
Tagung I/1984 vom 2. bis 4. 4. 1984 in München. 293 Seiten, DM 56,–/ÖS 409,–/SFr. 50,–

Band 20: Gorny/Kilian, Computer-Software und Sachmängelhaftung
Workshop am 29./30. 11. 1984 in Hannover. 208 Seiten, DM 48,–/ÖS 350,–/SFr. 43,–

Band 21: Kölsch/Schmidt/Schweiggert, Wirtschaftsgut Software
Tagung I/1985 am 26./27. 3. 1985 in Ulm. 318 Seiten, DM 58,–/ÖS 423,–/SFr. 52,–

Band 22: Molzberger/Zemanek, Software-Entwicklung: Kreativer Prozeß oder formales Problem?
Seminar am 20. 3. 1985 in Neubiberg. 176 Seiten, DM 42,–/ÖS 307,–/SFr. 38,–

Band 23: Klopcic/Marty/Rothauser, Arbeitsplatzrechner in der Unternehmung
Tagung II/1985 am 12./13. 9. 1985 in Zürich. 355 Seiten, DM 66,–/ÖS 482,–/SFr. 59,–

Band 24: Bullinger, Software-Ergonomie '85 Mensch-Computer-Interaktion
Tagung III/1985 am 24./25. 9. 1985 in Stuttgart. 482 Seiten, DM 78,–/ÖS 569,– SFr. 70,–

Band 25: Wedekind/Kratzer, Büroautomation '85
Tagung IV/1985 vom 2. bis 4. 10. 1985 in Erlangen. 280 Seiten, DM 56,–/ÖS 409,–/SFr. 50,–

Band 27: Remmele/Sommer, Arbeitsplätze morgen
Tagung II/1986 vom 11. bis 14. 3. 1986 in Marburg. 431 Seiten, DM 78,–/ÖS 569,–/SFr. 70,–

Band 28: Balzert/Heyer/Lutze, Expertensysteme '87
Tagung I/1987 am 7./8. 4. 1987 in Nürnberg. 493 Seiten, DM 82,–/ÖS 599,–/SFr. 74,–

Band 29: Schönpflug/Wittstock, Software-Ergonomie '87
Tagung II/1987 vom 27. bis 29. 4. 1987 in Berlin. 512 Seiten, DM 82,–/ÖS 599,–/SFr. 74,–

Band 30: Winkler, Proceedings of the International Workshop on Software Version and Configuration Control
January 27 – 29, 1988 Grassau. 478 Seiten, DM 78,–/ÖS 569,–/SFr. 70,–

Band 31: Dillmann/Swiderski, WIMPEL '88
Tagung I/1988 vom 28. bis 30. 6. 1988 in München. 479 Seiten, DM 78,–/ÖS 569,–/SFr. 70,–

Band 32: Maaß/Oberquelle, Software-Ergonomie '89
Fachtagung vom 29. bis 31. 3. 1989 in Hamburg. 509 Seiten, DM 88,–/ÖS 642,–/SFr. 79,–

Band 33: Ackermann/Ulich, Software-Ergonomie '91
Fachtagung vom 18. bis 20. 3. 1991 in Zürich. 383 Seiten, DM 84,–/ÖS 613,–/SFr. 76,–

Fortsetzung 3. Umschlagseite

Berichte des German Chapter
of the ACM   53

U. Arend / E. Eberleh /
K. Pitschke (Hrsg.)
Software-Ergonomie '99

Berichte des German Chapter of the ACM

Im Auftrag des German Chapter
of the ACM herausgegeben durch den Vorstand

Chairman
Wolf-Rüdiger Gawron, BMW AG, Petuelring 130, 80788 München

Vice Chairman
Prof. Dr. Günter Riedewald, Universität Rostock, Einsteinstraße 21,
18052 Rostock

Treasurer
Eckhard Jaus, CSC Ploenzke Consulting GmbH, Zettachring 2,
70567 Stuttgart

Secretary
Roland Dürre, Interface Connection GmbH, Leipziger Straße 16,
82008 Unterhaching

Band 53

Die Reihe dient der schnellen und weiten Verbreitung neuer, für die Praxis relevanter Entwicklungen in der Informatik. Hierbei sollen alle Gebiete der Informatik sowie ihre Anwendungen angemessen berücksichtigt werden.

Bevorzugt werden in dieser Reihe die Tagungsberichte der vom German Chapter allein oder gemeinsam mit anderen Gesellschaften veranstalteten Tagungen veröffentlicht. Darüber hinaus sollen wichtige Forschungs- und Übersichtsberichte in dieser Reihe aufgenommen werden.

Aktualität und Qualität sind entscheidend für die Veröffentlichung. Die Herausgeber nehmen Manuskripte in deutscher und englischer Sprache entgegen.

Software-Ergonomie '99

Design von Informationswelten

Herausgegeben von

Dr. Udo Arend
Dr. Edmund Eberleh
Dipl.-Inform. Knut Pitschke
SAP AG, Walldorf

B. G. Teubner Stuttgart · Leipzig 1999

Die Deutsche Bibliothek – CIP-Einheitsaufnahme

Software-Ergonomie '99 :
[gemeinsame Fachtagung des German Chapter of the ACM,
der Gesellschaft für Informatik (GI) und der SAP AG, Walldorf vom 8. bis 11. März 1999
in Walldorf/Baden] / Hrsg. von Udo Arend ... – Stuttgart ; Leipzig : Teubner, 1999
   Beitr. teilw. dt., teilw. engl.
   1999. Design von Informationswelten. – 1999
   (Berichte des German Chapter of the ACM ; Bd. 53)
ISBN 978-3-519-02694-5   ISBN 978-3-322-99786-9 (eBook)
DOI 10.1007/978-3-322-99786-9

Programmkomitee

Udo Arend, SAP AG, Walldorf
Astrid Beck, GUI Design, Stuttgart
Udo Bleimann, FH Darmstadt
Edmund Eberleh, SAP AG, Walldorf (Vorsitz)
Ulrich Eisenecker, FH Heidelberg
Peter Fach, RWG GmbH, Stuttgart
Jürgen Friedrich, Universität Bremen
Peter Gorny, Universität Oldenburg
Andreas Heinicke, FH Dortmund, hi-soft Hamburg
Michael Herczeg, Universität Lübeck
Christoph Kasten, DLR e.V., Bonn
Reinhard Keil-Slawik, Universität-GH Paderborn
Ulrich Klotz, IG Metall, Frankfurt
Hans-Guenter Lindner, humanIT GmbH, Sankt Augustin
Rüdiger Liskowsky, TU Dresden
Susanne Maaß, Universität Bremen
Rolf Molich, Dialog Design, Stenløse DK
Horst Oberquelle, Universität Hamburg
Reinhard Oppermann, GMD GmbH, Sankt Augustin
Ina Pitschke, Deutsche Telekom AG, Darmstadt
Knut Pitschke, SAP AG, Walldorf
Matthias Rauterberg, IPO Eindhoven
Harald Reiterer, Universität Konstanz
Karlheinz Sonntag, Universität Heidelberg
Friedrich Strauß, sd&m GmbH & Co. KG, München
Norbert Streitz, GMD GmbH, Darmstadt
Gerd Szwillus, Universität-GH Paderborn
Christoph Thomas, GMD GmbH, Sankt Augustin
Axel Viereck, Hochschule Bremen
Jens Wandmacher, TU Darmstadt
Wolfgang Wünschmann, TU Dresden
Volker Wulf, Universität Bonn
Jürgen Ziegler, FhG-IAO, Stuttgart

Organisationskomitee

Katrin Floegel, SAP AG, Walldorf
Leif Jensen-Pistorius, SAP AG, Walldorf
Khatoun Shahrbabaki, SAP AG, Walldorf
Jan Zwickel, SAP AG, Walldorf
Mette Lindvall Grüner, SAP AG, Walldorf
Markus Heilig, IST, Potsdam

Vorwort der Herausgeber

Erstmals in ihrer Geschichte findet die Fachtagung "Software-Ergonomie" bei einem Softwarehersteller statt. Hiermit wird die Praxisorientierung der software-ergonomischen Arbeiten in Deutschland konsequent fortgesetzt und betont, nachdem bereits die vorausgegangene Tagung 1997 in Dresden den Schwerpunkt auf die Gebrauchstauglichkeit von Softwareprodukten und die Methodik bei deren benutzerzentrierten Entwicklung legte. Die Software-Ergonomie ist damit auch sichtbar "in der Industrie angekommen". Hohe ergonomische Qualität von Software-Produkten stellt im internationalen Wettbewerb mittlerweile eine unabdingbare Voraussetzung für den Markterfolg dar und wird von den Benutzern als selbstverständlich erwartet. Diese praktische Bedeutsamkeit ergonomischer Themen spiegelt sich in der Tagung in vielen Vorträgen, Tutorien, Workshops, Postern und der Podiumsdiskussion wider.

Zugleich verändern sich viele der technischen und organisatorischen Rahmenbedingungen, die bisher wesentliche Bezugspunkte software-ergonomischer Arbeiten des letzten Jahrzehnts waren, welches von dem Bild des "desktop-orientierten" persönlichen Computers geprägt war.

Durch das Internet und WWW ist die Verbindung und Integration weltweit verteilter Information und der direkte Zugriff darauf von beliebigen Orten Realität geworden. Bislang voneinander getrennte Anwendungen und Daten wachsen zu einem globalen Netz ohne deutlich erkennbare Grenzen zusammen. Neue Arbeits- und Organisationsformen wie Telearbeit oder weltweit verteilte virtuelle Teams werden möglich. Grenzen verschwinden auch im Bereich der Mensch-Computer Benutzungsschnittstelle selbst: Agenten, Avatare, virtuelle Räume und neue Interaktionsgeräte verwischen den Unterschied zwischen der Welt "vor" und "hinter" der Monitorglasscheibe. Auf verschiedenste Geräte des täglichen Gebrauchs verteilte Mikroprozessoren erzeugen durch ihre möglich werdende Integration neue Formen und Probleme der Informations- und Gerätesteuerung. Dieser Wechsel bedingt eine ständige Weiterbildung von Individuen und Organisationen in der Nutzung der Möglichkeiten, eröffnet aber zugleich neue Möglichkeiten des Lernens, Arbeitens und Kommunizierens. Inwieweit diese neuen Möglichkeiten Chancen oder Risiken darstellen, hängt zu einem beträchtlichen Teil von der menschengerechten Gestaltung der technischen Optionen ab.

Die Benutzer zu Beginn des nächsten Jahrhunderts werden die Informations- und Kommunikationstechnologie stärker als bisher im ursprünglichen Sinne der Wortbedeutung nutzen. Sie werden nicht mehr bloße "Bediener" von Arbeitssystemen mit Schaltern, Knöpfen und Tasten sein. Arbeit am und mit dem Computer wird in vielen Arbeitsbereichen und für viele Benutzer immer weniger nur das Abarbeiten vorgegebener Teilaufgaben an einem festen Arbeitsplatz sein, sondern mehr und mehr ein Sich-Bewegen in künstlichen, anregend gestalteten Informationswelten, an verschiedenen realen und virtuellen Orten. Wichtigstes Arbeitsziel wird immer mehr das Suchen, Analysieren, Auswerten, Bewerten und Anwenden von Information sein, ermöglicht und unterstützt durch vielfältige Techniken der Mensch-Computer Interaktion. Die Benutzungsschnittstelle in Form einer explizit wahrnehmbaren Schicht zwischen Benutzer und Anwendung tritt in den Hintergrund zugunsten eines unmittelbareren Arbeitens mit Informationen. Es stellen sich Fragen nach der angemessenen Strukturierung und Darstellung dieser Informationen, nach der Navigation in der Informationsmenge und nach dem Nutzungspotential der damit verbundenen neuen

Möglichkeiten. Die Tagung soll sich unter dem Motto "Design von Informationswelten" mit diesen neuen Herausforderungen an die Software-Ergonomie beschäftigen.

Im Zentrum dieser 9. Tagung "Software-Ergonomie" steht daher die Diskussion über neue Möglichkeiten der Informationstechnologien (z.B. Virtual Reality, mobile Computing, Software-Agenten), über die benutzerorientierte Gestaltung von Informationswelten (z. B. WWW-Design, Verhältnis Design zu Funktionalität), sowie über die Nutzung und Anwendung von Informationssystemen (z.B. Erwartungen verschiedener Nutzergruppen, Veränderung von Kommunikation, Arbeit und Organisation).

Diese Themen werden in Form von Vorträgen, Tutorials, Workshops und Postern diskutiert: 4 Tutorien vermitteln den Stand der Technik sowohl bezüglich aktueller technischer Optionen und deren ergonomischer Gestaltung (WWW-Design, Virtuelle Welten) als auch bezüglich der Methodik und Qualitätssicherung benutzerzentrierter Softwareentwicklung. 5 Workshops zur Informationsstrukturierung, -visualisierung und zu ergonomischen Verfahren kommen dem immer wieder geäußerten Wunsch nach verstärktem direktem Austausch unter Fachkollegen entgegen. Eine Podiumsdiskussion behandelt das kontrovers diskutierte aktuelle Thema der Zertifizierung von Benutzungsoberflächen im Rahmen der EU-Bildschirmrichtlinie. Drei eingeladene Vorträge internationaler Fachexperten (Alan Cooper http://www.cooper.com, Dr. JoAnn Hackos http://www.comtech-serv.com, Prof. Bernhard E. Bürdek, http://www.rz.uni-frankfurt.de/hfg/asta/bernhard_buerdek/) beleuchten das Tagungsthema von verschiedenen Sichten und zeigen zukunftsweisende Entwicklung im Bereich des Designs multimedialer Informationswelten auf. An zwei Terminen außerhalb des Tagungsprogramms können sich interessierte Teilnehmer über die SAP AG und die ergonomischen Aktivitäten innerhalb der SAP-Produktentwicklung informieren.

Um die Vielfalt der Sicht- und Herangehensweisen zum Tagungsthema angemessen zu berücksichtigen und den Austausch über die weitere Entwicklung der Software-Ergonomie in Deutschland zu fördern, wurde das Programmkomitee deutlich umfangreicher als bisher – mit Vertretern aus Forschung und Praxis - zusammengesetzt. Die lebhaften Diskussionen innerhalb des Programmkomitees zur Tagungskonzeption und zu den einzelnen Beiträgen bestätigten dieses Konzept. Es bleibt zu wünschen, daß diese integrative Tagunskonzeption und Bündelung der verschiedenen software-ergonomischen Ansätze auch zukünftig bestehen bleibt, verstärkt und ausgebaut wird. Den Mitgliedern des Programmkomitees sei an dieser Stelle herzlich für ihr kritisches und konstruktives Engagement bei der Konzeption und Zusammenstellung des wissenschaftlichen Programms gedankt! Eine weitere Besonderheit dieser Tagung besteht in der ausdrücklichen Ermutigung zu Praxisbeiträgen im Call for Papers und durch ein Auswahlverfahren, welches die jeweiligen besonderen Bedingungen von Forschungs- und Praxisbeiträgen angemessen berücksichtigt. Auch hier konnte nur ein Anfang gemacht werden, von dem wir hoffen, daß er bei folgenden Tagungen fortgesetzt wird und die wachsende praktische Bedeutung der Software-Ergonomie in der Industrie sich dadurch auch in den zukünftigen Tagungsprogrammen angemessen widerspiegelt.

Wir möchten den Autoren dieses Tagungsbandes für ihren Beitrag zum wissenschaftlichen Austausch danken und für ihre große Kooperation bei der organisatorischen Abwicklung der Beitragseinreichung und -erstellung. Der Gesellschaft für Informatik und dem German Chapter of the ACM möchten wir für ihre finanzielle und organisatorische Unterstützung bei der Tagungsvorbereitung danken, namentlich besonders Frau Gabriele Wischnewski und Herrn Eckhard A. Jaus für die Unterstützung bei Fragen der Finanzierung und Organisation.

Der SAP AG sei für die Bereitschaft gedankt, diese Tagung in ihren Räumen durchführen zu können und sie bei der Vorbereitung durch personelle und finanzielle Ressourcen zu unterstützen. Hier gebührt besonders Frau Christiane Kretz, Frau Anne Goertz und Frau Mette Lindvall Gruener Dank für ihre Unterstützung bei vielen Fragen der Tagungsorganisation. Unseren Kolleginnen und Kollegen Frau Khatoun Shahrbabaki, Frau Katrin Floegel und Herrn Leif Jensen-Pistorius vom Organisationskomitee sei ganz herzlich für ihre beständige Mitarbeit von Anfang an gedankt. Ohne sie hätten wir es nicht geschafft! Herrn Jan Zwickel und Herrn Markus Heilig gebührt Dank für ihre Mitwirkung bei der Erstellung der Web-Seiten, des Buches und der Organisation der Tagungsdurchführung. Und schließlich möchten wir dem Teubner Verlag für die problemlose Zusammenarbeit bei der Erstellung des Tagungsbandes danken.

Wir wünschen und hoffen, daß durch diese Tagung und den vorliegenden Tagungsband die software-ergonomische Fachdiskussion in Deutschland am Ende des 20. Jahrhunderts einen neuen Impuls und eine gute Ausrichtung auf anstehende Themen des neuen Jahrtausends erhält.

Walldorf, Dezember 1998

Edmund Eberleh

Udo Arend

Knut Pitschke

Inhalt

Vorwort der Herausgeber 7

Eingeladene Vorträge

Cooper, A.:

The Inmates are Running the Asylum 17

Hackos, J. T.:

An Application of the Principles of Minimalism to the Design
of Human-Computer Interfaces 19

Bürdek, B. E.:

Beyond Interfaces 27

Angenommene Vorträge

Baumüller, H.; Beringer, J.; Eberleh, E.:

Empirical Evaluation of three Hierarchical Representation
Techniques 29

Brennecke, A.; Keil-Slawik, R.; Roth, W.:

Designorientierung und Designpraxis –Entwicklung und Einsatz
von konstruktiven Gestaltungskriterien 43

Burmester, M.; Komischke, T.:

Nutzeranforderungen als Grundlage für die Entwicklung innovativer
User Interfaces in der industriellen Prozeßführung 53

Fach, P. W.:

Der software-ergonomische Stellenwert von Rahmenwerken Erfahr-
ungen mit einem bankfachlichen Anwendungssystem 63

Frings, S.; Weisbecker, A.; Lahr, W.; Reinsch, V.:

Rollenkonzept in der Software-Entwicklung 73

Gronski, A.; Haller, H.:

GUIfizierung umfangreicher Hostanwendungen mit Java 85

Hamborg, K.-C.; Gediga, G.; Döhl, M.; Janssen, P.; Ollermann, F.:

Softwareevaluation in Gruppen oder Einzelevaluation:
Sehen zwei Augen mehr als vier? 97

Hassenzahl, M.; Prümper, J.:

„Benutzererwartung eingebaut": Gestaltungsempfehlungen für
Suchfunktionen auf der Basis einer empirischen Benutzerbefragung 111

Herrmann, T.:

Flexible präsentation von Prozeßmodellen 123

Holmer, T.; Streitz, N.:

Neue Möglichkeiten der Analyse der Mensch-Computer-Interaktion
zur Evaluation von computerunterstützten Gruppensitzungen 137

Hornecker, E.; Schäfer, K.:

Gegenständliche Modellierung virtueller Informationswelten 149

Hubwieser, P.; Schlichter, J.:

HyperCons: Eine Informationswelt für Bildungseinrichtungen 161

Iqbal, A.; Oppermann, R.; Patel, A.; Kinshuk:

A Classification of Evaluation Methods for Intelligent
Tutoring Systems 169

Kahler, H.; Stiemerling, O.; Wulf, V.; Hoepfner, J.-G.:

Gemeinsame Anpassung von Einzelplatzanwendungen 183

Kleinen, B.:

Ein Werkzeug zur Moderationsunterstützung 195

Mletzko, F.:

Designleitlinien und Bewertungskriterien für die Strukturgeometrie technischer Informationswelten 205

Nissler, J.; Thoma, V.:

Gestaltung von Software Agenten aus der Sicht des Benutzers 215

Nissler, J.; Thoma, V.; Manz, S.:

Online-Dienste für Alle:
Usability und Design einer PC/ITV Oberfläche 227

Oppermann, R.; Kinshuk; Kashihara, A.; Rashev, R.; Simm, H.:

Supporting Learner in Exploratory Learning Process in an
Interactive Simulation based Learning System 241

Oppermann, R.; Specht, M.:

Adaptive Information for Nomadic Activities. A process
oriented approach 255

Pfister, H.-R.; Wessner, M.; Beck-Wilson, J.:

Soziale und kognitive Orientierung in einer computergestützten
kooperativen Lernumgebung 265

Pitt, I.; Preim, B.; Schlechtweg, S.:

An Evaluation of Interaction Techniques for the Exploration of
3D-Illustrations 275

Rolles, R.; Schmidt, Y.:

Kontinuierliche Prozeßverbesserung durch Integration von
Workflow und Intranet 287

Schneider, J.; Strothotte, T:

Virtuelle taktile Karten – digitale Stadtpläne für Blinde 299

Totter, A.; Stary, C.; Riesenecker-Caba, T.:

Methodengesicherte Validierung von EU-CON II 309

Wandke, H.; Dubrowsky, A.; Hüttner, J.:

Anforderungsanalyse zur Einführung eines Unterstützungssystems bei Software-Entwicklern 321

Wittenberg, C.:

Aufgabenorientierte Visualisierung eines komplexen verfahrenstechnischen Prozesses unter Verwendung dreidimensionaler
Computergrafik 335

Ziegler, S.:

Der automatisierte Wissenserwerb im Kontext der Kommunikation.
Ein Vorschlag zur Entwicklung von Expertensystemen durch die
Experten 345

Workshops

Bruns, F. W.; Robben, B.; Rügge, I.:

Gestaltung von virtuellen und beGreifbaren Mensch-Computer-Schnittstellen 353

Gronski, A.; Königer, P.:

Nutzerunterstützung durch Klassifikationssysteme im Unternehmen 359

Gediga, G.; Hamborg, K.-C.; Hampe-Neteler, W.:

Was leisten ergonomische und arbeits-psychologische Verfahren
im Software-Entwicklungszyklus? 367

Lindner, H.-G.; Thomas, C. G.:

Werkzeuge zur Visualisierung von Unternehmensinformationen 371

Bleimann U.; Reiterer H.; Mann T. M.; Mußler G.:

Visualisierung von entscheidungsrelevanten Daten 379

Poster/Video

Bannert, M.:

Design und Evaluation von EDV-Lernumgebungen 385

Brennecke, A.; Holl F.-L.; Keil-Slawik R.; Meier J.; Selke H.:

Hyper-Skript – Entwicklung und Nutzung von verteilten
Multimediaskripten 387

Degen, H.:

Ein Wirkmodell für eine marktorientierte Konzeption und
Produktion von Softwareprodukten 389

Gu, F.; Mitritz, A.; Bendig, T.; Henseler, N.:

ProVision3D - Eine Virtual Reality Workbench zur Visualisierung
geschäftsprozeßbezogener Organisationsinformationen im virtuellen
Raum(Videobeitrag) 391

Schumacher, P.; Kahler, H.:

Anwendbarkeit software-ergonomischer Kriterien auf Einsatz und
Gestaltung von Agententechnologie 393

Weinberg, I.; Leutner, D.:

Double-Fading-Support als Trainingskonzept für Softwaresysteme
vergleichbarer Komplexität mit unterschiedlich realisierter
Benutzerschnittstelle? 395

The Inmates are Running the Asylum

Alan Cooper

Cooper Interaction Design, Palo Alto

Summary

Are you an inmate? What if we switched the metaphor to, "the building contractors are telling the architects where to put the windows?" Strike a little closer to home? The mechanics of building an application often end up taking precedence over the aims of the project, to the point where nobody--user, designer, programmer or manager--ends up getting what they want. Alan Cooper, the "Father of Visual Basic" and author of About Face: The Essentials of User Interface Design, sees a cure for this craziness in a new way to design interaction. Applications created using his Goal-Directed ® Design process provide users with power and pleasure. His keynote presentation will give you some much-needed perspective on design issues and show a case study of how a leading vendor has adopted Cooper's approach. He'll also offer tips on how you can make the business case for effective design to your managers. Alan is a motivating, thought-provoking, and original speaker. Come prepared to toss out some old ideas, hear some new ones and perhaps even escape from the asylum.

Author´s address

Alan Cooper
Cooper Interaction Design
2345 Yale Street
Palo Alto, CA 94306
USA
Email: alan@cooper.com

An Application of the Principles of Minimalism to the Design of Human-Computer Interfaces

JoAnn T. Hackos, Ph.D.

Comtech Services, Inc., Denver CO

Minimalism in information design, specifically as applied to user tutorials and manuals, was introduced in the early 1980s through the work of Dr. John M. Carroll, then a cognitive psychologist at the IBM Watson Research Center. Since that time, theorists and practitioners have further elucidated the principles of minimalism and have attempted to apply it to a variety of situations in which people attempt to learn how to use a software application. Most recently, a new exposition of minimalist principles and practices was published by MIT Press. This work, *Minimalism Beyond the Nurnberg Funnel*, represents the work of leading theorists and practitioners in the field.

I have long been in the habit of describing the user interface as an element of information design and thus amenable to the basic design tenets underlying information design. And, I have frequently characterized the user interface as an expression of minimalist design because the interface combines the traditional design elements of text, graphics, and layout in a two-dimensional space, with the occasional addition of movement, in as compact a form as is practical. More recently, I have been intrigued by the possibility of applying minimalist principles to interface design in a more systematic manner than has been heretofore discussed by its proponents.

In this presentation, I will introduce the four basic principles of minimalism, as well as some of the research done to support the principles, and explain how they might contribute to our understanding of interface design. To the extent which time and space permit, I hope to present some examples of how a minimalist interface might be similar but also differ from more traditional interfaces.

1 Background to the minimalist debate

John Carroll's most developed presentation of the minimalist concept appeared in 1990 in *The Nurnberg Funnel* (MIT Press). In this work, he fully described what he and his team at IBM had learned from observing people trying to learn to use IBM's DisplayWriter word processor and how to learn the computer language, SmallTalk. Learners of DisplayWriter and SmallTalk were most successful when they were provided with brief instructional material that encouraged them to act rather than read. They learned best when the tutorials emphasized the goals they really wanted to achieve, rather than tasks defined by the computer software and the interface.

In one case, for example, the researchers created a cue card that explained to the secretaries how to "Type something" rather than how to "Create a document." "Creating a document" was the heading used on one of the on-screen menus (before Windows) and in the actual DisplayWriter documentation. They had discovered that the secretaries did not want to create documents and were unable to relate their goal, which was to type something, to the name of the task in the interface. This mismatch between goal and execution of the task was a significant impediment to their learning.

In the design of the SmallTalk tutorial, the researchers discovered that programmers were best able to learn the computer language when they were engaged in the actual tasks of writing and debugging code rather than reading about the concepts underlying the language. Users who are encouraged to act, rather than read about acting, are more successful in their learning. Users who are encouraged to perform tasks that are directly related to their goals in using the software are more able to formulate realistic plans, the set of steps they envision will take them through a series of actions to their goals.

2   Minimalist Principles

In 1996, John Carroll and Hans van der Meij summarized the principles of minimalist design of documentation and training in an article in *Technical Communication*, the journal of the Society for Technical Communication. This article, reprinted in *Minimalism Beyond the Nurnberg Funnel*, most clearly states the four minimalist principles and how they are represented in the design of documentation and training.

The four basic principles of minimalism are

- Principle 1: Choose an action-oriented approach
- Principle 2: Anchor the tool in the task domain
- Principles 3: Support error recognition and recovery
- Principle 4: Support reading to do, study, and locate

My purpose here is to examine to what extent these principles, used to define goals for the development of documentation and training, apply to the design of the human-computer interface.

3   Principle 1: Choose an action-oriented approach

The first principle of minimalism points to the basic concept that underlies interface design. An interactive interface implies action; users are provided with opportunities for action. In a Windows environment, users select commands from a task bar, make selections in dialog boxes, type information into data-entry fields, and so on. Interfaces ordinarily provide many opportunities for action, even if that action is simply looking at an on-screen report. The intent of basic interface design is to allow users to do something.

However, there is more to the first principle than simply action. An action-oriented approach implies that the user is able to accomplish something. Users are most satisfied, it appears, when they have an immediate opportunity to act and when they believe that the actions they take will lead them toward their goals. Simply clicking on buttons or typing something into a field does not imply that the actions are anything more than random attempts to make progress. In many flawed interfaces, the actions that users take often have little purpose. They are simply attempts to see what, if anything, will result. We observe users trying to guess, unsuccessfully in many cases, which actions will lead to the results they want. The interfaces provide them with little or no guidance about where to start.

I recently reviewed an interface design in which the users are presented with a blank screen and a task bar when they enter the program. Their first cognitive task is to examine the task bar and attempt to guess what item they need to select to accomplish something. Users are immediately confused, unable to take the immediate and purposeful actions that they want to take, because the interface fails to lead them in a clearly defined direction. A more successful design, following minimalist principles, would present an immediate opportunity for action in the context of the users' goals. For example, if the users need to select a patient in order to view that patient's record, the first action available immediately in the interface should be patient selection. Providing an immediate opportunity to act means making the desired activities immediately apparent and reducing the number of choices that the users must make before taking action.

In addition to an immediate opportunity to act, the first principle of minimalism also points to the need for users to explore as a way of learning. Most graphic user interfaces provide more than enough opportunities for exploration, at least in principle. The standard Windows approach is to make many actions available through task bars, pull-down menus, and dialog boxes. Users can browse through the menus, select and review dialog boxes, and make different choices to see how those choices might affect what is happening on the screen. The problem with this approach is that the exploration is undirected. We have found that when users explore at random in an interface, they frequently become increasingly confused, unable to differentiate among explorations that lead to their goals and explorations that lead away from their goals. An interface that guides exploration in productive ways leads the users by using information on the screen that makes clear the direction to take.

Finally, the first principle implies that we must, in creating successful designs, respect the integrity of the users' actions. When we present message boxes that are unrelated to the users' actions we violate this aspect of the first principle. When we present error messages that fail to inform users how to correct the problem, we also fail to respect what the users are attempting to do. Tips that appear on the screen, unasked and unwanted, distract the users from their tasks. Tips that attempt to track user actions have the promise of maintaining the users' original intent, but, thus far, seem mostly clumsy and overbearing.

One of the researchers at Xerox PARC remarked to me a few years ago that they were having a difficult time making sense of users' seemingly random keystrokes to approach a task. They could not anticipate the users' intent simply from monitoring key strokes. The researcher suggested that a much more sophisticated approach to interpreting the users' logic was needed before such tracking might be successful.

For an interface to comply with the first minimalist principle, to take an action-oriented approach means the following:

- make critical actions immediately apparent when the users enter the interface, especially as users enter the interface for the first time
- provide opportunities for exploration through the interface, but guide those explorations using appropriate text and graphics so that the connection between actions and goals are easily apparent

- respect the integrity of the users' actions by eliminating annoying and often gratuitous invasions in the middle of tasks the user is attempting to pursue

4   Principle 2: Anchor the tool in the task domain

The second principle is most significant for the design of user-centered systems. Too often, interfaces reflect the underlying database structure rather than the users' goals. To anchor the tool in the task domain means to design the interface from the users' perspective rather than designing an interface that is functionally correct but disembodied. We have all seen interfaces that fail to resonate with the users' goals. These interfaces have functional names that make no sense to the users. They require sequences of actions that are hidden from view, requiring the users to figure out which task structure will lead them to their goals.

For example, we recently reviewed a web site designed by a major American university. We took the point of view of a user who wanted to find out if the university offered courses toward a degree in technical communication. We believed this goal to be typical of many individuals coming to the web site from outside the university. Unfortunately, the web site was not designed from the users' perspective, but rather used the organizational structure of the departments and schools in the university to organize the information. Unless the user already knew which school and department offered the courses she wanted to fit, she could not get the information she needed. In fact, she needed to know the solution to her problem in order to find the solution--a rather circular approach unlikely to produce success.

In order to anchor the tool in the task domain we must first be well-informed about the goals that the users want to achieve. Do they want to use the task bar or send a fax to a colleague? Do they want to complete the dialog box or find out how many days of vacation they have left? In many interfaces, the users' real goals are obscured, if they were ever understood in the first place. We need to use all of the user-centered design tools to ensure that we truly understand the world from the users' perspective if we hope to build successful interfaces.

The second principle of minimalism exhorts us to build on the users' prior skills, knowledge, and experience in the design of product. As a consequence, we must know the users' goals and relate them to tasks already known in the users' domain to achieve those goals. For example, the software program, Quicken, uses a metaphorical construct on screen of a check register, similar in appearance to the check register provided by banks with packets of checks. By using this visual metaphor of the task, the designers are able to build upon the users' knowledge of performing the task with the physical artifact. The affordances of the artifact are transferred to the interface. The users are able to bring their prior experience of recording checks and balancing their checkbooks to performing the task in the new environment.

Metaphoric constructs alone, however, are not sufficient to reinforce the users' prior skills, knowledge, and experience. In many more complex system designs, the overall relationship among tasks must reflect the users' experience. For example, consider a system to support the task of call tracking, in which the overall flow of information from the primary screen, through secondary screens, and into dialog boxes, is made to resemble the task sequence performed by the users' in the physical environment. If we are able to take advantage of known sequences, we spare the users the need to unlearn their previous behaviors and learn new ones. Their cognitive tasks are simplified, the learning curve is reduced, and they are able to accomplish their goals.

In building an interface that respects the second principle of minimalism, we must place information in the interface that supports task performance and enables the users to link the tasks to their goals. To do so often requires more than cryptic terminology. For some reason, we appear to have chosen to preserve the limited language present in the original teletype interfaces. Menu systems often consist of single words when phrases might be more meaningful. Dialog boxes reduce text to field labels and one-word descriptions of radio buttons or check boxes, often making it difficult for the users to know what actions to take.

Given the space we have to work with and the legibility of a graphic user interface, we appear to be constrained by unnecessary brevity, afraid to use words to help the users know what is going on. If we are to support the users' task performance, we should consider adding meaningful language to the screen, using text in phrases, sentences, or even paragraphs. In the past few years, interface designers have introduced the concept of the wizard or coach to assist users in performing certain tasks. Sometimes those tasks have been designed so that novice users are better supported, in effect, supplying an alternative to the more cryptic standard interface. In other cases, we have designed wizards to become the primary interface for complex tasks that users have difficulty performing successful with the cryptic assistance given by the "ordinary" interface. Performance support of these types have proven inordinately successful in improving task performance and enabling users to learn how best to perform tasks and achieve their goals.

It is interesting to note that, in most instances, the performance assistants (wizards and coaches) carry more text explanations than do the "ordinary" interfaces. If we have learned that more text can result in better performance, in the context of a well-designed task environment, why not adopt such devices most of the time instead of making them special cases?

In general, on-screen text should provide users with clues about the task structure in the software. Terminology used on the screen should reflect users' goals and lead them through the structure of the tasks. The text on-screen, as well as the layout of the information (reflecting reading order or other structures), should help users make the connection between what they want to do and how they accomplish the tasks within the software application.

For an interface to comply with the second minimalist principle, to anchor the tool in the task domain means the following:

- ensure that the users' goals are well understood by the interface designers so that the interface makes a clear connection between the users' goals and the tasks required to achieve the goals
- build on the users' prior skills, knowledge, and experience by creating a metaphoric structure for the interface that enables the users to connect known abilities to new requirements for interaction
- place information into the interface that supports task performance. Consider using more rather than less text in the interface, along with layout, color, and white space to assist the users in learning how to perform the tasks within the tool.

5   Principle 3: Support error recognition and recovery

The third principle of minimalism emphasizes the users' need to detect the problems when they occur, diagnose what has happened, and find effective ways to overcome those problems. At the heart of this principle is the need to prevent mistakes before they occur. Often, preventing mistakes requires careful use of default values for fields and choice buttons and error trapping. For example, I was reviewing an interface that required the users to type the hyphens in their U.S. Social Security number (the standard format is xxx-xx-xxxx). If the users failed to include the hyphen, a message appeared stating that the hyphens must be typed. In addition, all the numbers the user had already typed were deleted from the field and replaced with a blank space. Such an approach is insulting to the users and reflects sloppy programming by the developer. First, there is no need to require the hyphens at all. If they need to be added, they can be programmed to appear automatically as the users type. Second, if a particular sequence of characters is required in a field for some reason, and the users make an obvious mistake, the cursor should be returned to the position to correct the mistake following the appearance of the error message.

In many cases, as we have all seen, error messages are both insulting and belligerent toward users. The developers, in effect, accuse the users of being careless and stupid. The developers, in many instances, have chosen to blame the users rather than do the additional programming to prevent the error in the first place.

I am frequently annoyed when I enter a dialog box and have to establish the keyboard focus by clicking the cursor in the first field. Some additional programming is required to place the cursor in the first field, but someone considered that additional programming a waste of time. A decision to waste the time of hundreds, if not thousands of users, to save minutes of programming time should not occur.

In areas where errors are likely to occur or when error correction proves difficult, we need to design in cautions and warnings to alert the users about the potential problems and how to recover. If we provide error messages in such cases, the error messages must be careful worded to assist the users in detecting the error, diagnosing what happened and taking corrective action. Too often, error messages simply state the nature of the error, often in rather cryptic language, rather than providing assistance for correcting the error.

One way of assisting users to avoid errors is to provide many default values and explanations within the interface about what might be expected if new values are selected. Warning messages that point to the possibility of future problems are also useful additions, especially when they provide sufficient information to allow the users to make choices about their course of action.

For an interface to comply with the third minimalist principle, to support error recognition and recovery means the following:

- assist users in preventing the error in the first place by using error trapping techniques as thoroughly as possible
- assist users in making the right decisions by providing default values and complete information about the choices available in the interface itself

- create informative, helpful, and courteous error messages that enable users to detect, diagnose, and correct problems as soon as they occur.

6 Principle 4: Support reading to do, study, and locate

The fourth principle of minimalism, to support reading to do, study, and locate, applies, primarily, to the help system and documentation that you design to support the interface. We know that most of the time users' want to read to do (Redish 1988), seeking information that helps them complete the tasks in the software application and reach their goals. Help systems and documentation that emphasize reading to do, providing task-oriented instructional text, are most likely to assist users in navigating through the application interface.

Users who focus on reading to do profit from context-sensitive help and embedded help systems. Context-sensitive help is designed to anticipate the information the users may need at a particular point in the interface. If the help designer guesses correctly, the help that appears at the users' initial request will solve the users' problem and halt the search for information. Of course, guessing what help is desired is a difficult process and often requires that help developers observe user problems during usability testing of the interface. In some cases, we have recommended developing help text directly in response to information queries during testing. The users ask for assistance; the help developer responds with the minimal information necessary to move the users along; that text becomes the source of the help system.

Embedded help, a system more closely related to wizards and coaches, provides performance support in the context of the application. Since the help is always present, the users do not have to locate the information they need to complete a task. The information is provided to them immediately in the context of task performance. Development tools are now available to embed help within the application screens and to highlight the appropriate step in a process as the users proceed through the activity. (For more information, see the HelpXtender from Wextech Software.)

Well-designed help systems and user guides should be as brief as possible to encourage the users to engage with the software application and apply their previous knowledge and experience to the task. However, it is especially important that the text in the interface, the information on the screen, and be closely linked to the information in the instructional material. This approach to interface design recognizes that the help system and the user guides are integral parts of the interface and must be designed in association with the interface design.

For an interface to comply with the fourth minimalist principle, to support reading to do, study, and locate means the following:

- provide minimalist help and instructional text in user guides to support the users' task performance
- develop context-sensitive and embedded help systems so that the transition between actions in the interface and supporting information is as effortless for the users as possible
- carefully integrate the development of all aspects of the interface, including the development of help and additional instructional text so that all elements work together to support the users.

7   Summary

Minimalism provides us with insight into the design of user-centered interfaces. It de-emphasizes the construction of pleasing screen layouts, placing emphasis more properly on user goal-oriented design. Minimalism suggests, in some ways, that interfaces need to be designed to provide more information to users than they do today. That information might come in the form of metaphoric interface structures, assistance in moving users from their goals through the specific tasks that support the goals. It might come in the form of a richer textual content in screen design and certainly with a focus on selecting the right words and phrases to ensure that users make intelligent choices.

Minimalism also points to the need for close integration among the graphic interface, the help systems, and other forms of instruction. Close integration has been proven successful in the development of performance assistants such as wizards and coaches, which provide more text and more task direction than the typical menu-based design.

Finally, minimalism directs us to be aware of the real goals of our users and to explicitly support the achievement of those goals. Minimalism suggests that we anchor the tool in the task domain, rather than in the database. Such a perspective means that we must spend time understanding the users rather than designing screens.

8   References

Carroll, J. M. 1990. The Nurnberg Funnel: Designing Minimalist Instruction for Practical Computer Skill. Cambridge, MA: MIT Press.

Carroll, J. M., ed. 1998. Minimalism Beyond the Nurnberg Funnel. Cambridge, MA: MIT Press.

Lewis, C. and D. A. Norman. 1986. "Designing for error." In User Centered System Design: New Perspectives on Human-Computer Interaction, ed. D. A. Norman and S. W. Draper, pp. 411-432. Hillsdale, NJ: Erlbaum.

Redish, J. C. 1988. "Reading to learn to do." Technical Writing Teacher 15:223-233.

Author´s address

Dr. JoAnn Hackos, Comtech Services, Inc.
710 Kipling Street, Suite 400
Denver, Colorado 80215, USA
email: joann.hackos@comtech-serv.com

Dr. JoAnn Hackos is the president of Comtech Services, Inc., an information and interface design consulting firm headquartered in Denver, Colorado, USA. Dr. Hackos works with companies worldwide to assist them in designing more effective products and information, as well as to develop the organizational structures needed to promote user-centered design. She frequently speaks and teaches workshops on the design of user interfaces, documentation, and training on strategic planning, project management, and process maturity. She is the author of *User and Task Analysis for Interface Design* (1998, with Dr. Janice Redish), *Standards for Online Communication* (1997, with Dawn Stevens), and *Managing your Documentation Projects* (1994), all published by John Wiley & Sons, Inc. She is a Fellow of the Society for Technical Communication, where she served as international president from 1992 to 1993.

Beyond Interfaces

Prof. Bernhard E. Bürdek

Hochschule für Gestaltung, Offenbach/Main

Kurzfassung

Sir Charles Perry Snow (1905-1980) war Schrifsteller und Physiker, hoher Staatsbeamter und kritischer Beobachter gesellschaftlicher Entwicklungen. Im Mai 1959 hielt er in Cambridge eine Rede mit dem Titel: „Die Zwei Kulturen" - Literarische und naturwissenschaftliche Intelligenz, die für viele Jahre eine kontroverse Diskussion auslöste. Ich greife diesen in Deutschland erstmals 1967 publizierten Text wieder auf, da er mir im übertragenen Sinne auf eine Problematik anwendbar zu sein scheint, die auch Gegenstand dieses Symposiums ist:

Software-Ergonomie versus Software-Design.

Gerne wird ja - nicht nur in Ihren Kreisen hier - der Satz kolportiert:

- wenn es gut aussieht - aber nicht zu bedienen ist - dann ist es Design.

Diesen Satz kann man natürlich auch gerne umkehren:

- wenn es vermeintlicherweise gut zu bedienen ist - aber grausam aussieht - dann ist es Konstruktion, Informatik, Ergonomie usw.

In diesem Spannungsfeld bewegen sich auch heute noch die Debatten: einerseits die nach DIN und ISO entwickelten Programme, bei denen sodann die unvermeintlichen GUI-Style-Guides grüßen lassen, andererseits die überschäumenden Web-Seiten-Gestaltungen, CD-ROM´s, die Welt der Spiele usw.; eklatanter könnten die Brüche gar nicht sein. Eine Parallele drängt sich auf: in der Produktentwicklung des 20. Jahrhunderts dominierte über Jahrzehnte hinweg die Funktion vor der Form, da das Dogma lautete sogar: „form follows function". In den 80er Jahren geriet diese Bewegung an ein natürliches Ende: die sich immer mehr ähnlicher werdende Technik bedarf zunehmend der Differenzierung, um für den Kunden überhaupt noch unterscheidbar und damit erkennbar zu werden. Design ist also ein Instrument, den technologischen Fortschritt anschaulich zu machen, Innovationen zu visualisieren. Heute werden beispielsweise in der Automobilindustrie neue Fahrzeugkonzepte zuerst vom Design her bestimmt (Smart, New Beetle, AUDI TT etc.) und dann technologisch umgesetzt. In vielen anderen Produktbereichen ist es ähnlich, nicht mehr das Sein bestimmt unser Bewußtsein - sondern das Design.

Das Wort von den zwei Kulturen meint heute also einerseits die technisch-konstruktive und andererseits die ästhetisch-gestalterische Realität der Produktentwicklung. Seit den 80er Jahren wird auch propagiert, daß der Umgang mit Computern als eine neue Kulturtechnik begriffen werden müsse. Von „Kultur" jedoch kann man vielleicht gerade einmal bei dem einen oder anderen Hardware-Hersteller sprechen (wie z.B. Apple), in Bezug auf Software fallen mir dazu eigentlich nur ganz wenige Beispiele ein.

Woran dies im einzelnen liegt, was dagegen getan werden könnte und worauf es dabei überhaupt ankommt - darum soll es in diesem Vortrag gehen, der so gesehen vielleicht zu einem „Interface" werden könnte, zwischen den zwei Kulturen der Informatik und des Design. Dabei soll auch gezeigt werden, daß in vielen Fällen an den Menschen vorbei entwickelt wird, nicht die Bedürfnisse, Wünsche und Sehnsüchte der Benutzer stehen im Vordergrund, sondern die teilweise absurden Logiken der Entwickler. Die gesamte Digitalisierung ist möglicherweise einer jener gigantischen Irrtümer, der uns über Jahrzehnte hinweg bereits in Schach hält und auch noch rund zwanzig Jahre halten wird. Beyond Interface - das wäre eine Kultur von Hard- und Software nach dem Ende des mikroelektronischen Zeitalters.

Adresse des Autors

Prof. Dr. Bernhard E. Bürdek
P.O. Box 100 823
D-63008 Offenbach a.M
Fon: +49 (69) 80 059 (0)-74
Fax: +49 (69) 80 059 66
buerdek@em.uni-frankfurt.de
http://www.rz.uni-frankfurt.de/hfg

Vision & Gestalt
Design-Kommunikation
Human Interface Design
Darmstädter Str. 26a
D - 63179 Obertshausen
Fon: +49 (6104) 97 10 31
Fax: +49 (6104) 97 10 32

Empirical Evaluation of three Hierarchical Representation Techniques

Heike Baumüller, Jörg Beringer und Edmund Eberleh

SAP AG, Walldorf

Jens Wandmacher

TU Darmstadt

Zusammenfassung

Drei hierarchische Repräsentationen, der Scrolling Browser, der zweidimensionale Baum und der Hexagon Browser, wurden miteinander verglichen. In einer Einfachaufgabe sollten die Versuchspersonen Objekte in einer Hierarchie suchen. Dabei wurden die Reaktionszeit und die Zahl der Fehler erfaßt. In einer Doppelaufgabenbedingung wurden die Versuchspersonen häufig durch eine Zweitaufgabe unterbrochen. Wie erwartet war die Performanz beim Finden der Objekte am besten mit dem Scrolling Browser. Jedoch waren in der Doppelaufgabenbedingung die Reaktionszeit und die Fehlerrate für die Zweitaufgabe am niedrigsten mit dem Hexagon Browser. Dies wurde auf eine geringere mentale Beanspruchung durch den Hexagon Browser zurückgeführt, vermutlich aufgrund der hier geringeren Zahl der sichtbaren Begriffe und der verfügbaren räumlichen Information. Mit einer apriorischen GOMS-Analyse ohne Verwendung der beobachteten Daten konnten die durchschnittlichen Zeiten für den Zugriff auf die Objekte mit den drei Browsern einigermaßen genau vorhergesagt werden.

Abstract

This study compared three different hierarchical representations, the scrolling browser, the 2-dimensional flexible tree and the Hexagon Browser. In the single task condition, subjects were asked to retrieve items from a hierarchy. The retrieval time for each item and the number of errors were recorded. In the dual task condition, subjects were frequently interrupted by a secondary task. As expected, subjects using the scrolling browser performed better in terms of retrieval time and error rate. However, the users of the Hexagon Browser showed faster reaction times and less errors in the secondary task which might be due to the reduced mental effort required by the Hexagon Browser because of the limited number of shown objects and the location information displayed by the Hexagon Browser. With a 'blindfold' GOMS analysis of the retrieval task without use of any observed data the item retrieval times for the three browsers could be predicted fairly accurately.

Acknowledgement

Special thanks to Oliver Keim, Dr. Udo Arend and Peter Ebert for their help with the implementation, and to Nicola Spielmann and Stephen Corbett for their advice concerning the use of the usability lab. Finally, we are very grateful to all subjects who volunteered in the experiments.

1   Introduction

Due to growing computational resources increasingly large and complex information structures are stored on computers. Traditionally, simple lists and two-dimensional (2d) trees are frequently used to display the information. However, due to restricted screen space, these representations usually require scrolling or zooming because either a general overview or only a section of the hierarchy is displayed. Hence, with large information structures navigation becomes more difficult and the relationships between nodes are less obvious. Therefore, the need for finding new methods to effectively represent these structures arises.

In this study, the users' performance with three hierarchical representation techniques was compared. We used an information retrieval task with three different browsers to evaluate the advantages and disadvantages which arise from the different representation techniques. The three browsers were:

1. the *scrolling browser* which represents the hierarchical levels as vertically arranged scrollable lists,

2. the *2d flexible tree* (MS Windows Explorer) which utilises windows to represent hierarchical levels whose branches can be extended and pruned by the user, and

3. the *Hexagon Browser* (Ebert & Arend, 1997) which uses nested hexagons to represent the levels where each hexagon can contain other hexagons and associated level objects (for details see section 2.3).

A brief evaluation of the three browsers in terms of some usability aspects is given in Table 1.

	Scrolling Browser	2d Flexible Tree	Hexagon Browser
Metaphors	none, but most users familiar with lists	nodes connected by branches	concept of nested objects, but hexagons and 'rounded' rectangles not intuitive representations
Visual Complexity	low, only limited number of items, but increased due to the lack of differentiation between objects	variable depending on number of expanded branches	low, only few objects visible and the number of nested hexagons limited by the available display space
Backtracking	available for current path and to some objects in current top-level category, otherwise limited	available for current path, alternative paths variable depending on number of expanded branches	available for current path, difficult for alternative paths, only current path and limited number of objects visible
Preview	several lower levels, but only some objects	next lower level if branch expanded for the first time, otherwise dependent on paths opened before	only few objects on next lower level
Shortcuts	limited to currently visible objects and top-level nodes	variable depending on number of expanded branches	very limited, only few objects visible and current path
Context vs. Detail	limited context, top-level nodes always displayed	limited context, user gets lost when many branches expanded	only detail

Table 1 - Overview of the three browsers

Chimera & Shneiderman (1997) compared a stable, i.e., fully expanded, interface with the scrolling browser (which they referred to as 'multi-pane interface') and the 2d tree ('expand/contract interface'). They predicted the best performance for the scrolling browser since it provides a fisheye view of the information with the top-level nodes always and together visible. In addition, it does not require the undoing of operations to retain the global view. In contrast, the stable interface was expected to perform worst due to the large amount of scrolling required and because of the separation between the higher level nodes which resulted in a poor overview. The scrolling browser and 2d tree generally produced faster performance times than the stable interface, but the effectiveness of either of these two browsers depended on the type of task.

In the present study, the task was restricted to information retrieval from a static hierarchy. Each subject was given one of the three browsers. In the single task condition, subjects were asked to retrieve a set of items. It was tested which of the three representation techniques would allow the fastest retrieval and require least learning. The scrolling browser was expected to show the best performance because of its more comprehensive preview and a better overview of the structure. In the dual task condition, the mental effort required by the browsers was seized by means of a secondary task. No predictions were made for the dual task condition.

2 Behavioural Testing

2.1 Design

Three different representation methods were compared, the SAP Session Manager as a scrolling browser, the MS Windows 95/Windows NT Explorer as a 2d flexible tree, and the SAP Hexagon Browser (Ebert & Arend, 1997). A hierarchy of items in a supermarket was created. The scenario used in the study was an 'Online-Supermarket', and the subjects were asked to retrieve items and to place them in a separate 'Shopping Basket'. Each subject served in a single task condition and a dual task condition.

In the single task condition, all subjects received the same 'Shopping List' of 41 items to retrieve from the hierarchy represented by the browsers. The first item was given as a practice item, and its results were not recorded. The order of the remaining items was varied. To this end, the list was divided up into groups of four items and the groups were rotated, i.e., the first subject assigned to a browser received the original list, the second subject received the same list but with the last four items placed at the beginning of the list, the third subject received the same list as the second subject but again with the last four items placed at the beginning, etc. The retrieval times for each item and the number of errors, i.e., choosing an incorrect path, constituted the dependent variables.

In the dual task condition, subjects were given a second 'Shopping List' which was the same for all subjects. They were interrupted at random intervals averaging to 15 sec to perform a secondary task of searching and answering to a one digit number word among the distractor words 'SAP' and 'AG'. Subjects had to categorise the number as odd or even and compare its category to that of the previously shown number. The answer was given by pressing one of two buttons positioned on a separate pad and labelled with 'same' (gleich) and 'different' (ungleich). Thus, the secondary task induced a working memory load by requiring the subject to memorise the category of the number. Furthermore, it distracted the subject from viewing the browser by presenting the digit word at a random location among the distractor words. Each number word was accompanied by a tone which would only stop when the correct answer was given. The task lasted exactly five minutes. The dependent variables for the dual task condition were the number of items that could be retrieved in the given time, the reaction time to the secondary stimulus and the number of errors committed in the secondary task.

2.2 Subjects

Thirty employees of the Administration Department at SAP took part in this study. Before the experiment, subjects were asked about their familiarity with the three browsers They were allocated to the three browsers on the basis of their familiarity such that all subjects had equally little experience with the browser they were assigned to. However, it could not be avoided that subjects had at least some experience with the Microsoft Explorer, although subjects with the least experience were chosen.

2.3 Material

The hierarchy used in the experiment was supposed to represent an online-supermarket. It was made up of 633 nodes labelled with German expressions with 423 nodes representing categories of products and 210 nodes representing retrievable items. The top level consisted of five nodes, labelled 'Beverages', 'Fruit and Vegetables', 'Sweets', 'Farinaceous Products' and 'Animal Products'. The hierarchy had seven levels though not all items were located on the lowest level, but could be found somewhere on the 3rd to 7th level.

The three browsers were presented to the subjects on a PC in the SAP usability laboratory. Subjects' actions on the screen by using a mouse were recorded on video-tape. The time taken to retrieve each item was recorded by the computer. The experimenter recorded the errors by hand and noted down the different strategies used and items the subjects needed help with. For the secondary task, a laptop was placed to the right of the subjects together with a pad with two buttons attached to it. The Experimental Run Time System was used for the secondary task (Iwanek, 1994; Dutta, 1995).

2.3.1 Scrolling Browser

The SAP Session Manager (SM, Figure 2.1) was used as the scrolling browser. The nodes located on the first level are shown as buttons arranged vertically. This list of top-level nodes can be scrolled if the number of these nodes exceeds the available space. The lower levels are presented as vertical, scrollable lists arranged in parallel in descending order. When a node on some level is chosen, that node and the first nodes of the lower levels are highlighted. As long as the user stays within a category, i.e., the same top-level node, the path extending from a level remains highlighted even if the user moves between nodes on that level. The user can drag items, i.e., leaves of the tree but not nodes, into the lower part of the window in order to retrieve items of interest.

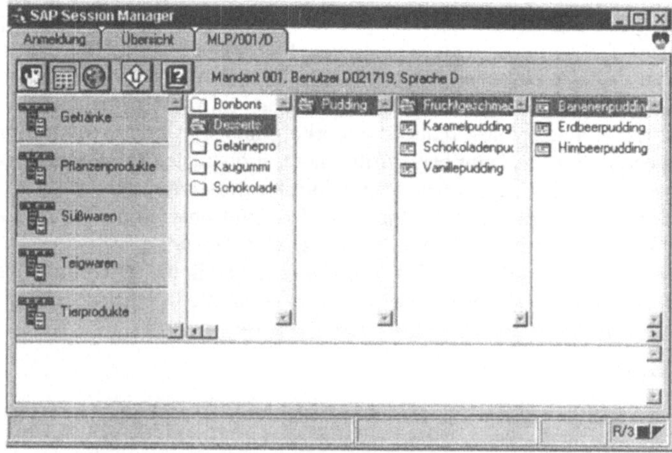

Figure 2.1 - A section of the supermarket-hierarchy in the SM

2.3.2 Explorer

The Microsoft Explorer is used in MS Windows 95/NT for the organisation of files (Figure 2.2). The Explorer window is split into two parts. On the left, a 2d flexible tree is shown with the files arranged in folders as the nodes of the tree. Each branch can be extended or contracted depending on which information the user is interested in. The tree grows vertically, i.e., the branches extend down and to the right. In order to expand a branch the user can either double-click on the appropriate node or click on the [+]-icon on the left of each node. The [+] then changes to a [−] to make apparent that the branch has been opened. To contract the branch the user can click on the [−]-icon or double-click on the node. When several nodes of the same branch on different levels are expanded, closing the branch by clicking on a higher-level or parent node will not close the branches that extend from child nodes. Thus, if the branch is expanded again all previously expanded nodes will be shown again.

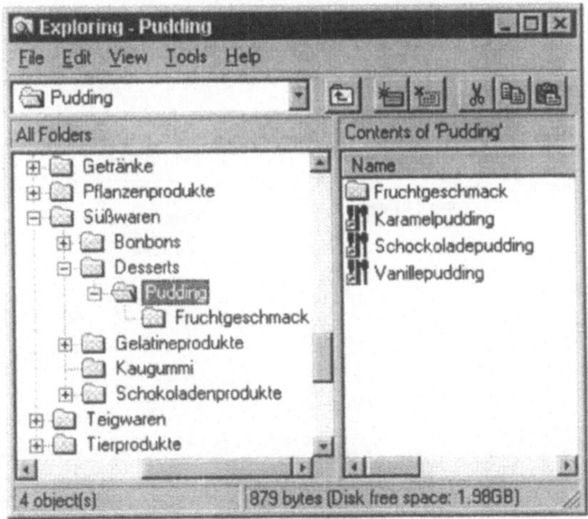

Figure 2.2 - A section of the supermarket-hierarchy in the Explorer

The right half of the window shows the items and nodes that are contained in the active (high-lighted) node in the left half. This part of the window can also be used to expand a branch by double-clicking on the appropriate node. The Explorer offers some additional features, such as a find-function or the adjustment of certain parameters, which were not used here. The user could drag an item into another window located on the desktop which was titled 'Shopping Basket'. Only items but no category nodes were allowed to be placed in this second window. Since 'moving' is the default action for dragging items between windows that are stored on the same hard disk, the second window was retrieved from a different source in order to ensure that the hierarchy remained unchanged.

2.3.3 Hexagon Browser

The Hexagon Browser (HB, Figure 2.3) was designed as a component of the Business Information Warehouse developed at SAP. The levels are shown as nested hexagons which are split into two parts. The right half contains the individual items of the current level in form of rounded rectangles. A maximum of four rounded rectangles can be displayed at any time. If there are more items, an alphabetical slider can be used to scroll through the list of items. Additionally, the user can click to the right or left of the items to scroll through them. In order to indicate whether more items are available the letters displayed on the slider will turn black if more items are present.

On the left, on or more other hexagons are shown which represented subcategories of the current level. These hexagons can again contain a hexagon or a rounded rectangle to indicate whether the levels they represent contain catgory or item nodes. Comparable to the right half, only three nodes can be displayed and an alphabetical slider or clicking within the left half can be used for scrolling. The user could retrieve a required item by dragging a rounded rectangle, yet not a hexagon, onto the icon in the top right-hand corner of the window.

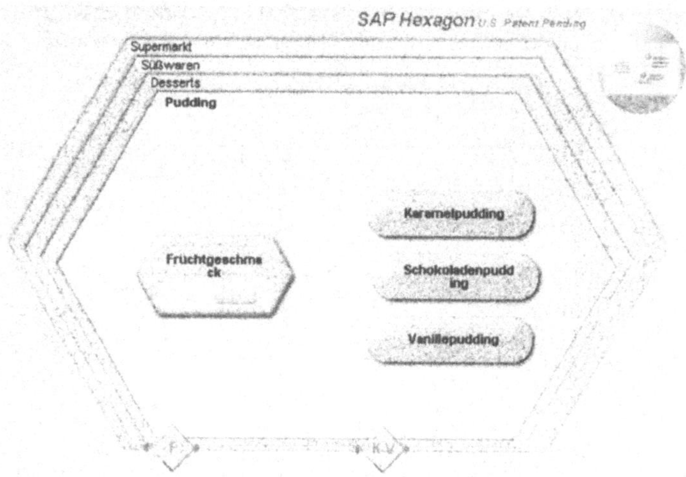

Figure 1.3 - A section of the supermarket-hierarchy in the Hexagon Browser

2.4   Procedure

The 30 subjects were split into three groups, one for each browser. Each session lasted between 30 and 45 minutes depending on the speed with which the subject could retrieve the items. At the beginning of the session each subject was asked to fill in the pre-test questionnaire. They were then given a brief description of the experiment. The usage of the browser assigned to the subject was explained by the experimenter on the computer. Subjects were asked to retrieve the items quickly but without feeling pressured, and with as little errors as possible. Additionally, they were advised to retrieve the items according to the order of the shopping list, to retrieve all the items and exactly those items on the list rather than similar ones. Even though they were given the possibility to ask questions, they were encouraged to find the items without help whenever possible.

After the single task condition was completed, the experimenter explained the dual task condition. A brief demonstration of the secondary task was given to familiarize the subject with this task. Subjects were asked to retrieve the items of the second list just like in the single task condition, but to perform the secondary task whenever a tone was emitted by the laptop. Subjects were then given five minutes to retrieve the items. After the completion of the dual task condition, subjects were asked to fill out the post-test questionnaire.

2.5 Results

2.5.1 Single Task Condition

Figures 3.1 and 3.2 give the overall mean retrieval times and error rates for each browser. The mean retrieval time was lowest for the SM with a difference of 8.5 sec to the Explorer and 17.6 sec to the HB. Both differences were significant by the Wilcoxon T-test ($p < 0.05$). The subjects using the SM showed the least average number of errors with a difference of 3 errors to the Explorer and of 6.1 errors to the HB. However, the differences between the error rates tested with the Mann-Whitney U-test did not reach significance. Likewise, the positive correlation between retrieval time and error rate did not reach significance.

Only five subjects using the HB completed the task while the remaining subjects retrieved 34 to 39 of the 40 recorded items.

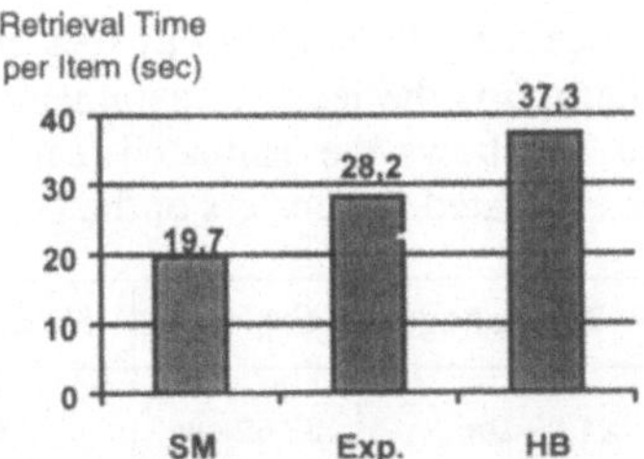

Figure 3.1 - Mean retrieval time for each browser

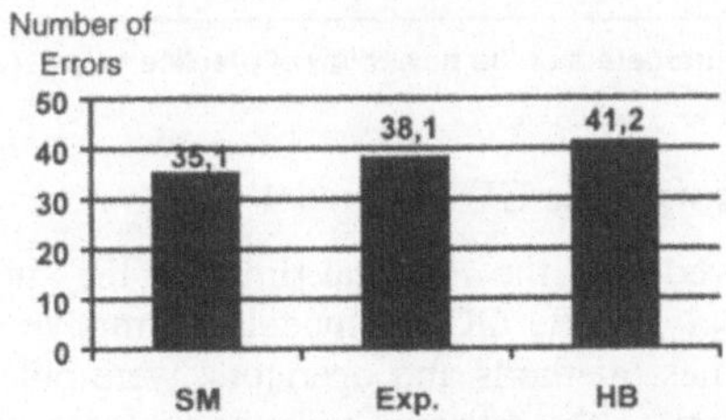

Figure 3.2 - Mean error rate

Besides, a learning curve was calculated. The retrieval times of the ten subjects per browser were averaged over the first eight items, the 9th to 16th item, the 17th to the 24th item, the 25th to the 32th item, and the 33th to the 40th item, respectively, yielding five data points of the learning curve for each browser. By these averages the differences due to the different levels of the hierarchy on which a particular item can be found, and due to difficulties in finding a particular item had about the same influence across the five data points. Figure 3.3 shows the learning curves. The data points of the SM lie below the values of the two other browsers, and the data points of the Explorer lie below those of the HB.

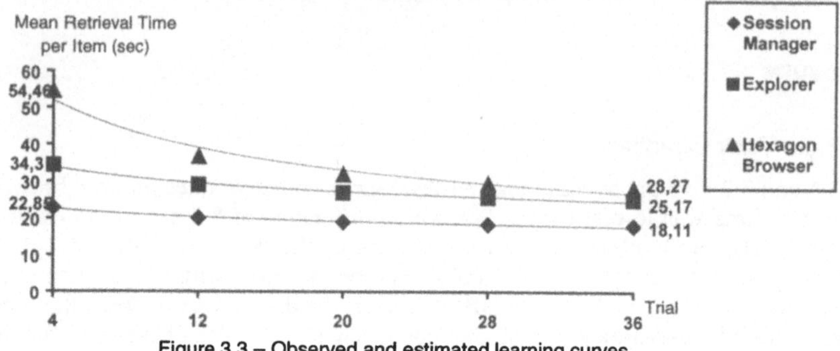

Figure 3.3 – Observed and estimated learning curves

The power law of practice (Newell & Rosenbloom, 1981) was fitted to the five data points of each browser. The power law of practice equation is $t = cx^b + d$, where t is the predicted retrieval time, x is the amount of practice, i.e., the number of executed item retrievals, c is the expected retrieval time for the first item, b is the learning parameter (learning rate), and d is the asymptotic retrieval time. Figure 3.3 shows the estimated learning curves of the power law of practice, and Table 2 gives the estimated parameters of the power law of practice.

	Session Manager	Explorer	Hexagon Browser
Expected Retrieval Time for the first item (c)	21.09 sec	37.52 sec	85.72 sec
Learning Parameter (b)	0.16	0.18	0.65
Asymptotic Time (d)	5.9 sec	5.6 sec	19.8 sec

Table 2 – Estimated parameters of the power law of practice $t = cx^b + d$

2.5.2 Predicting retrieval times with the GOMS model

Without using any of the observed data the retrieval times of the single task condition of Figure 3.1 were predicted by applying the GOMS model. Normative task structures of the GOMS model, i.e., goal hierarchies, methods and operators, were obtained by applying the functions of the respective browser to the retrieval tasks as an ideal user would have done. The GOMS analyses and the predicted retrieval times were obtained with GOMSED, a tool for editing and evaluating GOMS analyses (Wandmacher, 1998). Operator times, e.g., mouse click, positioning of coursor with mouse or screen search, were adapted from Olson and Olson (1990) and other published results (cf. Wandmacher, 1998). Since the HB displayed a new hexagon with a rather long delay a system response time operator roughly estimated as 0.7 sec became necessary. This estimation might have been too small. The GOMS predictions given in Table 3 are the means of the predicted retrieval times over ten item retrievals.

	Session Manager	Explorer	Hexagon Browser
Observed Retrieval Time (Figure 3.1)	19.7	28.2	37.3
GOMS Prediction	7.2	10.4	11.3
Scaled GOMS Prediction	21.2	30.7	33.3
Δ	−1.5	−2.5	4.0
Δ %	−7.7	−8.7	10.7

Table 3 - GOMS predictions of retrieval times in sec. Δ = observed retrieval time − scaled GOMS prediction.
Δ % = relative prediction error = 100Δ /observed retrieval time

The obtained GOMS predictions match the rank order of the observed retrieval times with the three browsers. However, the GOMS predictions are relative performance times which cannot be used as estimates of absolute performance times. For comparing observed and predicted retrieval times scaled GOMS predictions had to be calculated. For this the GOMS predictions were multiplied by a unit constant, which is the sum of the observed retrieval times divided by the sum of GOMS predictions. These scaled GOMS predictions can be used as predictions of absolute retrieval times. The average absolute relative prediction error of the scaled GOMS predictions is about 9 %. These results support the assertion that the GOMS model together with an efficient tool for generating GOMS analyses like GOMSED can serve as a discount method for coarse estimation of performance times without laborious empirical observations.

2.5.3 Dual Task Condition

The data of the dual task condition were analysed by calculating the mean number of items retrieved in the time allowed (5 min), the mean reaction time and error rate in the secondary task. The subjects using the SM were able to retrieve more items than those using the Explorer and the HB with a difference of 4.4 and 5.6, respectively (Figure 3.5). These differences were significant (Mann-Whitney U-test, $p < 0.05$). The number of retrieved items was negatively correlated with the error rates and retrieval times in the single task condition, and these correlations were significant with $p < 0.04$ and $p < 0.01$, respectively.

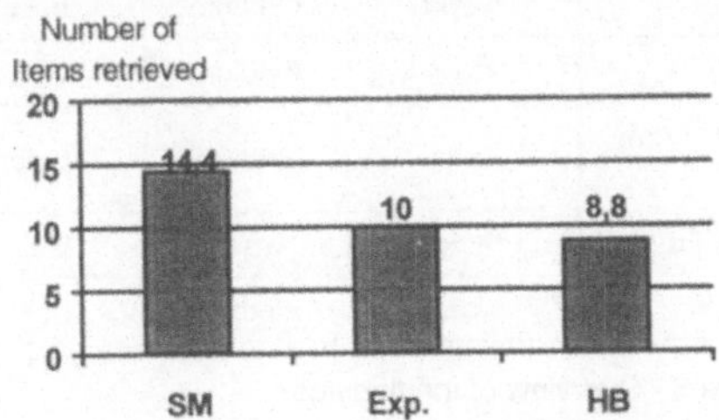

Figure 3.5 - Mean number of retrieved Items

Table 4 gives the mean reaction times and error rates to the secondary task. The users of the HB showed the shortest reaction times differing by 678 msec from the SM and by 1450 msec from the Explorer. The differences between the HB and the other two browsers were significant according to the Mann-Whitney U-test (p < 0.05), whereas the SM and the Explorer did not differ significantly. The highest error rate to the secondary task stimulus was observed for the SM with a difference of 1.1 to the Explorer and 1.6 to the HB. None of these differences turned out significant.

	SM	Explorer	HB
Reaction Time in sec	4.05	4.82	3.37
Error Rate	3.6	2.5	2.0

Table 4 - Mean reaction times and error rates in the secondary task

2.6   Discussion

Table 5 summarizes the results of our experiment.

2.6.1 Single Task Condition

The findings in the single task condition support the hypothesis that users of the SM can retrieve the required items faster and commit fewer errors than subjects using the two other browsers. Subjects using the Explorer showed faster retrieval times and fewer errors than the users of the HB, even though these differences were not significant. The SM users needed the least learning, closely followed by the subjects using the Explorer, while the HB required considerably more learning. However, the HB enabled the fastest learning with an estimated learning parameter of 0.65 compard to 0.16 and 0.18 obtained for the SM and the Explorer (see Table 2).

	Session Manager	Explorer	Hexagon Browser
Retrieval Time	best	medium	worst
Error Rate	best	medium	worst
Amount of Learning Required	best	medium	worst
Learnability (Learning Rate)	worst	medium	best
Self-descriptiveness	best	medium	worst
Number of Items Retrieved in Dual Task Condition	best	medium	worst
Reaction Time in Secondary Task	medium	worst	best
Error Rate in Secondary Task	worst	medium	best

Table 5 - Overview of the findings

Users of the SM and Explorer generally found the usage of the browser and the finding of items easy, whereas most of the HB users regarded the usage as 'fairly easy' and the finding of items as in between easy and difficult. The hierarchical structure was obvious to the majority of the users of each browser, most of all to the users of the Explorer. This could be expected for a 2d tree since branches are used to make the relationships between the nodes explicit. The layout of the three browsers was generally considered as 'good' and using them as 'fun'.

Interestingly, the performance of the Explorer users strongly depended on their strategy. Three subjects almost never contracted a branch, thus being forced to frequent scrolling in the left half of the window thereby quickly loosing the overview of the structure. Consequently, these subjects showed by far the worst performance, differing in their mean retrieval time from the remaining subjects by 9 sec. This finding corresponds to the results of Chimera and Shneiderman (1997) since the three subjects effectively converted the flexible tree into a stable interface, which was shown to produce the worst performance. To overcome this problem, some subjects suggested integrating an additional button into the interface for contracting an entire subtree. This facility would allow users to quickly regain the overview of the top level categories without having to close each branch individually. Other subjects suggested the use of colour to increase the discriminability of the different levels.

The fastest retrieval time was exhibited by one subject who closed almost every branch after retrieving an item. The subject with the second fastest retrieval time adopted another strategy after 'getting lost' due to the large number of expanded branches: the subject clicked on the icon representing the hard disk thereby displaying the top level folders in the right half of the window and then following the path by double-clicking on the nodes in the right half. However, as mentioned above, this subject encountered difficulties when following an incorrect path and was forced to search for the higher level nodes in the left half. The remaining Explorer users did not show a consistent strategy, but instead used both halves of the window and contracted the branches whenever the left half became too cluttered.

2.6.2 Dual Task Condition

In the dual task condition, SM users retrieved significantly more items in the time given than users of the Explorer and the HB (see Figure 3.5). According to the results of the single task condition, this outcome could be expected since subjects using the SM were generally faster in retrieving items and could thus retrieve more items.

In the secondary task, the HB users showed shorter reactions times and committed less errors than the users of the Explorer and SM (see Table 4). The HB seems to have required less mental effort since only a very limited amount of information is displayed at a time. Therefore less scanning of the screen is required, and the required decisions are fairly simple. Furthermore, the subjects did not need to remember their current location in the hierarchical structure since it is clearly displayed by the HB. Thus, while the limited information presented by the HB might be one of the reasons for the worse performance in the first task, the users appeared to benefit from it in the secondary task.

This hypothesis is also confirmed by other findings. First, subjects using the SM and Explorer showed a tendency to 'mark' their present position on the structure despite the tone emitted by the laptop. Second, subjects using the SM showed a slightly faster reaction time compared to Explorer users. This could be due to the additional information about the current location available to the SM users since only the information of the current top level category is displayed. Finally, the three Explorer users who left most of the branches expanded and were thus presented with the most information show slower reaction times than the remaining subjects (the difference averaging to 777 msec) and committed slightly more errors (3.7 vs. 2 for the remaining subjects).

2.6.3 Problems encountered by the Subjects

Some SM users were confronted with two problems even though these problems did not seem to slow down the subjects substantially:

- The medium levels were sometimes overlooked and the subjects moved from the higher levels directly to the lower levels disregarding the levels in between.

- Many subjects frequently clicked on folders even though they were already highlighted, thus apparently not realising that the required path was already expanded.

Several users of the Explorer were faced with two problems which might account for the slower performance compared to the SM users:

- Some subjects frequently overlooked items that were contained in nodes which not only held retrievable items but also category nodes. This problem affected in particular those subjects who expanded the branches using the [+]-icon since this technique does not highlight the nodes that are opened but merely expands the branch in the left half of the Explorer window. In contrast, subjects who expanded the branches by double-clicking on the nodes automatically highlighted these nodes. Thus, they were able to view not only the nodes, but also the items contained in the highlighted nodes.

- Subjects who extended the branches by double-clicking on the nodes in the right half of the window were confronted with problems when following an incorrect path since they were not able to easily backtrack and follow a different path but had to search for a higher level node in the left half of the window. Only one subject used the appropriate button which allowed step by step backtracking in the right half of the window.

The relatively slow performance of subjects using the HB can be explained by a bunch of reasons:

- The HB itself slowed the subjects down since the changing from one hexagon or category to the next is fairly slow. Thus, even the fastest users could not reach the same times as the users of the SM. When predicting the retrieval times with the GOMS model (see section 2.5.2) and reducing the system response time operator to 0.5 sec, the predicted retrieval time of the HB is about equal to that of the Explorer. With a system response time operator equal to zero the predicted retrieval time of the HB is 8 sec and still above that of the SM, viz., 7.2 sec (see Table 3). Note that the GOMS model does not predict the performance time of beginners but of practiced subjects. The GOMS predictions are not affected by the larger amount of learning required for the HB. Therefore, according to the GOMS predictions even an optimized HB would be less efficient then the SM.

- Several users failed to notice the slider that can be used to scroll through the list of objects and often searched in the visible hexagons first before realising that more nodes or items were available. This problem could be solved by rendering the sliders more visible or by placing additional buttons inside the hexagon for scrolling through the objects.

- Several subjects scrolled through the list of hexagons even though the correct option was already visible. If more objects could be displayed at any time, this problem could possibly be reduced.

- The subjects who clicked inside the hexagon in order to scroll through the list of categories or items sometimes clicked in the wrong half of the hexagon. This could be avoided by rendering the separation between the two halves visually more distinctive.

- Several subjects clicked too close to the hexagons which caused them to involuntarily move down a level. This problem was caused by the sensitive zone around a hexagon which had a more or less rectangular envelope.

- Several subjects seemed to think that on some levels the items in the right half belonged to the hexagon in the left half. Thus, they searched through the list of items first before realising that they had to move down a level.

- Some subjects pointed out that navigation became more difficult by the lack of context since only a few objects and one path were visible. Others criticised that they (almost) always had to go back to the top level instead of being able to move between subtrees.

- The additional information about the contents of a hexagon which was given to the subjects in form of hexagon- and 'rounded rectangle'-symbols within the hexagon icon did not seem to be utilised by the subjects since most subjects also clicked on empty hexagons when looking for an item.

- Some subjects followed the same path several times when looking for an item, apparently not realising that they had been there before, possibly because of the lack of context.

3 Conclusions

The experiment revealed that users of the SM performed substantially better at retrieving a set of items from a pre-defined static hierarchy, both in terms of retrieval time and error rate. Subjects using the HB showed the worst performance. Furthermore, the SM required the least learning, closely followed by the Explorer, while users of the HB had to to learn most. This finding was also supported by the length of explanation that was required prior to the task with the HB needing more explanations than the other two browsers. On the other hand, the HB allowed the fastest learning. Above, several reasons were mentioned that might have caused the worse performance of the HB users, among them the lack of context and the rather limited amount of information displayed which increased the need for scrolling. The subjective evaluation of the three browsers did not seem to differ considerably even though the usage of the HB was regarded as slightly more difficult than of the other two browsers.

The SM seems to be best suited for the retrieval of information and can also be easily used by beginners. However, due to the limited depth and width of the hierarchy no scrolling was required for the SM, and research should be done to determine its suitability for deeper and wider hierarchical structures. Further research is required to investigate the suitability of the three browsers for other tasks. These tests could include similar tasks as used by Chimera and Shneiderman (1997) such as textual searching tasks or sibling comparison tasks at different levels of the hierarchy.

In the secondary task, the users of the HB showed the best performance, both in terms of reaction time and error rate. This result could be explained by the reduced amount of information displayed by the HB at any given time, such that the user is not required to scan through a large list of items but instead has to chose among a rather small number of options. Furthermore, the current location does not need to be remembered since it is immediately visible. This explanation is confirmed by the longer reaction times of the Explorer users who are presented with even more information than the SM users. Apart from that, the current location is at least partly displayed by the SM in form of the activated category, while the current branch in the Explorer window is not necessarily highlighted. In order to remember the present location, users of the SM and Explorer frequently 'marked' the current location while the HB users could immediately turn their attention to the secondary task. However, it can also be argued that due to the generally slow performance of the HB itself while changing between categories or hexagons, the subjects had more time to concentrate on the secondary task. More experiments should be done to test the hypothesis of the reduced mental effort for users of the HB. If this hypothesis were confirmed, the actual benefit of the reduced mental effort should be evaluated in order to determine whether the reduction of the required mental effort is sufficiently beneficial to justify the longer retrieval times, which can be shortened by optimizing the HB.

4 References

Chimera, R., Shneiderman, B. (1997) An explanatory evaluation of three interfaces for browsing large tables of content. Technical Report UMCP-CSD:CS-TR-2620, University of Maryland.

Dutta, A. (1995). Experimental RunTime System: Software for Developing and Running Reaction Time Experiments on IBM-compatible PCs. Behavior Research Methods, Instruments, & Computers, 27 (4), 516-519.

Ebert, P., Arend, U. (1997) Business Explorer Browser. SAP-AG. Internal Paper.

Iwanek, R. (1994). Review of the Experimental RunTime System. Psychology Software News, 5(2), 65-67, CTI Centre for Psychology, University of York.

Newell, A., Rosenbloom, P.S. (1981) Mechanisms of skill acquisition and the law of practice. In J.R. Anderson (ed.) Cognitive skills and their acquisition. Hillsdale. Lawrence Erlbaum Associates.

Olson, J.R., Olson, G.M. (1990) The growth of cognitive modeling in human-computer interaction since GOMS. *Human Computer Interaction*, 5, pp. 221-265.

Wandmacher, J. (1998) GOMS-Analysen mit dem GOMS-Editor GOMSED. Http://www.tu-darmstadt.de/fb/fb3/kogpsy/gomsed/gomsed.html

Authors´ addresses

Dr. Jörg Beringer und Dr. Edmund Eberleh
Usability Engineering Center
SAP AG
Postfach 1461
69185 Walldorf
Email: joerg.beringer@sap-ag.de
Email: edmund.eberleh@sap-ag.de

Prof. Dr. Jens Wandmacher
TU Darmstadt
Institut für Psychologie
Angewandte Kognitionspsychologie
Hochschulstr. 1
64289 Darmstadt
Email: wandmacher@psychologie.tu-darmstadt.de

Designorientierung und Designpraxis – Entwicklung und Einsatz von konstruktiven Gestaltungskriterien

Andreas Brennecke, Reinhard Keil-Slawik und Werner Roth

Heinz Nixdorf Institut, Universität-GH Paderborn

Zusammenfassung

Zur Gestaltung von interaktiven Systemen gibt es eine Fülle von Kriterien. Was bei deren praktischer Anwendung jedoch häufig fehlt, ist eine verbindende Orientierung, mit der unterschiedliche Anforderungen abgewogen und Designkonflikte aufgelöst werden können. Anhand von theoretischen Überlegungen motivieren wir einen Gestaltungsansatz, der eine solche Designorientierung aufweist. Basierend auf dem Leitprinzip „Reduzierung erzwungener Sequentialität" haben wir durch viele praktische Beispiele einen Satz von Kriterien entwickelt, mit denen bei der Gestaltung verschiedene Designalternativen bewertet werden können. Am Beispiel der Gestaltung einer Museumsanwendung zeigen wir, wie der vorgestellte Ansatz konstruktiv umgesetzt werden kann.

Abstract

Many criteria exist for the design of interactive systems. However, most of these criteria neither provide a design orientation for practical purposes nor do they offer help in situations where different requirements collide or where design conflicts arise. In this paper, we present a design guideline that is motivated by a theoretical framework. Centred around the general principle to "mimimize the amount of enforced sequentiality" we have developed a set of criteria that allows us to evaluate different alternatives in the design process. A practical example is presented to demonstrate the applicability of our approach for the design of a multimedia terminal in a museum.

1 Einleitung

Die Gestaltung interaktiver Systeme ist ein Problem, bei dem situationsunabhängige und situationsabhängige Faktoren angemessen miteinander kombiniert werden müssen. Bei der Fülle der dabei zu berücksichtigenden Aspekte ist es nicht möglich, durch einen methodisch geleiteten systematischen Konstruktionsprozeß alle relevanten Faktoren im vorhinein zu erheben und dann gewissermaßen nacheinander abzuarbeiten. Abgesehen davon, daß es keine universellen und allgemein anwendbaren Gestaltungsregeln gibt, treten häufig Designkonflikte auf (vgl. Brennecke, Keil-Slawik [1]), bei denen es darauf ankommt, gleichermaßen berechtigte, aber einander zuwiderlaufende Anforderungen auf Kosten der jeweils anderen umzusetzen. Die Auflösung solcher Designkonflikte stellt die eigentliche Herausforderung für die Entwickler dar. Sie erfordert klare Kenntnisse des Einsatzumfeldes, grundlegende Gestaltungskompetenz und einen Designansatz, bei dem operationalisierbare Gestaltungskriterien mit einer Designorientierung derart verknüpft sind, daß es möglich ist, verschiedene Gestaltungsanforderungen miteinander in Beziehung zu setzen und unter einem Leitprinzip gegeneinander abwägend umzusetzen.

Einen solchen Ansatz haben wir in den letzten Jahren aufbauend auf theoretischen Grundannahmen entwickelt (siehe Keil-Slawik [7]) und anhand vieler praktischer Beispiele aus Forschung und Lehre (siehe Brennecke, Keil-Slawik [1] oder Engbring, Keil-Slawik, Selke [5]) zunehmend verfeinert. Im vorliegenden Beitrag werden wir nach der Darstellung unseres Gestaltungsansatzes ein Praxisbeispiel vorstellen, bei dem die Gestaltungsprinzipien produktiv angewendet worden sind.

Ausgangspunkt ist die Entwicklung einer multimedialen Museumsinstallation zum Rechnen mit einer historischen Rechenmaschine für das Heinz Nixdorf MuseumsForum in Paderborn. Die Anforderungen, die sich aus diesem Einsatzumfeld ergeben, werden zunächst kurz skizziert, um dann eine Lösung vorzustellen, die in spezifischer Weise die Umsetzung der Anforderungen des Einsatzumfeldes vor dem Hintergrund des zugrundegelegten Gestaltungsansatzes verdeutlicht. Eine kritische Würdigung sowohl der erzielten Lösung als auch des Gestaltungsansatzes zeigt zum einen die Grenzen der Gestaltung auf, eröffnet zum anderen aber auch spezifische Perspektiven für die zukünftige Arbeit.

2   Konstruktives Design

Für den Bereich der Software-Ergonomie gibt es mittlerweile eine Fülle nationaler und internationaler Normen, die von der Gestaltung der Arbeitsaufgabe bis zur Festlegung der Prinzipien der direkten Manipulation oder der Spezifikation von Zeichengrößen reichen (siehe DIN/EN 9241 [4]). Das Problem dabei ist, daß die Kriterien meist beziehungslos nebeneinanderstehen, häufig sehr allgemein formuliert sind und durch Analyseverfahren teilweise unterschiedlich interpretiert und abgeändert werden (vgl. Stary [11]). Weiterhin sind sie im Hinblick auf die Evaluation bestehender Systeme ausgerichtet und weniger auf die Erfordernisse des Gestaltungsprozesses selbst. Erst durch die Einbettung in einen umfassenden Gestaltungsansatz, der es auf Grund einer speziellen Designorientierung gestattet, verschiedene Anforderungen gegeneinander abzuwägen und so Designkonflikte partiell aufzulösen, können die verschiedenen Kriterien und Gestaltungsregeln im Gestaltungsprozeß fruchtbar gemacht werden. Die Designorientierung ergibt sich dabei aus der Fragestellung, worin denn die spezifische kognitive Unterstützungsfunktion technischer Artefakte besteht.

Ausgehend von der Feststellung, daß sinn- und bedeutungsschöpfende Prozesse des Menschen selbstorganisiert sind und damit von außen weder gelenkt noch vorherbestimmt werden können, zugleich aber solche Prozesse auf sinnlicher Wahrnehmung beruhen und damit auf physische Gegebenheiten Bezug nehmen, läßt sich die primäre Funktion technischer Artefakte bestimmen. Sie besteht zum einen darin, z. B. durch Visualisierung oder Symbolisierung etwas wahrnehmbar zu machen. Wichtig dabei ist, daß durch den tätigen Umgang mit diesen Artefakten Veränderungen sichtbar werden, die es erlauben, durch Verknüpfung von Wahrnehmen und Handeln den Erkenntniskreis zu schließen. Erst durch die Gruppierung zusammenhangloser Sinnesreize entsteht eine Gestalt, eine bedeutungsvolle Form. Um aber verschiedene Sinnesreize miteinander verbinden zu können, müssen zum Zweiten die entsprechenden Artefakte möglichst gleichzeitig ins Wahrnehmungsfeld gebracht werden, denn wenn die jeweils in Beziehung zu setzenden Aspekte zeitlich und räumlich zu weit auseinanderliegen, können sie nicht miteinander verknüpft werden. Schließlich soll drittens eine einmal hergestellte Beziehung auch physisch abgebildet werden, so daß man sich zu einem späteren Zeitpunkt auf das entsprechende Wahrnehmungsmuster beziehen kann, ohne es zuvor wieder aufwendig rekonstruieren zu müssen. Diese drei Kernaspekte des Erzeugens, Verknüpfens und Speicherns verkörpern die primäre Unterstützungsfunktion technischer Artefakte (vgl. Keil-Slawik, Selke [8]). Sie sollen an einem Beispiel kurz erläutert werden.

Ohne physische Hilfsmittel wie Rechengeräte oder Papier und Bleistift kann beispielsweise ein durchschnittlich begabter Mensch kaum mehr als zwei zweistellige Zahlen im Kopf multiplizieren. Sobald im Verlaufe einer solchen Berechnung mehr als zwei Zwischenergebnisse entstehen, sind die meisten von uns überfordert. Mit einem Abakus oder Taschenrechner ist es leichter möglich, denn jetzt kann der gesamte Rechenprozeß in Teilschritte aufgeteilt werden. Man muß den jeweiligen Algorithmus beherrschen, dessen Regeln sich auf die jeweils zu verschiebenden Kugeln oder zu drückenden Tasten beziehen und man muß die elementaren Rechenschritte beherrschen, um die Kugeln bzw. Tasten in der entsprechenden Anzahl und Reihenfolge zu bewegen bzw. zu drücken. Der Vorteil ist, jeder Rechenschritt bezieht sich auf eine sinnlich wahrnehmbare Konstellation von Kugelposition bzw. Tastenarrangements mit einer aktuellen Anzeige in einem Display. Der Nachteil ist, jeder weitere Rechenschritt überschreibt den vorherigen Zustand. Deshalb sprechen wir hier von sequentiellen Rechenmitteln. Rechnet man dagegen mit Papier und Bleistift, bleibt die Spur des Rechenprozesses erhalten, d. h. alle Zwischenergebnisse sind gleichzeitig im Wahrnehmungsfeld präsent und dauerhaft gespeichert. Um ein Rechenergebnis zu überprüfen, müssen nicht mehr alle Zwischenergebnisse durch manuelle Operationen neu erzeugt werden. Auch können einzelne Rechenschritte jetzt unabhängig voneinander durchgeführt werden. Und schließlich ist es sogar möglich, verschiedene Rechnungen miteinander in Beziehung zu setzen und auf Gemeinsamkeiten und Unterschiede hin zu vergleichen. Aus diesem Grund bezeichnen wir die hierbei verwendeten Hilfsmittel und Verfahren auch als räumliche Rechenmittel.

Analog zum Übergang von den sequentiellen zu den räumlichen Rechenmitteln kann der Übergang von kommandoorientierten zu grafischen Benutzungsoberflächen aufgefaßt werden. Die sequentielle Folge von Eingabe und Ausgabe (mißverständlich meist als Dialogsystem bezeichnet), bei der jeweils die Ausgabe die Eingabe überschreibt (und umgekehrt), wird zu einer Arbeitsumgebung, bei der Handlungs- und Wahrnehmungsraum möglichst eng gekoppelt sind. Das Ziel, daß jede Handlung sich möglichst an wahrnehmbaren Objekten vollzieht und daß jede Veränderung der Objekte auch sichtbar und einsehbar ist, wird durch das Leitprinzip der „Reduzierung erzwungener Sequentialität" verdeutlicht: Es gilt, jede Sequenzierung manueller Operationen im Wahrnehmungsfeld zu vermeiden, die nicht der Aufgabenstruktur inhärent sind oder aus Gründen der Erlernbarkeit erforderlich sind. Die Designorientierung lautet folglich nicht, ein Dialogsystem zu entwickeln, sondern eine Arbeitsumgebung zu entwerfen, bei der durch die direkte Kopplung von Wahrnehmungs- und Handlungsraum der manuelle Aufwand zum Verstehen und Ausführen von Aufgaben minimiert wird. Zur Umsetzung dieses Leitprinzips haben wir einen Satz kanonischer Gestaltungsregeln entwickelt, der in Abbildung 1 skizziert ist.

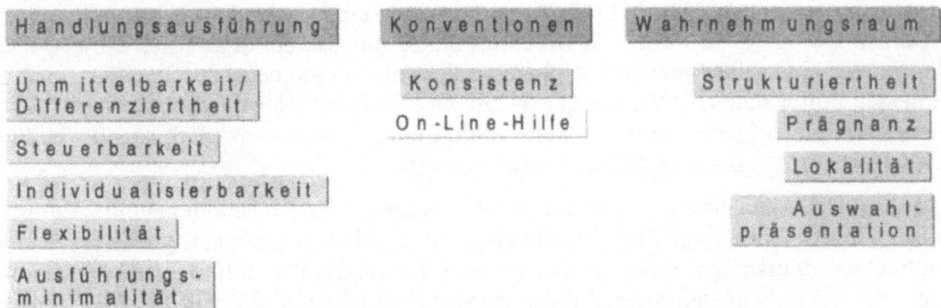

Abb. 1: Kriterien zur Reduzierung erzwungener Sequentialität.

Die Dreiteilung der Kriterien spiegelt in gewisser Hinsicht eine Priorisierung einzelner Aspekte bezüglich bestimmter Designkonflikte wider. Die Kriterien unter der Rubrik Wahrnehmungsraum beziehen sich jeweils auf ein aktuelles Wahrnehmungsfeld; die Kriterien der

Handlungsausführung beziehen sich unmittelbar auf einen Interaktionsschritt. Erst wo diese Kriterien nicht unmittelbar greifen (können), weil sich z. B. eine Handlung über mehrere Wahrnehmungsräume bzw. unterschiedliche Systeme erstreckt, werden die Kriterien der dritten Rubrik entscheidend. Insofern ist im Unterschied zur gängigen Literatur (vgl. z. B. Shneiderman, 1998 [10]) das Kriterium der Konsistenz nicht das wichtigste Gestaltungsprinzip, sondern eher ein wichtiges, aber nachgeordnetes Prinzip.

Wie sich das Leitprinzip „Reduzierung erzwungener Sequentialität" sowie einzelne Kriterien im Gestaltungsprozeß umsetzen lassen, soll nachfolgend anhand der Gestaltung einer Multimedia-Anwendung für den Einsatz in einem Museum verdeutlicht werden.

3 Rahmenbedingungen für das Praxisbeispiel

Museumsanwendungen stellen besondere Anforderungen an die Gestaltung, da Benutzer mit unterschiedlichstem Vorwissen die Anwendung nutzen können sollen (vgl. Nielsen, S. 110 [9]). Das Ziel der Multimedia-Anwendung ist es, den Besuchern des Computermuseums das Rechnen an einem Original-Comptometer zu ermöglichen. Das Comptometer ist eine mechanische Rechenmaschine, die von der Firma Felt&Tarrant ab der Jahrhundertwende in Amerika produziert wurde (siehe Darby, 1968 [3]). Obwohl man sie als „Volltastatur Addiermaschine" bezeichnet, kann man auf ihr alle Grundrechenarten oder auch andere Rechenarten wie beispielsweise das Ziehen von Quadratwurzeln direkt berechnen (siehe Boys, 1901 [2]). Das Rechnen mit einer Volltastaturmaschine ist heute kaum noch jemandem vertraut und die Maschine unterscheidet sich in Aussehen und Bedienung so sehr von einem heute üblichen Taschenrechner (siehe Abbildung 2), daß von keinem Vorwissen bzgl. der Bedienung eines solchen oder vergleichbaren Gerätes ausgegangen werden kann. Das einzige Wissen, das vorausgesetzt wird, ist die Kenntnis der Grundrechenarten.

Die Multimedia-Anwendung soll den Museumsbesuchern ermöglichen, eine selbst gewählte Rechenaufgabe mit dem Comptometer durchzuführen. Dabei werden nicht die Maschine und auch nicht die verwandten Rechenalgorithmen erklärt – hierfür müßte erst einiges an Hintergrundwissen z. B. zur Verwendung des Neunerkomplementes vermittelt werden – aber die Besucher sollen erfahren, daß man mit einer einfachen Maschine, die mechanisch nur die Addition ausführt, auch kompliziertere Rechnungen ausführen kann. Die verwendeten Algorithmen zur Durchführung der Grundrechenarten wurden aus didaktischen Erwägungen vereinfacht, was mit einer erhöhten Zahl an Arbeitsschritten einhergeht. Die Notwendigkeit, überhaupt einen Computer in diesen Ausstellungsteil zu integrieren, ergibt sich daraus, daß die Erklärung wie eine selbst gewählte Rechnung mit dem Comptometer durchgeführt wird, nicht angemessen durch statische Medien (Text- und Grafiktafeln) oder durch einmaliges Vorrechnen einer Beispielaufgabe – beispielsweise mittels eines Videos – möglich ist, so wie dies bei anderen Rechenmaschinen im Museum realisiert wird.

Die Multimedia-Anwendung wurde mit der Programmiersprache Delphi erstellt. Autorenprogramme wie Toolbook oder Macromedia-Director sind besser geeignet, grafische Animationen interaktiv zu erstellen, es ist mit ihnen aber schwieriger, interaktive Simulationen zu realisieren, die auf errechneten Werten beruhen (vgl. Freibichler, 1995 [6]). Bei der Comptometer-Anwendung kommt es aber gerade darauf an, die Simulation für alle Rechenarten mit unterschiedlichen Zahlenwerten durchzuführen. Weiterhin war die Anfangs für diesen Ausstellungsteil zur Verfügung gestellte Hardware so leistungsschwach, daß mit einer Programmiersprache gearbeitet werden mußte, die systemnahe Entwicklungen ermöglicht.

An dem Entwurf der Multimedia-Anwendung waren Informatiker, ein Historiker, eine Museumspädagogin sowie ein Designer beteiligt, für die es auch das erste gemeinsame Projekt war. Problematisch stellte sich im Nachhinein heraus, daß bei der arbeitsbereichübergreifenden Teamarbeit keine Entscheidungsstrukturen festgelegt waren. So mußten an einigen Stellen Kompromisse geschlossen werden, so daß die endgültige Multimedia-Anwendung nicht an allen Stellen der oben beschriebenen Designorientierung entspricht.

Im folgenden zeigen wir einige Aspekte bei der Gestaltung der Multimedia-Anwendung auf, vernachlässigen dabei aber die eben beschriebenen Faktoren, die auf das Endprodukt ebenfalls maßgeblichen Einfluß hatten.

4 Designpraxis

Die Anwendung der oben skizzierten Designorientierung hatte unmittelbare Konsequenzen auf die Grundkonzeption. Anders als bei einer Multimedia-Anwendung für den Heimbereich, die in wechselnden Kontexten eingesetzt wird, steht bei einer Museumsanwendung die Integration in das Ausstellungskonzept im Vordergrund. Dies bedeutet insbesondere, daß die Gestaltung nicht auf den Computer beschränkt sein muß, sondern den gesamten zur Verfügung stehenden Raum nutzen sollte.

Abb. 2: a) Aufbau der Multimedia-Installation, bestehend aus einem Terminal, zwei Rechenmaschinen und einer Hinweistafel b) Startbildschirm der Anwendung.

Die gewählte Lösung besteht aus zwei Original-Rechenmaschinen, zwischen denen ein Bildschirm und über denen eine Hinweistafel angeordnet sind (siehe Abbildung 2a). Die zwei Rechenmaschinen sind nicht für einen Benutzer erforderlich, sie sollen es zwei Personen ermöglichen, gleichzeitig zu rechnen und die Multimedia-Installation kommunikativ zu nutzen. Die Bedienung des Rechners erfolgt über einen Touch-Screen; es sind keine zusätzlichen

Eingabegeräte vorhanden. Die Multimedia-Anwendung selbst enthält keine Hintergrundinformation über den Aufbau und die historische Einbettung der Rechenmaschine; diese wird in der umgebenden Ausstellung gegeben. Beim Rechnen mit dem Comptometer erstreckt sich der Handlungsraum auf die Rechenmaschine und den Bildschirm; der Wahrnehmungsraum umfaßt zusätzlich auch die Hinweistafel. Die Informationen auf der Hinweistafel könnten auch auf dem Bildschirm zugänglich gemacht werden. Dieses würde jedoch zusätzlich Naviagtionselemente erfordern und zugleich Handlungssequenzen erzwingen, die mit der Ausführung einer Rechenaufgabe nichts zu tun haben.

Bei der gewählten Lösung hat ein Benutzer dagegen alle Informationen gleichzeitig im Wahrnehmungsraum präsent und kann schnell zwischen den jeweils vermittelten Informationen wechseln. Ein weiterer Vorteil besteht darin, daß sich die umstehenden Personen unabhängig von jeweiligen Benutzern mit wichtigen Informationen vertraut machen und so eine mögliche Wartezeit produktiv nutzen können. Die Multimedia-Installation soll als Teil des Ausstellungskonzeptes der historischen Rechenmaschine Comptometer, die Brücke zwischen Rechenmaschine und Rechnen mit der Maschine schlagen. Sie steht jedem Besucher zur Verfügung und ist nicht auf bestimmte Zielgruppen ausgerichtet. Dies bedeutet, daß man weder auf Erfahrungen des Benutzers mit grafischen Oberflächen zurückgreifen kann, noch bestimmte Konventionen in der Bedienung einführen kann.

Das didaktische Konzept sieht für die Multimedia vor, daß größtmögliche Freiheit bei der Wahl der Rechenaufgabe im Rahmen des didaktisch Sinnvollen gewährt werden soll. Diese Rahmenbedingung der Museumspädagogik erfordert die freie Eingabe einer Aufgabe mit der Schwierigkeit, daß bestimmte Aufgaben nicht zugelassen werden können, weil das Ergebnis beispielsweise die vorhandene Stellenzahl überschreitet. Andere Aufgaben sind aus didaktischen Gründen nicht angemessen, weil für deren sinnvolle Durchführung Hintergrundwissen z. B. über das Neunerkomplement bei negativem Rechenergebnis erforderlich ist.

Nachdem eine Beispielaufgabe eingegeben wurde, soll für das Rechnen der Aufgabe am Original-Comptometer jeder Rechenschritt textuell erläutert und auf der virtuellen Rechenmaschine der Multimedia-Anwendung vorgeführt werden.

Ausgehend von einem Startbildschirm (siehe Abbildung 2b), welcher nach einer längeren Zeit ohne Aktivitäten der Benutzer auf dem Touch-Screen automatisch aufgerufen wird, kann durch Druck auf eine beliebige Stelle des Bildschirms zum Vorführungsmodus der Multimedia weitergeschaltet werden. In einer speziellen Maske wird die vorzuführende Rechenaufgabe, die aus zwei Operanden und einem Operator besteht, eingegeben. Dabei sind bei der Subtraktion Aufgaben mit negativem Ergebnis nicht erlaubt, bei Multiplikation und Division ist die Anzahl der Stellen beschränkt. Dieses ist nicht in der Rechenmaschine begründet, sondern wurde aus didaktischen und feinmotorischen Gründen bei der Bedienung des Comptometers eingeführt. Diese Beschränkungen sollen dem Benutzer transparent gemacht werden und sie sollen seine Herangehensweise beim Eintippen der Aufgabe so wenig wie möglich beeinflussen.

Nach der Festlegung einer Aufgabe wird diese schrittweise textuell erläutert und wenn vom Benutzer gewünscht auf der schematischen Darstellung eines Comptometers (siehe Abbildung 3a) animiert. Danach soll jeder einzelne Schritt auf einer neben dem Bildschirm stehenden Originalmaschine nachvollzogen werden. Dazu kann die Animation gegebenenfalls wiederholt werden. Einen Erklärungsschritt rückgängig zu machen, verbietet sich, da man auf dem Comptometer im Gegensatz zu typischen Vierspezies-Rechenmaschinen keine Eingaben widerrufen kann. Hat sich der Benutzer verrechnet oder ist er aus einem anderen Grund aus

dem Tritt geraten, so wird ihm empfohlen, die gesamte Vorführung und seine Rechnung auf
dem Comptometer neu zu beginnen.

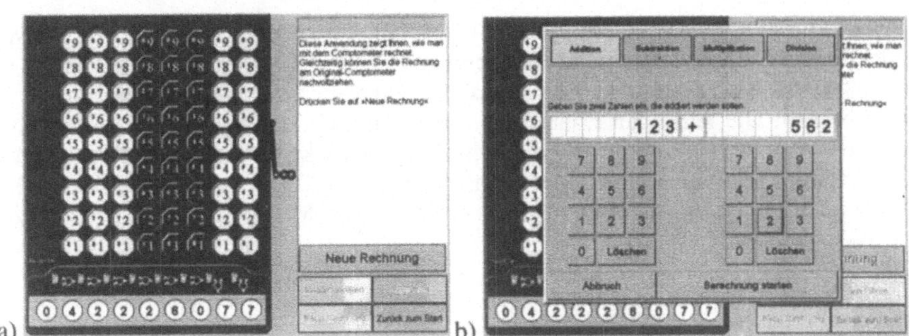

Abb. 3: a) Vorführungsmodus und b) Eingabemaske für die Rechenaufgabe in der aktuellen Version.

Auf ein Hilfesystem innerhalb der Multimedia wurde verzichtet, da durch die Integration einer mechanischen Rechenmaschine mit einem elektronischen Erklärungs- und Vorführsystem
viele Fragen entstehen, die beide Geräte bzw. deren Integration betreffen. Statt dessen wurde
der zur Verfügung stehende Raum genutzt, um eine Hinweistafel über der Arbeitsstation an-
zubringen. Diese Konzeption löst das Integrationsproblem, steht wartenden Besuchern zur
Verfügung und entkoppelt von räumlichen und zeitlichen Abhängigkeiten, die durch ein Hilfesystem innerhalb der Multimedia entstehen würden.

Als Beispiel für die Gestaltung gemäß des Leitkriteriums „Reduzierung erzwungener Sequentialität" beschreiben wir nachfolgend die Entwicklungsarbeit einer Maske zur Eingabe
der gewünschten Rechenaufgabe. Das Designziel war die Schaffung eines Handlungsraums,
in dem sämtliche Möglichkeiten und dabei die Konsequenzen verdeutlicht werden. Bei der
Gestaltung der Eingabemaske geht es in erster Linie um die Bewältigung der beschriebenen
Beschränkungen.

Ein naheliegender und von der Museumspädagogik präferierter Ansatz war, die Aufgabe
mittels eines nachempfundenen Taschenrechners einzugeben. Als Vorteil einer solchen Konzeption wurde die Ähnlichkeit mit dem bekannten Taschenrechnermodell und damit die Vertrautheit vieler Besucher mit dieser Anordnung herausgestellt. Auch die geringe Anzahl von
Bedienelementen wurde als Argument angeführt.

Die Nachbildung eines Taschenrechners, wie sie in Abbildung 4a zu sehen ist, erfordert jedoch eine sequentielle Eingabe der Aufgabe. Dies bedeutet, daß bei der Eingabe der erste
Operand während der Erfassung des zweiten Operanden überschrieben wird. Dieses Problem
tritt noch deutlicher bei Fehlersituationen in den Vordergrund. Diese können frühestens nach
Eingabe des Operators gefunden und behandelt werden. Bei der Multiplikation ist beispielsweise der Multiplikand, also der erste einzugebende Operand, auf zwei Stellen beschränkt. Da
die Rechenart aber erst durch Eingabe des Operators nach dem ersten Operanden festgelegt
wird, muß bei eingegebenem längerem Multiplikanden im nachhinein eine Reaktion erfolgen.
Wenn also eine unzulässige Eingabe auftritt, werden zusätzliche Meldungen und Abfragen
erforderlich. Die Korrektur der Eingabe muß dabei als zusätzliche sequentielle Handlung
durchgeführt werden. Die Handlungssequenzen bei fehlerhaften Eingaben sind jedoch, wie
unsere spätere Lösung zeigt, weder aufgabeninhärent, noch tragen sie etwas zur Einsicht in

die Aufgabenstruktur bei. Es handelt sich demnach um eine erzwungene Sequentialität, die sich in der Verletzung des Kriteriums Ausführungsminimalität niederschlägt.

Ein weiteres Problem, welches zusätzliche Sequentialität erzwingt, ist, daß bei dieser Anordnung der Elemente ohne zusätzliche Erläuterungen nicht klar wird, welche Aufgaben überhaupt zulässig sind und daß die Eingabe aus zwei Operanden und einem Operator in Infix-Notation besteht. Die fehlenden Erläuterungen können traditionellerweise durch Erklärungstext auf einem separaten Fenster, innerhalb der Eingabemaske oder grafisch dargeboten werde. Ein separates Fenster bedeutet aber einen zusätzlichen Modus, also Sequentialität in der Handlungsausführung; ein Erläuterungstext innerhalb der Maske ist das Pendant dazu im Wahrnehmungsraum.

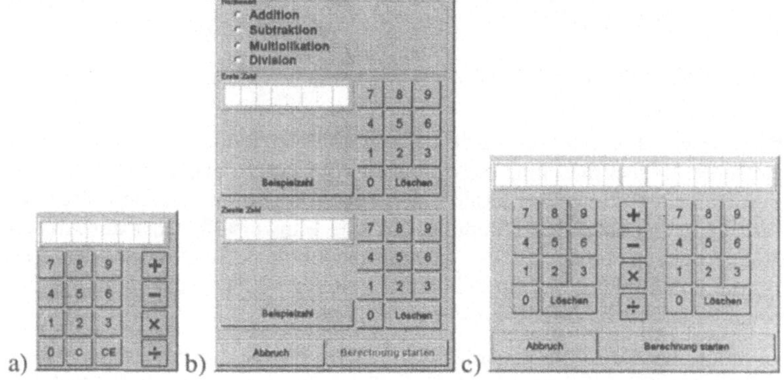

Abb. 4: Verschiedene Entwürfe für die Eingabemaske für die Rechenaufgabe.

Eine mögliche Lösung beider Problemfelder ist, die zeitlich sequentielle Eingabe in eine räumliche Aufteilung der Eingabemaske zu überführen (siehe Abbildung 4b). Diese gibt die Benutzereingaben durch ihre räumliche Struktur vor, schreibt jedoch keine Reihenfolge vor. Der Benutzer kann im Sinne des Kriteriums der Steuerbarkeit jedes Element der Beispielaufgabe zu jedem Zeitpunkt eingeben. Diese Reduzierung erzwungener Sequentialität erforderte die Einführung eines zweiten Tastenfeldes inkl. Anzeige der eingetasteten Zahl und eines Radiobuttons zur Auswahl des Operators. Neben der Erhöhung der Steuerbarkeit legen diese neuen Eingabefelder die Elemente der Aufgabe optisch ohne externe Hinweise fest, lösen also gleichzeitig das oben aufgeworfene Problem der Erläuterung der Aufgabenstruktur. Abhängig von der gewählten Rechenart ändert sich die Stellenanzahl in den Eingabebereichen. Das bedeutet die Elimination möglicher Fehleingaben bei Multiplikation und Division, bei denen die Stellenzahl eingeschränkt ist. Das bedeutet, die der Aufgabe inhärenten Abhängigkeiten werden modusfrei gelöst und transparent gemacht und dies genau dann, wenn sie auch auftreten. Durch die Verfügbarkeit von einzelnen Eingabebereichen für die zwei Operanden und den Operator löst man sich sowohl von der sequentiellen Eingabe als auch von der sequentiellen Korrektur.

Die von der Museumspädagogik vermuteten Vorteile des Taschenrechneransatzes gehen auch bei diesem Entwurf nicht verloren. Durch die Gruppierung der Eingabefelder der Operanden nutzt man einen eventuellen Wiedererkennungswert auch in diesem Dialog, durch die Gruppierung und Gleichheit der Elemente spielt die absolute Anzahl der Einzelelemente nur noch eine untergeordnete Rolle. Für das Problem, den Aufbau der Rechenaufgabe erläutern zu müssen und somit ein weiteres gestalterisches Element zu haben, wurde demnach mit dem Kriterium Strukturiertheit eine Lösungsalternative gefunden.

Die eingegebene Aufgabe wird in dem eben präsentierten Entwurf jedoch nicht in der gewohnten Weise – sequentiell nebeneinanderstehende Operanden mit dem Operator in der Mitte – dargestellt. Der Entwurf ist so gestaltet, daß die Reihenfolge der Benutzereingaben durch die räumliche Anordnung geleitet wird. Verfolgt man diesen Grundsatz weiter und entkoppelt die Repräsentation der Beispielaufgabe von der gewohnten Schreib-Lese-Richtung, so entsteht der Dialog, der jetzt in der Multimedia zu finden ist. Die Anzeige der Aufgabe als Operand, Operator und Operand, spiegelt die geforderte Aufgabenstruktur ohne zusätzliche Erklärung wider. Die Darstellung der Abhängigkeit der Stellenzahl vom gewählten Operator wird durch das Anordnen der Operatorauswahl über der Darstellung der Aufgabe ähnlich wie in Abbildung 4b gelöst. Sämtliche Beschränkungen innerhalb der Eingabe werden nach Auswahl der Rechenart unmittelbar sichtbar gemacht. Die verbleibende Fehlersituation, daß bei der Subtraktion zweier eingegebener Zahlen das Ergebnis negativ ist, wird ebenfalls durch eine unmittelbare Rückmeldung gelöst: Liefert die aktuell erfaßte Aufgabe ein solches negatives Ergebnis, so wird eine Textfläche der Eingabemaske für eine Fehlermeldung genutzt, der Fehlertext wird gegenüber dem normalen Text hervorgehoben und die „Berechnung starten"-Taste gesperrt. So wird das Kriterium der Unmittelbarkeit umgesetzt, d. h. auf Fehleingaben wird durch eine zeitlich unmittelbare Reaktion des Systems hingewiesen und diese müssen nicht im nachhinein beispielsweise nach Quittierung der Eingabe in zusätzlichen Masken korrigiert werden. Die situative Rückmeldung und der Hinweistext erfüllen zudem das Kriterium einer differenzierten Rückmeldung, bei der sofort die Konsequenzen der Eingabe verdeutlicht werden.

In der letztendlich umgesetzten Maske, wie sie in Abbildung 3b zu sehen ist, wurden die Operatortasten wie beschrieben über die Anzeige der Aufgabe gesetzt. Diese Gestaltung hebt, durch die räumliche Anordnung, die Auswahl der Rechenart hervor. Da die Rechenart für eine Beschränkung der Stellenzahl sorgen kann, hat sich diese Anordnung als aufgabenangemessener herausgestellt als die Maske in Abbildung 4c, die durch ihre räumliche Anordnung eher die Sequenz Operand, Operator und Operand unterstreicht. Unmittelbar unter den Operatortasten befindet sich ein Textfeld, in dem einzelne Rechenarten und deren Beschränkungen erläutert werden. Diese räumliche Bindung entspricht dem Kriterium Lokalität und sorgt dafür, daß die Zuordnung zu den Rechenarten keine weiteren Handlungsschritte erfordert.

Das Textfeld wird erst einige Zehntelsekunden nach Erscheinen des Dialogs eingeblendet, da sich bei der Evaluation mit einer Schulklasse herausgestellt hat, daß auf diese Weise die Erklärungstexte eher die gewünschte Beachtung finden und trotzdem im Gegensatz zu einer anderen Hervorhebung während der Eingabe keine unnötige optische Anziehung haben, wie dies bei der Verwendung von Farbe oder blinkendem Text der Fall wäre. Neben den eben benutzten Kriterien zur „Reduzierung erzwungener Sequentialität" spielen somit auch noch andere zum Beispiel empirisch ermittelte Faktoren bei der Gestaltung eine Rolle.

5 Zusammenfassung und Bewertung

Die vorgestellten Kriterien decken natürlich nicht alle Aspekte ab, die hinsichtlich der Gestaltung multimedialer Systeme denkbar und erforderlich sind. Durch die zugrunde gelegte Designorientierung soll insbesondere in den Bereichen Hilfestellung gegeben werden, wo Designkonflikte auftreten oder wo Entwickler sich Neuland erschließen. Wo immer bestimmte Standards allgemein akzeptiert und gefordert werden oder wo klare empirische Befunde vorliegen, sollen und müssen diese natürlich berücksichtigt werden. Die Kriterien zur „Reduzierung erzwungender Sequentialität" erheben insofern nicht den Anspruch, alle gestaltungsrelevanten Tatbestände zu erfassen oder abzudecken. Vielmehr zielen sie darauf, eine gewisse Orientierung in das Dickicht der vielfältigen Normen und Gestaltungsanforderungen zu bringen und, soweit das möglich ist, kanonische Regeln zu formulieren, die die Entwickler bei ihrer Arbeit unterstützen.

Insofern bietet der vorgestellte Gestaltungsansatz sowohl praktische Hilfestellung bei konkreten Designproblemen als auch eine Forschungsperspektive, die darauf zielt, möglichst viele eher analyseorientierte Gestaltungskriterien in einem gestaltungsorientierten Ansatz vollständig zu integrieren. Ob das gelingen kann, ist fraglich, aber je mehr es gelingt, desto besser werden Anforderungen nach Prinzipien der Software-Ergonomie auch für Informatiker und Systementwickler lehrbar und vermittelbar und damit auch unter alltäglichen Bedingungen der Systemgestaltung nutzbar.

6 Literatur

[1] A. Brennecke, R. Keil-Slawik: Alltagspraxis der Hypermediagestaltung – Erfahrungen beim Einsatz des World Wide Web und Mosaic in der Lehre. In: H.-D. Böcker (Hg.): Software-Ergonomie´95 – Mensch-Computer-Interaktion – Anwendungsbereiche lernen voneinander. Stuttgart, 1995: B.G. Teubner. S. 107-123.

[2] C. V. Boys: The Comptometer. Nature London, No. 1654, Vol. 64, 11 Juli 1901. S. 265-268.

[3] E. Darby: It All Adds Up – The Growth of the Victor Comptometer Corporation. 1968: Victor Comptometer Corporation.

[4] DIN/EN 9241: Ergonomische Anforderungen für Bürotätigkeiten mit Bildschirmgeräten – Teil 1-17. Berlin, unterschiedliche Erscheinungsdaten: Beuth Verlag.

[5] D. Engbring, R. Keil-Slawik, H. Selke: Neue Qualitäten in der Hochschulausbildung – Lehren und Lernen mit interaktiven Medien. Paderborn, 1995: Technischer Bericht Nr. 45, Heinz Nixdorf Institut.

[6] H. Freibichler: Werkzeuge zur Entwicklung von Multimedia. In: L. J. Issing, P. Klimsa (Hg.): Information und Lernen mit Multimedia. Weinheim, 1995: Psychologie Verlags Union. S. 221-240.

[7] R. Keil-Slawik: Konstruktives Design – Ein ökologischer Ansatz zur Gestaltung interaktiver Systeme. Berlin, 1990. Habilitation, Forschungsberichte des Fachbereichs Informatik, Bericht Nr. 90-14, TU Berlin.

[8] R. Keil-Slawik, H. Selke: Forschungsstand und Forschungsperspektiven zum virtuellen Lernen von Erwachsenen. Erscheint in der Reihe: Schriften zur beruflichen Weiterbildung. Berlin, 1998: QUEM Qualifikations-Entwicklungs-Management.

[9] J. Nielsen: Multimedia, Hypertext und Internet – Grundlagen und Praxis des elektronischen Publizierens. Braunschweig u.a., 1996: Vieweg.

[10] B. Shneiderman: Designing the user interface – strategies for effective human computer interaction, 3. Auflage. Reading, Mass. u.a., 1998: Addison-Wesley.

[11] C. Stary: Interaktive Systeme – Software-Entwicklung und Software-Ergonomie, 2. Auflage. Braunschweig u.a., 1996: Vieweg.

Adressen der Autoren

Andreas Brennecke
Heinz Nixdorf Institut
Universität-GH Paderborn
Fürstenallee 11
33102 Paderborn
Email: anbr@uni-paderborn.de
http://hyperg.uni-paderborn.de/~anbr

Prof. Dr.-Ing. Reinhard Keil-Slawik
Heinz Nixdorf Institut
Universität-GH Paderborn
Fürstenallee 11
33102 Paderborn
rks@uni-paderborn.de
http://hyperg.uni-paderborn.de/~rks

Werner Roth
Heinz Nixdorf Institut
Universität-GH Paderborn
Fürstenallee 11
33102 Paderborn
tiberius@uni-paderborn.de
http://hyperg.uni-paderborn.de/tiberius

Nutzeranforderungen als Grundlage für die Entwicklung innovativer User Interfaces in der industriellen Prozeßführung

Michael Burmester und Tobias Komischke

Fachzentrum User Interface Design, Siemens AG

Zusammenfassung

Dieser Artikel (1) stellt das Siemens-Forschungsprojekt „MediaPlant" vor, das den Nutzen für Anlagenbetreiber, Projekteure und Operateure fördern will; (2) beschreibt anhand erster Untersuchungsergebnisse den aktuellen Stand der Führung exemplarischer Produktionsprozesse, sowie Nutzeranforderungen an User Interfaces in Leitwarten und (3) nennt neue Technologien, die diesen Nutzeranforderungen gerecht werden könnten.

Abstract

This article (1) presents the Siemens Research-Project „MediaPlant", which aims at improving the benefits for industrial companies, operators and automation engineers; (2) describes, using primary results, the current state of the control of exemplary production processes as well as user interface requirements and (3) gives recent technologies which could fulfill with these user requirements.

1 Einführung

Ein Prozeß ist der Vorgang des Umformens und/oder Transports von Materie, Energie oder Daten [1]. Die Anzahl der beobachtbaren Prozeßvariablen kann als grobes Maß für die Komplexität eines Prozesses angesehen werden [2]. In großen Energieversorgungsnetzen gibt es bis zu 200.000, in Walzwerken weniger als 100 beobachtbare Prozeßvariablen, was zeigt, daß Prozesse sehr unterschiedlich komplex sein können. Um Komplexität beherrschbar zu machen, werden Prozesse automatisiert. Zudem wird automatisiert, um Produkte mit höherer Qualität herzustellen oder den Einsatz von Material und Energie zu optimieren [1]. Die Verantwortung und Überwachung über den Prozeß trägt jedoch noch immer der Mensch. Wie Abbildung 1 zeigt, braucht er in vollautomatisierten Anlagen nicht in den Prozeß eingreifen [3]. Die Ausgangsgrößen des technischen Prozesses werden über Sensoren gemessen, die diese Daten an das Informationsverarbeitungssystem weitergeben, das wiederum auf Grundlage dieser Daten über Aktoren auf den Prozeß einwirkt. Ist die Anlage nicht vollautomatisiert, übt der Mensch als sogenannter Operateur eine leitende Kontrolle aus [2]. Die Prozeßdaten werden ihm durch das Mensch-Maschine-Kommunikationssystem („User Interface") vermittelt, er nimmt diese Daten auf und interpretiert sie. Schließlich reagiert der Operateur auf die Daten, indem er über das Mensch-Maschine-Kommunikationssystem Eingaben tätigt, die den Prozeß steuern.

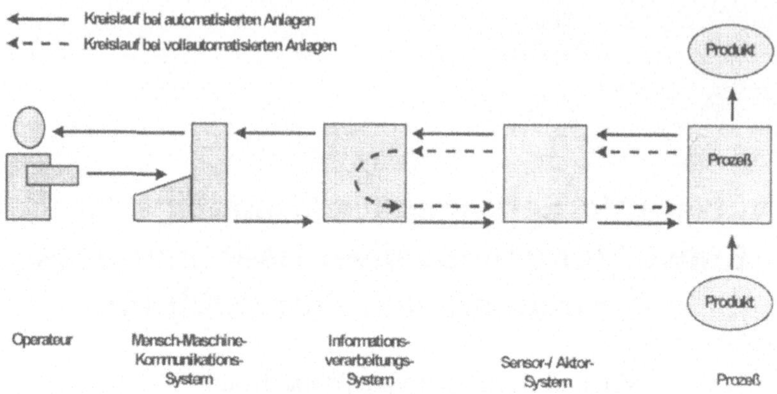

Abb. 1: Modell der Prozeßführung (nach Oberquelle, 1997) [3]

In Mensch-Maschine-Systemen arbeiten verschiedene Klassen von Nutzern [2]:

- Anlagenfahrer („Operateure"): arbeiten über das User Interface mit dem technischen System und führen den Prozeß, indem sie bedienen und beobachten
- Automatisierungsingenieure („Projekteure"): gestalten das User Interface und konfigurieren das Automatisierungs-System
- Betriebsingenieure: verfügen über fundiertes technisch-wissenschaftliches Funktionswissen und können zur Unterstützung der Operateure angefordert werden
- Servicepersonal: führt notwendige Reparaturen aus
- Wartungsmannschaft: wartet die Anlage in regelmäßigen Abständen
- Management: leitet die Anlage und ist für die Erfüllung, der übergeordneten Ziele wie Sicherheit, Wirtschaftlichkeit, Umweltverträglichkeit und Sozialverträglichkeit verantwortlich

Die Prozeßführung ist einem ständigen Wandel unterworfen, so daß sich auch heute folgende Fragen stellen:

- Wie entwickelt sich die Prozeßführung weiter?
- Welches sind die Trends bei User Interfaces in den nächsten 5-10 Jahren?
- Welche verfügbaren User Interface-Technologien haben das Potential für den praktischen Einsatz?
- Wo liegen die derzeitigen Probleme in der Mensch-Maschine-Interaktion?

Das im folgenden Kapitel vorgestellte Forschungsprojekt von Siemens will Antworten auf diese Fragen liefern.

2   Das Siemens Projekt „MediaPlant"

Bezüglich der User Interfaces in der industriellen Prozeßführung liegen noch Potentiale für den nutzergerechten Einsatz neuer Technologien, worunter wir neuartige, moderne Technologien, wie z.B. Multimedia, Virtual Reality, dreidimensionale Informationsaufbereitung oder Spracheingabe verstehen. Als „innovativ" bezeichnen wir Technologien, die nicht nur neu sind, sondern auch soweit entwickelt sind, daß sie in Anlagen eingesetzt werden können.

Das Siemens-Forschungsprojekt „MediaPlant" verfolgt folgende Ziele:

- Neue Informations- und Kommunikationstechnologien sollen für die Prozeßführung nutzbar gemacht werden.
- Dazu sollen innovative User Interface Lösungen für die Optimierung des Bedienen & Beoachtens prototypisch entwickelt werden. Es soll geprüft werden, ob diese Lösungen branchenübergreifend eingesetzt werden können.
- Da ebenso Projekteure bei der Projektierung und Gestaltung von User Interfaces optimal unterstützt werden sollen, werden ferner ihre Anforderungen ermittelt und neue Konzepte der Projektierung entwickelt.

Zur Erreichung dieser Ziele wird ein benutzer- und aufgabenzentriertes Vorgehen gewählt, wie es in der Forschung u.a. von Alty & Bergan [4] gefordert wird und das mittlerweile auch in der neuen ISO-Norm 13407 „Benutzer-orientierte Gestaltung interaktiver Systeme" verankert ist [5]. Das im Projekt verfolgte Vorgehen umfaßt folgende Arbeitsschritte:

1. Erhebung von Anforderungen an User Interfaces, sowie Aufgaben und Einsatzumgebungen von Operateuren und Projekteuren durch branchenübergreifende Interviews mit in der Prozeßführung erfahrenen Experten sowie Arbeitsanalysen in verschiedenen Anlagen, in denen automatisierte Produktionsprozesse ablaufen. Die Interviews dienen zur Wissensakquisistion über Ist-Zustände, Trends und Probleme in der Prozeßführung und Projektierung, während die Arbeitsanalysen zu einem detaillierten Verständnis der Bedienziele und -handlungen der Nutzer dienen.
2. Ableitung von Benutzungsanforderungen in unterschiedlichen Branchen für das Bedienen & Beobachten sowie die Projektierung
3. Entwicklung von User Interface Prototypen unter Berücksichtigung von kognitionsergonomischen Anforderungen und relevanten Normen sowie technischen Randbedingungen (z.B. Plattform-Beschränkungen)
4. Empirische Überprüfung der Prototypen anhand von Prozeßführungsaufgaben mit Operateuren und Überprüfung der Projektierungskonzepte in Kooperation mit Projekteuren
5. weitere Iterationsschritte von Prototypenentwicklung und empirischer Evaluation bis die Benutzungsanforderungen erfüllt sind

Im folgenden Kapitel werden erste Untersuchungsergebnisse präsentiert.

3 Der gegenwärtige Stand der industriellen Prozeßführung und Nutzeranforderungen an das User Interface

3.1 Fragestellung

Um den gegenwärtigen Stand in der Führung industrieller Prozesse, sowie Anforderungen an das User Interface und an Projektierungssoftware zu erheben, wurden Experteninterviews durchgeführt. Von Interesse waren dabei vor allem folgende Punkte:

- die Mensch-Maschine-Funktionsteilung
- Tätigkeiten und Probleme der Operateure
- Tätigkeiten und Probleme der Projekteure
- Trends in der Prozeßführung
- Anforderungen an User Interfaces und Projektierungs-Software

3.2   Erhebungs- und Auswertungsmethode

Der aktuelle Entwicklungsstand in der Prozeßführung wurde anhand einiger ausgewählter automatisierter Produktionsprozesse analysiert:

- Energieerzeugung
- Energieverteilung
- Papier- und Zellstoffindustrie
- Stahlstraßen
- Automobilindustrie
- chemische Industrie

Die halbstrukturierten Interviews bestanden aus:

- 9 allgemeine Fragen zur Prozeßführung in der jeweiligen Anlage (Bsp.: *„Was ist die Rolle des Menschen in der Leitwarte?"*)
- 5 Fragen zu den Operateuren (Bsp.: *„Welche Ausbildung haben die Operateure?"*)
- 5 Fragen zu den Projekteuren (Bsp.: *„Wie groß ist der Gestaltungsrahmen beim Projektieren?"*)

Es wurden bislang Interviews mit 11 Personen in 8 führenden deutschen Firmen durchgeführt, weitere sind in Vorbereitung. Die Interviewpartner waren Experten in der Prozeßführung bzw. Prozeßleittechnik und hatten auf diesem Gebiet eine durchschnittliche Berufserfahrung von 13½ Jahren. Ihre ausgeübten Tätigkeiten waren:

- Bereichs-Management
- Projektierung
- Software-Entwicklung
- Vertrieb

Die Bereichs-Manager und Vertriebsleute hatten einen umfassenden Überblick über das Untersuchungsthema. Projekteure und Software-Entwickler konnten sehr detailliert ihre eigenen Tätigkeiten und die der Operateure schildern. Die Interviews mit Projekteuren waren für das Projekt sehr wichtig, weil durch deren spezielles Wissen einerseits Szenarien als Input für Beobachtungen determiniert werden können, an denen zu einem späteren Zeitpunkt in dem Projekt die neu entwickelten Bausteine mit Hilfe von Operateuren experimentell validiert werden. Andererseits dienten die Interviews mit den Projekteuren als Anforderungserhebung für neue Konzepte für Projektierungssoftware. Die Interviews wurden auf Audio-Kassetten mitgeschnitten und anschließend exzerpiert. Im folgenden werden Aussagen getroffen, die für alle Branchen gelten und somit einen möglichst einheitlichen, branchenübergreifenden Überblick zu geben.

3.3 Ergebnisse

3.3.1 Wartentypen und -belegung

Es werden sowohl zentrale wie auch dezentrale Leitwarten angetroffen. Letztere werden auch als örtliche Leitstände bezeichnet und dort eingesetzt, wo man sich neben Kosteneinsparungen eine Entlastung des Servicepersonals erhofft, indem die Arbeit des Warten-Operateurs angereichert wird und er bei Störungen und Pannen die Leitwarte verläßt und selbst Hand an die Anlage legt. Dies ist vor allem in fertigungsorientierten Anlagen der Fall.

In der konventionellen Energieerzeugung, in der ausschließlich zentrale Leitwarten im Einsatz sind, wird für die Zukunft erwartet, daß sich der Operateur räumlich weiter entfernt vom laufenden Prozeß befinden wird als bisher. Über eine zentrale Fernleitwarte könnten mehrere Kraftwerke überwacht und per Fernzugriff gesteuert werden. Nur noch das Servicepersonal wäre vor Ort stationiert, um im Bedarfsfall schnell eingreifen zu können.

Die Anzahl von Operateuren in den Warten ist in den letzten Jahren durch den steigenden Automatisierungsgrad gesunken. Heute sind im Normalfall zwischen einer und drei Personen in der Warte. Obwohl in einigen Warten zur Vermeidung von Monotonie bewußt mehrere Operateure eingesetzt werden, weißt der Trend in Richtung Ein-Mann-Belegung.

3.3.2 Typische Bedienhandlungen und -fehler der Operateure

Die Aufgabe des Operateurs in der Leitwarte wird von den Interviewpartnern mit „viel" beobachten und „wenig" bedienen umschrieben. Der Mensch soll von den reinen Prozeßführungsaufgaben entlastet werden und soll statt dessen Prozeßoptimierung betreiben.

In allen Branchen müssen

- Prozeßbilder angewählt und analysiert,
- binäre Schaltungen durchgeführt,
- Meldungen aufgenommen und interpretiert,
- Kurven analysiert und
- die Instandhaltung über Störungen informiert werden.

In anderen Prozessen können zusätzliche Bedienhandlungen hinzukommen, wie z.B. Rezeptänderungen in chemischen Anlagen.

Typische Fehler des Personals bei der Prozeßführung treten nur im Handbetrieb auf und sind dementsprechend selten. Es handelt sich dabei um:

- falsche Interpretation von Meldungen
- falsche analoge Eingaben
- falsche Schalthandlungen
- Wahl einer falschen Strategie
- übersehen, daß Prozeß aus dem Ruder läuft
- mehrmaliges Auslösen von Funktionen bei verzögertem Feedback (z.B. aufgrund geringer Performanz des User Interface Management Systems)

Die Steigerung des Automatisierungsgrades bewirkt die Reduktion von Bedienhandlungen und verlagert die Fehlerverteilung auf den drei Handlungsebenen nach Rasmussen [6] von unten nach oben. Das bedeutet, daß eher Strategie-Fehler, wie beispielsweise an falscher

Stelle im Prozeß per Hand eingreifen, in der täglichen Arbeit eines Operateurs vorkommen, als klassische Fehlleistungen wie z.B. versehentlich den falschen Knopf zu drücken.

3.3.3 Anforderungen an User Interfaces

Die folgenden Anforderungen an User Interfaces in der Prozeßführung gehören teilweise zu den Standard-Anforderungen der Software-Ergonomie [7] und können durch Modifikation der bestehenden Software gelöst werden, ohne daß neue Technologien integriert werden müßten. Andere Forderungen verlangen jedoch nach Innovationen.

- **Selbstbeschreibungsfähigkeit**: die Benutzer des User Interfaces haben sehr unterschiedliche Qualifikationen. Die Spannbreite reicht von Universitätsingenieuren (kerntechnischer Bereich) bis zu Hilfsarbeitern (chemische Industrie). Es kann also nicht in jedem Fall davon ausgegangen werden, daß die Operateure EDV-Erfahrungen oder Programmierkenntnisse haben. Ist das User Interface selbsterklärend, kommt dies den Anlagen-Betreibern entgegen, die darauf achten, den Schulungsaufwand möglichst gering zu halten.

- **Erwartungskonformität** sowohl innerhalb des Systems als auch zu MS-Windows als meistverwendetes Betriebssystem für PCs: Konsistenz innerhalb eines Systems gehört zu den ergonomischen Standardforderungen, die an Software gestellt werden [8]. Mit der zunehmenden Verbreitung von PCs im privaten Bereich, wächst die Wahrscheinlichkeit, daß die Bediener mit der Windows-Oberfläche und deren grundlegenden Funktionen vertraut sind, so daß durch eine konsistente Gestaltung des User Interfaces zu Windows ein positiver Lerntransfer ausgenutzt werden kann.

- **Sicherheit** bezieht sich mehr darauf, die Verfügbarkeit der Anlage nicht zu gefährden als darauf, die Gefahr von Störfällen zu reduzieren. Störfälle werden durch die Leittechnik, die unabhängig vom Operateur im Bedarfsfall in den Prozeß eingreift, weitgehend ausgeschlossen. Durch Fehlbedienung können jedoch Störungen provoziert werden, die es möglicherweise nötig machen, die Anlage anzuhalten, oder die sehr kostenintensive Schäden verursachen.

- **Individualisierbarkeit**: die Oberfläche des User Interfaces soll für verschiedene Nutzerklassen individualisierbar sein, einzelne Nutzer einer Nutzerklasse brauchen im Normalfall keine individualisierten Darstellungen. Weiterhin will der Anlagenbetreiber Prozeßbilder bequem nach seinen eigenen Vorstellungen modifizieren können.

- **effiziente Navigationsstrukturen**: der Operateur muß in kürzester Zeit die benötigten Informationen und Eingriffsmöglichkeiten in Störungssituationen erhalten und die notwendigen Handlungen ausführen können. Eine Anforderung bezüglich Effizienz lautet heute: der Anlagenfahrer muß mit zwei Auswahlaktionen (z.B. Mausklicks) zu der Störung gelangen können. Zukünftig wird eine Auswahlaktion genügen müssen.

- **innovative Darstellungstechnologien**: ein immer noch aktuelles Problem in der Führung von Prozessen stellt die Datenflut dar, die dem Operateur präsentiert wird und von ihm interpretiert werden muß. Da das Beobachten automatisierter Prozesse den größten Anteil der Aufgaben von Operateuren einnimmt, spielt die schnelle Einschätzung des Anlagenzustands und der Überblick über die Prozeßparameter in Störungssituationen eine entscheidende Rolle

- **starke Nutzerführung**: da Eingriffe in Störfallsituationen selten und die Operateure zum Teil nur angelernt sind, ist eine starke Benutzerführung zur Lösung dieser Situationen notwendig. Die Integration des Betriebshandbuchs in das User Interface wird von den Experten als wünschenswert eingestuft. Ein Anwendungsszenario ist das Anfahren der Anlage per Hand, wenn die Automatik ausgefallen ist.

- **Hilfestellungen für die Nutzer**: den Operateuren sollen Hilfestellungen bei der Anlagendiagnose und Entscheidungen gegeben werden. Dabei sollen auch ergonomische Aspekte und die kognitiven Verarbeitungsvorgänge beim Bediener in den Spezifikationen der Systeme mit berücksichtigt werden.

3.3.4 Anforderungen an Projektierungssoftware für User Interfaces

Um effizient User Interfaces für die immer komplexer werdenden Prozesse gestalten zu können, benötigen Projekteure geeignete Projektierungssoftware. Aus den Erfahrungen mit Projektierungstools unterschiedlicher Hersteller wurden in den Interviews folgende Forderungen erhoben:

- **Selbstbeschreibungsfähigkeit und Konsistenz zu Windows**, um die Einarbeitungszeit in die Programme zu reduzieren (s. auch Kapitel 3.3.3)
- **verbesserte Hilfesysteme**: bei den heutigen Hilfesystemen wird beklagt, daß sie oft nicht durchgängig kontextsensitiv sind, zum Teil falsche Informationen enthalten und oft unvollständig sind
- **zentrale Änderbarkeit**: die Modifikation eines sich in mehreren Bildern befindlichen Objekts an einer Stelle soll automatisch in allen Bildern übernommen werden
- **Standards / Bausteine**: es gibt heute schon eine Reihe von vordefinierten Bausteinen, die modular zu einem System zusammengefügt werden können und somit eine effizientere Arbeit der Projekteure ermöglichen, allerdings müssen noch oft Bildelemente selbst konfiguriert werden.

4 Lösungsansätze durch neue User Interface Technologien

Den im vorigen Kapitel formulierten Anforderungen stehen Lösungsansätze in Form neuer User Interface Technologien gegenüber, welche das Potential haben, die Prozeßführung sicherer, einfacher, effizienter und benutzergerechter zu gestalten. Im folgenden werden einige dieser Technologien im Kontext mit den Nutzer-Anforderungen beschrieben.

- **effizientere Navigationsstrukturen und neue Darstellungstechnologien**: eine mögliche Lösung könnten 3-dimensionale Informationsdarstellungen sein. In einem Überblicksbild können tiefergehende Informationen in der dritten (Tiefen-) Dimension dargestellt werden, so daß der Operateur nicht Bilder wechseln muß. Auch Störungen können 3-dimensional dargestellt werden, indem in der Tiefe die Wertigkeit und in der Fläche die Information über die Lokalisation der Störung dargestellt wird. Eine neue Technologie ist KOAN (Kontext-Analysator) [9], ein Werkzeug, das Daten nach dem Grundprinzip „kontextuelle Ähnlichkeit" = „räumliche Nähe" dreidimensional in einem Netzwerk aus Gegenständen, Merkmalen und den zugehörigen Relationen präsentiert. Der Nutzer kann in diesem 3D-Raum navigieren, selektieren, Aktionen auslösen oder Beschreibungen und Hintergrundinformationen abrufen.
  Von Griem und Oberquelle wurde ein Oberflächenkonzept entwickelt, das sie „Leitstandsmetapher" nennen [3]. Bei diesem Ansatz wird bewußt die Verwendung der bekannten und verbreiteten Desktop-Metapher vermieden und durch eine andere, die Leitstandsmetapher, ersetzt. Desktop-Metaphern sind in der Prozeßführung insofern problematisch, als daß durch die frei verschiebbaren und überlappenden Fenster wichtige Infor-

mationen verdeckt werden können. Als Alternative wurde in Analogie zu realen Leitwarten eine Metapher mit folgenden Komponenten vorgestellt:

- der Bedienbereich (in Analogie zum Bedienpult) löst direkt auf den Prozeß einwirkende Funktionen aus

- der Anzeigenbereich (in Analogie zum Informationsbord) zeigt qualitative Informationen an, die ständig sichtbar sein müssen

- der Tafelbereich (in Analogie zur Wandtafel) bietet mittels unterschiedlicher anwählbarer Darstellungen eine Übersicht über den Prozeßzustand

- der Alarmbereich (in Analogie zu Signallampen in klassischen Leitwarten) zeigt Alarme an, wobei der Alarm mit höchster Priorität zusätzlich textuell angezeigt wird. Durch Klicken auf ein Alarmsymbol wird das entsprechende Übersichtsbild angezeigt.

Auf weitere interessante Visualisierungstechnoogien wird hier nur kurz verwiesen:

- Bildpyramide (GRADIENT-Projekt) [10]
- Fisheye View [11] [12]
- Massendatendisplay [13]
- Dreidimensionale Prozeßvisualisierung [14]

- **Hilfestellungen für die Nutzer** sollen durch intelligente und integrierte Expertensysteme gegeben werden, die ihr in Datenbanken gespeichertes Wissen den Operateuren zur Anlagendiagnose und für Entscheidungen zur Verfügung stellen. Den Beitrag, den Expertensysteme für die Prozeßführung leisten können, wird in den durchgeführten Interviews als sehr groß eingeschätzt. Ein intelligenter Assistent kann unter Verwendung einer aufgabenbezogenen Datenbank speichern, welche Bedienhandlungen eines Operateurs zu einer Störung geführt hatten. Sollte ein Bediener ein weiteres Mal diese Eingaben tätigen und somit auf eine Störung zusteuern, kann er von dem System auf diesen Umstand hingewiesen werden. Gleichzeitig würden ihm - etwa durch ein integriertes Betriebshandbuch - alternative Schritte vorgeschlagen werden, welche Eingaben er tätigen sollte, um den Prozeß zu stabilisieren. Eine existierende Anwendung eines Echtzeit-Expertensystems ist MIP, das nach seiner Entwicklung in einer spanischen Anlage eingesetzt wurde [15]. Es gibt Informationen und Empfehlungen in Echtzeit, um eine effiziente und stabile Prozeßführung zu erreichen.

- **Virtual Reality & Multimedia**: in den Interviews wurde Interesse an Virtual Reality und Multimedia nur in bezug auf Simulationen, Präsentationen und Schulungen [vgl. auch 16] geäußert. Der Einsatz dieser Technologien in Leitwarten ist jedoch auch in Prozeßführungs-Szenarien, wie z.B. die Prozeß-Diagnose und Ursachenforschung, interessant. Der Vorteil von Virtueller Realität liegt in der Freiheit der Betrachtung. So können der Betrachtungswinkel und - im Gegensatz zu Kameras - die Detailstufen der Darstellung verändert werden. Multimedia könnte durch die Verteilung von Informationen auf mehrere Sinneskanäle und durch die optimale Ausnutzung medialer Informationsaufbereitung die kognitive Belastung und Beanspruchung der Operateure reduzieren.

- **Ergonomische Hilfestellung für Projekteure**: Kolski stellte das in LISP programmiertes Design-Expertensystem SYNOP vor, das mit Produktionsregeln eine ergonomische Evaluation von graphischen Darstellungen und im Anschluß daran geeignete Modifikationen vornehmen kann [17]. Dadurch wird ergonomischen Anforderungen Rechnung getragen, auch wenn kein Ergonom bei der Gestaltung der Automatisierungssoftware zur Verfügung steht. Das „MediaPlant"-Projekt verfolgt eine andere Strategie zur Lösung des gleichen Problems. Dadurch, daß die Bausteine ergonomisch gestaltet und usability-getestet sind, sollte sich ein evaluierendes Expertensystem wie SYNOP erübrigen. Für den Gebrauch der Bausteine beim Projektieren ist ein Assistent denkbar, der Entscheidungshilfen für die Kombination dieser Module gibt.

5 Ausblick

Wie in Kap. 2 berichtet, ist das Ziel von „MediaPlant", Anforderungen von Operateuren und Projekteuren in innovative User Interfaces zu übersetzen. Dazu sollen die im vorigen Kapitel vorgestellten von Hochschulen und Industrieforschungszentren entwickelten Technologien für den Einsatz in Leitwarten nutzbar gemacht werden. Durch die Interviews wurden die Anforderungen, wie in Kap. 3 beschrieben, branchenübergreifend erhoben.

Nach der Durchführung der restlichen Interviews werden als nächster Schritt Arbeitsanalysen in Leitwarten verschiedener Branchen, in denen automatisierte technische Prozesse ablaufen, vorgenommen. Durch die Analyse der Aufgaben und Handlungen der Operateure lassen sich einerseits die Nutzeranforderungen, die in den Interviews formuliert wurden, anhand von Endbenutzern validieren. Zudem soll überprüft werden, ob branchenübergreifende Kernabläufe, also Handlungen des Beobachten und Bedienens, die in allen Branchen vorkommen, identifiziert und beschrieben werden können, für die - unter der möglichen Verwendung neuer Technologien - innovative Baustein-Prototypen gestaltet werden sollen. Weiterhin dient die Analyse von Aufgaben und Handlungen der Operateure dazu, detaillierter als in den Interviews branchenspezifisch Szenarien zu ermitteln, in denen sich die Baustein-Prototypen empirisch bewähren sollen.

Ein entsprechendes Vorgehen ist für die neuen Konzepte der Projektierung vorgesehen. Auch hier sind als nächster Schritt Arbeitsanalysen geplant, die das Vorgehen bei der Gestaltung von User Interfaces detailliert untersuchen sollen, um letztlich diese Arbeit durch innovative Projektierungs-Tools optimal zu unterstützen.

6 Literatur

[1] H.J. Charwat: Lexikon der Mensch-Maschine-Kommunikation. München, 1992: R. Oldenbourg Verlag

[2] G. Johannsen: Mensch-Maschine-Systeme. Berlin, 1993: Springer-Verlag.

[3] U. Griem, H. Oberquelle: Die Gestaltung der Benutzerschnittstelle von Prozeßleitsystemen nach der Leitstandsmetapher. In R. Liskowsky, B.M. Velichskovsky & W. Wünschmann (Hrsg,): Usability Engineering: Integration von Mensch-Computer-Interaktion und Software-Ergonomie. Tagungsband der Software-Ergonomie '97.Stuttgart, 1997: Teubner.

[4] J.L. Alty, M. Bergan: The Design of Multimedia Interfaces for Process Control. In: H.G. Stassen (Ed.): Analysis, Design and Evaluation of Man-Machine Systems 1992. Oxford, 1992: Pergamon Press.

[5] DIN EN ISO 13407 (Normentwurf), Ausgabe 1998-02, Benutzer-orientierte Gestaltung interaktiver Systeme. Berlin: Beuth Verlag.

[6] J. Rasmussen, K. Duncan, J. Leplat (Eds.): New Technology and Human Error. Chichester, 1987: Wiley & Sons.

[7] DIN EN ISO 9241 Teil 10, Ausgabe 1996-08, Ergonomische Anforderungen für Bürotätigkeiten mit Bildschirmgeräten. Grundsätze der Dialoggestaltung. Berlin: Beuth Verlag.

[8] G.I. Johnson, C.W. Clegg, S.J. Ravden: Towards a practical method of user interface evaluation. Applied Ergonomics 20 (1989), 4, 255-260.

[9] B.W. Kolpatzik, L. Pfefferer, A. Schappert: Content Analysis and Visualization of Epidemiological Documents on the Internet. In: L. Gierl, A. D. Cliff, A.-J. Flahault (Eds.): GEOMED'97 International Workshop on Geomedical Systems (Proceedings of the GEOMED'97 in Rostock, 4.- 6. September 1997). Leipzig, Stuttgart, 1998: Teubner-Verlag, S. 238-248.

[10] C. Weisang, K. Zinser: GRADIENT, ein intelligentes, wissensbasiertes System zur sicheren Prozeßführung. ABB Technik (1992), 5, 349-361.

[11] K. Zinser: Fisheye-views - gezielte Informationsdarstellung zur Beherrschung komplexer Systeme. ABB Technik (1995), 3, 10-15.

[12] K. Zinser: Fisheye Views - Interaktive, dynamische Visualisierungen. atp -Automatisierungstechnische Praxis 37 (1995), 9, Seiten 42, 44-45, 48-50.

[13] C. Beuthel u. a.: Advantages of Mass-Data-Displays in Process S&C. In: Department of Mechanical Engineering and Department of Aeronautics, Massachusetts Institute of Technology (MIT), 1995.

[14] C. Beuthel: Dreidimensionale Prozeßvisualisierung zur Führung technischer Anlagen am Beispiel eines Kohlekraftwerks. Dissertation, Technische Universität, Clausthal.

[15] C. Aguirre, E. de Pablo, J.L. Zaccagnini, X. Alaman: The user interface in expert systems for real-time process control: the MIP system experience. IFIP Transactions A (Computer Science and Technology) A-14 (1992), 266-272.

[16] M. Akiyoshi, S. Miwa, T. Ueda, S. Nishida: A learning environment for maintenance of power equipment using virtual reality. In: Image Processing and its Application, Conference Publication No. 410. Edinborough, 1995: IEE.

[17] C. Kolski: Formalization approaches of ergonomic knowledge for 'intelligent' design, evaluation and management of man-machine interface in process control. In: A. Ollero, E.F. Camacho (Eds.): Intelligent Components and Instruments for Control Applications. Selected Papers from the IFAC Symposium. Oxford, 1993: Pergamon Press.

Adressen der Autoren

Dipl.-Psych. Michael Burmester
Siemens AG
Fachzentrum User Interface Design (ZT IK 7)
Otto-Hahn-Ring 6
81730 München
Email: michael.burmester@mchp.siemens.de

Dipl.-Psych. Tobias Komischke
Siemens AG
Fachzentrum User Interface Design (ZT IK 7)
Otto-Hahn-Ring 6
81730 München
Email: tobias.komischke@mchp.siemens.de

Der software-ergonomische Stellenwert von Rahmenwerken
Erfahrungen mit einem bankfachlichen Anwendungssystem

Peter W. Fach

RWG, Stuttgart

Zusammenfassung

Rahmenwerke sind Strukturierungsmittel in der modernen, objektorientierten Software-Entwicklung und verfolgen den Sinn, die Wiederverwendung ganzer Entwurfsstrukturen zu ermöglichen. Dabei ist der software-technische Aspekt der Wiederverwendung nur eine Seite der Medaille, die andere, ob und ggf. wie sich ein benutzungsfreundliches Design wiederverwenden läßt, ist nach wie vor eine interessante software-ergonomische Fragestellung. In diesem praxisorientierten Bericht werden Methoden und Erfahrungen präsentiert, wie mit Hilfe des Reifungsprozesses in Rahmenwerken und der sog. „Entwurfsmuster" benutzungsfreundliche und wiederverwendbare Anwendungssysteme für eine Universalbank erstellt werden können.

Abstract

Frameworks are structural aids in the development of modern, object-oriented software with the objective to facilitate the reuse of already existing design structures. However, the technical aspect of software reuse is only one side of the coin; the other question which presents an interesting challenge is if and how a user-friendly design can be reused. This report describes methods and experiences concerning how to develop reusable and user-friendly application systems for a universal bank, the central issues being the process of framework maturation and the so-called "design patterns".

1   Einleitung

An den Arbeitsplätzen großer Dienstleister wie Versicherungen oder Banken verändert sich die Rolle der Anwendungssoftware kontinuierlich vom ehemaligen Datenerfassungsterminal hin zu multimedialen Informations- und Beratungsinstrumenten. So sind beispielsweise Serviceinseln im Bankenbereich immer stärker gefragt, an denen einerseits eine umfassende Anlageberatung andererseits aber auch Reiseschecks ausgestellt oder Versicherungen organisationsnaher Institute abgeschlossen werden können. Dieses breite Spektrum an „Diensten", das Bankberater und -beraterinnen für ihre alltägliche Arbeit benötigen (analysieren und bewerten von Wertpapierverläufen, Auszahlen von Fremdwährungen, elektronisches Abschließen von Versicherungsverträgen), stellt allerhöchste Anforderungen an deren Benutzungsfreundlichkeit. Besonders hervorzuheben sind dabei:

- Aufgabenangemessenheit
- Performanz
- Durchgängig konsistente Bedienung

Weil sich dieser Trend, Kunden einen „Rundumservice" anzubieten, branchenweit verfestigt, und die Anforderungen an eine typische Bank-Anwendungssoftware zunehmend ähnlicher werden, entsteht auch das Potential für standardisierte, branchenweite Lösungen. So sehr sich aber einerseits die Geschäftsbereiche gleichen, so sehr müssen andererseits die einzelnen Arbeitsplätze an die jeweiligen Arbeitsplatzanforderungen adaptierbar sein. Rahmenwerke und Entwurfsmuster, das Thema dieses Beitrags, stellen eine software-technische Antwort auf solche und ähnliche, im Grunde sich gegenseitig ausschließende, Standardisierungs- und Flexibilisierungsanforderungen dar.

Die RWG als zentrales Dienstleistungsunternehmen für Daten- und Informationsverarbeitung der württembergischen Genossenschaftsorganisation erstellt für über 350 württembergische Raiffeisen- und Volksbanken Rahmenwerke für banktypische Anwendungsbereiche. Das auf diesen Rahmenwerken aufbauende Anwendungssystem GEBOS wird seit über acht Jahren erfolgreich an zwischenzeitlich über 12.000 z.T. stark divergierenden Arbeitsplätzen eingesetzt (vgl. z.B. [1]). In diesem Beitrag sollen nicht nur die zahlreichen Vorteile von Rahmenwerken aus software-ergonomischer Sicht dargelegt werden, sondern auch Probleme und Hindernisse, die auftreten können. Denn Rahmenwerke werden evolutionär entwickelt, sie „reifen" oder besser festigen sich erst im Laufe von Jahren. Deshalb ist es extrem wichtig, Probleme insbesondere software-ergonomischen Ursprungs in den frühen Stadien zu erkennen und zu bereinigen, denn die Korrektur wird mit zunehmendem Reifegrad aufwendiger und kostspieliger. Darüber hinaus soll auch deutlich werden, welche zusätzlichen Anforderungen auf Software-Ergonominnen und -Ergonomen zukommen, wenn sie die Entwicklung derartiger Rahmenwerke betreuen.

2   Rahmenwerke - eine kurze Einführung

Das Schlüsselkonzept beim Einsatz von Rahmenwerken lautet „Wiederverwendbarkeit des Designs". Im Gegensatz zu bisherigen software-technischen Lösungsansätzen, wo sich Wiederverwendbarkeit auf einzelne Software-Funktionen bezog, z.B. mathematische Funktionen (Sinus, Exponentialfunktionen), setzen Rahmenwerke auf die Wiederverwendbarkeit kompletter fachlicher Einheiten, wie Kundenakten, Konten oder Wertpapierdepots. Mit anderen Worten versucht man, die Entwicklungsarbeiten wie Aufgabenanalyse, fachlicher Klassenentwurf und Werkzeugdesign zu konservieren, indem man ein Anwendungsskelett bzw. -rahmen erstellt, der beispielsweise die software-technische Implementierung eines typischen Kontos darstellt. Ein solcher Rahmen muß von Anwendungsentwicklern dann nur noch auf die Besonderheiten des jeweiligen Anwendungsbereichs zugeschnitten werden. Um den Unterschied auf den Punkt zu bringen: In herkömmlichen Entwicklungsansätzen erstellt man ein Anwendungsprogramm *neu* unter Verwendung *bestehender* Klassenbibliotheken für Datenbankzugriffe, mathematische Funktionen oder die Benutzungsoberfläche. Im Sinne der Wiederverwendung des Designs kommt ein *fertiges*, ausgereiftes Rumpf-*Anwendungssystem* zum Einsatz, das von Programmierern nur an die Besonderheiten des jeweiligen Anwendungsbereichs angepaßt werden muß. Rahmenwerke geben also bereits die „Programmlogik" vor.

2.1   Schichtenbildung durch Sedimentation

Zunächst wird die Anwendungsentwicklung mit einem relativ kleinen und überschaubaren Tätigkeitsbereich begonnen, z.B. mit der Abwicklung von Daueraufträgen. Durch Aufgabenanalysen werden die typischen Tätigkeitsfelder erhoben, also Arbeitsgegenstände und Handlungsabläufe, sowie die an eventuellen Kooperationen beteiligten Rollen beschrieben. Daraus lassen sich beispielsweise mit Hilfe von CRC-Karten die fachlichen Klassen (Dauerauftrag, Konten, Auftraggeber und Empfänger, etc.) modellieren. Auf der Basis dieser fachlichen Klassen wird schließlich damit begonnen, die Benutzungsoberfläche zu entwickeln. Prototypen und Autor-Kritiker-Zyklen unterstützen bei dieser Vorgehensweise die Benutzerpartizipation. Dieses oder ein vergleichbares Vorgehen hat sich im Laufe der letzten Jahre bewährt und wurde auch in verschiedenen Arbeiten dokumentiert (siehe z.B. [2], [3]). Nach der gleichen Vorgehensweise werden nun Zug um Zug Softwarelösungen für weitere Teile des Anwendungsbereichs erstellt; diese werden teilweise Überlappungen mit den bereits existierenden Lösungen aufweisen, teils aber auch neue Aspekte hinzufügen.

Vereinfacht dargestellt besteht nun die „Kunst" beim Erstellen von Rahmenwerken darin, während der Entwicklung des Anwendungssystems diejenigen Implementierungsteile (im OO-Jargon: Klassen) zu finden, die beispielsweise für alle Konto-Implementierungen immer wieder neu erstellt werden müßten. Diese werden dann als Rahmenwerk in einer eigenständigen Schicht zusammengefaßt, dem *Gegenstandsbereich* (siehe Abb. 1). Am Beispiel des Kontos führt diese Vorgehensweise dazu, daß alle Klassen, in denen das Ein- und Auszahlen von Beträgen implementiert ist, in eine tiefere Schicht sedimentieren. Das Hinzufügen und Prüfen eines „Kreditlimits" ist dagegen eine Besonderheit von Girokonten, oder das Anwenden gestaffelter Zinssätze eine von bestimmten Sparkonten. Solche konkreten Ausprägungen von Konten gehören immer auch zu einem bestimmten Typ von Tätigkeit, wie etwa die Kreditsachbearbeitung oder die Anlageberatung. Daher werden alle Implementierungsteile, die zur Unterstützung einer solchen Tätigkeit dienen, in einem Rahmenwerk zusammengefaßt, das in der Schicht *Tätigkeitsbereich* angesiedelt ist (siehe Abb. 1). Alle Rahmenwerke dieser Schicht können dann gemeinsam die Rahmenwerke aus dem Gegenstandsbereich nutzen. Im Verlaufe mehrerer Entwicklungszyklen, die sich u.U. über mehrere Jahre erstrecken können, sedimentieren gewisse Implementierungsteile in die tiefere Schicht ab und bilden dort stabile Rahmenwerke heraus [4], die in ihrer Gesamtheit alle diejenigen Arbeitsgegenstände (Konten, Kunden, Formulare, usw.) repräsentieren, die typisch für den Geschäftsbereich einer Universalbank sind [5].

2.2 Hohes Wiederverwendungspotential durch Sedimentation

Beachtenswert ist, daß diese Gegenstände nicht unmittelbar durch eine Aufgabenanalyse erhoben werden können, weil es beispielsweise das Arbeitsmittel „Konto" auf einer Bank zunächst einmal überhaupt nicht gibt, sondern nur Sparkonten, Girokonten, Darlehenskonten usw. Die Summe der Gemeinsamkeiten, die alle Kontoarten zusammen nutzen können, kristallisiert sich erst durch den Sedimentationsprozeß heraus.

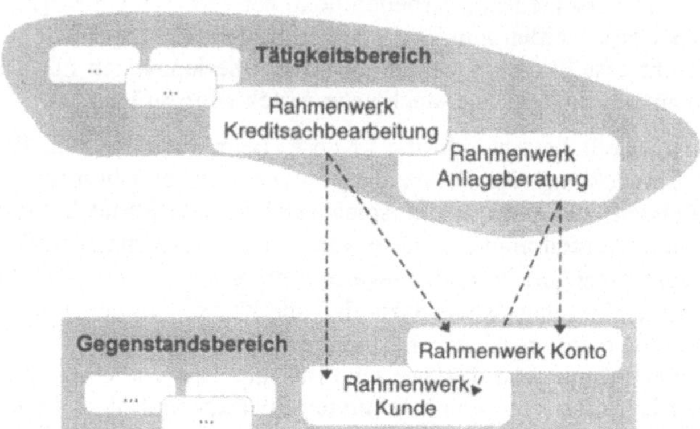

Abbildung 1: Stabile und wiederverwendbare Rahmenwerke durch den Sedimentationsprozeß. Rahmenwerke des Tätigkeitsbereichs verwenden diejenigen des Gegenstandsbereichs wieder. Eigenschaften, die in mehreren Rahmenwerken auftreten, sedimentieren so in den Gegenstandsbereich.

Genauso, wie Rahmenwerke für die „Gegenstände" eines Anwendungsbereichs entstehen, bilden sich auch solche für den Interaktionsbereich heraus. Das Ausfüllen eines Kontoeröffnungs-Antrags am Bildschirm unterscheidet sich im Prinzip nur unwesentlich vom Erfassen der Kontodaten, wie Kontonummer, Dispolimit, usw. Beide Gegenstände aus dem Anwendungsbereich lassen sich mit einem Editor bearbeiten. Auch hier kommt das gleiche Prinzip zum Einsatz, das oben bereits beschrieben wurde: Im Laufe mehrerer Entwicklungszyklen „setzen sich" die Funktionalitäten, die allen Editoren gemein sind, wie z.B. die Möglichkeit Felder anzuordnen, fehlerhafte Werte hervorzuheben, Art und Weise der Darstellung von Rückmeldungen. Auf diese Weise entsteht ein Rahmenwerk, mit dem Editoren erstellt werden können, die das Bearbeiten aller gängigen Gegenstände des Anwendungsbereichs ermöglichen. Auf die gleiche Weise bilden sich weitere Rahmenwerke heraus, die immer wiederkehrende Formen der Kooperation unterstützen, wie Versende-Mechanismen, elektronisches Unterschreiben im Rahmen einer Kompetenzverteilung, usw. Diese Rahmenwerke werden dann ebenfalls dem Gegenstandsbereich zugeordnet (siehe Abb.1).

3 Entwurfsmuster - Transparenz zwischen Benutzungsoberfläche und Implementierung

Software-technisch ist ein Rahmenwerk eine Sammlung kooperierender Klassen (z.B. in Java oder C++), die „von außen" als in sich geschlossene Einheiten genutzt werden. Die Anzahl solcher Klassen und deren Kooperation untereinander kann allerdings sehr umfassend werden, so daß es häufig schwierig ist, die Funktionsweise eines Rahmenwerkes zu dokumentieren oder auch neuen Mitarbeitern zu vermitteln. Aus dieser Problematik heraus entstanden erstmals die sog. *Entwurfsmuster* (z.B. [6]). Einerseits sind Entwurfsmuster im wesentlichen Metaphern, die Verhalten und Struktur eines einzelnen oder das Zusammenspiel mehrerer Rahmenwerke beschreiben und somit deren Funktionsweise transparent machen. Andererseits bieten sie aber auch einzigartige Möglichkeiten, um Anwenderinnen und Anwender in den evolutionären Prozeß der Entwicklung von Rahmenwerken miteinzubeziehen und somit die

Voraussetzungen für ein benutzungsfreundliches System entscheidend zu verbessern. Dieser Sachverhalt wird im folgenden anhand zweier Entwurfsmuster erörtert, die im GEBOS-Anwendungssystem erfolgreich zum Einsatz kommen.

3.1 Das Werkzeug-Material-Muster

Ein bekanntes Entwurfsmuster, welches das Zusammenspiel zwischen Objekten aus dem Anwendungsbereich (z.B. Konto) und Objekten aus dem Interaktionsbereich (z.B. Editor) beschreibt, leitet sich aus der Werkzeug-Material-Metapher ab [7]. Diese Metapher, daß nämlich mit Werkzeugen Materialien bearbeitet werden, ist ein Bild dafür, wie zwei (software-technische) Objekte zueinander in Beziehung stehen: Ein Werkzeug (Editor) wird für die Bearbeitung spezieller Materialien (Konten) hergestellt, um beispielsweise Ein- und Auszahlungen zu bewerkstelligen. Materialien haben somit keinen direkten Bezug zur Benutzungsoberfläche, dieser wird ausschließlich durch Werkzeuge hergestellt. Die Metapher wird nun für die Implementierung handlungsleitend, weil sie die Abhängigkeiten zwischen Objekten definiert: Für die Implementierung eines Werkzeugs ist Kenntnis über die Beschaffenheit der relevanten Materialien von Nöten; dagegen bleibt die Implementierung eines Materials völlig unabhängig von der eines bestimmten Werkzeugs'. Damit ist gewährleistet, daß Änderungen an der Benutzungsoberfläche keinerlei Auswirkungen z.B. auf die Rahmenwerke des Gegenstandsbereichs haben.

Handlungsleitend ist diese Metapher aber auch für die Strukturierung der Benutzungsschnittstelle. So „hantieren" qualifizierte GEBOS-Benutzer bewußt mit Materialien und wenden auf diese auch gezielt Werkzeuge an, z.B. einen Kundensucher, um anhand gegebener Kriterien eine gewünschte Menge von Kunden aufzulisten. Dies versetzt Benutzer einerseits in die Lage, auftretende Problembereiche samt Arbeitskontext präzise zu benennen (z.B. „der Kundensucher kann in dieser oder jener Situation nicht eingesetzt werden, weil ...") und ermöglicht Entwicklern anderseits derartige Anforderungen fachlich einzuordnen, zu lokalisieren und zu beurteilen (vgl. [4]). Anwender und Entwickler sprechen also dieselbe Sprache und beziehen sich dabei auch auf dieselben technischen Entitäten! Damit kommt diesem Entwurfsmuster eine zentrale Rolle hinsichtlich der Benutzerpartizipation in der evolutionären Entwicklung von Rahmenwerken zu, die ihrerseits wiederum eine notwendige Voraussetzung für rasch konvergierende Sedimentationsprozesse ist. So ist das Entwurfsmuster „Werkzeug-Material" komplett in den GEBOS-Entwicklungsprozeß integriert ([2], [3], [8]) und ein herausragendes Beispiel dafür, wie konzeptionelle Modelle und Entwurfsmodelle (Normans „conceptual model" und „design model" [9], [10]) durch den Einsatz einer geeigneten Metapher zur Konvergenz gebracht werden können.

Auch in diesem Beitrag wird im folgenden von *Materialien* gesprochen, wenn es sich um fachlich motivierte Implementierungen (Konten oder Daueraufträge) handelt und von *Werkzeugen*, wenn es um Implementierungen der Benutzungsoberfläche geht, mit denen Materialien angezeigt oder bearbeitet werden (Auflister, Editor). Daß dieses Entwurfsmuster, das seinen eigentlichen Ursprung in den Entwicklungs-Labors hat, auch in diesem Beitrag als Metapher handlungsleitend werden kann, ist ein weiteres und aussagekräftiges Indiz für das Potential so verstandener Entwurfsmuster. Ein zweites Muster, das in GEBOS ebenfalls erfolgreich eingesetzt wird, das sog. Rollenmuster, wird im nächsten Abschnitt beschrieben.

---

1 Für OO-Methodiker: Das Entwurfsmuster „Werkzeug-Material" entspricht dem sog. MVC-Paradigma (vgl. z.B. [11]) und wurde hier nur in Teilen dargestellt.

3.2 Das Rollenmuster

In einer Universalbank treten Kunden häufig in verschiedenen Rollen auf: als Anleger, Kreditnehmer, Bürge oder auch als Kunde organisationsnaher Institute wie Versicherungen oder Fondsgesellschaften. Entsprechend vielfältig sind auch die Anforderungen an die jeweiligen Arbeitsplätze der Bankangestellten. Bei der Kreditsachbearbeitung muß beispielsweise ein Gesamtüberblick über die wirtschaftlichen Verhältnisse eines Kunden gewährleistet sein, in den o.a. Serviceinseln, wo eine schnelle und effiziente Abwicklung eines Kundenwunsches im Vordergrund steht, genügen die wichtigsten Kundendaten und ggf. ein Kontenüberblick. Dagegen dürfen spezielle Kundendaten generell nur autorisierten Personen zugänglich sein, usw. Derartige Anforderungen an die Adaptierbarkeit der Arbeitsplätze ziehen u.a. software-technische Herausforderungen wie die folgenden nach sich:

- *Performanz.* In Serviceinseln müssen alle relevanten Kundendaten unmittelbar verfügbar sein, dennoch sollten einzelne Rollen so von einander entkoppelt werden, daß auch nur diejenigen „geladen" werden, die in der jeweiligen Situation erforderlich sind; kurze Antwortzeiten wäre sonst nicht leistbar.
- *Sedimentation.* Häufige Änderungen der Kundendaten, z.B. wegen gesetzlicher Anforderungen, oder Erweiterungen des Anwendungssystems, verlangsamen oder verhindern im Extremfall sogar die Reifung des betroffenen Rahmenwerks.

Anders als im Fall des Entwurfsmusters „Werkzeug-Material", das einen software-technischen Ursprung hat, dient in diesem Fall der bankfachliche Alltag als Metapher für die Strukturierung software-technischer Lösungen, woraus das sog. „Rollenmuster" entstand [5].

In Abb. 1 wurde dargestellt, auf welche Weise Rahmenwerke in unterschiedliche Schichten eingeteilt werden können. Die Idee beim Rollenmuster besteht nun darin, daß sich die Kernimplementierung eines Kunden, die von allen Kundenrollen (z.B. Bankkunde oder Bürge) genutzt werden, im Gegenstandsbereich setzt. Jene Kundenimplementierungen, die rollenspezifisch sind, verbleiben dagegen in Rahmenwerken des Tätigkeitsbereichs. Fallen nun Änderungen an einer bestimmten Rolle an, wirken sich diese nur in dem betroffenen Rahmenwerk aus. Der Kundenkern im Gegenstandsbereich sowie die Implementierungen der anderen Rollen bleiben davon unberührt. Denn die Idee besteht ja gerade darin, daß alle Gemeinsamkeiten im Gegenstandsbereich sedimentieren, und die Implementierungen im Tätigkeitsbereich voneinander disjunkt sind.

3.3 Adaptierbarkeit des Anwendungssystems

Wegen dieser Aufteilung in unterschiedliche Rahmenwerke, die alle denselben Kundenkern nutzen, untereinander jedoch völlig unabhängig sind, können auf sehr elegante Weise und ohne weiteren Implementierungsaufwand arbeitsplatz-spezifische Anwendungen erstellt werden[2]. Beispielsweise stellt Abb. 2 ein GEBOS-Werkzeug dar, das einen Gesamtüberblick über einen Bankkunden vermittelt. Im oberen Teil des Werkzeugs werden die invarianten Kundendaten präsentiert, die zum Kundenkern des Gegenstandsbereichs gehören. Dagegen listet der untere, linke Bereich des Werkzeugs nur diejenigen Rollen auf, die an dem jeweiligen Arbeitsplatztyp zur Verfügung stehen müssen. Die Auswahl einer bestimmten Rolle, z.B. Bank-

---

kunde, listet dann im rechten Bereich alle Informationen auf, die typischerweise mit dieser Rolle assoziiert werden.

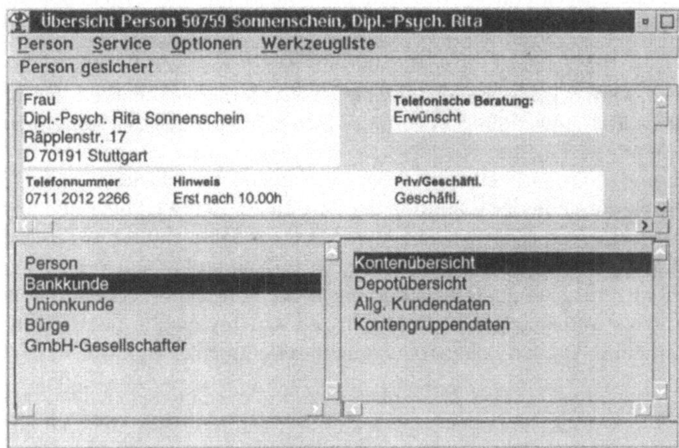

Abbildung 2: Struktur der Implementierung und der Benutzungsoberfläche fallen zusammen. Im unteren, linken Teil werden die arbeitsplatz-spezifischen Rollen und im rechten die dazugehörenden Informationen aufgelistet; letztere sind als Rahmenwerke des Tätigkeitsbereichs implementiert.

Wie auch schon das Werkzeug-Material-Muster, strukturiert das Rollenmuster die Benutzungsoberfläche und die software-technische Implementierung vor dem selben bankfachlichen Hintergrund und fördert damit die Konvergenz zwischen konzeptionellem Modell der Benutzer und dem Entwicklungsmodell. Darüber hinaus bietet das Rollenmuster fachlich sinnvolle und software-technisch elegante Möglichkeiten, wie ein Anwendungssystem auf bestimmte Arbeitsplatztypen zugeschnitten werden kann. Gleichwertige Möglichkeiten bietet übrigens auch das Werkzeug-Material-Muster, indem auf bestimmten Arbeitsplatztypen nur ein eingeschränkter oder aber ein erweiterter Satz an Werkzeugen und Materialien verfügbar gemacht wird.

Dieses Kapitel sollte verdeutlichen, welches software-ergonomische Potential in der Anwendung von Rahmenwerken steckt: beste Voraussetzungen für eine kompetente Benutzerpartizipation, fachlich adaptierbare Arbeitsplatztypen und vor allem anderen: die Wiederverwendbarkeit eines „guten" Designs (vgl. Abb. 1), was in der Deckungsgleichheit von technischer Implementierung (Rahmenwerk) und fachlicher Entität gründet. Ein weiterer entscheidender Faktor ist jedoch, daß nicht nur Materialien, sondern ganze Werkzeuge wiederverwendet werden können, und nicht wie bislang üblich nur einzelne Controls bzw. Widgets (z.B. Pushbuttons oder Listboxen). Dadurch ändert sich jedoch die Rolle des Styleguides bei der Entwicklung von Rahmenwerken drastisch.

4 Styleguides - Verwendungskataloge für bewährte Lösungen

Bei traditionellen Entwicklungsmethoden wurde die Rolle des Styleguides mit wachsendem Umfang der Anwendung immer wichtiger. Denn immer mehr Personen sind an der Entwicklung beteiligt, immer neue Komponenten kommen hinzu und die Gefahr, daß ein einheitliches

Look&Feel verloren geht, steigt kontinuierlich an. So umfaßt z.B. der Styleguide der Sparkassen-Organisation [12] über 800 Seiten. Auch die ursprüngliche, von der RWG als firmeninterner Styleguide ausgearbeitete Richtlinie umfaßte knapp 300 Seiten. Werke, die aller Erfahrung nach bei Anwendungsentwicklern auf Akzeptanzprobleme stoßen.

Beim Einsatz von Rahmenwerken verändert sich dagegen die Rolle des Styleguides gravierend, was seinen Grund in der bereits erwähnten Wiederverwendbarkeit der Werkzeuge hat. So werden im GEBOS-Anwendungssystem nur selten Werkzeuge neu konzipiert; in den meisten Fällen können bestehende wiederverwendet werden (die einfachsten Beispiele hierfür sind Auflister und Editoren). Dadurch konzentriert sich jedoch die Standardisierungsarbeit auf eine sehr frühe Phase in der Evolution der Anwendungsentwicklung und betrifft nur einen kleinen und überschaubaren Kreis von Entwicklern, so daß sich hier in Zusammenarbeit mit Software-Ergonomen das notwendige Know-how aufbaut und in den Werkzeugen selbst konserviert. Dadurch stellt sich auch die Frage nach der korrekten Verwendung einzelner Controls bzw. Widgets immer seltener. Dagegen muß ein Styleguide für die Entwicklung von Rahmenwerken die folgenden Schwerpunkte berücksichtigen:

- *Wiederverwendungskriterien.* Die Situationen, in denen ein spezielles Werkzeug wiederverwendet werden darf, müssen präzise definiert werden. Dies läßt sich am besten durch einen Katalog realisieren, der alle zur Wiederverwendung bereitstehenden Werkzeuge darstellt und deren Einsatzbedingungen definiert. Ansonsten besteht die Gefahr, daß Werkzeuge - weil sie bereits implementiert und deswegen mit verhältnismäßig geringem Aufwand nutzbar sind - außerhalb des vorgesehenen Kontexts eingesetzt werden. Aller Erfahrung nach sind die Werkzeuge dann nicht mehr handhabbar, müssen mittel- oder langfristig angepaßt werden, und machen letztlich die Vorteile der Wiederverwendbarkeit wieder zunichte.

- *Handhabungsmodelle.* Die Menge aller Metaphern (in diesem Zusammenhang als Handhabungsmodell bezeichnet), die für die Bedienung eines Anwendungssystems handlungsleitend sind, wie z.B. die Werkzeug-Material-Metapher oder die Rollenmetapher, muß frühzeitig standardisiert werden. Dies verhindert, daß bei der parallelen Entwicklung von Rahmenwerken alternative Handhabungsmodelle angewendet werden, deren spätere Vereinheitlichung zu Überarbeitungen der Rahmenwerke aus dem Gegenstandsbereich (vgl. Abb. 1) führen kann.

Nach wie vor muß ein Styleguide für Rahmenwerke z.B. Menüeinträge und damit assoziierte Aktionen standardisieren. Aller Erfahrung nach sind Rahmenwerke jedoch gegenüber derartigen Veränderungen sehr tolerant, weil nur durch wenige, lokal begrenzte Änderungen (z.B. im Gegenstandsbereich, vgl. Abb. 1), auf das systemweite Look&Feel Einfluß genommen werden kann (vgl. auch [4]).

5 Diskussion

Wie ein roter Faden zog sich der Begriff der Sedimentation durch diesen Beitrag und letztlich zielte alles darauf ab, diesen evolutionären Reifungsprozeß von der software-ergonomischen Seite von Beginn an zu unterstützen, sei es durch geeignete Metaphern für Entwurfsmuster, die ein hohes Maß an Benutzerpartizipation sicherstellen, sei es durch spezielle Kataloge, die die Wiederverwendung reglementieren. Natürlich ist dieses „vorauseilende" Wesen der Software-Ergonomie nicht erst durch das Aufkommen der Rahmenwerke aktuell geworden, sondern wird schon seit fast zehn Jahren immer wieder hervorgehoben [13]. Gereifte Rahmenwerke tragen jedoch ein extrem hohes Trägheitspotential in sich, und Änderungen am Design von Rahmenwerken haben oft weitreichende Folgen und destabilisieren im ungünstigsten Fall das gesamte System. Rein software-ergonomisch bedingte Änderungen finden deshalb in den späten Phasen der Rahmenwerke tendenziell nur noch wenige Realisierungschancen. Dies bewirkt natürlich eine Veränderung im Tätigkeitsfeld der Software-Ergonominnen und -Ergonomen. Zu den typischen Aufgabenfeldern kommt noch die Notwendigkeit hinzu, den technischen Entwurfsprozeß zu verstehen, im Rahmen des Möglichen auf die Entwicklung passender Metaphern für Entwurfsmuster hinzuwirken und vor allen Dingen darauf zu achten, daß deren Realisierung im Entwicklungsprozeß konsequent eingehalten wird.

Die in diesem Beitrag immer wieder hervorgehobenen Entwurfsmuster sind zunächst einmal Metaphern, um technische Lösungen transparent zu machen und den Grad der Wiederverwendbarkeit zu erhöhen, mit dem Ziel, mittel- und langfristig die Entwicklungskosten zu minimieren. So gibt es zwischenzeitlich ganze Kataloge [6] mit technischen Lösungen, die anhand von Metaphern als Entwurfsmuster dokumentiert sind. Viele dieser Entwurfsmuster werden die Entwicklungslabors sicherlich niemals verlassen; andere aber, und hierzu gehören das Werkzeug-Material-Muster und das Rollenmuster, haben das Potential, Entwicklungsmodelle und konzeptionelle Modelle zur Konvergenz zu führen. So entstehen transparente und auch für Anwender leicht nachvollziehbare technische Lösungen, die in vergleichbaren Kontexten wiederverwendet werden und so einen wertvollen Beitrag zur Erwartungskonformität von Anwendungssoftware leisten.

6   Literatur

[1]   RWG & Volksbank Herrenberg: Nutzungspotentiale von GEBOS-Passiv. Diebold-Studie. Stuttgart,1996: RWG.

[2]   U. Bürkle, G. Gryczan & H. Züllighoven: Object-Oriented System Development in a Banking Project: Methodology, Experience, and Conclusions. Human Computer Interaction, 10, 1995, 4, pp. 293-336.

[3]   H. Züllighoven: Das objektorientierte Konstruktionshandbuch. Nach dem Werkzeug- und Materialansatz. Heidelberg, 1998; dpunkt.verlag.

[4]   W.A. Kellogg et al.: NetVista: Growing an Internet solution for schools. *http://www.almaden.ibm.com/journal/sj/371/kellogg.txt*, 1998.

[5]   D. Bäumer et al.: Framework Development for Large Systems. Commun. ACM, 40, 1997, 10, pp. 52 - 95.

[6]   E. Gamma et al.: Design Patterns: Elements of Reusable Software. Reading, Massachusetts, 1994: Addison Wesley.

[7]   R. Budde & H. Züllighoven: Software-Werkzeuge in einer Programmierwerkstatt. München, 1990: Oldenbourg.

[8]   RWG: Die GEBOS-Methode, Interner Entwicklungsleitfaden. Stuttgart, 1996: RWG.

[9]   D.A. Norman: Cognitive engineering. In: D.A. Norman & S.W.Draper (eds.): User centered system design. Hillsdale, NJ, 1986: Lawrence Erlbaum.

[10] M.B. Rosson & S. Alpert: The cognitive consequences of object-oriented design. Human-Computer Interaction, 5, 1990, 4, pp. 345-379.

[11] T. Reenskaug: Working with Objects. New York, 1996: Prentice-Hall.

[12] Sparkassen-Organisation: SIZ Gestaltungsleitlinien für Grafische Oberflächen. Stuttgart, 1997: Deutscher Sparkassenverlag.

[13] A. Grünupp & K.-P. Muthig: Software-Ergonomie als prospektive Gestaltung der Funktionalität von Werkzeugen. In: A. Reuter (Hrsg.): GI - 20. Jahrestagung: Informatik auf dem Weg zum Anwender, Band 1. Berlin, 1990: Springer Verlag, S. 532-542.

Adresse des Autors

Dr. Peter Fach
RWG GmbH
Produktplanung / Software-Ergonomie
Räpplenstr. 17
70191 Stuttgart

Tel: +49 711 2012 2266
Fax: +49 711 2012 6266
Email: peter.fach@rwg.de
http://www.rwg.de

Rollenkonzept in der Software-Entwicklung

Sandra Frings, Anette Weisbecker, Wilhelm Lahr, Volker Reinsch

Fraunhofer IAO, Stuttgart; Fraunhofer IAO, Stuttgart; Bonndata, Bonn, GRAINsoft, Freiberg

Zusammenfassung

Die notwendige Nutzung innovativer Technologien in der Software-Entwicklung bedingt neue Tätigkeiten und stellt neue Anforderungen an die Qualifikation der Mitarbeiter. Um diese Anforderungen genauer spezifizieren, definieren und transparenter machen zu können, wurde ein Rollenkonzept entwickelt, das die Rollen in der Software-Entwicklung, deren Beziehungen zueinander und deren Integration in den Software-Entwicklungsprozeß beschreibt. Ein an die Unternehmensspezifika angepaßtes Rollenkonzept kann zu folgenden potentiellen Nutzen führen:

- Transparenz der Tätigkeiten und Verantwortlichkeiten durch Definition von Rollen
- Unterstützung beim Software-Projektmanagement (z.B. bei der Projektteamzusammensetzung und der Personalbedarfsabschätzung)
- Unterstützung bei der Ermittlung des Qualifizierungsbedarfs und Ableitung von Qualifizierungsmaßnahmen

Die gemachten Umsetzungserfahrungen bei dem Aufbau und der Einführung eines unternehmensspezifischen Rollenkonzepts in einem großen und in einem sehr kleinen Software-Haus (innerhalb des Verbundprojekts PROMPT) bestätigen diese möglichen Potentiale mit unterschiedlichen Ausprägungen und werden in diesem Beitrag näher beleuchtet (Kapitel 6).

Abstract

The necessary use of new technologies in software engineering asks for new activities and demands new personnel qualification requirements. A role concept was developed to specify these requirements in detail and to define and make them more transparent. The role concept describes the roles needed in software engineering, their relationships between them and the integration into the software development process itself. Only a company specific role concept leads to the potentials summaries in the following:

- more transparency of activities and responsibilities through the definition of roles
- support in software project management (e.g. team composition and resource estimation)
- support in determination of the qualification demands and derivation of qualification measures

The experiences made when setting up and introducing a company specific role concept in a large and a very small software house (within the research project PROMPT) confirm these possible potentials in different ways and are discussed in more detail in chapter 6.

1 Einleitung

Innovative Branchen zeichnen sich durch rasante Technologiewechsel und sich schnell entwickelnde Märkte aus. Auch im Bereich des Software-Engineerings führt dies zu einem andauernden Qualifizierungsbedarf der Mitarbeiter und einer laufenden Anpassung bzw. Verbesserung der Prozesse. Verwendet ein Unternehmen Werkzeuge und Vorgehensweisen, um die verfügbaren Wissenspotentiale effizient und flexibel in neuen Organisationsstrukturen einzusetzen, bietet sich ihm ein Wettbewerbsvorteil.

Die komplexer werdenden Informationssysteme aus Sicht der Anwendungsentwicklung führen zu der Notwendigkeit eines immer stärker ansteigenden Grads der Spezialisierung in der

Software-Entwicklung [7]. Aufgrund der daraus entstehenden Wichtigkeit, eine bestmögliche Zusammensetzung eines Projektteams zu erhalten, wurde ein Rollenkonzept entwickelt, in dem die Tätigkeiten, Ziele und der Qualifizierungsbedarf dem Mitarbeiter und dem Management transparent gemacht werden. Durch die Definition der Qualifikationsanforderungen an die Rollen wird effektiv die Personalplanung unterstützt. Die über die Beschreibung der Rollen vermittelte Transparenz motiviert nicht nur den Mitarbeiter sondern gestaltet Projektplanungen zuverlässiger.

Aufgrund der vielen verschiedenen, aber nur teilweise transparenten Qualifikationen von Mitarbeitern und aufgrund der notwendigen Flexibilität eines größeren Software-Hauses, ist die Planung des zukünftigen Qualifizierungsbedarfs der Mitarbeiter mit großem zusätzlichen Aufwand verbunden. Solch eine konkrete Planung kann nur durch die Aufnahme der vorhandenen Qualifikationen der Mitarbeiter und einem Vergleich dieses mit den gewünschten Qualifikationen für die Zukunft erreicht werden. Das hier dargestellte Rollenkonzept im Bereich der Software-Entwicklung beschreibt die Notwendigkeit nach einem solchen Konzept, die Vorgehensweise, wie dieses aufgebaut und eingeführt wird, die einzelnen Bestandteile und eine mögliche Integration des Konzepts in ein Personalinformationssystem.

Das hier beschriebene Rollenkonzept wurde im Rahmen des vom BMBF geförderten Verbundvorhaben PROMPT[1] (Organisationsgestaltung und Methoden für menschengerechte Software-Entwicklungsprozesse) [4, 10] entwickelt und gemeinsam mit Projektpartnern erprobt. Das Projekt wurde vom Fraunhofer Institut für Arbeitswirtschaft und Organisation IAO und der GSM Gesellschaft für Software-Management bmH gemeinsam mit sieben Industriepartner durchgeführt.

Bei PROMPT wurde von Projektstart an ein ganzheitlicher Ansatz verfolgt, den Software-Entwicklungsprozeß unter Beteiligung von software-entwickelnden Unternehmen zu gestalten.

Durch das Einbinden der verschiedenen Firmen, die das vorhandene Spektrum möglicher Unternehmenstypen in der Software-Entwicklung gut abdeckten, erfolgte im Rahmen von PROMPT erstmals neben der Entwicklung auch in breiterer Form die Evaluation arbeitsgerechter Organisationskonzepte, Methoden und Werkzeuge. In PROMPT wurden explizit die immer wichtiger werdenden Organisationsformen der Software-Entwicklung unter Zusammenarbeit von Softwarehäusern und Anwenderunternehmen berücksichtigt.

Neben Schwerpunkten, die den gesamten Software-Entwicklungsprozeß betreffen, wurde innerhalb von PROMPT ein Prozeßmodell entwickelt, welches sich aus einem prozeßorientierten Vorgehensmodell, einem Qualifizierungsmodell und darauf aufbauend und vervollständigendem Rollenkonzept zusammensetzt. Aufgrund der flexiblen Anpassungs- und der vielseitigen Umsetzungsmöglichkeiten des Rollenkonzepts konnte insbesondere dieses im Rahmen der Betriebsprojekte aufgebaut und praktische Maßnahmen zur Verbesserung des Software-Entwicklungsprozesses abgeleitet werden [4].

---

1 "Das diesem Beitrag zugrundeliegende FE-Vorhaben wurde im Auftrag des BMBF unter dem Förderkennzeichen 01 HP 904/4 durchgeführt. Die Verantwortung für den Inhalt dieser Veröffentlichung liegt bei den Autoren."

2 Prozeßorientierte Vorgehensweise bei der Software-Entwicklung

Von zentraler Bedeutung für die Software-Entwicklung sind Vorgehensmodelle, die deren wesentliche Prozesse beschreiben und die mit Hilfe des Tailorings an Firmen- und Projektspezifika angepaßt werden können [2, 3]. Im Rahmen von PROMPT wurde ein Vorgehensmodell für den ganzheitlichen Software-Entwicklungsprozeß entwickelt und erprobt, welches unter expliziter Berücksichtigung der Wiederverwendung und Benutzerpartizipation auch die kaufmännischen Prozessen abdeckt [10].

Das PROMPT-Vorgehensmodell ist hierarchisch aufgebaut und besteht aus fünf Bereichen (siehe Abb. 1). Jeder Bereich ist in Kernprozesse, die aus Subprozessen bestehen, untergliedert. Aktivitäten (bzw. Tätigkeiten) und zugeordnete Rollen beschreiben die Subprozesse.

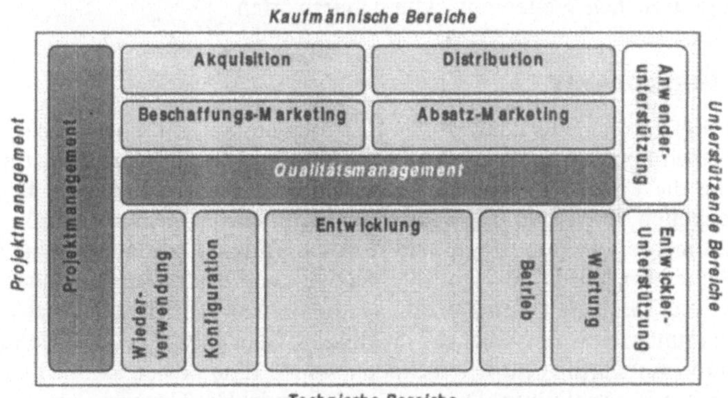

Abb. 1: Kernprozesse im PROMPT-Vorgehensmodell

Zur Identifikation und Beschreibung der notwendigen Rollen war eine strukturierte Vorgehensweise erforderlich, die das Grundgerüst für ein Rollenkonzept bildete, welches das Vorgehensmodell ergänzt. Das PROMPT-Rollenkonzept beschreibt das Zusammenspiel aller Rollen in der Software-Entwicklung und die Vorgehensweisen der Definitionen, Ermittlungen und Integration der Rollen. Somit ist bei dem Aufbau eines Rollenkonzepts eine unternehmensspezifische Definition zum Ziel gesetzt.

Abb. 2: Bestandteile des Rollenkonzepts

Diese beinhaltet die Integration von Rollen in die prozeßorientierte Vorgehensweise bei der Software-Erstellung, Aspekte im Management von Software-Projekten, welche z.B. die Teamzusammensetzung und die Bedarfsabschätzung einschließt [4], und die Integration von Rollen in ein Qualifizierungskonzept für die Software-Entwicklung. Die Ermittlung des Qualifizierungsbedarfs und die Ableitung von Qualifizierungsmaßnahmen werden durch das Rollenkonzept unterstützt.

Aufgrund verschiedener Organisationsformen und unterschiedlicher Projekttypen, kann das Rollenkonzept keinem verkaufsfertigen Rezept zur Verbesserung der Softwareprojektplanung entsprechen. Es muß an die unternehmensspezifischen Anforderungen, sogar an verschiedene projektspezifische Bedingungen, angepaßt werden. Diese Vorgehensweise bzw. die Vorschläge, Hinweise und Empfehlungen sind Bestandteil des Rollenkonzepts. Ferner ist das Rollenkonzept durch den Einsatz neuer Technologien und daraus entstehenden neuen Prozessen einer kontinuierlichen Weiterentwicklung unterworfen.

3 Das Rollenkonzept

Im Rollenkonzept wird eine Rolle durch die Erfahrungen, Kenntnisse und Fähigkeiten, die für einzelne Aufgaben notwendig sind, definiert wird [5 und 6]. Sie sagt jedoch nichts über die Person aus, die diese Rolle wahrnimmt. Eine Rolle wird mit Tätigkeiten, Ergebnissen und Verantwortlichkeiten verbunden. Sie kann verantwortlich für mehrere Tätigkeiten sein, aber ebenso können mehrere Rollen für ein und dieselbe Tätigkeit zuständig sein. Abhängig von den Gegebenheiten im Projekt oder im Unternehmen, vom Aufgabengebiet und von den Prioritäten werden die Tätigkeiten den Rollen zugeordnet. Der Rolleninhaber ist die Person, die die Rolle einnimmt, somit entsprechend qualifiziert sein muß, die geforderten Tätigkeiten durchführt und verantwortlich für deren Ergebnisse ist. Eine Person kann dabei mehrere Rollen wahrnehmen z.B. Analytiker, Designer und Entwickler. Genauso kann eine Rolle von mehreren Rolleninhabern eingenommen werden, z.B. mehrere Entwickler in einem Projekt.

Das Rollenkonzept besteht grundlegend aus einer Vorgehensweise (siehe Abb. 3), die beschreibt, wie es an ein Unternehmen angepaßt und weiterentwickelt werden kann. Neue Geschäftsfelder, Technologien oder Wissen, durch die neue Prozesse entstehen würden, erfordern die Ergänzung von neuen Rollen. Folgende Schritte sollten somit bei der Erweiterung des Rollenkonzepts bedacht werden:

1. Identifikation von Rollen
2. Integration in das Prozeßmodell
3. Analyse des Aufgabenbereichs
4. Analyse der Rollenart

5. Analyse der Beziehungen zu anderen Rollen
6. Erstellung der Rollenbeschreibung
7. Kennzahlen (zur Ressourcenabschätzung aufgrund der neu identifizierten Rollen, z.B. Größe der Projekte)

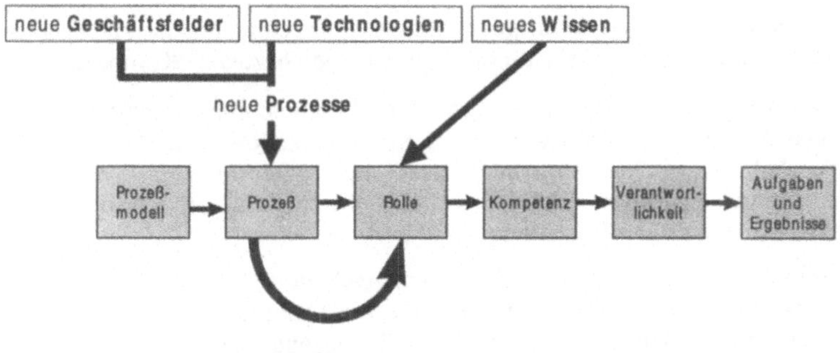

Abb. 3: Vorgehensweise bei der Entwicklung und Weiterentwicklung des Rollenkonzepts

Weiterhin gibt die Rollenstruktur den identifizierten Rollen einen festgelegten Aufbau, die in unterschiedliche Rollenarten unterteilt ist. Phasenabhängige Rollen beziehen sich auf eine bestimmte Phase des Software-Entwicklungsprozesses. Die Rolle "Analytiker" ist beispielsweise eindeutig der Phase "Analyse" zugeordnet. Wohingegen die Rolle "Projektleiter" einer phasenübergreifenden Rolle entspricht. Weiterhin gibt es die domänenspezifischen Rollen, die sich auf eine Domäne, d.h. einem Themenbereich in der Software-Entwicklung, beziehen. Z.B. ist der "Ergonomie-Experte" eine domänenspezifische Rolle.

Solch ein Aufbau ist für die Identifikation aller notwendigen Rollen und späteres Hinzufügen von Rollen hilfreich.

Zuletzt gibt die Rollenbeschreibung vor, wie eine Rolle geeignet definiert werden soll, um so z.B. den Qualifizierungsbedarf möglichst einfach bestimmen zu können.

Eine Rollenbeschreibung besteht im wesentlichen aus:

- einem leicht verständlichen, möglichst bereits im Unternehmen geläufigen Namen als Bezeichnung für die Rolle,
- einer stichwortartige Auflistung der vorwiegenden bzw. wichtigsten Aufgaben und zu erreichenden Ergebnisse,
- den Qualifikationsanforderungen an die Rolle, untergliedert in vier verschiedene Kompetenzarten (Fach-, Methoden-, Sozial- und Medienkompetenzen),
- einer Angabe zur notwendigen Qualifikationsausprägung der einzelnen Kompetenzen und
- einer Angabe zur Beziehung der Rolle zu anderen Rollen, die beschreibt, welche Abhängigkeiten zwischen Rollen bestehen oder welche Konflikte hervorgerufen werden könnten, wenn ein und dieselbe Person verschiedene Rollen einnimmt.

Die Kompetenzarten [1, 4] der Rollenbeschreibung sind folgendermaßen definiert.

Unter **fachlicher Kompetenz** versteht man berufsbedingt erworbene Qualifikationen und Erfahrungen sowie das fachspezifische und fachübergreifende Wissen. Die **Methodenkompetenz** entspricht der Fähigkeit, dieses Fachwissen zu nutzen, zu kombinieren und zu ergänzen. **Soziale Kompetenzen** umfassen die persönlichen Ausprägungen bzw. Grundverhaltensmuster. Sie können in zwei Bereiche untergliedert werden. Zum einen sind Kompetenzen im

Umgang mit anderen Personen damit gemeint, wie z.B. Teamfähigkeit und Kooperationsfähigkeit und zum anderen sind es rein auf die eigene Person bezogene Kompetenzen.

Die **Medienkompetenz** schließlich hat aufgrund des Internetbooms zunehmend an Bedeutung gewonnen. Das Nutzen von Informations- und Kommunikationstechnologien umfaßt die Informationsbeschaffung, -aufbereitung und -darstellung. Weiterhin ist das Verwalten von Wissen und das Filtern von Informationen nach deren Wichtigkeit sowie das Beherrschen verschiedener Medien von großer Bedeutung.

Für die einzelnen Kompetenzarten werden Themengebiete für das benötigte Wissen festgelegt. Eine Aufstellung dieser Themengebiete, die die Qualifikationsanforderungen der vier Kompetenzarten im Bereich der Software-Entwicklung abdecken (siehe Abb. 4), sollte abhängig von der Vorgehensweise bei der Software-Entwicklung durchgeführt werden. Hierfür ist entweder die Übernahme der Angaben der bereits definierten Prozesse (aus dem dokumentierten Vorgehensmodell) oder eine Analyse der entsprechenden Prozesse notwendig. Die zweite Möglichkeit hat, neben des Aufbaus eines Rollenkonzepts, den Nutzen, daß diese Prozesse untersucht, definiert und transparent gemacht werden.

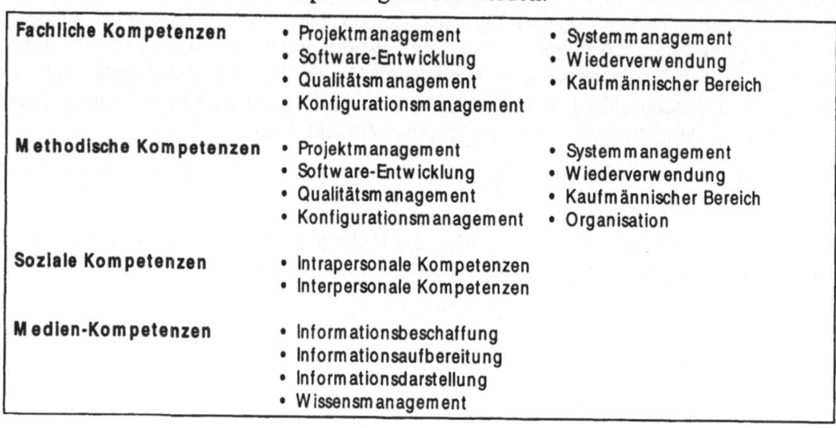

Abb.4: Qualifikationsanforderungen der vier Kompetenzarten

Als Grundvorgabe könnten die Rollen in Abbildung 5 zu verstehen sein. Abhängig von der Anzahl der Mitarbeiter im Software-Entwicklungsbereich und der Schwerpunkte des Unternehmens werden die passende Rollen mit Hilfe oben angegebener Vorgehensweise daraus ausgesucht oder weitere hinzugenommen.

<table>
<tr><td valign="top">

Phasenabhängige Rollen
- Analytiker
- Designer
- Entwickler
- Tester

Domänenspezifische Rollen
- C/S-Experte
- Datenbank-Experte
- Ergonomie-Experte
- Groupware-Experte
- Internet-Experte
- Multimedia-Experte
- Netzwerk-Experte
- Reuse-Experte
- Datenschutz-Experte
- Softwaremethoden-Experte
- Softwaretechnologie-Experte
- Workflow-Experte

</td><td valign="top">

Phasenübergreifende Rollen
- Projektleiter
- Projektassistent
- Qualitätsmanager
- QM-Assistent
- Qualitätsprüfer
- Reuse-Leiter
- Konfigurationsmanager
- KM-Assistent
- KM-Administrator
- Softwarearchitekt
- Wissensmanager
- Datenschutzbeauftragter
- Release-Manager
- Dokumentations-Manager
- Technischer Autor
- Systemadministrator
- Wartungsleiter
- Schulungsleiter
- Trainer
- HW-Berater
- Reuse-Support
- Support

</td><td valign="top">

Beteiligte Rollen
- Anwendungsbereich-Experte

Kaufmännische Rollen
- Vertriebsleiter
- Einkaufsleiter
- Marketingleiter (Akquisition)
- Verkaufsleiter

Externe Rollen
- Benutzer / Anwender
- Lieferant
- Kunde

</td></tr>
</table>

Abb. 5: Beispielrollen in der Software-Entwicklung

4 Qualifizierung

Im PROMPT-Prozeßmodells stellt das Rollenkonzept und die definierte Rolle die Schnittstelle zwischen dem Vorgehensmodell und dem Qualifizierungsmodell dar.

Ausgehend von dem Soll-Profil einer Rolle kann der Qualifizierungsbedarf ermittelt werden. Für die Besetzung einer Rolle durch einen Mitarbeiter kann relativ einfach ein Ist-Profil seiner Qualifikation erarbeitet werden. Dazu können verschiedene Methoden, z.B. Befragungen oder Selbsteinschätzung eingesetzt werden. Das ermittelte Ist-Profil wird mit dem Soll-Profil verglichen. Auf diese Weise werden Übereinstimmungen, Defizite und Überqualifizierungen aufgedeckt. Aus den aufgezeigten Defiziten können der Qualifizierungsbedarf abgeleitet und entsprechende Qualifizierungsmaßnahmen initiiert werden. Somit kann das Rollenkonzept zur gerichteten Qualifizierung der Mitarbeiter eingesetzt werden.

In Abbildung 6 sind eine graphische Darstellung eines Rollenprofils der Rolle "Entwickler" und die Themengebiete beschränkt auf die eigentliche Software-Entwicklung zu sehen. Es stellt einen Ist-Soll-Vergleich der notwendigen und vorhandenen Ausprägungen der Qualifikationsanforderungen dar.

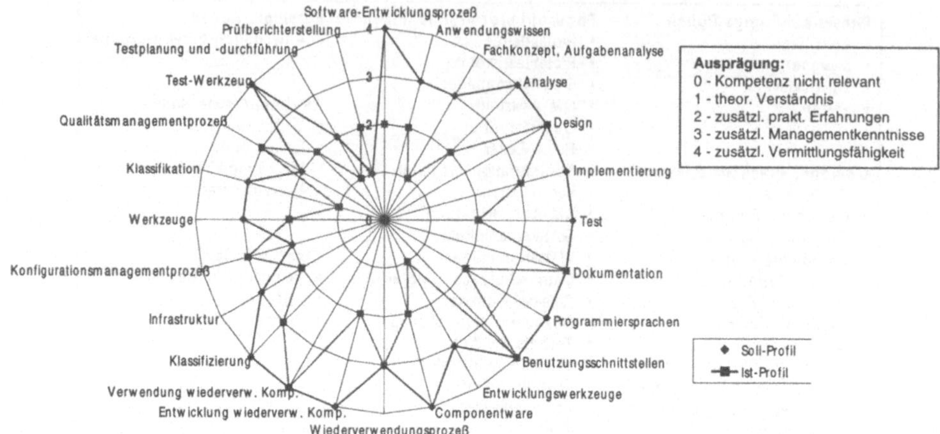

Abb. 6: Graphisches Rollenprofil der Rolle "Entwickler" aus Ist-Soll-Vergleichssicht

5 Umsetzungserfahrungen und Fazit

Die Erfahrungen bei der Einführung des PROMPT-Rollenkonzepts in größeren Softwarehäusern zeigen, daß zuerst die etablierten Prozesse und damit verbundenen, bereits vorhandenen Rollen und Tätigkeiten identifiziert werden mußten, um diese dann in ein einheitliches Rollenkonzept zu integrieren. Die Analyse führte zu einem besseren Verständnis der Prozesse in der Software-Entwicklung und vereinfachte die Zuordnung der vorhandenen Rollen in das PROMPT-Rollenkonzept. Daraus konnte dann eine Rollenbeschreibung erstellt werden.

Konkrete Umsetzungserfahrungen bei dem Aufbau eines Rollenkonzepts wurden bei der Bonndata Gesellschaft für Datenverarbeitung mbH gemacht, die mit über 500 Mitarbeitern zu den führenden Beratungs- und Softwareunternehmen. im deutschsprachigen Raum zählt. Sie betreibt als Kompetenzzentrum Versicherungswirtschaft die vollständige Informationsverarbeitung für die Versicherungsgruppe der Deutschen Bank.

In immer stärkeren Maße beeinflussen neue Technologien die Qualifikationsanforderungen der Mitarbeiter, die diese einführen, integrieren und umsetzen. Die Ausbildung der Mitarbeiter für zukünftige, neue Technologien benötigt eine viel stärker am aktuellen Bedarf orientierte, strukturierte Vorgehensweise zur spezifischen Qualifizierung, die sich an den betrieblichen Anforderungen und der Firmenstrategie orientiert.

Aus diesem Grund wurde ein Rollenkonzept zusammen mit Bonndata entwickelt, für dessen Präsentation eine Argumentenliste für die Zielgruppen Mitarbeiter und Projektleiter erstellt wurde. So konnten die damit verbundenen Nutzen und Potentiale geeignet dargestellt werden.

Um genau zu bestimmen, welche Rollen bei Bonndata notwendig waren, wurden die etablierten Software-Entwicklungsprozesse analysiert (vgl. Vorgehensweise aus Abb. 3). Es hat sich gezeigt, daß das Bonndata-Vorgehensmodell zusammen mit der Erweiterung der Kernprozesse "Prozeßmanagement" und "Know-How-Management" des innerhalb von PROMPT entwickelten Vorgehensmodells eine geeignete Grundlage für das Rollenkonzept darstellt.

Auf den Kernprozeß "Know-How-Management" mit den Subprozessen "Wissensakquisition, -darstellung und –verbreitung" und der zugeteilten Rolle "Know-How-Manager" wurde besonderer Wert gelegt, da dadurch explizit im Vorgehensmodell der Software-Entwicklung der Aspekt "Wissensmanagement" berücksichtigt wird.

Bei der Analyse der Prozesse wurde eine große Anzahl von Rollen zusammengetragen, die nach Betrachtung der unterschiedlichen Rollenarten zusammengefaßt 21 Rollen ergaben und alle Phasen und Prozesse abdecken. Für eine spätere Ergänzung der Rollen gibt die beschriebene Vorgehensweise zur Erweiterung des Rollenkonzepts eine genaue Anleitung.

Aufgrund bereits etablierter Aktivitäten im Bereich der Unternehmensentwicklung bei Bonndata konnten zur einheitlichen Strukturierung der Rollenbeschreibungen fünf Kompetenzarten den Bonndata Förderprogrammen entnommen werden, die sich in fachliche, methodische, persönliche, soziale und strategische Kompetenzen untergliedern. Weiterhin wurden die in Abbildung 4 dargestellten Themengebiete der Kompetenzarten diese Gegebenheiten angepaßt.

Empfehlungen zur Zuordnung der Rollen zu Rolleninhaber und Projekten basieren auf einer Charakterisierung von Projekten, die vorwiegend bei Bonndata durchgeführt werden. Daher wurden Projekttypen identifiziert und für diese unterschiedliche Vorgehen bei der Zuordnung von Rollen zu Prozessen in der Linie und in den Projekten vorgeschlagen.

Aufgrund neuer Technologien ist es sowohl für das Unternehmen Bonndata als auch für die Mitarbeiter in Bezug auf ihre Karriereplanung von großer Bedeutung, frühzeitig die in Zukunft anstehenden Qualifizierungen zu planen. Dies bedeutet, daß es wichtig ist zu wissen, wie viele Rolleninhaber für eine Rolle qualifiziert oder umgeschult werden müssen, um somit die erforderliche Qualifikation zu gewährleisten. Für diesen Fall wurde anhand einer hypothetischen Beispielberechnung gezeigt, auf welche Faktoren zu achten ist und welche Voraussetzungen gemacht und welche Randbedingungen angenommen werden können bzw. müssen.

Zusammenfassend ist in Abbildung 7 dargestellt, wie das Know-How-Management bei Bonndata in der Software-Entwicklung aufgebaut ist und wie sich dessen Potentiale mit denen eines Rollenkonzepts verbinden läßt.

Auf Kernprozesse konzentrieren	• Rollenbeschreibungen	– Aufgaben / Eigenschaften – Fachlich / Methodisch – Qualifikationsausprägung
Know-How-Einsatz organisieren	• Rollenzuordnung	– Projekttypen → Client/Server, Host, ... – Linie → Entwicklung, Betrieb, ...
Know-How wirtschaftlich einsetzen	• Kennzahlen	– Qualitativ – Quantitativ
Know-How aktualisieren, erweitern	• Bedarfsanalyse • Qualifikationspfade	– Soll / Ist
Know-How aktiv verwalten	• Rollensystem • Mitarbeiter-Überblick	– Pflege, Ausbau – Know-How-Datenbank – aktives Wissensmanagement

Abb. 7: Nutzung des Rollenkonzepts zum Know-How-Management bei Bonndata

Die Umsetzung des Rollenkonzepts bei Bonndata ist zeitnah geplant.

Bei kleineren software-entwickelnden Unternehmen, wie zum Beispiel bei der GRAINsoft GmbH in Freiberg / Sachsen, wird nicht die gesamte Bandbreite des Rollenkonzepts von der Ermittlung und Beschreibung über die Projektplanung bis hin zu der Ableitung von Qualifizierungsmaßnahmen genutzt. Hier liegt der Schwerpunkt auf der transparenten Darstellung der Tätigkeiten einer Rolle und der dafür benötigten Fähigkeiten. Das Rollenkonzept trägt dazu bei, sich der gerade aktuell wahrgenommenen Rolle und den damit zu erzielenden Ergebnissen bewußt zu werden.

Wie in Abbildung 8 zu sehen ist, wurden bei GRAINsoft die in der Software-Entwicklung vorhandenen Rollen identifiziert und sowohl den Prozessen als auch den Rolleninhabern zugeordnet. Es zeigt sich, daß ein Rolleninhaber hier immer mehrere Rollen wahrnimmt.

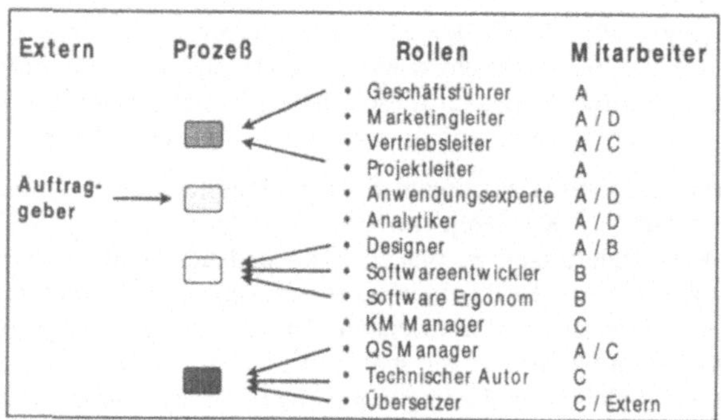

Abb. 8: Identifizierte Rollen für die Auftragsabwicklung bei GRAINsoft und exemplarische Zuordnung bei Rolleninhaber (A-D und Extern sind spezifische Rolleninhaber bei GRAINsoft)

Sowohl in einem großen als auch in einem kleinen software-entwickelnden Unternehmen kann die Verwendung eines Rollenkonzepts diesem helfen, sein bedeutendstes Potential, die Mitarbeiter, effizient und flexibel innerhalb der Unternehmensprozesse einzusetzen.

Das beschriebene Konzept stellt jedoch keine computerunterstützte Lösung dar, wie sie für eine möglichst flexible, leicht erweiterbare und änderbare Realisierung sinnvoll wäre. Die Umsetzungserfahrungen haben gezeigt, daß eine technische Unterstützung unabdingbar ist. In der Aufbauphase kann diese Datenbank zur reinen Verwaltung der Rollen dienen und später in ein Skill-Management- oder Personalinformationssystem [8] integriert werden.

6 Literatur

[1] H. J. Bullinger: Arbeitsgestaltung. Stuttgart, 1995: Teubner

[2] A.-P. Bröhl, W. Dröschel: Das V-Modell. München, 1995: Oldenbourg

[3] R. Singh: Information Technology – Software Lif-Cycle Processes, Committee Draft, ISO/IEC(JTC1)-SC7, 12.2.1993

[4] A. Weisbecker, G. Groh: PROMPT – Organisationsgestaltung und Methoden für menschengerechte Software-Entwicklungsprozesse. Stuttgart, 1998: Fraunhofer IRB

[5] H. Oberquelle: Sprachkonzepte für benutzergerechte Systeme. Berlin, 1987: Springer

[6] H. Balzert: Lehrbuch der Software-Technik. Heidelberg, 1998: Spektrum

[7] H. Weber: Die Software-Krise und ihre Macher. Berlin,1992: Springer

[8] H. J. Bullinger: Software-Management-Forum. Stuttgart, 1998: Fraunhofer IRB
[9] S. Frings, A. Weisbecker: Für jeden die passende Rolle. In: it-Management, Ausgabe Juli 1998
[10] PROMPT – Organisationsgestaltung und Methoden für menschengerechte Software-Entwicklungsprozesse. Tagungsband zur Veranstaltung "Software-Technologien in der Praxis", Fraunhofer IAO, Stuttgart, 22.4.97

Adressen der Autoren

Dipl.-Inf. Sandra Frings
Fraunhofer Institut für Arbeitswirtschaft und Organisation
Competence Center Software-Management
Nobelstr. 12
70569 Stuttgart
Email: Sandra.Frings@iao.fhg.de
http://www.iao.fhg.de

Bonndata Gesellschaft für Datenverarbeitung mbH
Dipl.-Math.Wilhelm Lahr, EPV
Rochusstraße 4
53123 Bonn
Email: Willi.Lahr@db.com

Dr.-Ing. Anette Weisbecker
Fraunhofer Institut für Arbeitswirtschaft und Organisation
Competence Center Software-Management
Nobelstr. 12
70569 Stuttgart
Email: Anette.Weisbecker@iao.fhg.de
http://www.iao.fhg.de

GRAINsoft GmbH
Dipl.-Math. Volker Reinsch
Chemnitzer Straße 40
09599 Freiberg/Sachsen
Email: GRAINsoft@t-online.de

GUIfizierung umfangreicher Hostanwendungen mit Java

Andreas Gronski   Harald Haller

sd&m software design & management GmbH & Co. KG, München

Zusammenfassung

In diesem Vortrag wird ein in Produktion befindliches GUI-Frontend zu seit Jahren bei der Deutschen Bahn AG im Einsatz befindlichen Hostanwendungen vorgestellt. Die Realisierung erfolgte mit Java 1.1 und geht in ihrer Funktionalität deutlich über die automatischer GUIfizierung hinaus. Kernpunkte sind die konzeptionelle Integration bisheriger, bei den Benutzern bekannter Abläufe und Bedienelemente mit den neuen Möglichkeiten einer GUI sowie deren technische Umsetzung.

Abstract

This paper presents the conception and realisation of a GUI-frontend to a complex IBM-mainframe application, which is in use at the Deutsche Bahn AG for several years. It's implemented in java and offers much more functionality than a GUI-wrapper. The main challenge was to integrate the new opportunities of a modern GUI and the established functions in order to address users familiar with host applications as well as new users.

1 Einleitung

Zeichenorientierte Benutzeroberflächen wie beispielsweise der CICS-Transaktionsmonitor auf einem IBM-Mainframe sind nach wie vor in zahlreichen Unternehmen und an zentralen Stellen im operativen Betrieb. Durch den zunehmenden Einsatz von Intranet-Techniken bietet sich die Chance, auch die bestehenden Hostanwendungen in eine aus der PC-Standardsoftware gewohnte, GUIfizierte Form zu bringen. Letztere umfaßt neben der optischen Aufwertung insbesondere auch eine umfangreiche Erweiterung der Bedienungsfunktionalität. Die zentrale Aufgabe besteht nun in der Integration der gewohnten Bedienung von CICS-Dialogen und der aus der Standardsoftware gewohnten Handhabung einer GUI-Anwendung, sodaß beide Arbeitsweisen adäquat unterstützt werden.

1.1 Zur Vorgeschichte

Die in diesem Vortrag vorgestellte Softwarelösung wurde von Dezember 1997 bis August 1998 entwickelt und ist von September 1998 bis voraussichtlich Januar 1999 in der Testphase für den produktiven Einsatz im zentralen Controlling der Deutschen Bahn AG. Diese Entwicklung entstand im Kontext des Wartungsprojektes „Transportleistungsrechnung" (TLR), das bereits sei März 1993 besteht, verschiedene Teilsysteme beinhaltet und durch einen hohen Verknüpfungsgrad mit Nachbarsystemen gekennzeichnet ist. Programmiersprache war und ist XCobol, ein hauseigener Cobolaufsatz, das Betriebssystem OS390, die Datenbank DB2. Zum

Projekt-Kickoff des GUI-Frontends existierten rund 300 Dialoge mit zusammen 1000 Masken
sowie Batches mit einer Gesamtlaufzeit von 10 Tagen.

1.2 Rahmenbedingungen

Derzeit nutzen rund 3000 Anwender im Netzwerk der Deutschen Bahn die TLR-Software.
Viele davon arbeiten schon seit 10 Jahren und mehr mit CICS-Dialogen, sodaß eine GUI für
die fachlich gleichen Vorgänge nicht „auf der grünen Wiese" anfangen darf. Gleichzeitig sind
die EDV-Arbeitsplätze der Deutschen Bahn AG seit mehreren Jahren nahezu vollständig mit
dem Betriebssystem Microsoft Windows standardisiert, sodaß die GUI-Bedienung für die Benutzer keine grundsätzlichen Neuerungen beinhaltet. Mit der geplanten Umstellung ergab sich
weiter die Chance, ein unternehmensweit einheitliches Erscheinungsbild und Verhalten der
Benutzeroberfläche für Hostanwendungen im beschriebenen Anwendungskontext zu etablie-
ren. Folglich mußte es also möglich sein, die neue Benutzerschnittstelle sowohl in ihren wesentlichen Funktionen über die Funktionstasten zu steuern, als auch zum unternehmensweiten
Styleguide für Applikationen kompatible Oberflächen, Bedienelemente, Abläufe, etc. bereitzustellen. Da die Konsistenzanforderungen in hohem Maße gewahrt werden konnten, besteht
mit der hier vorgestellten GUIfizierung auch die berechtigte Aussicht auf deutlich kürzere
Einarbeitungszeiten.

1.3 Vorhergehende Benutzerschnittstelle

Um die folgenden Gestaltungsentscheidungen nachvollziehen zu können, mag es hilfreich
sein, zunächst in einem kurzen Überblick die zuvor eingesetze Benutzerschnittstelle zu betrachten. Diese besteht für den Benutzer aus Dialogen und Masken. Ein Dialog ist in der Regel
die Präsentation einer Datenbanktabelle. Er ist über Einstiegsmenüs zugänglich und besteht im
wesentlichen aus Such- , Übersichts- und Einzelsatzmaske. Eine Maske ist dabei gleichbe-
deutend mit einer Bildschirmanzeige. Die folgende Abbildung zeigt einen solchen Dialog mit
seinen drei Masken.

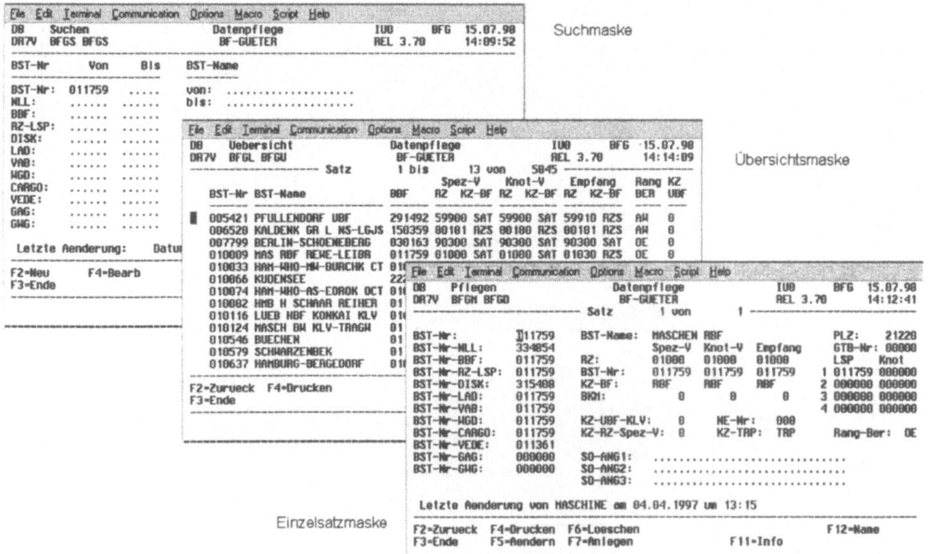

Abbildung 1: herkömmliche Hostmasken

Der typische Arbeitsablauf besteht aus folgenden Schritten:

a) Suchkriterien angeben
b) in gefundenen Datensätzen blättern
c) einzelne Datensätze lesen, pflegen oder drucken

Die im folgenden Abschnitt skizzierte Spezifikation der (graphischen) Benutzerschnittstelle unterstützt genau dieses Modell, ohne dabei das Vorgängermodell zu imitieren.

2 Gestaltung der Benutzerschnittstelle

2.1 Arbeiten mit Dialogen und Masken

Die Metapher des Dialogs bietet eine in der Zielgruppe etablierte Orientierungsmöglichkeit in den zur Verfügung stehenden Datenbeständen. Ähnlich wie ein Dokument, ein Foliensatz oder eine Gruppe von Hypertextdokumenten bündelt ein Dialog fachlich zusammengehörende Daten zu einer Einheit. Technisch ist dabei keineswegs festgelegt, ob die präsentierten Daten aus einer oder mehreren Datenbanktabellen stammen. Die Dialoge werden wie bei den CICS-Anwendungen hierarchisch angeordnet, d.h. zu Gruppen (Menüs) zusammengefaßt, welche aus Untermenüs und/oder Dialogen bestehen können. Diese hierarchische Struktur wird durch ein hierarchisch aufgebautes Menü „Dialog-Auswahl" im Hauptmenüpunkt „Dialog" abgebildet, woraus der Benutzer den gewünschten Dialog auswählt.

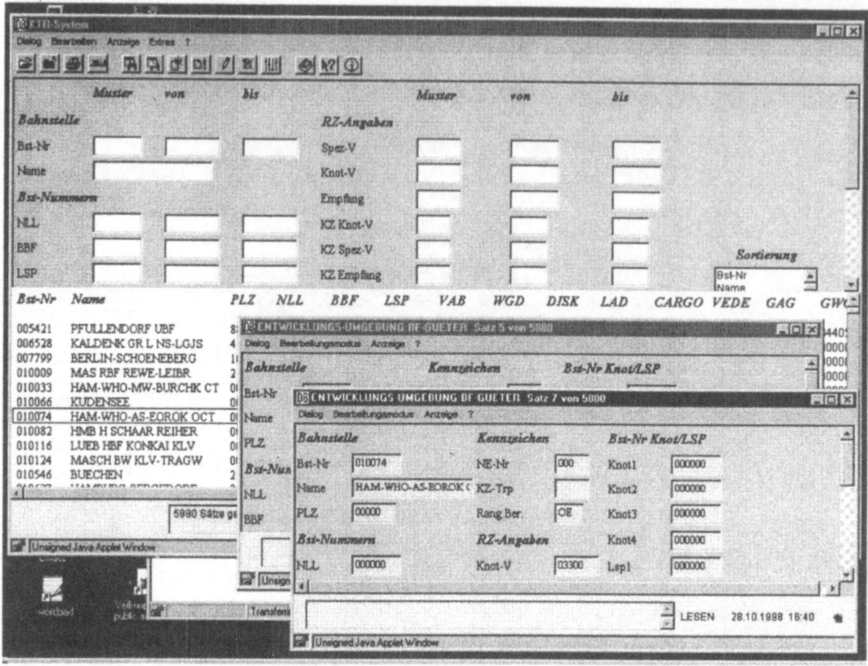

Abbildung 2: Oberfläche neuer Intra-/Internet-Dialoge

Ebenso wie das Dialogmodell sollte auch die Maskenmetapher beibehalten werden. Der erste Ansatz ist, jede Maske auf ein Fenster abzubilden, wie dies in vielen automatischen GUIfizierungstools für Hostanwendungen der Fall ist. Dadurch ist kaum etwas gewonnen, denn das Wechseln der Masken über Funktionstasten ist hier nur durch das wechselnde Anklicken von Fenstern ersetzt. Die zweite offensichtliche Variante, alle Masken eines Dialogs zugleich in ein Fenster zu setzten, würde ein intensives Scrolling erfordern, wodurch bei größeren Masken der Überblick verloren geht. Nach Erwägung dieser und zahlreicher weiterer Möglichkeiten entschieden wir uns für zwei Fensterarten: Ein Hauptfenster mit Such- und Übersichtsmaske und ein oder mehrere Fenster zur Darstellung von einzelnen Datensätzen (siehe Abbildung 2).

Grund für die Zusammenlegung von Such- und Übersichtsmaske war die enge funktionale Bindung. Die Suchmaske stellt quasi das Bedienelement der Datenbank bzw. deren Eingabe dar, während die Übersichtsmaske das Resultat des Programms anzeigt. Die so entstandene Vergleichsmöglichkeit von Eingabe und Ausgabe auf einen Blick ermöglicht eine Interaktion mit direktem Feedback und stellt somit insbesondere für neue Benutzer einen echten Mehrwert dar. Im Vergleich dazu bringt die Kopplung von Übersichtsmaske und Einzelsatzmaske für den Benutzer einen deutlich geringeren Mehrwert, da der Ausgangspunkt der Aktion nicht notwendig zur Orientierung beiträgt.

Das ausschlaggebende Argument für die Abbildung der Einzelsatzmaske in ein eigenes Fenster war die Vergleichsmöglichkeit verschiedener Datensätze. Diese ist zwar auch in der Übersichtsmaske gegeben (und dort sogar günstiger), aber durch Anzeige eines Datensatzes auf einer Zeile limitiert. Für jede Einzelsatzmaske ein neues Fenster zu öffnen, ohne andere zu schließen, bedeutet aber auch, einen Fensterwildwuchs zuzulassen. Es ist in zahlreichen Gesprächen mit verschiedenen Benutzergruppen deutlich geworden, daß der Umgang mit Fenstern individuell sehr unterschiedlich ist. Daher haben wir entgegen unserer generellen Entscheidung, Programmverhalten nicht zu individualisieren (siehe Abschnitt 0), die Anpassungsmöglichkeit integriert, die maximale Anzahl geöffneter Einzelsatzfenster zu wählen. Ist diese Anzahl erreicht, werden die Masken nach dem FIFO-Verfahren aktualisiert.

Schließlich stellt sich noch die Frage nach der automatischen Aktualisierung der Maskeninhalte. Wird die Änderung eines Datensatzes im Einzelsatzfenster in der Übersichtsmaske automatisch übernommen? Verschwinden Einzelsatzfenster, wenn die entsprechenden Datensätze in der Übersichtsmaske gelöscht werden? Wie verhalten sich zwei Einzelsatzfenster, die denselben Datensatz anzeigen? Die Lösungsmöglichkeiten spannen sich zwischen den Positionen „immer aktualisieren" und „nie aktualisieren" auf. Da die Vorgängerschnittstelle letztere Position implementiert hat, entschlossen wir uns dazu, Einzelsatzfenster zu „historisieren", also deren Inhalt nicht automatisch zu ändern. Auf diese Weise bleibt insbesondere die Möglichkeit erhalten, bereits gelöschte Datensätze erneut anzulegen, welches eine unter den bisherigen Benutzern verbreitete Vorgehensweise war und ist. Um trotzdem sicherzustellen, daß ein gelöschter Datensatz nicht irrtümlich als gültiger Datensatz interpretiert wird, kennzeichnet das Programm gelöschte bzw. geänderte Datensätze in der Übersichts- und in der Einzelsatzmaske jeweils farblich gleich. Die Aktualisierung der angezeigten Daten geschieht dann entweder explizit durch den Benutzer oder implizit bei Öffnen eines Einzelsatzfensters.

2.2   Individualisierung vs. Standardisierung

Ein nicht unerheblicher Klärungsbedarf bestand zwischen Standardisierung und Individualisierung. Sind die kennzeichnenden Farben, die Bedienelemente für Funktionen oder das Iteraktionsverhalten des Programms von Benutzer zu Benutzer verschieden, wird der Austausch über Fragen der Bedienung zwischen den Benutzern und die Unterstützungsmöglichkeit durch Onlinehilfe, Handbuch oder persönlichen Support erschwert. Andererseits ist die Individualisierung gerade bei einem großen und heterogenen Benutzerkreis ein wichtiges Mittel zur Akzeptanz und effizienten Arbeit mit der neuen Benutzerschnittstelle. Zudem sind grundlegende Individualisierungsmöglichkeiten wie Farben oder Schriften bereits in der verwendeten Emulationssoftware zur Anzeige der CICS-Masken gegeben.

Wir haben uns daher für eine auch für die Benutzer klar erkennbare Trennung von Programmverhalten und Erscheinungsbild entschieden. Letzteres ist individualisierbar, ersteres nicht. Um aber die bedeutungstragende Verwendung von Farben und Schriften trotzdem nicht aufzugeben, werden im Anpassungsmenü Farb- und Schriftgruppen angeboten, die jeweils eine gemeinsame Bedeutung besitzen. Beispiele sind die Gruppen „Farbe gelöschter Datensätze" oder „Schrift editierbarer Felder". Damit ist sichergestellt, daß z.B. ein gelöschter Datensatz in der Übersichtsmaske die gleiche Farbe besitzt wie eine Einzelsatzmaske, die einen gelöschten Datensatz anzeigt.

Eine weitere Individualisierungsmöglichkeit im Erscheinungsbild bietet die Übersichtsmaske. Hier kann der Benutzer für jeden Dialog individuell einstellen, welche Attribute in welcher Anordnung angezeigt werden sollen. Einige Spalten können darüberhinaus in der linken Maskenhälfte fixiert werden, um beim horizontalen Scrollen die Orientierung in der Tabelle zu verbessern. Auf diese Weise werden benutzerspezifische Datensichten auf den verfügbaren

Datenbestand umgesetzt. Diese Anpassungsoption ist eingerichtet worden, da der Arbeitskontext der Benutzer sehr unterschiedlich ist, diese also oft bezüglich eines Dialoges an ganz unterschiedlichen Attributen und Datenkonstellationen interessiert sind.

2.3 Verknüpfung der Datenbestände und Orientierung

Ein wichtiges Qualitätsmaß für die Benutzerschnittstelle auf große Datenbestände ist die Darstellung bedeutungstragender Zusammenhänge bei bleibender Orientierungsfähigkeit des Benutzers. Dazu ist keine graphische Benutzerschnittstelle nötig. De facto waren alle wesentlichen Konzepte bereits in der cobolbasierten Vorgängerschnittstelle implementiert. Diese galt es zu integrieren und zu standardisieren, um dem Benutzer gegenüber eine zuverlässige Erwartungshaltung bei der Benutzung dieser Funktionen zu ermöglichen.

Die neue Benutzerschnittstelle kennt daher drei Arten des Dialogwechsels: Das Schließen des einen und Eröffnen eines anderen Dialogs, den Wechsel in einen anderen Dialog mit automatischer Datenübernahme in die Suchmaske sowie die Anzeige eines Datenausschnitts in einer Listbox zur Wertewahl bei der Pflege eines Attributs. Besonders hervorzuheben ist die zweite Variante. Hier kann von einem konkreten Datensatz, beispielsweise ein Buchungskonto, in einen Dialog „Kontierung Region Nord" gewechselt und durch die Vorbelegung der Suchmaske erreicht werden, daß genau die Datensätze selektiert werden, die dieses Konto belasten. Diese Dialogwechsel können mit wenigen Parametern vom Programmierer festgelegt werden (siehe Abschnitt 0).

3 DV-Design und Integration

3.1 Viele Dialoge - ein Programm

Graphische Benutzerschnittstellen lassen sich bei normalen WAN-Bandbreiten im Netz der Deutschen Bahn AG kaum über das Netzwerk steuern, wie dies bei den CICS-Masken der Fall war. Somit stellt sich die Frage der Softwareverteilung. Die Rahmenbedingungen des Kunden waren dabei klar: Keine zusätzlichen Installationen auf dem Client, keine deutliche Performanceeinbuße, Fachlichkeit ausschließlich auf dem Server. Die technische Entscheidung viel auf ein schlankes Java-Applet und einen Javaserver (vgl. Abschnitt 4). Im Gegensatz zu den zahlreichen Cobolmodulen der Vorgängerschnittstelle wurde das Applet so konzipiert, daß es zwar durch nachladbaren Code erweiterbar ist, im Regelfall aber ohne diese Erweiterung einen beliebigen Dialog anzeigen kann. Dadurch reduziert sich nicht nur der Wartungsaufwand erheblich, sondern es wurde möglich, eine zuverlässige Standardisierung der Bedienelemente zu realisieren, die zuvor durch zahlreiche Änderungen in einzelnen Dialogen nur schwer umzusetzen war. Diese Standardisierung kennt drei Dialogtypen: "Anzeige", "Pflege" und "Anwendung". Dabei ist ein Anzeigedialog im wesentlichen ein Pflegedialog mit gesperrten Pflegefunktionen. Der Typ „Anwendungsdialog" wurde eingeführt, um in Spezialfällen die nötige Flexibilität zu gewährleisten. Er stellt quasi eine Art „Restklasse" mit minimalen Anforderungen dar.

3.2   Generischer Aufbau durch Metadaten

Um mit einem Programm alle Dialoge darstellen zu können, mußten zahlreiche Verhaltens- und Darstellungsweisen über Parameter einstellbar sein. Diese Parameter nennen wir Metadaten. Jeder Dialog hat eine eigene Metadatendatei, die beim Starten des Servers eingelesen wird und dem Client gesendet wird, wenn der entsprechende Dialog gewählt wurde.

Die Metadaten umfassen u.a. Parameter zu jedem Attribut (Herkunft / Berechnung / Bezeichnung / Prüfung / Zusatzfunktionen / Attributgruppe), zu jeder Maske (Auswahl und Reihenfolge der Attribute) sowie zur Entsperrung und Einrichtung von Funktionen (Hilfefunktionen, Pflegefunktionen, Verknüpfungsfunktionen, Sonderfunktionen). Die Metadaten bestehen aus zwei Hauptteilen: einem Dialogkopf, der allgemeine Informationen für den Dialog enthält, und einem Rumpf mit Angaben für jedes einzelne Attribut:

```
Kopf:

    DialogName    =EMPFANGSLAND              // Referenzname
    DialogDBName  =EMPFANGSLAND              // Datenbanktabelle

    /* Code, der beim Löschen ausgeführt wird */
    DeleteCode
        Wenn ( FREMDLEISTUNG(EMPFANGSLAND==DIES.EMPFANGSLAND) != NULL )
        Fehler (2,"FREMDLEISTUNG");

    /* Liste der Attribute in den einzelnen Masken (Suche,Liste,Pflege)
       Reihenfolge legt Anordnung in GUI fest */
    SelectAttr  =EMPFANGSLAND, KURZTEXT
    ListAttr    =EMPFANGSLAND, KURZTEXT, LANGTEXT
    DisplayAttr =EMPFANGSLAND, KURZTEXT, LANGTEXT, LOGIN_NAME, BEARB_DATUM

Rumpf:

    Name        =EMPFANGSLAND      // Referenzname
    DisplayName ="Länder-Code"     // Label auf der Maske
    Group       ="Empfangsland"    // Gruppierung mit anderen Attributen
    DB-Name     =EMPFANGSLAND      // Name in der Datenbank
    Writeable                      // Dieses Feld darf editiert werden
    STRING(2)                      // Der Typ ist ein String der Länge 2
    KeyAttr                        // Ist Schlüsselattribut der Tabelle
    TipText     ="Code für das Empfangsland"      // Hilfetext
```

Auf diese Weise ist die Erstellung von Standarddialogen sehr komfortabel und effizient möglich. Allerdings kann es ungewohnt sein, daß sich die Darstellung auf dem Bildschirm so nicht direkt bestimmen läßt, da diese aus den logischen Angaben in den Metadaten generiert wird. Unser Vorgehen wurde durch die Entkopplung von logischer und physischer Darstellung, wie sie (u.a.) dem AWT von Java 1.1 zugrundeliegt, erst ermöglicht und erweitert dieses Konzept.

4 Realisierung als Java-Applet und Java-Server

Von Seiten der Deutschen Bahn AG gab es zwei zentrale Anforderungen:

1. Entwicklung von neuen Intranetanwendungen, die direkt auf die Datenbank DB2 zugreifen.
2. Integration der bestehenden CICS-Anwendungen in die neuen Benutzeroberflächen, siehe Abbildung 2, ohne die Vielzahl der zum Teil sehr umfangreichen CICS-Anwendungen ändern oder neu entwickeln zu müssen.

Ferner sollte die Integration der bestehenden Anwendungen so erfolgen, daß der Benutzer möglichst keine Unterschiede zwischen den beiden Anwendungsarten feststellen kann, d.h. es gibt identische Oberflächen, identische Bedienelemente und Dialogabläufe sowie nahezu den identischen Funktionsumfang für beide Dialogtypen.

Durch dieses Vorgehen können bereits bestehende Batchläufe, Eingabeprüfungen, Nachbarsystemschnittstellen und Archivierungsprogramme unverändert bleiben. Die natürliche Schnittstelle zwischen Cobol und Java ist die Datenbank. In einigen Fällen - insbesondere bei aufwendigen Eingabeprüfungen - können Cobolmodule darüber hinaus auch direkt vom Java-Server aufgerufen und verwendet werden.

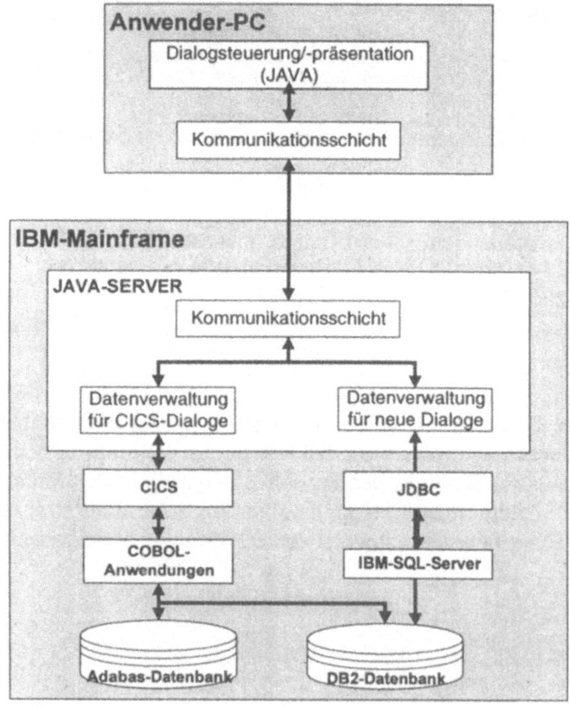

Abbildung 3: DV-Architektur der Intra-/Internet-Dialoge

4.1 Hostanbindung und Datenbankzugriff

Für die Anbindung der vorhandenen Datenbestände in den DB2-Datenbanken wurde die in Abbildung 3 dargestellte Architektur gewählt.

Für die Realisierung der neuen Dialoge, die nicht mehr auf CICS-Anwendungen zugreifen, ist die rechte Hälfte des Architekturbildes auf dem IBM-Mainframe relevant:[1]

Der Client besteht im wesentlichen aus der Präsentationsschicht, welche die Erzeugung der graphischen Benutzeroberfläche sowie die Auswertung der Benutzereingaben und die Ablaufsteuerung (Events) übernimmt, siehe auch Abbildung 2, indem die vom Benutzer gewünschten Aktionen angestoßen werden. Fachliche Informationen, Daten und Metadaten (letztere dienen sowohl dem Bildschirmaufbau als auch dem Datenbankzugriff) werden vom Server geliefert, der unter OpenEdition, einem UNIX-Aufsatz auf das OS390, auf einem IBM-Mainframe als Java-Applikation läuft.

Zur Kommunikation, die zur erhöhten Sicherheit mittels RSA und DES verschlüsselt wird, dient das Verfahren RMI (remote method invocation), mit dem Client und Server Objekte (Daten) austauschen können. Dabei kann der Client Aktionen auf dem Server auslösen, d.h. Methoden auf diesem aufrufen. Der Server öffnet dabei für jeden Client Threads, die die Anfragen des Clients ausführen.

Auf dem Server finden außer den Typprüfungen alle fachlichen Prüfungen, wie zum Beispiel Konsistenzprüfungen mit Datenbanktabellen statt. Die Typprüfungen erfolgen bereits auf dem Client, damit dem Benutzer falsche Eingaben ohne Umweg über den Server mitgeteilt werden können und so das Netz entlastet wird. Der Zugriff auf die DB2-Datenbank erfolgt mittels JDBC, einer für Java entwickelten Schnittstelle für den Datenbankzugriff.

4.2 Zum Vergleich: Automatische Umsetzung von CICS-Masken

Da eine Vielzahl bereits bestehender Masken möglichst schnell in der neuen Technik (Intranet/Internet) abrufbar sein sollten, gab es von Seiten der Deutschen Bahn AG die Anforderung, die CICS-Masken ebenfalls in einer GUI darzustellen. Wegen der Vielzahl der Masken, dem Umfang der bestehenden Anwendungen und der Verwendung von Adabas-Datenbanken, die weder eine ODBC- noch eine JDBC-Schnittstelle besitzen und daher nicht mit C++ oder Java-Anwendungen kommunizieren können, schied eine schnelle Umsetzung aller Masken analog zu 4.1 aus.

Um die vorhandenen 3270-Masken in graphische Benutzeroberflächen zu verwandeln, gibt es zahlreiche Produkte, die den vom Host gesendeten 3270-Datenstrom auswerten und in einzelne Masken umwandeln. Diese stellen dann eine normale Terminal-Emulation zur Verfügung, die jedoch in einem Browser dargestellt wird. Die Produkte erkennen die Bedeutung der einzelne Felder (Labels, Ein- und Ausgabefelder) und bieten die Möglichkeit, einzelne Felder entsprechend darzustellen. So werden beispielsweise Eingabefelder als Textfelder („Textfields") und Funktionstasten als Buttons dargestellt (siehe Abbildung 4).

---

1 Die linke Hälfte wird in den Abschnitten 4.2 und 4.3 erläutert.

Allgemein erhält man mit diesem Verfahren zunächst nur eine Terminal-Emulation, die mit zusätzlichem Arbeitsaufwand optisch aufbereitet werden kann. Will man die Funktionalität und die Bedienungsfreundlichkeit der Masken deutlich erhöhen, so ist der hierfür nötige Arbeitsaufwand sehr hoch, so daß wir uns für eine andere Alternative entschieden haben.

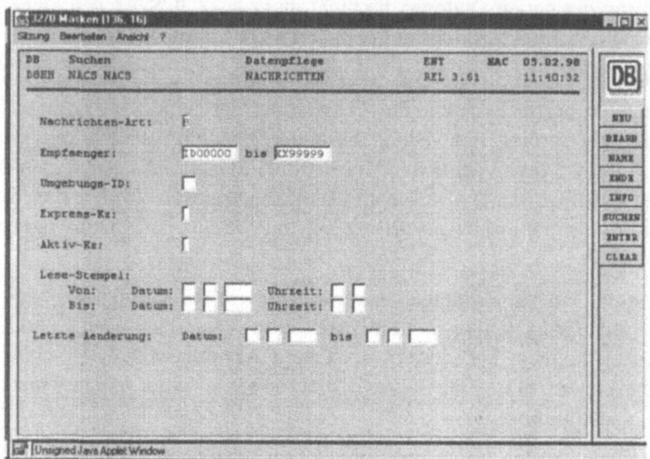

Abbildung 4: Automatische GUIfizierung von Hostmasken

4.3 Verbindung der alten und neuen Dialoge

Als Alternative bot sich an, für die CICS-Dialoge und die neuen Dialoge dieselbe neu graphische Benutzeroberfläche gemäß Abbildung 2 zu verwenden. Zusätzliche Vorteile gewinnt man, wenn man für beide Dialogtypen dasselbe Applet benutzt. Dazu ist es möglich, aus der bestehenden Spezifikation der CICS-Dialoge die in Abschnitt 0 vorgestellten Metainformationen weitgehend zu generieren, so daß ein Teil der für die erweiterte Funktionalität nötigen Eingaben ohne manuellen Arbeitsaufwand erzeugt werden kann. Insgesamt ergeben sich durch diese Lösungen folgende Vorteile:

- Der Benutzer kann über ein gemeinsames Auswahlmenü alle Dialoge unabhängig vom Typ auswählen. Es ist kein erneutes Einloggen und kein Wechsel des Applets erforderlich, wenn der Benutzer zwischen Dialogen mit neuer und alter Technologie wechselt.

- Der Benutzer merkt nicht, mit welcher Technik der Dialog arbeitet, da er eine einheitliche Benutzeroberfläche erhält. Er muß sich nicht auf unterschiedliche Bedienoberflächen oder Funktionsweisen für die verschiedenen Dialogtypen einstellen.

- Es werden eine erweiterte Funktionalität (z.B. eine frei gestaltbare Tabelle in der Übersichtsmaske), mehrere Einzelsatzfenster und eine moderner GUI-Oberfläche ermöglicht, die bei einer automatischen Umsetzung nicht geboten werden. Dem Benutzer steht hiermit nahezu die volle Funktionalität der neuen GUI-Dialoge zur Verfügung, siehe Abbildung 2, anstelle der eingeschränkten Möglichkeiten bei den herkömmlichen Hostmasken sowie bei der automatischen GUIfizierung.

Für die Realisierung dieses Konzeptes wurde die in Abbildung 3 dargestellte Architektur verwendet. Der Arbeitsablauf für den Benutzer sieht dann wie folgt aus:

Der Benutzer gibt nur einmal zu Beginn einer Session seinen Benutzernamen und sein Paß-
wort ein. Mittels Menüs wählt er den von ihm gewünschten Dialog aus. Der Typ des Dialogs
bestimmt das weitere Vorgehen. Handelt es sich um einen Dialog mit der neuen Intranet-
Technologie, so wird direkt die Verbindung zum Java-Server aufgebaut und der Dialog ge-
startet, die Verbindung erfolgt auf dem in Abschnitt 4.1 beschriebenen Weg. Hat der Benutzer
dagegen einen CICS-Dialog gewählt, so wird im Hintergrund eine Verbindung zum CICS
hergestellt und der gewünschte Dialog automatisch - ohne weitere Eingaben vom Benutzer -
aufgerufen. Für jeden Dialogtyp steht dem Benutzer somit der volle Funktionsumfang graphischer Benutzeroberflächen gemäß Abbildung 2 zur Verfügung.

5 Fazit

Durch die Einführung eines GUI-Frontends für Hostapplikationen ist es möglich, dem Benutzer ansprechendere und übersichtlichere Oberflächen mit vereinfachter Bedienung sowie um-
fangreicheren Funktionen als bei den herkömmlichen Hostmasken anzubieten, die der Benutzer spezifisch an seine Bedürfnisse und Wünsche anpassen kann. Dies ist auch für unverän-
derte, bereits seit Jahren im Einsatz befindliche Hostanwendungen möglich.

Durch die Anpassung der Bedienelemente an die der Standardsoftware der Zielplattform des
Cleints ist sowohl der Umstieg von der Arbeit mit den Hostmasken auf die graphische Benutzeroberfläche für erfahrene Benutzer als auch das Erlernen der Arbeitsweisen für neue Mitarbeiter nach ersten Erfahrungen rasch und problemlos zu vollziehen. Bei Pilotstudien stießen
wir auf Seiten der Benutzer auf ein sehr positives Echo und die Benutzer waren sehr schnell in
der Lage, mit den neuen Oberflächen umzugehen und ihre Arbeiten in der neuen Umgebung
durchzuführen.

Zusätzliche Vorteile ergeben sich durch das Metadatenkonzept, wodurch mit sehr geringem
Arbeitsaufwand neue Dialoge implementiert werden können. Dies erlaubt auch alte CICS-
Masken einfach zu integrieren, falls die Beschreibungen der CICS-Masken vorliegen. Die
Qualität der GUI für bestehende CICS-Dialoge ist abgesehen von den höheren Antwortzeiten
nahezu identisch mit der für Dialoge mit neuer Technologie.

Neuentwicklungen von Dialogen werden aber aufgrund des deutlich geringeren Entwick-
lungsaufwands nur noch auf neuen Technologie basieren.

6 Danksagung

Wir möchten uns bei der Deutschen Bahn AG, Projekt Controlling, für die langjährige und
sehr gute Zusammenarbeit und speziell für die Unterstützung bei der Entwicklung der hier
dargestellten Intranet-Anwendungen bedanken.
Unser Dank gilt ferner Herrn Dr. Friedrich Strauß, sd&m Technologiemanagement, für viele
wertvolle Diskussionen und Hinweise bei der Erstellung dieses Artikels.

7 Literatur

[1] *Nielsen, Jakob* (Hrsg.): Coordinating User Interfaces for Consistency. Academic Press: San Diego 1989
[2] *Microsoft Corporation*: Die Windows-Oberfläche. Leitfaden zur Softwarege-staltung. Microsoft Press:
Redmond 1995

Adressen der Autoren

Andreas Gronski
Sd&m
Software design & management GmbH & Co. KG
Thomas-Dehler-Straße 27
81737 München
Email: Andreas.Gronski@sdm.de

Dr. Harald Haller
sd&m
software design & management GmbH & Co. KG
Thomas-Dehler-Straße 27
81737 München
Email: Harald.Haller@sdm.de

Softwareevaluation in Gruppen oder Einzelevaluation: Sehen zwei Augen mehr als vier?

K.-C. Hamborg, G. Gediga, M. Döhl, P. Janssen & F. Ollermann

Fachbereich Psychologie, Universität Osnabrück

Zusammenfassung

Es finden sich Hinweise dafür, daß die Evaluation von Softwaresystemen effektiver in Gruppen- als in Einzel-settings ist. In drei Untersuchungen mit verschiedenen Varianten des IsoMetrics-Verfahrens werden diese Befunde überprüft. Die Ergebnisse zeigen Vorteile des Gruppensettings in Bezug auf die Qualität konstruktiver Anmerkungen, nicht jedoch in Bezug auf die Menge und inhaltliche Breite der Informationen und den Durchführungsaufwand. Spezifische Unterschiede der Methoden und Konsequenzen für die Praxis werden diskutiert.

1   Problemstellung

In diesem Beitrag wird ein möglicher Ansatz zur Effizienzsteigerung formativer Evaluation geprüft. Auf der Grundlage des IsoMetrics Fragebogens [10] wird die Einzelversion des Verfahrens mit einer gruppenbezogenen Variante verglichen. Ausgangspunkt für diese Untersuchung sind Befunde aus Vergleichsuntersuchungen von gruppen- und personenbezogenen Evaluationsmethoden, die darauf verweisen, daß sich die Effizienz von Evaluationsmethoden durch gruppenbezogene Settings steigern läßt. Zwei Arbeiten zu diesem Thema sollen im folgenden kurz dargestellt und durch allgemeine Argumente aus der Forschung von Gruppendiskussionsverfahren ergänzt werden.

Hackman und Biers [5] untersuchten die Methode des Lauten Denkens als Einzel- und als Gruppenmethode. Die Ergebnisse der Untersuchung zeigen, daß Evaluatoren im Zweier-Team insgesamt mehr Zeit mit dem Verbalisieren verbrachten als einzelne Evaluatoren. Im Team wurde mehr Zeit darauf verwendet, qualitativ hochwertige Verbalisierungen vorzunehmen. Wurden die Werte jedoch an der Personenzahl relativiert, gingen die Effekte verloren. Die Autoren halten die Ergebnisse, daß zwei Personen gemeinsam mehr Zeit mit dem Verbalisieren verbringen und dabei auch mehr hochwertige Informationen als eine Person generieren, für nicht trivial. Sie bewerten die Ergebnisse dahingehend, daß die Teammethode im gleichen Zeitraum mehr Informationen bringt als die Einzelmethode und daher zeiteffizienter funktioniert. Weiterhin verweisen sie darauf, daß im Team andere Anmerkungsqualitäten als bei der Einzelevaluation entstehen. Dazu zählen sie insbesondere auch Anmerkungen, die Unsicherheit mit dem Interface und dessen Bedienung ausdrücken.

In einer Untersuchung verglich Desurvire [1] Labortests mit Inspektionsmethoden. Ein System wurde mit der Methode des „Cognitive Walkthrough" und der „Heuristischen Evaluation" jeweils als Gruppen- und als Einzelmethode bewertet. Desurvire kommt zu dem Schluß, daß sowohl Experten als auch Nicht-Experten eine größere Anzahl Probleme, gemessen an dem Kriterium der Ergebnisse aus Labortests, vorhersagen, wenn die Evaluation in dem Gruppenparadigma durchgeführt wurde.

1.1 Vorteile gruppenorientierter Verfahren

Die Befunde der dargestellten Untersuchungen erscheinen aus methodischer Sicht plausibel. Teamorientierte Settings ähneln prinzipiell Gruppendiskussionsmethoden [2, 7]. Im Vergleich zu Einzelinterviews wird gruppenorientierten Verfahren zugeschrieben, daß sie einen größeren Bereich verschiedener Reaktionsweisen erfassen, Befragte zu detaillierten Meinungsäußerungen anregen und zur Aktualisierung und Explikation „tieferliegender Bewußtseinsinhalte" stimulieren, daß sie weiterhin zum Abbau psychischer Kontrollen beitragen und eher spontane, unkontrollierte Reaktionen, die den Schluß auf den latenten Inhalt geäußerter Meinungen zulassen, provozieren. Unter dem Eindruck der Auskunftsbereitschaft anderer Gesprächspartner sollten sie vor allem auch stärker gehemmte Teilnehmer zu Beiträgen ermutigen [2]. Weiterhin wird erwähnt, daß gruppenorientierte Verfahren auch ökonomischer seien. Ein geringerer sachlicher und personeller Aufwand erbringe soviel Material wie mehrere Einzeluntersuchungen [2, 7].

Die genannten Argumente beziehen sich in erster Linie auf den Vergleich von Einzelinterviews und Gruppendiskussionen. Nach unserer Auffassung gelten sie jedoch auch für den Vergleich schriftlicher Einzelbefragung und gruppenorientierter Befragungstechniken.

1.2 Mögliche Nachteile gruppenorientierter Verfahren

Jedoch wird auch eine Reihe möglicher Nachteile gruppenorientierter Verfahren im Vergleich zu Einzelinterviews genannt [2, 7]. Sie beziehen sich 1.) auf Beeinträchtigungen bei der Informationsgewinnung, 2.) auf gruppendynamische Prozesse und 3.) auf ökonomische Aspekte. Die Argumente werden im folgenden ebenfalls kurz dargestellt.

zu 1.) Relevante Informationen werden aus Gründen, die in der Person liegen, nicht genannt. Ursachen dafür können mangelnde Motivation oder Unkenntnis sein:

- Es liegt im Ermessen der TeilnehmerInnen, ob, wann und in welchem Umfang sie sich zu dem thematisierten Inhalt äußern. Es kann nicht ausgeschlossen werden, daß Personen zu einzelnen Punkten der Diskussion gänzlich schweigen oder Inhalten ausweichen.
- Einzelne Personen und Meinungen kommen nicht zum Zug, es kann zu einer ungleich–mäßigen Beteiligung einzelner Personen und einer mengenmäßig unterschiedlichen Verteilung der Beiträge kommen.

zu 2.) Das Zusammenspiel einzelner Personen in der Gruppe wirkt sich auf den Informationsgewinn aus:

- Gruppendynamische Effekte können sich möglicherweise ungünstig auf Atmosphäre und Gesprächsbereitschaft auswirken.
- Gesprächsmonopolsierung, Gruppendruck, Beeinflussungsversuche oder die frühe Verständigung über Gruppenstandards bzw. der Rekurs auf bestimmte Normen können verhindern, daß ein breites Spektrum von Perspektiven zu einem Thema entwickelt wird.

zu 3.) Die Erstellung und Verarbeitung von Diskussionsprotokollen ist zeitaufwendig und mit vielfachen Schwierigkeiten struktureller und inhaltlicher Art verbunden. Die Auswertungsökonomie kann daher als eher ungünstig bewertet werden. Weiterhin wird auf den hohen organisatorischen Aufwand bei der Vorbereitung von Gruppendiskussion hingewiesen.

Die referierten Forschungergebnisse zeigen widersprüchliche Befunde zu Vor- und Nachteilen von gruppen- und einzelpersonbezogenen Befragungsverfahren, während aus dem direkten Anwendungsgebiet der Softwareevaluation vorteilhafte Effekte, insbesondere mit Bezug auf die Menge und die Qualität gewonnener Informationen, berichtet werden.

In dem vorliegenden Bericht wird das Verhalten von gruppen- und einzelpersonbezogenen Evaluationsmethoden am Beispiel entsprechender Varianten des IsoMetrics-Verfahrens untersucht. Der Vergleich bezieht sich nach *formativen* Gesichtpunkten auf die Menge, die Qualität und das Spektrum erhobener Bemerkungen zu einem Softwaresystem als Ausgangspunkt für Verbesserungen und nach *summativen* Gesichtspunkten auf die Profilbewertung des Programms aufgrund von Bewertungsskalen.

2 Methoden

In diesem Abschnitt werden die beiden Varianten des IsoMetrics- Verfahrens dargestellt.

2.1 IsoMetrics^L

Der Fragebogen IsoMetrics^L umfaßt sieben Subskalen mit derzeit insgesamt 75 Items. Jede Subskala von IsoMetrics^L repräsentiert einen Gestaltungsgrundsatz gemäß ISO 9241/10. Die Skalen sind den Gestaltungsgrundsätzen entsprechend benannt: Skala 'A': *Aufgabenangemessenheit*, Skala 'S': *Selbstbeschreibungsfähigkeit*, Skala 'T': *Steuerbarkeit*, Skala 'E': *Erwartungskonformität*, Skala 'F': *Fehlerrobustheit*, Skala 'I': *Individualisierbarkeit*, Skala 'L': *Erlernbarkeit*.

Je Item werden drei unterschiedliche Daten erhoben (s. Abbildung 1):

1. Die Bewertung der Software auf einer 5-stufigen Ratingskala in Bezug auf das Item. Die Skalierung der Skala reicht von "stimmt nicht" (Rating 1) bis "stimmt sehr" (Rating 5).
2. Die Gewichtung des Items als Ausdruck der Bedeutung des durch das Item operationalisierten Aspekts der ISO 9241/10 in Bezug auf den Gesamteindruck der Software. Die Gewichtung wird ebenfalls auf einer 5-stufigen Ratingskala, von "nicht wichtig" (Rating 1) über "mittelmäßig wichtig " (Rating 3) bis "sehr wichtig" (Rating 5) vorgenommen.
3. Eine oder mehrere auf das Item bezogene Problemmeldungen, die der oder die Befragte in freiem Text selbst formuliert.

Bearbeitet wird der IsoMetrics von prospektiven Nutzern des zu evaluierenden Systems. Die Konstruktion, Validität und Reliabilität des Verfahrens als auch die Gebrauchstauglichkeit der von den Benutzern erzeugten Bemerkungen für die Software-Entwicklung wurden in [10, 11] dokumentiert.

	stimmt nicht	Stimmt Wenig	Stimmt Mittelmäßig	stimmt ziemlich	stimmt sehr	Keine Angabe
Wenn Menü-Optionen in bestimmten Bearbeitungsschritten nicht zur Verfügung stehen, wird mir die Sperrung sichtbar gemacht.	1	2	3	4	5	

	nicht wichtig	Wenig Wichtig	Mittelmäßig Wichtig	ziemlich wichtig	sehr wichtig	Keine Angabe
Wie wichtig ist dieser Aspekt für Ihren Gesamteindruck von der Software?	1	2	3	4	5	

Können Sie konkrete Beispiele nennen, bei denen Sie dieser Aussage nicht zustimmen können?

Abbildung 1: Ein Fragebogenitem aus IsoMetricsL (Erläuterungen im Text)

2.2 IsoMetrics^G

Als Modifikation bietet das IsoMetrics-Verfahren eine Gruppenversion an. Auch hier werden alle Items des Fragebogens benutzt, wobei jede Frage vom Befragungsleiter auf einem Overhead-Projektor vorgelegt wird. Jeder Teilnehmer erhält folgendes Format für die Beantwortung der Fragen (Abbildung 2).

	Stimmt nicht	Stimmt Wenig	Stimmt Mittelmäßig	stimmt ziemlich	Stimmt Sehr	Keine Angabe
A.1 Die Software zwingt mich Arbeitsschritte durchzuführen, die für meine Arbeit nicht sinnvoll sind.	1	2	3	4	5	

	Nicht wichtig	Wenig wichtig	Mittelmäßig wichtig	ziemlich wichtig	Sehr wichtig	Keine Angabe
Wie wichtig ist dieser Aspekt für Ihren Gesamteindruck von der Software?	1	2	3	4	5	

Kreuzen Sie bitte von denen auf der Overhead-Folie aufgelisteten Problemen diejenigen an, die Sie als relevant empfinden.
(1)   (2)   (3)   (4)   (5)   (6)   (7)   (8)   (9)

Abbildung 2: Ein Item aus dem IsoMetricsG-Fragebogen

Die Vorgehensweise in der Gruppenbefragung gestaltet sich wie folgt:

- Die Befragung der Gruppe wird von einem Moderator durchgeführt.
- Der Moderator benutzt pro IsoMetrics-Frage eine Overhead-Folie, trägt den Text des Kriteriums vor und erläutert u.U. diesen Text weiter – gibt Beispiele, etc.
- Die beteiligten Personen notieren jeder für sich das Rating für das Zutreffen der Aussage.
- Danach wird die Gewichtung von den Personen eingeschätzt (zweites Rating im Fragebogen).
- In der Gruppe werden Schwachpunkte der Software generiert, die der Moderator auf der Folie festhält.

- Die Schwachpunkte werden durchnumeriert.
- Am Ende der Generierung der Schwachpunkte für ein Item kreuzt jede beteiligte Person für sich an, welche der festgehaltenen Schwachpunkte ihrer Meinung nach relevant sind.

3 Versuchsplan und Methodik

3.1 Die Untersuchungen

Es wurden drei Untersuchungen durchgeführt, deren Gegenstand das Online Recherche Programm (OPAC) der Universitätsbibliothek Osnabrück war. In der ersten Untersuchung wurde das Programm mit IsoMetrics^L ($N = 15$), in der zweiten und dritten Untersuchung mit IsoMetrics^G (jeweils $N = 7$) bewertet. Die „Evaluatoren" waren Studierende der Universität Osnabrück, entstammten also der typischen Nutzerpopulation des Systems. Die Gruppen waren in Bezug auf Alter und Vorerfahrung nach der Anzahl genutzer Programme homogen. In Bezug auf die Vorerfahrung nach zeitlichen Aspekten unterscheiden sich die IsoMetrics^L- und die zweite IsoMetrics^G Gruppe, nicht aber die anderen Gruppen voneinander (siehe Tabelle 1). Die Evaluatoren verfügten über keine Vorbildung in Bezug auf Fragen der Software-Ergonomie.

Vor der Bewertung explorierten die Evaluatoren das Programm ca. 30 Minuten. Daraufhin bearbeiteten sie mit dem Programm vier Aufgaben. Für die Aufgabenbearbeitung standen weitere 15 Minuten zur Verfügung. Bei den Aufgaben handelte es sich um vier Rechercheaufträge, bezogen auf die Bestände der Universitätsbibliothek Osnabrück (s. Kasten 1).

1) Wie viele Einträge finden sich unter dem Titelstichwort „Organisationspsychologie"?
_______Treffer.

2 a) Schauen Sie sich den Eintrag zu Gros, Eckhardt: *Anwendungsbezogene Arbeits-, Betriebs- und Organisationspsychologie an*, der unter dem Titelstichwort „Organisationspsychologie" auftaucht.
b) Erkunden Sie, ob das Buch ausgeliehen ist.
c) Drucken Sie den Anzeigetext in Langdarstellung aus.
d) Übernehmen Sie den Titel in ein Speicherset.
e) Löschen Sie bitte das Speicherset.

3a) Suchen Sie alle Bücher mit dem Titelstichwort „Organisationspsychologie" von Siegfried Greif.
b) Speichern Sie die ersten drei Titel in einem Speicherset.

4a) Fügen Sie dem Speicherset die Arbeiten von Eberhard Ulich unter dem Titelstichwort „Arbeitspsychologie", die seit 1990 erschienen sind, hinzu.
b) Lassen Sie sich das gesamte Speicherset anzeigen.
c) Drucken Sie das Speicherset in Kurzdarstellung aus.

Kasten 1: Testaufgaben

Die Aufgaben wurden von den UntersuchungteilnehmerInnen eigenständig bearbeitet. Die VersuchsleiterInnen konnten notfalls um Hilfe gebeten werden. Dies geschah jedoch nur in wenigen Fällen und überwiegend bei Problemen mit dem Telnet Programm, das Voraussetzung für die Arbeit mit dem OPAC-System war. Sowohl für die Explorationsphase als auch für die Aufgabenbearbeitung konnten die Evaluatoren ein Kurzmanual, wie es Nutzern des Programms in der Regel zur Verfügung steht, benutzen.

	Alter	Vorerfahrung, Zeit (Index[*])	Vorerfahrung, Anzahl genutzter Programme (N)			
			T	D	K	P
IsoMetrics^L ($N = 15$)	$M = 25{,}4$; $S = 5{,}12$	$M = 4{,}5$; $S = 1{,}13$	12	3	4	1
IsoMetrics^G 1 ($N = 15$)	$M = 23{,}6$; $S = 3{,}87$	$M = 5{,}0$; $S = 1{,}0$	7	1	1	1
IsoMetrics^G 2 ($N = 15$)	$M = 24{,}6$; $S = 3{,}41$	$M = 6{,}14$; $S = 1{,}4^*$	7	2	2	2

Erläuterungen:

M = Mittelwert; SD = Standardabweichung, T = Textverarbeitung, D = Datenbank, K = Kalkulation, P = Programmiersprache; Index = aggregiert aus Vorerfahrung in Jahren, Nutzungshäufigkeit in Tagen/Monat und Stunden/Sitzung; * Differenz (Scheffé) zu IsoMetrics^L p = ,021.

Tabelle 1: Charakterisierung der Untersuchungsgruppen nach Vorerfahrung und Alter

3.2 Abhängige Variablen und Datenaufbereitung

Mögliche Effekte der Methodenvarianten können nach formativen Gesichtspunkten bezüglich der Menge, der Qualität und des Spektrums der erhobenen Anmerkungen, sowie nach summativen Gesichtspunkten in Bezug auf die Bewertungsergebnisse der IsoMetrics-Skalen erwartet werden. Für die summative Bewertung wurden die Mittelwertsprofile und die Varianzen der Subskalen des IsoMetrics berechnet.

Voraussetzung für die Auswertung des Datenmaterials der formativen Analyse war als erster Schritt die Explikation [8] der durch die Evaluatoren vorgenommenen Anmerkungen zu dem Programm. Die Explikation konzentrierte sich darauf, die Anmerkungen gegebenenfalls so zu ergänzen, daß sie durch Dritte ohne zusätzliche Informationen verständlich waren, d.h. es wurden unvollständige Sätze ergänzt, Bezüge zu den Fragebogenitems hergestellt etc..

In einem nächsten Schritt wurden redundante, d.h. inhaltlich übereinstimmende Aussagen aus dem Datensatz herausgefiltert und gekennzeichnet. Für die unterschiedlichen Auswertungsschritte bildete der so aufbereitete Datensatz die Grundlage. Die weitere Datenauswertung wird aus Gründen der Übersichtlichkeit im Zusammenhang mit den Ergebnissen dargestellt.

4 Ergebnisse

4.1.1 Summative Bewertung

Grundlage für die summative Bewertung der Software durch IsoMetrics sind die Ausprägungen auf den Subskalen des Verfahrens. Abbildung 3 zeigt die Varianzen um die Mittelwerte der Ratings für die drei Untersuchungen. In Abbildung 4 sind die aus den Subskalenwerten resultierenden Mittelwertsprofile für IsoMetrics^G aus den Untersuchungen dargestellt und in Abbildung 5 ist das Mittelwertsprofil der zusammengefaßten Werte von IsoMetrics^G zu dem Profil aus der IsoMetrics^L Untersuchung in Beziehung gesetzt.

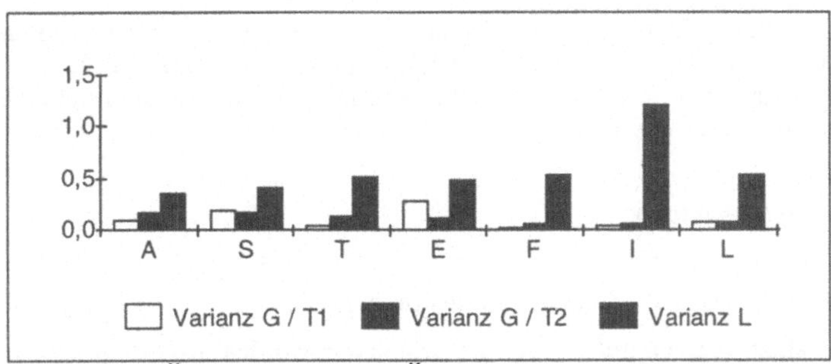

Es ist zu bemerken, daß die Bewertungen der Einzelversion deutlich stärker streuen als die der Gruppenversion. Der Vergleich der Werte der beiden Untersuchungen mit IsoMetrics^G (Abbildung 4) zeigt bei relativ geringen Skalenunterschieden (mit Ausnahme der Skala T) bedeutsame Unterschiede (Effektgröße > 0,8) außer bei der Skala Individualisierbarkeit (I). Aus statistischer Sicht wird dies durch die geringe Streuung der Skalenwerte der IsoMetrics^G Untersuchungen begünstigt (s. Abbildung 3).

Der Vergleich der Bewertungsergebnisse aus der Gruppen- mit denen der Einzelversion (Abbildung 5), zeigt, daß die Gruppenversion zu einer „strengeren" Bewertung als die Einzelversion führt. Der Unterschied der Bewertungsergebnisse ist auf allen Subskalen signifikant und bedeutsam (Effektgröße > 0,8). Es ist jedoch hervorzuheben, daß sich die Profilverläufe nicht strukturell unterscheiden: Werden die aggregierten Mittelwertsprofile aus den Untersuchungen mit der Gruppenversion um einen Skalenpunkt erhöht, unterscheiden sie diese nicht mehr von dem Profil, das aus der Einzeluntersuchung resultiert.

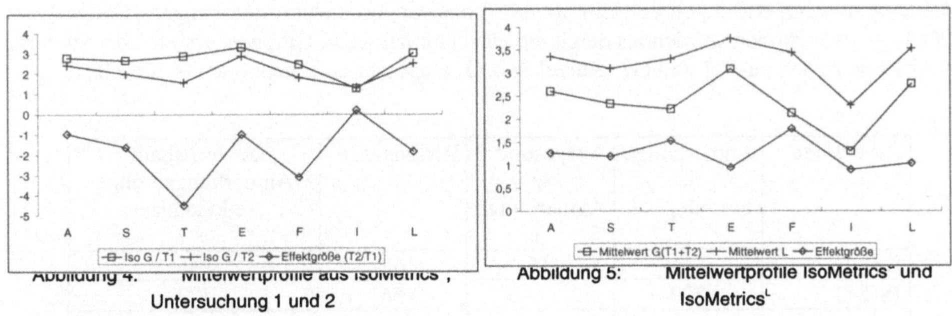

Abbildung 4:   Mittelwertprofile aus IsoMetrics^G, Untersuchung 1 und 2

Abbildung 5:   Mittelwertprofile IsoMetrics^G und IsoMetrics^L

Zusammenfassung

Die Ergebnisse der Gruppenversion zeigen im Vergleich zu der Einzelversion eine deutlich geringere Varianz bei der summativen Bewertung des zu evaluierenden Systems. Dies kann als Indikator dafür gewertet werden, daß durch den Gruppenprozeß extreme Bewertungen ausgeschlossen werden und eine Art Gruppenstandard homogenisierende Wirkung zeigt.

Weiterhin ist zu beobachten, daß die Gruppenversion zu einer strengeren Beurteilung des Systems führt. Die strengere Beurteilung erfolgt systematisch, d.h. die zusammengefaßten Profile aus der Gruppenversion und der Einzelversion unterscheiden sich um eine Konstante, unabhängig von der Wahl des Kriteriums.

4.1.2 Formative Bewertung

4.1.2.1 Vergleich der Anzahl erhobener Anmerkungen

Von den 15 Evaluatoren, die mit IsoMetrics^L arbeiteten, wurden insgesamt 345 Anmerkungen zu dem System formuliert.

Für die weitere Auswertung wurden davon 35 Bemerkungen nicht berücksichtigt. Dies waren Bemerkungen, die sich auf den Fragebogen (z.B. dessen Verständlichkeit) bezogen, (3), Bemerkungen, die sich auf das Betriebssystem oder auf die Programmumgebung richteten (11), oder die auf andere Bemerkungen verwiesen (9) sowie unvollständige, nicht explizierbare oder unverständliche Anmerkungen (12). Die resultierenden 310 Items enthielten 65 redundante, d.h. thematisch übereinstimmende Anmerkungen, die in der weiteren Auswertung ebenfalls nicht berücksichtigt wurden. Es verblieben also insgesamt 245 redundanzfreie und weiterhin auswertbare Anmerkungen. Das entspricht im Durchschnitt 16,3 Anmerkungen pro Teilnehmer der Evaluation.

In der ersten Untersuchung mit der Gruppenversion des IsoMetrics wurden 129, in der zweiten 110 Anmerkungen erhoben. Darin sind in der ersten Untersuchung 37, in der zweiten 29 Redundanzen enthalten. Ohne diese Redundanzen verblieben für die erste Untersuchung 92 und für die zweite 81 Anmerkungen. Durchschnittlich sind dies 13,1, respektive 11,6 Anmerkungen pro Evaluator.

Der Häufigkeitsvergleich der mit der Gruppen- und der Einzelversion erhobenen Anmerkungen zeigt, daß durch den Einsatz der Einzelversion mehr Anmerkungen (Chi^2 = 12,4, df =1, p<1%) erhoben wurden als durch die einzeln eingesetzte Gruppenversion, aber auch im Verhältnis zu den aus G1 und G2 (Kürzel ISO-G) aggregierten Ergebnissen (s. Tabelle 2).

Methode	Anmerkungen ursprünglich	Ausgesonderte Anmerkungen	Redundanzen	Auswertbare Anmerkungen ohne Redundanzen
L	345	35	65	245
ISO-G	239		66	173
G1	129		37	92
G2	110		29	81

Tabelle 2: Anmerkungshäufigkeiten IsoMetrics L und G

Zusammenfassung

Die Annahme, daß durch gruppenbezogene Evaluationsverfahren mehr Anmerkungen erhoben werden, bestätigt sich in der vorliegenden Untersuchung bei etwa gleicher Anzahl Evaluatoren nicht.

4.1.2.2 Vergleich der Anmerkungen nach Relevanz

Zur Bestimmung der Relevanz der erhobenen Anmerkungen wurden die von den Untersuchungsteilnehmern erzeugten redundanzfreien Anmerkungen zunächst verschiedenen für Entwickler relevanten Kategorien zugewiesen (Tabelle 3, s.a. [3]).

Kategorie	Art der Bemerkung
Irr	Die Aussage bezieht sich nicht auf die Software.
Pos	Die Aussage ist positiv und zeigt keine Schwächen des Systems auf.
X	Die Bemerkung drückt ein generelles Unverständnis aus. Sie handelt mehr vom Fühlen, Denken und Erleben der Person als von den auslösenden Schwächen der Software. Es muß aus der Bemerkung ersichtlich sein, daß es sich bei der Software um den auslösenden Faktor handelt.
S	Die Aussage bezieht sich auf eine lokalisierbare Stelle oder eine bestimmte Funktion oder einen bestimmten Befehl innerhalb der Software. Der Nutzungskontext muß in der Anmerkung genannt sein.
S+	Wie Kategorie S, die Aussage enthält jedoch mindestens einen konkreten Vorschlag *wie* das Problem behoben werden könnte.
G	Die Aussage bezieht sich auf ein Problem, das die gesamte Software durchzieht. Es ist nicht genau angegeben, welche Stellen betroffen sind. Beispiele, die zur Illustration dienen, machen aus einer generellen Anmerkung keine spezielle Anmerkung.
G+	Wie Kategorie G, die Aussage enthält jedoch zusätzlich mindestens einen konkreten Vorschlag *wie* das Problem behoben werden könnte.

Tabelle 3 Kategoriendefinition für die Analyse der Bemerkungen

Die Zuordnung der Anmerkungen zu den Kategorien erfolgte durch 4 Rater. Drei der Rater waren Studierende der Universität Osnabrück, Fachbereich Psychologie, die spezielle Kurse zu den Themen Evaluation und Software-Ergonomie besucht hatten, der vierte Rater ist Lehrender in diesem Gebiet. Vor dem Rating wurde ein Trainingsdurchgang durchgeführt. Bei dem Training und dem nachfolgenden Rating standen ein Informationsblatt mit Erläuterungen zu den Kategorien und mit Abgrenzungsbeispielen zur Verfügung.

Für die Auswertung der in diesem Abschnitt verfolgten Frage wurden nur die Anmerkungen berücksichtigt, die von wenigstens drei Ratern übereinstimmend kategorisiert wurden. Nach diesem Kriterium konnten 356 der insgesamt 418 redundanzfreien Anmerkungen (ca. 85%) ausgewertet werden. Neben der Tatsache, daß bei Kategorisierungen so gut wie nie eine völlige Übereinstimmung erreicht wird, können einige allgemein formulierte Anmerkungen, die eine trennscharfe und übereinstimmende Zuordnung zu den Kategorien teilweise erschwerte, als Ursache für die fehlende Übereinstimmung der Rater in 15% der Fälle gesehen werden.

Tabelle 4 zeigt die Häufigkeitsverteilung dieser Anmerkungen zu den Kategorien in Abhängigkeit von der eingesetzten Methodenvariante.

Kategorie	G1		G2		ISO-G		L	
	N	%	N	%	N	%	N	%
Irr, Pos, X	0 (-)	0 %	1 (-)	1,4%	1 (-)	,7%	27 (+)	12,7%
S	26 (+)	33,8%	14	20,0%	40 (+)	28%	29 (-)	13,6%
S+	3	3,9%	1	1,4%	4	2,8%	5	2,3%
G	47	61%	51	72,9%	94	65,7%	145	68,1%
G+	1	1,3%	3	4,3%	4	2,8%	7	3,3%
Gesamt	77	100 %	70	100 %	143	100 %	213	100 %

Tabelle 4: Häufigkeitsverteilung der Anmerkungen zu den Relevanzkategorien

Nach Berechnung einer Konfigurationsfrequenzanalyse zeigen die mit (+) gekennzeichneten Werte bedeutsam positive bzw. die mit (-) gekennzeichneten Werte auf dem 5% Niveau bedeutsam negative Abweichungen von dem erwarteten Wert aus dem Vergleich von IsoMetrics^L und IsoMetrics^G (Untersuchung G1, G2, bzw. ISO-G gegen Untersuchung L). Die Verteilung der Anmerkungen auf die unterschiedlichen Relevanzklassen zeigt, daß der Anteil irrelevanter, positiver und der Anmerkungen aus Kategorie X bei der Einzelversion bedeutsam überwiegt, während die Gruppenversion verhältnismäßig mehr spezielle Anmerkungen in Untersuchung G1 als auch bei Berücksichtigung der Anmerkungen aus ISO-G (Kategorie S), jedoch nicht in Untersuchung G2, evoziert.

Zusammenfassung

Durch die Gruppenvariante des Verfahrens werden weniger nicht konstruktiv verwendbare Anmerkungen erhoben. Weiterhin zeigen sich Unterschiede bei speziellen Anmerkungen (Kategorie S). Durch die Gruppenversion wurden in Untersuchung G1 und bei Berücksichtigung der aggregierten Anmerkungen aus G1 und G2 mehr Anmerkungen dieser Kategorie als mit der Einzelversion erhoben.

4.1.2.3 Vergleich des Anmerkungsspektrums

Weiterhin wurde in der vorliegenden Untersuchung das Anmerkungsspektrum in Abhängigkeit von der Methodenvariante untersucht. Ausgangspunkt für diese Analyse war die Verteilung der Anmerkungen auf die Subskalen des IsoMetrics.

Hierzu wurden die redundanzfreien Anmerkungen pro Skala für die Methodenvarianten ausgezählt. Die Ergebnisse sind in Tabelle 5 dargestellt. Sie zeigen, daß sich die mit IsoMetrics^L und IsoMetrics^G erhobenen Anmerkungen relativ zu der Gesamtmenge unterschiedlich auf die Subskalen verteilen (Chi^2=38,7, df=6, p<1%). Die mit (+)/(+*) und (-)/(-*) gekennzeichneten Werte markieren bedeutsam positive bzw. negative Abweichungen vom erwarteten Wert auf dem 1% bzw. 5% Niveau.

	Aufgabenangemessenheit	Selbstbeschreibungsfähigkeit	Steuerbarkeit	Erwartungskonformität	Fehlertoleranz	Individualisierbarkeit	Erlernbarkeit	gesamt
ISO-G	103 (+)	23	15	8 (-*)	16	0 (-)	8 (-)	173
L	79 (-)	43	29	22 (+*)	30	16 (+)	26 (+)	245
Gesamt	182	66	44	30	46	16	34	418

Tabelle 5: Häufigkeitsverteilung der Anmerkungen über die Subskalen des IsoMetrics

In der Gruppenversion finden wir eine starke Konzentration auf die erste Skala (Aufgaben-angemessenheit) bei weitgehendem Abfall auf den folgenden Subskalen. Bei der Einzel-version werden ebenfalls die meisten Anmerkungen durch die erste Skala (Aufgaben-angemessenheit) erhoben, die folgenden Skalen tragen aber in einem stärkeren Maße als bei IsoMetrics^G zur Erhebung weiterer Anmerkungen bei.

Zusammenfassung

Die Anmerkungen der gruppenorientierten Variante konzentrieren sich auf die zuerst bearbei-tete Skala (Aufgabenangemessenheit). Vermutlich wurde die Auskunftbereitschaft der Evalu-atoren für die folgenden Skalen dadurch weitgehend erschöpft. Hierdurch ensteht die Gefahr, daß sich die Evaluationsergebnisse auf ein eingeschränktes Spektrum der zugrundegelegten Gestaltungsgrundsätze reduziert. Bei der Einzelversion des Fragebogens verteilen sich die Anmerkungen dagegen auch auf die in der Bearbeitungsfolge später dargebotenen Skalen.

4.1.2.4 Durchführungsaufwand

Die Einzel- und die Gruppenversion unterscheiden sich nach ihrem Durchführungs- und Aus-wertungsaufwand.

In Bezug auf die Durchführung erfordert die Gruppenversion mehr organisatorische Vorbe-reitungen, da die TeilnehmerInnen auf einen Termin hin koordiniert werden müssen. Hier er-möglicht die Einzelversion mehr Flexibilität, da prinzipiell Personen unabhängig voneinander den Fragebogen bearbeiten können. Da für die Durchführung der Gruppenversion ein ge-schulter Moderator und zusätzlich ein Protokollant, für die Aufzeichnung der Anmerkungen der TeilnehmerInnen, benötigt werden, ist nach personellen Gesichtspunkten diese Ver-fahrensvariante aufwendiger. Die Untersuchung mit der Einzelversion kann dagegen von einer angelernten Hilfskraft durchgeführt werden. Der Zeitaufwand für die Gruppenversion betrug in den hier dargestellten Untersuchungen ca. 3 Stunden, für die Durchführung der Einzelversion 1,5 bis max. 2 Stunden (inkl.Pausen, ohne Explorationsphase und Aufgaben-bearbeitungen).

Der Auswertungsaufwand der summativen Ergebnisse ist versionsspezifisch nicht unter-schiedlich. Dies ist jedoch anders bei der Auswertung der formativen Ergebnisse. Bei den Ergebnissen der Gruppenversion besteht weniger Explikationsaufwand, da die Anmerkungen durch den Prozeß der Informationsvermittlung, der die Verbalisierung und Niederschrift der Anmerkungen in einer für alle GruppenteilnehmerInnen verständlichen Weise umfaßt, zum größten Teil besser ausformuliert sind. Ebenfalls ist die Auseinandersetzung mit nicht relevanten Aussagen bei der Gruppenvariante geringer als bei der Einzelvariante.

Zusammenfassend ist ein höherer Durchführungsaufwand nach Zeit- und Personalkosten für die Gruppenvariante festzuhalten, der insgesamt die Ersparnisse bei der Auswertung überstei-gen dürfte.

5 Diskussion

In der vorliegenden Untersuchung wurde die Gruppen- und Einzelversion des IsoMetrics, eines Verfahrens zur formativen und summativen Evaluation von Software, untersucht. Ausgangspunkt für die Untersuchung waren Befunde, die auf den effizienzsteigernden Charakter gruppenorientierter Verfahren zur Evaluation von Software hinweisen.

Der Methodenvergleich zeigt, daß auf der summativen Ebene die Bewertungsergebnisse der Gruppenvariante (IsoMetrics^G) weniger streuen und um einen Skalenpunkt strenger bewerten als die der Einzelvariante (IsoMetrics^L). Der letztgenannte Aspekt ist insbesondere bei der vergleichenden Untersuchung von Softwaresystemen zu beachten, um den Einfluß methoden-spezifischer Bewertungstendenzen kontrollieren zu können.

Auf der formativen Bewertungsebene zeigte sich, daß mit der Einzelvariante (IsoMetrics^L) insgesamt mehr Anmerkungen zu der evaluierten Software erhoben wurden und die erhobenen Anmerkungen die Gestaltungsgrundsätze der zugrundeliegenden Norm breiter abdecken. Demzufolge können die Befunde von Desurvire [1], nach denen gruppenorientierte Evaluationsvarianten mehr Informationen liefern als einzelpersonorientierte, nicht gestützt werden.

Die Stärke der Gruppenvariante besteht darin, daß mehr spezielle Anmerkungen erhoben werden, diesbezüglich die Anmerkungsqualität im Unterschied zu der Einzelversion also besser ist. Dieser Effekt stimmt mit den in der Literatur genannten Befunden überein, daß Gruppendiskussionsverfahren zu detaillierteren Meinungsäußerungen anregen.

Ein weiterer Vorteil der Gruppenversion kann darin gesehen werden, daß weniger nichtver-wertbare Informationen erhoben werden. Eine Erklärung hierfür ist, daß es in der Gruppen-situation erforderlich ist, die Anmerkungen für die weiteren Mitglieder und den Moderator verständlich formulieren zu müssen. Die Hemmschwelle, wenig aussagekräftige Hinweise zu formulieren, dürfte in der Gruppensituation höher sein als bei der alleinigen Bearbeitung eines Fragebogens. Zum geringeren Anteil nicht verwertbarer Anmerkungen dürfte ebenfalls bei-tragen, daß diese in der Gruppenversion durch den Moderator weiterverarbeitet werden, wenn er die Aussagen auf Overheadfolie festhält, ggf. nachfragt und reformuliert.

Es kann zusammengefaßt werden, daß die untersuchten Methodenvarianten weder bei der summativen noch bei der formativen Evaluation von Software äquivalent funktionieren. Die summativen Bewertungsergebnisse der Gruppenvariante sind strenger und streuen weniger. Befunde in bezug auf einen besseren Mengenertrag von Information zur Optimierung evalu-ierter Programme durch gruppenbezogene Verfahren können nicht gestützt werden. In Bezug auf die Informationsqualität zeigen sich teilweise Vorteile der Gruppenvariante, nicht jedoch in Bezug auf das thematische Anmerkungsspektrum und den Durchführungsaufwand.

Vor dem Hintergrund der vorliegenden Ergebnisse stellt sich die Frage, welche Variante in der Praxis bevorzugt werden sollte. Der geringere Durchführungsaufwand, die größere An-merkungsmenge und die inhaltliche Anmerkungsbreite sprechen für die Einzelversion, für die Gruppenversion die höhere Anmerkungsqualität, die wesentlich darin besteht, daß die An-merkungen konkreter auf Funktionen der untersuchten Software gerichtet sind.

Verschiedene Autoren empfehlen, im Anschluß an die Systembewertung Usability Reviews, an denen ggf. Anwender, Nutzer und Systementwickler beteiligt sind, durchzuführen [6, 11, 12]. Hier lassen sich erhobene Funktions- und Gestaltungsprobleme an Hand vorsortierer An-merkungen aus Evaluationsuntersuchungen priorisieren und Maßnahmen für das (Re-) Design planen [4, 11].

Im Rahmen solcher Reviews ließe sich auch die niedrigere Spezifität der Anmerkungen aus IsoMetrics^L kompensieren. Diesem Vorgehen folgend, wäre die Einzelvariante die effektivere und zu empfehlende Methode.

Der Einsatz des gruppenorientierten Settings ist jedoch dann angezeigt, wenn aufgrund von Vorbildung oder Erfahrung die Anwender des Verfahrens individuelle Unterstützung im Umgang mit den Bewertungsskalen und der Generierung von Problempunkten benötigen. Unter diesen Bedingungen ist die Gruppenvariante die effizientere und vorzuziehende Methode.

6   Literatur

[1] Desurvire, H. (1994). Faster, Cheaper!! Are Usability Inspection Methods as Effective as Empirical Testing?. In: J. Nielsen & R.L. Mack (eds.). *Usability Inspection Methods*. New York: J. Wiley & Sons.

[2] Dreher, M. & Dreher, E. (1994). Gruppendiskussion. In: G.L. Huber & H. Mandl (Hrsg.) *Verbale Daten. 2. Auflage*. Weinheim: Psychologie Verlags Union.

[3] Gediga, G. & Hamborg, K.-C. (1997). Heuristische Evaluation und IsoMetrics: Ein Vergleich. In: R. Liskowsky, B.M. Velichkovsky & W. Wünschmann (Hrsg.). *Software Ergonomie '97, Usability Engineering: Integration von Mensch-Computer-Interaktion und Software-Entwicklung*. Stuttgart: Teubner.

[4] Gediga, G. & Hamborg, K.-C. (1997). *Das IsoMetrics-Manual*. Osnabrücker Schriftenreihe Software-Ergonomie 2.

[5] Hackmann & Biers (1992).Team usability testing: Are two heads better than one? *Proceedings of the 36th annual meeting of the Human Factors society*, 36, S. 1205-1209.

[6] Karat, C.-M. (1994). A Comparison of User Interface Evaluation Methods. In: J. Nielsen & R.L. Mack (eds.). *Usability Inspection Methods*. New York: J. Wiley & Sons.

[7] Mangold, W. (1973). Gruppendiskussionen. In. R. König (Hrsg.). *Handbuch der empirischen Sozialforschung, Bd. 2. Grundlegende Methoden und Techniken der empirischen Sozialforschung, Erster Teil*. Stuttgart: Enke.

[8] Mayring, P. (1997). *Qualitative Inhaltsanalyse. 6., durchgesehene Auflage*. Weinheim: Beltz Deutscher Studienverlag.

[9] Nielsen, J. (1993). *Usability Engineering*. Boston, AP Professional.

[10] Gediga, G., Hamborg K.-C. & Düntsch, I. (in print). The IsoMetrics Usability Inventory: An operationalisation of ISO 9241-10. *Behaviour and Information Technology*.

[11] Willumeit, H., Gediga, G. & Hamborg, K.-C (1996) IsoMetrics[L]: Ein Verfahren zur formativen Evaluation von Software nach ISO 9241/10. *Ergonomie & Informatik*, 27, 5 - 12.

[12] Wixon, D., Jones, S., Tse, L. & Casady, G. (1994). Inspections and Design Reviews: Framework, History and Reflection. In: J. Nielsen & R.L. Mack (eds.). *Usability Inspection Methods*. New York: J. Wiley & Sons.

Adressen der Autoren

Dr. Günther Gediga,
Universität Osnabrück,
FB Psychologie
Fachgebiet Methodenlehre
Seminarstr. 20
49069 Osnabrück

Dr. Kai-Christoph Hamborg
Cand. Psych. Meike Döhl
Cand. Psych. Philip Janssen
Cand. Psych. Frank Ollermann
Universität Osnabrück,
FB Psychologie
Fachgebiet Arbeits- und Organisationspsychologie
Seminarstr. 20
49069 Osnabrück

„Benutzererwartung eingebaut": Gestaltungsempfehlungen für Suchfunktionen auf der Basis einer empirischen Benutzerbefragung[1]

Marc Hassenzahl und Jochen Prümper

bao - Büro für Arbeits- und Organisationspsychologie, Berlin
FHTW - Fachhochschule für Technik und Wirtschaft, Berlin

Zusammenfassung

Suchfunktionen sind zentral und allgegenwärtig. Sie stehen zwischen dem Benutzer und der von ihm gewünschten Information. Die vorliegende Arbeit präsentiert praktische Gestaltungsempfehlungen für Suchfunktionen auf der Basis einer empirischen Benutzerbefragung. Ziel war es, bisherige Erfahrungen der Benutzer beim Suchen von Information abzufragen, um sie in einem neu zu gestaltenden System explizit berücksichtigen zu können. Eine solche Form der „externen Konsistenz" mit den Erfahrungen der Benutzer ist ein Schlüssel zur Gebrauchstauglichkeit eines Systems (z.B. Erwartungskonformität). Die zentralen Aspekte der Benutzerbefragung sind Suchstrategien und logische Operatoren. Des weiteren wurde Computer- und Interneterfahrung berücksichtigt.

Abstract

Searching is a primary and ubiquitous task when working with the computer. Most of the time, the only access to the information required is a search functionality. The present paper introduces practical guidelines for designing search functions. These guidelines were derived from an empirical user survey, which was aimed at the individual experiences of users with searching. Considering former experiences when designing a new system is one way to assure „external consistency". Consistency in turn is a key to the system's usability. The topics of the survey were generalised search strategies and the use of boolean operators with regard to differences in computer and internet experience.

1   Einleitung

Bei der Arbeit mit dem Computer ist das Suchen von Information eine zentrale Aufgabe, die sich nicht nur auf Datenbankanwendungen oder neuerdings auf das Internet (bzw. WorldWideWeb) beschränkt. In fast jedem Anwendungsprogramm und Betriebssystem findet sich eine Suchfunktion (z.B. „Suchen und Ersetzen" in einer Textverarbeitung oder die Suche von Dateien auf Betriebssystemebene).

Gerade mit Verbreitung des Internets und damit auch der zunehmenden Menge an bereitstehender Information werden optimal gestaltete Suchfunktionen immer wichtiger: sie sind das Bindeglied zwischen dem Benutzer und der für ihn nützlichen Information. Relativ viel Energie wird dabei in die Entwicklung neuer Suchalgorithmen investiert, weniger in die Entwicklung empirisch gestützter Gestaltungsempfehlungen, die bei der Auswahl aus der Vielzahl möglicher Gestaltungsvarianten helfen.

---

1 Wir danken Edmund Buchbinder, Tobias Ley und Uta Sailer für ihre hilfreichen und anregenden Kommentare zu früheren Versionen der vorliegenden Arbeit.

Wir verstehen solche allgemeingültigen Gestaltungsempfehlungen als einen Beitrag zur Sicherung der *Gebrauchstauglichkeit* [8] betroffener Systeme. Wie notwendig dies ist, zeigt eine Untersuchung der „User Interface Engineering"-Gruppe [1]. In Gebrauchstauglichkeitsstudien (usability tests) mit Web-Sites zeigte sich, daß das Verwenden einer „On-Site"-Suche (im Gegensatz zum Verwenden von Verweisen, sog. „links") die Chance des Benutzers, gewünschte Information zu finden, von 53% auf 30% reduzierte.

Ein Schlüssel zum Ableiten von Gestaltungsempfehlungen für Suchfunktionen sind die bisherigen Erfahrungen der Benutzer mit den Suchfunktionen *anderer* Softwaresysteme. Eine solche Form der *externen Konsistenz* [10] ist ein wichtiger Teilfaktor der *Erwartungskonformität* [7]. Externe Konsistenz ermöglicht es dem Benutzer, Wissen aus anderen, häufiger genutzten Softwaresystemen auf das neue System zu transferieren [16] und es so schneller zu erlernen.

Das Abfragen von Erwartungen bezüglich Suchfunktionen scheint zwar ein wünschenswertes, aber auch fast unmögliches Unterfangen zu sein, bedenkt man die große Anzahl unterschiedlicher Suchfunktionen und Suchaufgaben.

Ein möglicher Ansatz ist es, davon auszugehen, daß Benutzer auf der Basis konkreter Erfahrungen einen Satz allgemeinerer Strategien herausbilden. Eine solche *Suchstrategie*, wie z.B. „Ich verwende Synonyme", kann sich entweder durch technische Unterstützung („Ich verwende Systeme, die eine Suche nach Synonymen auf Wunsch automatisch durchführen können") oder durch den Bedarf entwickelt haben („Wenn ich im Internet unter einem bestimmten Begriff nichts finde, verwende ich ein Synonym, da ich nicht genau weiß, unter welchem Begriff der Autor der Information diese abgelegt hat").

Wichtig ist, daß es sich dabei um eine *generalisierte* Strategie handelt, die der Benutzer bei der Verwendung einer neuen Funktion parat hat und anwenden will. Dementsprechend ist es das Ziel der vorliegenden Arbeit, aus der Menge der Gestaltungsmöglichkeiten solche herauszufinden und bevorzugt anzubieten, die den Erwartungen eines Großteils der Benutzer entsprechen. Dabei soll hier über die konkreten Systeme, mit denen die Erfahrung gesammelt wurde, und mögliche Kategorien von Suchaufgaben [15] generalisiert werden. Es wird davon ausgegangen, daß sich sowohl häufig genutzte Systeme und deren spezielle Möglichkeiten als auch häufig vorkommende Aufgaben in den verwendeten Suchstrategien widerspiegeln.

Sobald mehr als ein Suchwort gleichzeitig verwendet wird, muß implizit oder explizit mit *logischen Operatoren* (z.B. UND, ODER, NICHT etc.) gearbeitet werden. Es zeigt sich allerdings, daß Benutzer - auch Experten! - Schwierigkeiten im Umgang mit diesen Operatoren haben [11]. Diese Schwierigkeiten machen logische Operatoren zu einem Aspekt, der - neben den Suchstrategien - in Gestaltungsempfehlungen berücksichtigt werden muß.

Der Anspruch, allgemeingültige Gestaltungsempfehlungen zu formulieren, macht es notwendig, die allgemeine *Computer-* und *Interneterfahrung* explizit zu berücksichtigen. Es liegt nahe, daß *Erfahrung* eine wichtige Determinante für verwendete Suchstrategien und den Umgang mit logischen Operatoren ist.

Die vorliegende Untersuchung verfolgte das Ziel, alle oben angesprochenen Aspekte bei Computerbenutzern zu erheben, um daraus generalisierte Gestaltungsempfehlungen abzuleiten. Eine solche Generalisierung verlangt nach einer ausreichenden Anzahl Befragter und damit nach einer ökonomischen Erhebungsmethode. Aus diesem Grund wurde eine schriftliche Benutzerbefragung eingesetzt.

Im folgenden werden zunächst Umstände und Art der Benutzerbefragung näher erläutert (Abschnitt 0). Danach werden zentrale Ergebnisse vorgestellt und diskutiert (Abschnitt 0). Im letzten Abschnitt (Abschnitt 0) werden die Ergebnisse zu Gestaltungsempfehlungen verdichtet.

2 Methode

2.1 Befragungsdurchführung und Befragte

Die hier vorgestellte schriftliche Befragung war Teil eines software-ergonomischen Beratungsprojekts zur Gestaltung eines web-basierten „Informationssystem zur Gesundheitsberichterstattung des Bundes (IS-GBE)" [4-6]. Ein großer Teil der schriftlichen Befragung (79 von 97 Personen) fand auf dem „Forschungsforum '97" (Messe Leipzig, 16. - 20.9.1997) im Rahmen einer Präsentation des IS-GBE statt. Zusätzlich nahmen noch 18 Personen aus der Fachabteilung „Gesundheitsberichterstattung" des statistischen Bundesamtes teil.

Das Durchschnittsalter der Gesamtstichprobe von 97 Befragten (32 Frauen, 65 Männer) betrug etwas mehr als 32 Jahre (Min=15, Max=55). In der Stichprobe fanden sich Schüler, Studenten, Wissenschaftler, Techniker, Informatiker, Sachbearbeiter, Sekretärinnen, Experten aus der Fachabteilung und Journalisten. Sie kann als relativ heterogen bezeichnet werden. Nur 8% der Befragten gaben an, über keinen Internetzugang zu verfügen.

2.2 Maße

2.2.1 Befragtenmerkmal „Computererfahrung"

Das Merkmal „allgemeine Computererfahrung" wurde mit den in Tabelle 1 dargestellten fünf Items erfaßt (in Anlehnung an [3]) und setzt sich additiv aus den Aspekten *Nutzungsintensität* (Items 1,2,3) und *Nutzungsbreite* (Items 4,5) zusammen.

	Item	**Antwortvorgabe**	
1	Wie viele Jahre haben Sie schon Erfahrung mit Computern?	0 bis 1 Jahr / 1 bis 3 Jahre / 3 bis 6 Jahre / mehr als 6 Jahre	
2	An wieviel Tagen in der Woche sind Sie durchschnittlich mit Arbeiten am Computer beschäftigt?	Überhaupt nicht / weniger als 2 Tage / 2 bis 5 Tage / mehr als 5 Tage	
3	Wie lange sind Sie im Durchschnitt täglich mit dem Computer beschäftigt?	Überhaupt nicht / weniger als 2 Stunden / 2 bis 6 Stunden / mehr als 6 Stunden	
4	Mit welcher Art von Software haben Sie bereits gearbeitet? *(keine oder Mehrfachnennung möglich)*	**Lokal:** Textverarbeitung (z.B. Word) / Datenbank (z.B. Access) / Statistikprogramm (z.B. SPSS) / Werkzeug (z.B. Dateimanager)	**Netzwerk:** Internet-Suchmaschine (z.B. Altavista) / Browser (z.B. Netscape Navigator) / Elektronische Post (z.B. Exchange) / „Chat"-Programm (z.B. IRC)
5	Haben Sie bereits programmiert?	Nie / selten / häufig	

Tabelle 1: Items und entsprechende Antwortvorgaben zur Einschätzung des Befragtenmerkmals „Computererfahrung". Die beiden Softwarekategorien „Lokal" (Item 4, linke Spalte) und „Netzwerk" (Item 4, rechte Spalte) dienten zur Einschätzung der „Interneterfahrung"

Jeder Antwortvorgabe ist ein bestimmter Wert zugewiesen (z.B. „Haben Sie bereits programmiert?" nie=0, selten=1, häufig=2). Der Kennwert *Computererfahrung* ergibt sich als Summenwert aller fünf Items. Er reicht von 0 bis 19, wobei höhere Werte einem höheren Grad an Erfahrung entsprechen.

Der Median der Stichprobe liegt mit 12 (Min=6, Max=19) etwas über dem theoretischen Median der Skala von 9,5. Es zeigen sich allerdings keine Deckeneffekte und damit auch keine fehlende Differenzierung im oberen Skalenbereich. Ein Kolmogorov-Smirnov-Anpassungstest zeigte keine signifikante (K-S Z=0,93, n.s.) Abweichung von der zu erwarteten Normalverteilung des Merkmals *Computererfahrung* in der Population der Computernutzer.

Offensichtlich ist die Stichprobe bezüglich des Kennwertes *Computererfahrung* in ausreichender Weise normalverteilt. Auch der Wertebereich der Skala wird durch die Stichprobe gut ausgenutzt. Man kann daher von einer - in bezug auf Computererfahrung - *repräsentativen* Stichprobe ausgehen, die eine gewisse Generalisierung der Ergebnisse zuläßt.

Um die Betrachtung von Unterschieden in Abhängigkeit von der Computererfahrung zu vereinfachen, wurde die Stichprobe in drei gleich große Gruppen aufgeteilt: *niedrige* (Md=9), *mittlere* (Md=12,5) und *hohe Computererfahrung* (Md=16).

2.2.2 Befragtenmerkmal „Interneterfahrung"

Die *Interneterfahrung* ist ein Teil der generellen Computererfahrung. Dies drückt sich in Item 4 („Mit welcher Software haben sie bereits gearbeitet?", siehe Tabelle 1) aus. Ein Befragter mit hoher Nutzungsintensität, der selbst häufig programmiert, aber nie Erfahrung mit Internetanwendungen sammeln konnte, kann dementsprechend den maximalen Wert für Computererfahrung nicht erreichen.

Item 4 ermöglicht - in begrenztem Maße - die Einschätzung der Interneterfahrung unabhängig von der Computererfahrung. Dazu werden Personen, die mit keiner der unter Item 4 angebotenen Internetanwendungen Erfahrungen haben, der Gruppe *internetnaiv* und Personen, die mit mindestens einer der angebotenen Internetanwendungen Erfahrungen haben, der Gruppe *interneterfahren* zugeordnet.

Ein einwandfreier Rückschluß auf Effekte durch Interneterfahrung ist nur dann möglich, wenn Unterschiede zwischen *internetnaiven* und *interneterfahrenen* Personen auf einer ähnlichen Stufe der Computererfahrung untersucht werden. Diese finden sich nur im mittleren Bereich des Kennwerts Computererfahrung. Dementsprechend wird zur Betrachtung der *Interneterfahrung* eine Teilstichprobe von 28 Personen mit mittlerer Computererfahrung (Kennwertebereich 8 bis 11) herangezogen, von denen 12 Personen keine Erfahrung mit Internetanwendungen (*internetnaiv*) und 16 Personen Erfahrung mit mindestens einer Internetanwendung haben (*interneterfahren*). Ein Mann-Whitney-U-Test bestätigt, daß sie sich bezüglich ihrer *Computererfahrung* nicht unterscheiden (U=68,5; n.s.). In beiden Gruppen beträgt der Median der *Computererfahrung* jeweils 10.

2.2.3 Suchstrategien

Zur Abfrage der von den Befragten eingesetzten *Suchstrategien* verwendeten wir ein Mehrfachwahl-Item („Bei der Suche mit dem Computer verwende ich folgende Vorgehensweisen:"). Jeder Befragte konnte aus den in Tabelle 2 aufgeführten acht Möglichkeiten die von ihm verwendeten auswählen.

Die angebotenen Strategien stellen eine willkürliche Auswahl dar. Um sicherzustellen, daß keine entscheidenden Strategien unberücksichtigt blieben, bekam jeder Befragte zusätzlich die Möglichkeit, eigene Strategien anzugeben. Von dieser Möglichkeit machten nur sechs Personen (6 %) Gebrauch. Dies spricht für die Angemessenheit der Auswahl angebotener Strategien. Zusätzlich angegebene Strategien waren beispielsweise „Ich frage Kollegen" oder „Ich kenne den URL" (uniform resource locator).

Suchstrategie
Ich verwende ein Suchwort.
Ich achte auf exakte Rechtschreibung.
Ich verwende Ausdrücke (z.B. „Statistisches Bundesamt").
Ich verschaffe mir einen Überblick.
Ich verwende Synonyme.
Ich suche nach bestimmten Dateiformaten.
Ich suche mit mehreren Worten gleichzeitig.
Ich verwende Platzhalter (z.B. * oder ?).

Tabelle 2: Suchstrategien

2.2.4 Logische Operatoren

Logische Operatoren wurden mit einer offenen Frage erfaßt („Eine Frage für 'Experten': Welche Operatoren zur Verknüpfung von Suchworten halten Sie für sinnvoll?"). Die offene Form der Frage wurde gewählt, um „Scheinantworten" von unerfahrenen Befragten zu verhindern. Durch den Zusatz „halten Sie für sinnvoll?" sollte die *Bewertung* der Operatoren erfragt werden. Alle Antworten wurden als UND-, ODER- bzw. NICHT-Operator klassifiziert. Eine vierte Kategorie beinhaltet die Kombination von Operatoren, im weiteren als „Klammerung" bezeichnet.

3 Ergebnisse und Diskussion

3.1 Suchstrategien

3.1.1 Suchstrategien und Computererfahrung

Beurteiler mit *niedriger Computererfahrung* geben im Mittel 3,2 (s=1,40), Beurteiler mit *mittlerer Computererfahrung* 3,8 (s=1,48) und Beurteiler mit *hoher Computererfahrung* 4,1 (s=1,41) unterschiedliche Suchstrategien an. Der Unterschied von nur rund einer Suchstrategie zwischen wenig erfahrenen und sehr erfahrenen Befragten ist wesentlich kleiner als zu erwarten war. Die Vermutung, daß Computererfahrung keinen allzu großen Einfluß auf die Anzahl der eingesetzten Suchstrategien hat, wird durch die zwar hoch signifikante, aber relativ niedrige Korrelation von .29 (Spearmans ρ, p<0,01, zweiseitig) zwischen dem Kennwert

Computererfahrung und der *Anzahl eingesetzter Suchstrategien* bestätigt. Der geringe Unterschied in der Zahl der Strategien ist konsistent mit einer Beobachtung von Canter et al. [2]. In einer experimentellen Untersuchung zur Navigation in komplexen Datenbasen zeigte sich, daß Benutzer, die eine erfolgreiche Strategie entdeckt haben, an dieser unter Ausschluß anderer Strategien festhalten (S. 253, vgl. auch [9]). Dementsprechend gaben die Befragten also eher besonders *erfolgreich* angewendete als alle beherrschten Suchstrategien an.

Tabelle 3 zeigt die Prozentzahlen für die Nutzung einzelner Suchstrategien in Abhängigkeit von der Computererfahrung. Betrachtet man die drei am häufigsten genannten Strategien für jede Stufe der Computererfahrung (siehe Rahmen) zeigt sich ein relativ konsistentes Bild: Auf jeder Stufe der Computererfahrung wird eines oder mehrere Suchworte verwendet. Personen mit *niedriger* Computererfahrung bevorzugen einen Überblick (vgl. „Answers First, Then Questions"-Paradigm [12]). Personen mit *mittlerer* oder *hoher Computererfahrung* verwenden eher explizite Platzhalter („wildcards"). Es sollten also mindestens diese vier Suchstrategien vom System optimal unterstützt werden.

		Computererfahrung			niedrig vs. hoch	
	gesamt	*niedrig*	*mittel*	*hoch*		
Suchstrategie	%	%	%	%	χ^2	sig.
ein Suchwort	89	91	78	97	,02	n.s.
Exakte Rechtschreibung	39	36	38	44	,15	n.s
Ausdrücke	40	36	44	41	,04	n.s.
Überblick	42	52	47	28	2,46	n.s.
Synonyme	35	21	44	41	1,80	n.s.
Dateiformate	18	12	22	19	,40	n.s.
Mehrere Worte	56	39	53	75	3,27	p≤0,10
Platzhalter	53	30	59	69	4,50	p<0,05
	n=97	n=33	n=32	n=32		

Tabelle 3: Suchstrategien und Computererfahrung (Prozentwerte sind ganzzahlig gerundet, Rahmen kennzeichen die drei am häufigsten genannten Strategien)

χ^2-Tests zeigen bei der Verwendung von Platzhaltern einen deutlichen Unterschied zwischen den beiden Extremgruppen *niedrige* und *hohe Computererfahrung*. Je höher die Computererfahrung, desto eher wird mit expliziten Platzhaltern bei der Suche gearbeitet.

Nur marginal signifikant, aber dennoch deutlich, ist der Unterschied zwischen den Extremgruppen bezüglich des Verwendens mehrerer Suchworte gleichzeitig. Dies wird von erfahrenen Befragten häufiger angegeben als von weniger erfahrenen. In diesem Zusammenhang schlägt Nielsen [11] vor, die Suche nach mehreren Worten auf einer getrennten Experten-Suchmaske zu „verstecken". Damit soll verhindert werden, daß der nicht so erfahrene Benutzer mit logischen Operatoren umgehen muß (was zwangsläufig eine Folge des Verwendens mehrerer Suchworte gleichzeitig ist) und dabei mit unlösbaren Problemen konfrontiert wird.

Uns erscheint dieser Vorschlag nicht angebracht, da immerhin noch 40% der Befragten mit niedriger Computererfahrung mit mehreren Worten suchen. Wir meinen, daß der Schritt vom Verwenden nur eines Suchwortes zum Verwenden mehrerer Suchworte automatisch gemacht wird, bevor überhaupt ein Verständnis für logische Operatoren entsteht (vgl. Abschnitt 3.2). Die Folge ist eine Art „Vakuum", in dem zwar vom Suchenden mehrere Suchworte gleichzeitig verwendet werden und er auch eine implizite Vorstellung von der Verknüpfung dieser Worte durch das System hat, ohne daß diese allerdings der Realität entsprechen muß. Anstatt dem wenig erfahrenen Benutzer die Möglichkeit des Verwendens mehrerer Suchworte zu nehmen, sollte ein logischer Operator als Standardvoreinstellung („default") gewählt werden, der den Erwartungen des Benutzer entspricht.

3.1.2   Suchstrategien und Interneterfahrung

Betrachtet man Suchstrategien in Abhängigkeit von der *Interneterfahrung*, so zeigt sich ein deutlicher Unterschied beim Verwenden mehrerer Suchworte bei internetnaiven und internetfahrenen Befragten: *Internetnaive* verwenden selten mehrere Suchworte (17% vs. 69%; χ^2=4,01, df=1, p<0,05). Gerade bei der Gestaltung von Suchfunktionen im Internet muß also Wert auf die Unterstützung der offenbar besonders notwendigen Suchstrategie „mehrere Worte gleichzeitig" gelegt werden. Gleichzeitig muß für den Interneteinsteiger (Übergang von internetnaiv zu interneterfahren) auf eine verständliche Präsentation des Konzepts „logischer Operator" geachtet werden, da es dem Benutzer durch die bisher bevorzugte Suche mit nur einem Suchwort wahrscheinlich eher unbekannt ist.

Ein weiterer, marginal signifikanter Unterschied (Binomialtest, p<0,10) zwischen internetnaiven und interneterfahrenen Benutzern zeigt sich beim Verwenden von Synonymen. Nur zwei internetnaive Befragte (17%) geben an, Synonyme als Suchstrategie zu verwenden. Dem stehen sechs interneterfahrene Befragte (38%) gegenüber.

Dieser Anstieg beim Verwenden von Synonymen bei *interneterfahrenen* Befragten macht auf eine Besonderheit der Information im Internet (bzw. Netzwerk) aufmerksam. Information im Internet ist im Gegensatz zu Information auf lokalen Systemen oft nicht vom Benutzer selbst erstellt worden. Der Benutzer sucht also nach Information, ohne genau zu wissen, ob der Autor ähnliche Worte bzw. Begriffe verwendet wie er selbst. Dies macht eine Suche mit Synonymen notwendig. Ein Beispiel dafür ist eine Nutzerin, die in einem Gesundheitsinformationssystem im Internet nach Information zum Thema „Brustkrebs" sucht und nichts findet, weil die relevante Information unter dem Begriff „Mamma-Karzinom" abgelegt wurde. Dementsprechend ist im Internet die Suche nach Synonymen eine zentrale Suchstrategie, die von einer Suchmaschine oder einer „On-Site"-Suche unbedingt unterstützt werden sollte.

Weiterhin deutet sich an, daß *interneterfahrene* Befragte bei der Formulierung ihrer Anfragen eher auf Rechtschreibung achten (63% vs. 25%, χ^2=2,08, df=1, p=0,15). Trotzdem bleibt gerade die Rechtschreibung ein großes Problem [1;13], das durch entsprechende Funktionalität, z.B. eine Rechtschreibhilfe, gemildert werden kann. Weiterhin verwenden *interneterfahrene* Beurteiler weniger oft explizit Platzhalter (31% vs. 50%, χ^2=0,61, df=1, n.s.). Dies ergibt sich wahrscheinlich aus der Tatsache, daß viele Internet-Suchsysteme auf die explizite Eingabe von Platzhaltern verzichten.

3.2   Logische Operatoren

Im Rahmen der vorliegenden Befragung gaben 50,5% der Befragten keine Antwort auf die Frage „Eine Frage für 'Experten': Welche Operatoren zur Verknüpfung von Suchworten halten Sie für sinnvoll?". Diese Befragten können entweder mit dem Konzept bzw. Begriff des „logischen Operators" nichts anfangen oder halten keinen Operator für sinnvoll.

Es zeigt sich, daß 19,6% der Befragten, die die Suchstrategie „mehrere Worte gleichzeitig" angegeben haben, **keinen** Operator für sinnvoll halten, obwohl dieser bei mehreren Suchworten technisch gesehen zwingend notwendig ist. Es liegt die Interpretation nahe, daß hier immerhin rund ein Fünftel der Befragten logische Operatoren zwar implizit verwenden, ohne allerdings Auskunft über logische Operatoren geben zu können.

Betrachtet man nun die 49,5% der Befragten, die mindestens einen Operator als sinnvoll erachten, findet man folgende Verteilung: 90% erachten den UND-Operator, 69% den ODER-Operator, 42% den NICHT-Operator und nur noch 8% die Kombination, d.h. „Klammerung" von Operatoren als sinnvoll (Prozentwerte beziehen sich auf die Teilstichprobe der Befragten, die mindestens einen Operator für sinnvoll halten, n=48).

Die Befragten, die Auskunft über logische Operatoren geben können, bevorzugen den UND-Operator. Dies macht deutlich, daß das Verwenden mehrerer Suchworte von den Befragten hauptsächlich als eine weitere *Spezifizierung* der Suchanfrage verstanden wird. Getreu dem Motto „Je mehr ich erkläre, desto genauer soll die Antwort meine Wünsche treffen" wird der UND-Operator zu einem wichtigen Werkzeug für die Suche in einer großen Datenbasis. Auf der Basis der Erfahrungen und Erwartungen rund einer Hälfte der Befragten ist es daher sinnvoll, die UND-Verknüpfung von Suchworten als Standardvoreinstellung („default") einer jeden Suchfunktionalität einzusetzen, wie es z.B. bei der Internet-Suchmaschine „Hotbot" (www.hotbot.com) realisiert wurde. Der ODER-Operator wird zwar seltener als sinnvoll erachtet, ist aber dennoch als wichtiges Werkzeug anerkannt und sollte ebenso unterstützt werden.

Nur noch 42% der Befragteb halten den NICHT-Operator für sinnvoll. Dies reflektiert die Tatsache, daß man für einen sinnvollen Einsatz des NICHT-Operators sehr viel über die Struktur und Inhalte der durchsuchten Datenbasis wissen muß. Er macht es möglich, große Treffermengen gezielt zu reduzieren. Eine solche Form der interaktiven, mehrstufigen Suche, bei der gezielt unpassende Fundstellen herausgefiltert werden, ist besonders in großen, unbekannten Datenbasen, wie das Internet eine darstellt, sinnvoll. Bei der Gestaltung von Suchmaschinen sollte man allerdings eher neue Formen der Filterung finden, als auf den schwer zu handhabenden NICHT-Operator zu vertrauen.

Nur noch 8% halten die Kombination von Operatoren für sinnvoll. Eine solche „Klammerung" logischer Ausdrücke ist oft selbst für den Experten nicht immer auf Anhieb zu verstehen. Bei der Gestaltung einer einfachen Suchfunktion kann diese Funktionalität getrost auf eine Expertenseite verschoben werden oder sogar gänzlich entfallen.

3.2.1 Logische Operatoren, Computererfahrung und Interneterfahrung

Bis jetzt wurden logische Operatoren noch nicht in Abhängigkeit von der *Computererfahrung* betrachtet. Hier zeigt sich ein sehr konsistentes Bild: Im allgemeinen nimmt die Anzahl der Befragten, die keinen Operator für sinnvoll erachten, mit zunehmender *Computererfahrung* ab. Rund 88% aller Befragten mit *niedriger Computererfahrung* geben keinen Operator als sinnvoll an, bei *mittlerer Computererfahrung* sind es noch 41% und bei *hoher Computererfahrung* nur noch 22%. Die Rangkorrelation zwischen der *Anzahl angegebener Operatoren* und dem Kennwert für *Computererfahrung* beträgt .55 (Spearmans ρ, n=97, p< 0,001, zweiseitig).

Die festgestellte Reihenfolge der Operatoren UND, ODER, NICHT und „Klammerung" bleibt auf jeder Stufe der *Computererfahrung* (niedrig, mittel, hoch) bestehen. Nur die Anzahl der Personen, die den entsprechenden Operator für sinnvoll halten, nimmt zu. Dementsprechend ist die Empfehlung, den UND-Operator als Standardvoreinstellung („default") zu verwenden, unabhängig von der Computererfahrung und damit allgemeingültig.

In bezug auf Interneterfahrung zeigt sich ein ähnliches Bild. Elf von 12 *internetnaiven* Befragten (92 %) halten keinen Operator für sinnvoll. Nur ein Befragter hält den UND-Operator für sinnvoll. Elf von 16 *interneterfahrenen* Befragten (69%) halten keinen Operator für sinnvoll. Die „Klammerung" von Operatoren findet kein *interneterfahrener* Befragter sinnvoll.

Alles in allem hat Interneterfahrung bei einer niedrigen bis mittleren Computererfahrung keine große Auswirkung auf das Wissen über logische Operatoren. Dies ist besonders interessant, da das Verwenden logischer Operatoren im Internet mit seinen großen Mengen an Information immer wichtiger wird, aber selbst interneterfahrene Befragte das Konzept „logischer Operator" nicht unbedingt kennen. Es liegt wiederum der Schluß nahe, daß Benutzer intuitiv mehrere Suchworte verwenden und auch eine gewisse Funktionalität damit verbinden, ohne die genaue Funktionsweise verschiedener logischer Operatoren verstanden zu haben.

4 Gestaltungsempfehlungen und Ausblick

Aus den hier dargestellten Ergebnissen der Benutzerbefragung lassen sich die folgenden Gestaltungsempfehlungen ableiten:

Gestaltungsempfehlung 1 (vgl. Abschnitt 0): Eine Suchfunktion muß die vier häufigsten Suchstrategien unterstützen. Sie sollte die *Suche mit einzelnen* und *mehreren Suchworten gleichzeitig* ermöglichen, ohne daß spezielle Bildschirmseiten oder -masken aufgerufen werden müssen. Für Benutzer mit geringer Erfahrung ist eine *Übersicht* wichtig, die es ihnen möglich macht, etwas über die Struktur und die Inhalte der Datenbasis zu erfahren. Für erfahrenere Benutzer haben *Platzhalter* ein wichtige Funktion. Das System kann hier den Benutzer unterstützen, indem es z.B. ein eingegebenes Suchwort automatisch als Wort*anfang* interpretiert. Unsere Erfahrungen zeigen, daß Platzhalter häufig als Substitut für das Wortende verwendet werden. Allerdings muß das System den Benutzer auf das automatische Verwenden von Platzhaltern aufmerksam machen.

Gestaltungsempfehlung 2 (vgl. Abschnitt 0): Eine Suchfunktion für das **Internet** muß die *Suche nach Synonymen* unterstützen. Das System sollte hier von der Möglichkeit Gebrauch machen, dem Benutzer die automatisch verwendeten Synonyme transparent zu machen. Weiterhin muß eine Suchfunktion auf *Rechtschreibfehler bzw. Tippfehler* in der Suchanfrage aufmerksam machen [13] oder eine Funktion zur Suche nach ähnlich klingenden Suchworten bereitstellen.

Gestaltungsempfehlung 3 (vgl. Abschnitt 0): Die optimale Standardvoreinstellung („default") für die Verknüpfung mehrerer Suchworte ist das logische *UND*. Diese Standardvoreinstellung muß dem Benutzer deutlich signalisiert werden. Dabei sollte neben den konkreten logischen Operatoren auch verbale Umschreibungen der Art „alle Worte" o.ä. angeboten werden, die es dem Benutzer ermöglicht, ihre Funktionsweise nachzuvollziehen. Auf die „Klammerung" und ggf. den NICHT-Operator kann verzichtet werden, wenn andere Mechanismen der Filterung verwendet werden. Diese sollten kein explizites Wissen über die Inhalte und Struktur der Datenbasis voraussetzen. Gute Möglichkeiten sind hier die Kombination von gezielter Suche mit anschließendem „Browsen", das Unterstützen einer iterativen Suche durch erneutes, auf die Treffermenge beschränktes Durchsuchen oder wiederverwendbare und miteinander kombinierbare Treffermengen (z.B. [14] S. 108).

Die hier vorgestellten Gestaltungsempfehlungen bilden praktische Minimalanforderungen an Suchfunktionen auf der Basis technischer Realisierbarkeit. Sie stellen einen Ausgangspunkt für die Gestaltung von Suchfunktionen dar. Mit ihrer Hilfe kann der Forderung von Pollock und Hockley [13] entsprochen werden, nämlich zuerst die einfache Suche gut zu gestalten, bevor erfahrenere Benutzer mit exotischen Funktionalitäten bedacht werden.

Die Gestaltungsempfehlungen haben den Charakter von Hypothesen, die in zukünftigen Arbeiten empirisch geprüft werden sollten. Gebrauchstauglichkeitsstudien mit realen Systemen müssen klären, mit welchen Nutzungsproblemen trotz Berücksichtigung der Gestaltungsempfehlungen gerechnet werden muß.

Alles in allem erscheint es wünschenswert, Gestaltungsempfehlungen aus empirischen Daten abzuleiten. Dabei sollten die bereits häufig zur Anwendung kommenden experimentell-orientierten Verfahren (z.B. Gebrauchstauglichkeitsstudien) um schriftliche Benutzerbefragungen ergänzt werden. Benutzerbefragungen können an einer weitaus größeren Stichprobe ökonomisch durchgeführt werden. Dies ermöglicht eine „Makro"-Perspektive auf das Ge-

staltungsproblem, die eine wichtige Ergänzung zur eher detaillierten, konkreten „Mikro"-
Perspektive experimenteller oder quasi-experimenteller Studien darstellt.

5   Literatur

[1] Why On-Site Searching Stinks. In: User Interface Engineering (1998),
     http://world.std.com/~uieweb/searchart.htm.

[2] Canter, D.; Powell, J.; Wishart, J. & Roderick,C.: User navigation in complex database systems. In: Beha-
     viour & Information Technology 5 3 (1986), 249-257.

[3] Hamborg, K.-C.: Zum Einfluß der Komplexität von Software-Systemen auf Fehler bei Computernovizen
     und Experten. In: Zeitschrift für Arbeits- und Organisationspsychologie 40 1 (1994)

[4] Hassenzahl, M. & Prümper, J.: Software-ergonomische Beratungsprojekte in der Praxis: Probleme und Kon-
     zepte am Beispiel eines Informationssystems im WorldWideWeb (WWW), in Vorbereitung.

[5] Hassenzahl, M., Prümper, J. & Schulz, J.: Die software-ergonomische Gestaltung von WorldWideWeb-
     Seiten: Benutzerbefragungen und „Usability"-Tests. In: G. W. Himmelmann (Hg.): Oracle in Theorie und
     Praxis: Vortragsband zur 10. Jahrestagung der DOAG-Konferenz. Stuttgart, 1997: Deutsche Oracle Anwen-
     der Gruppe e.V. 309-323.

[6] Himmelmann, G. W., Prümper, J., Hassenzahl, M., Schulz, J., Eberhardt, W., Brückner, G. und Dubrow,
     M.: A General Health Information System in the Web: What is its Impact? - The German Federal Health In-
     formation System IS-GBE as a Sociocybernetic Experience. World Congress of Sociology, Montreal, July
     26th - August 1st, 1998.

[7] ISO 9241: Ergonomic requirements for office work with visual display terminals. Part 10: Dialogue prin-
     ciples. 1994: International Organization for Standardization.

[8] ISO 9241: Ergonomic requirements for office work with visual display terminals. Part 11: Guidance on
     usability. 1996: International Organization for Standardization.

[9] Luchins, A. S.: Mechanization in problem-solving. In: Psychological Monographs 54 6 (1942).

[10]   Maaß, S., Rosson, M. B. & Kellog, W. A.: Benutzerfreundlichkeit, Systemkonsistenz und andere
     schwer definierbare Prinzipien: Interviews mit Systementwicklern. In: W. Schönpflug und M. Wittstock
     (Hg.): Software-Ergonomie '87: Nützen Informationssysteme dem Benutzer? Stuttgart, 1987: B.G. Teubner.
     417-427.

[11]   Nielsen, J.: Search and You May Find. In: Jakob Nielsen's Alertbox July (1997),
     http://www.useit.com/alertbox/9707b.html.

[12] Owen, D.: Answers First, Then Questions. In: D. A. Norman und S. W. Draper (Hg.): User Centered System
     Design. Hillsdale (NJ), 1986: Lawrence Erlbaum Associates. 361-375.

[13]   Pollock, A & Hockley, A.: What's Wrong with Internet Searching. In: D-Lib Magazine, March (1997).

[14]   Rosenfeld, L. & Morville, P.: Information Architecture for the World Wide Web. Cambridge, 1998:
     O'Reilly.

[15]   Shneiderman, B.: Designing information-abundant web sites. In: International Journal of Human-
     Computer Studies 47 (1997), 5-29.

[16]   Wandmacher, J.: Software-Ergonomie. Berlin, 1993: de Gruyter.

Adressen der Autoren

Marc Hassenzahl
bao - Büro für Arbeits- und Organisationspsychologie
Gutzmannstraße 38
14165 Berlin
Email: hassenzahl@aol.com

Prof. Dr. Jochen Prümper
FHTW - Fachhochschule für Technik und Wirtschaft
Treskowallee 8
10313 Berlin
Email: j.pruemper@fhtw-berlin.de

Flexible Präsentation von Prozeßmodellen

Thomas Herrmann

Fachbereich Informatik, Universität Dortmund

Zusammenfassung

Geschäftsprozeßmodelle müssen für alle Beteiligte nachvollziehbar sein, insbesondere für diejenigen, die die Aufgaben im Rahmen dieser Prozesse bearbeiten. Deshalb sollte man sich Diagramme, die Geschäftsprozesse darstellen, mit Hilfe software-basierter Präsentationswerkzeuge ansehen können. Diese Werkzeuge sollen verschiedene Ebenen von Details präsentieren können. Es wird erklärt, wie man Mechanismen des Ein- und Ausblendens verwenden kann, wie man Gruppierung von Elementen, Kontextualisierung oder die Nachvollziehbarkeit für Laien unterstützt und wie sich Visualisierungselemente einsetzen lassen.

Abstract

Models of business processes must be comprehensible for all participants, also for those people who have to carry out tasks in the course of the process. Therefore, the diagrams representing the models should be used with the help of software-based viewers, which can provide different levels of details. It is explained how hide-and-show-mechanisms can be supported, how grouping of elements, availability of context or comprehensibility for non-experts can be achieved and how elements of visualization can be used.

1 Einleitung

In dem interdisziplinären Forschungsprojekt MOVE wird das Ziel verfolgt, Workflow Management Systeme mitarbeiter-orientiert einzuführen. Dazu gehört, daß künftige Benutzer partizipieren können und daß ihre Interessen im Verlauf einer stetigen Verbesserung der zu unterstützenden Geschäftsprozesse berücksichtigt werden. Hierzu sind verschiedene Instrumente und Methoden erforderlich: partizipative Erhebungsverfahren, Verfahrensvorschläge für die Partizipation, Methoden der Soll-Konzept-Gewinnung, Darstellung von Geschäftsprozessen, Flexibilisierung und Behandlung von Sonderfällen, Evaluation und Ermittlung von Feedback zu den implementierten Geschäftsprozessen (Herrmann et al., 1998). Es ist nicht unumstritten, ob Workflow Management Systeme den ausführenden Mitarbeitern überhaupt Vorteile bringen können. Allerdings ist davon auszugehen, daß im Falle ihrer Einführung durch partizipative Verfahren die besseren Effekte erzielt werden. Ferner ist davon auszugehen, daß sich die hier präsentierten Anforderungen auch auf die Darstellung und Analyse semistrukturierter Kooperationsprozesse übertragen lassen, wie sie von Groupware im allgemeinen unterstützt werden.

Es wird die Frage behandelt, wie sich Geschäftsprozesse so darstellen lassen, daß folgende Anforderungen erfüllbar sind:

- Die verwendeten Diagramme müssen für Laien auf dem Gebiet der Modellierung nachvollziehbar sein oder ihnen nachvollziehbar präsentiert werden können.
- Die Diagramme müssen auch Informationen beinhalten, die für die Verbesserung aus Mitarbeitersicht benötigt werden (z.B. sollten die Entscheidungsspielräume einzelner Aufgaben nachvollziehbar sein, damit qualitative Mischarbeit realisiert werden kann).
- Zusätzliche Informationen müssen leicht ergänzbar sein, z.B. im Laufe eines Workshops.

Diese Anforderungen sind widersprüchlich, da zum einen eine Reduktion dargestellter Information zwecks leichterer Nachvollziehbarkeit relevant ist und zum anderen die Informationsvielfalt noch zu erweitern ist. Dieser Widerspruch kann mit Hilfe computergestützter Editoren und Präsentationswerkzeuge zumindest teilweise aufgefangen werden, indem Aus- und Einblendungsmechanismen angeboten werden. Sie spielen unseres Erachtens eine zentrale Rolle, wenn Diagramme, die reale und damit umfangreiche Prozesse darstellen, noch nachvollziehbar sein sollen. Deshalb konzentriert sich diese Analyse auf solche Mechanismen und ergänzt sie um Möglichkeiten des flexiblen Gruppierens.

Hauptaufgabe der vorliegenden Analyse ist es, die Anforderungen an solche Mechanismen in der Domäne der mitarbeiter-orientierten Darstellung von Geschäftsprozessen aufzuarbeiten. Ferner ist zu diskutieren, inwieweit die hier vorgeschlagenen Möglichkeiten zur flexiblen Generierung von Darstellungen automatisch erfolgen können oder in hohem Maße interaktives Arbeiten erfordern. Die hier vorgestellten Anforderungen gehen davon aus, daß der Wechsel zwischen einzelnen Darstellungen durch Verwendnung geeigneter Funktionen unterstützt werden kann, die zum Teil durch Schaltflächen in den Diagrammen aktivierbar sind. Zielgruppen, die diese Funktionen nutzen sollen, sind zum einen die Betrachter der Diagramme selbst und zum anderen diejenigen, die solche Diagramme präsentieren. Entsprechend dem hier behandelten Thema erfolgt die Darstellung der Anforderungen vor allem mittels Abbildungen, da sie aussagekräftiger sind als textliche Erläuterungen. Die einzelnen Beispiele wirken für sich genommen recht einfach, die Komplexität der Anforderungen ergibt sich jedoch aus der Kombinierbarkeit und Auswählbarkeit des gesamten Spektrums der hier zu beschreibenden Möglichkeiten. Dabei wird auch deutlich, daß die Funktionen zum schrittweise Einblenden von Elementen bei gängigen Präsentationstools diesen Anforderungen nicht genügt. Es wird nicht auf 3D-Effekte eingegangen, da zusätzliche die Anforderung besteht, daß die verwendeten Zeichungssymbole im Bedarfsfalle auch mittels Stift und Papier festgehalten und kombiniert werden können. Aus ähnlichen Gründen wird hier nicht auf animierte Simulationen von Prozessen eingegangen.

Zur Behandlung der Grundlagen der skizzierten Fragestellung sind vor allem die wahrnehmungspsychologische Literatur und kommunikationstheoretische Schriften relevant. Zur Wahrnehmungspsychologie finden sich als Beiträge zur Software-Ergonomie zahlreiche Zusammenfassungen (z.B. Glaser, 1994), die jedoch dem eigentlichen, schon lange bekannten Reichtum an Details des Themas (s. z.B. Metzger, 1953) nicht gerecht werden können. Auch neuere Arbeiten, die sich mit der Gestaltung von Diagrammen befassen, fokussieren sich meistens auf einen Ausschnitt, wie etwa der Wirksamkeit der Nähe, der Symmetrie oder der „Guten Gestalt" (vergl. z.B. Blythe, 1996 et al. oder Purchase et al. 1995). Auf diese neueren Beiträge sowie auf die Schriften von Moody (1996) und Lohse et al. (1995) wird bei der Beschreibung der Anforderungen Bezug genommen. Allerdings müssen ihre Ergebnisse auf die hier behandelte Domäne übertragen werden, da sich in der Literatur auf das Problem der Darstellung von Geschäftsprozessen und der interessen- bzw. kontext-orientierten Hinzunahme und Ausblendung von Information kaum Hinweise befinden. Ausnahmen bilden zum Beispiel Hoffmann (1998) oder Rosemann (1996), der in seinen Erläuterungen zum ordnungsgemäßen Modellieren auch Anmerkungen zur geeigenten Darstellung von Geschäftsprozessen aus betriebswirtschaftlicher Sicht macht. Hinsichtlich des kommunikationstheoretischen Hintergrundes muß hier aus Platzgründen auf die Analyse von Herrmann (1997) verwiesen werden.

Der empirische Hintergrund der hier entwickelten Anforderungen ergibt sich aus der Kooperation mit 10 Unternehmen, bei denen Geschäftsprozesse sowohl analysiert als auch modelliert wurden. Bei diesen Fallstudien stellte sich mehrfach die Frage, wie mitarbeiter-orientierte Aspekte in die Modelle und in die Optimierung integriert werden können. Hierzu wurden Workshops und Präsentationen durchgeführt und Erfahrungen gesammelt, die in die unten beschriebenen Anforderungen eingehen. Diese empirische Grundlage zeigt insbesondere die Defizite und Probleme, für die die folgenden Vorschläge eine Lösung darstellen. Die Lösungen selbt konnten noch nicht empirisch überprüft werden. Ferner wurde durch eigene Praxis bei der Erstellung von ca. 200 Einzeldarstellung im Rahmen der Lehre zu kommunikations- und kooperationsunterstützenden Systemen ein Fundus von Erfahrungen gebildet. Hierzu wurde eine eigens entwickelte Methode (SeeMe) verwendet, die den folgenden Ausführungen als Grundlage dient (Kap. 2). Die sich anschließenden Kapitel vermitteln die Anforderungen zur flexiblen Präsentation. Kapitel 7 geht abschließend auf Möglichkeiten zur kontrollierten empirischen Validierung der getroffenen Anforderungen ein und auf ihre software-technischen Unterstützung.

2   Die Modellierungsmethode SeeMe

Abb. 1 zeigt einige Basiskonzepte der semi-strukturierten, sozio-technischen Modellierungsmethode SeeMe am Beispiel eines Geschäftsprozesses zur Vertragserstellung. In den Ovalen werden Rollen dargestellt. Die von den Rollen ausgehenden Pfeile zeigen auf Aktivitäten (abgerundete Rechtecke), die von den Rollen ausgeführt werden können. Aktivitäten können Entitäten (Rechtecke) erzeugen oder verändern. Dies wird durch einen Pfeil ausgedrückt, der von der Aktivität auf das Rechteck zeigt (z.B. erzeugt Vertragsanfrage stellen den Vertragsentwurf). Pfeile in umgekehrter Richtung drücken aus, daß eine Aktivität eine Entität als Bearbeitungsmittel benutzt. Pfeile stellen also Relationen dar. Zwischen Aktivitäten drücken sie z.B. den Ablauf aus. Ferner gibt es logische Verknüpfungen, die mit gedrehten Quadraten dargestellt werden, wobei zwischen UND, ODER, exklusivem ODER und Sondersymbolen unterschieden wird. Wie in den noch folgenden Abbildungen zu sehen ist, werden auch Sechsecke verwendet, um Ereignisse oder Bedingungen auszudrücken. Diese sogenannten Modifikatoren geben an, wie mit logischen Verzweigungen umzugehen ist. Ferner besteht die Möglichkeit, die Eigenschaften von Rollen, Aktivitäten, Entitäten oder Relationen durch Attribute näher zu beschreiben.

SeeMe wurde speziell für die Darstellung semi-strukturierter, sozio-technischer Prozesse und

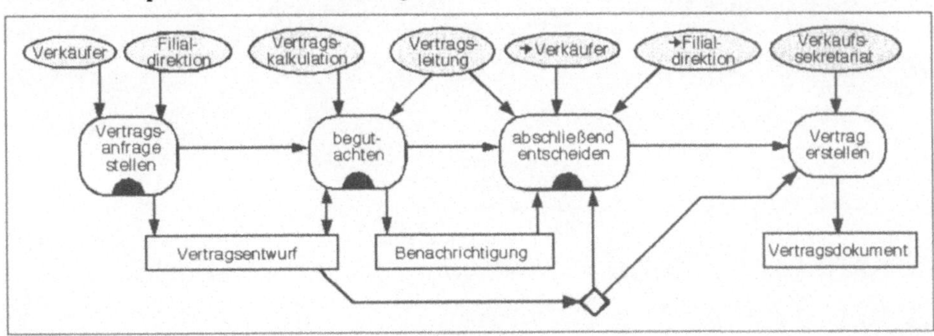

Abbildung 1

Systeme entwickelt. Dabei wurde auf Konzepte der ereignis-gesteuerte Prozeßketten, der objekt-orientierten Modellierung, State-Charts nach Harel und andere Bezug genommen. Einen detaillierteren Überblick zu SeeMe geben Hoffmann et al. (1998). Hauptziel war es, die vielfältigen Besonderheiten sozialer Systeme berücksichtigen zu können. Der derzeit Verbreitung findende Standardisierungsversuch mit UML reicht hierzu nicht aus. Auch Petrinetz basierte Ansätze sind mit Hinblick auf ihre Abfolgebedingungen zu starr, da in semi-strukturierten Prozessen Folgeaktivitäten auch starten können, wenn die vorangegangenen noch nicht abgeschlossen sind. SeeMe geht insbesondere auf Möglichkeiten der Kennzeichung von unvollständiger und vager Information ein (s. Herrmann; Loser, 1998). Die hierfür eingeführten Darstellungsmöglichkeiten werden im folgenden nur begrenzt benutzt, etwa die Verwendung von drei Punkten in einem Halbkreis, um Auslassungen anzuzeigen, zu denen keine nähere Spezifizierung bekannt ist. Der nicht ausgefüllte Konnektor in Abb.1 (leere Raute) ist ein weiteres Beispiel. Die schwarzen Halbkreise zeigen dagegen an, daß die Information absichtlich unvollständig dargestellt ist. Solche Symbole können im Sinne einer Schaltfläche genutzt werden, um mehr Information abzurufen (solche Zusatzinformation wird auf den folgenden Abbildungen dargestellt).

Der dargestellte Geschäftsprozeß behandelt die Erstellung von Verträgen, bei denen insbesondere zu berücksichtigen ist, wie hoch die zu gewährenden Rabatte ausfallen können. Daher besteht der Prozeß in erster Linie aus Prüf- und Begutachtungsschritten, in die auch Nachverhandlungen eingebettet sein können. Um zunächst einen Überblick zu geben, werden die einzelnen Schritte zum Zweck der vorliegenden Ausführungen in vier abstrakteren Einheiten zusammengefaßt. Auf die Problematik solcher Abstraktionen wird unten eingegangen.

3 Allgemeine Grundsätze zur Darstellung von Diagrammen

Wir gehen davon aus, daß Modellierer zunächst ein möglichst vollständiges Modell bzw. Diagramm eines Geschäftsprozesses erzeugen. Dieses vollständige Diagramm würde aber nur in den seltensten Fällen den Interessenten gezeigt. Vielmehr wird zunächst durch Reduktion der dargestellten Elemente ein Überblick gegeben (vergl. Abb. 1). Die Betrachter können dann entsprechend ihrem Informationsbedürfnis weitere Elemente einblenden (Kap. 4), größere Mengen von Elementen übersichtlich gruppieren (Kap. 5) und bei sich änderndem Erkenntnisinteresses Elemente ausblenden (Kap. 6), um an anderer Stelle wieder zusätzliche Informationen abzurufen. Diese Strategie ergibt den Hintergrund zu der hier verwendeten Gliederung.

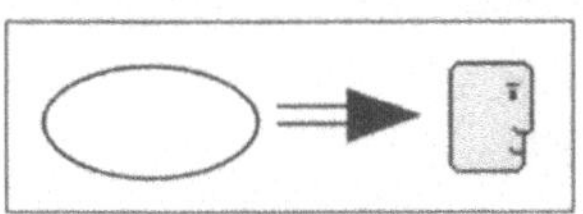

Abbildung 2

Bei der Gestaltung des vollständigen Diagramms sind bereits wahrnehmungspsychologische Grundsätze zu berücksichtigen. Hierzu gehört die Verwendung von Visualisierungselementen, wobei insbesondere solche von Bedeutung sind, die man parallel anstatt sequenziell wahrnimmt, wie z.B. Schrift und Farbe (s. Lohse et al., 1995). Beispiele hierfür ist die Verwendung eines Grautones, um die Einbettung auf verschiedenen Ebene zu verdeutlichen (s. Abb. 3) oder auch die Ersetzung abstrakter Symbole durch Ikone, wie in Abb. 2 gezeigt. Dabei sollten möglichst solche Ikone verwendet werden, die in der Domäne der potentiellen Betrachter bereits vertraut sind. Ferner können Mittel der Hervorhebung zum Einsatz kommen, wie besondere Linienstärke etc. Nach Purchase (1997) sind sich überschneidende Linien, gekrümmte Linien und spitze Winkel zu vermeiden. Hoffmann (1998) bündelt Gestaltungsempfehlungen unter den Gesichtspunkten Verständlichkeit, Übersichtlichkeit und Deutlichkeit.

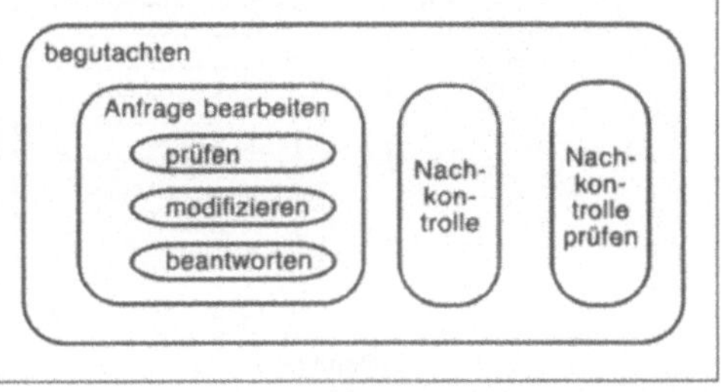

Abbildung 3

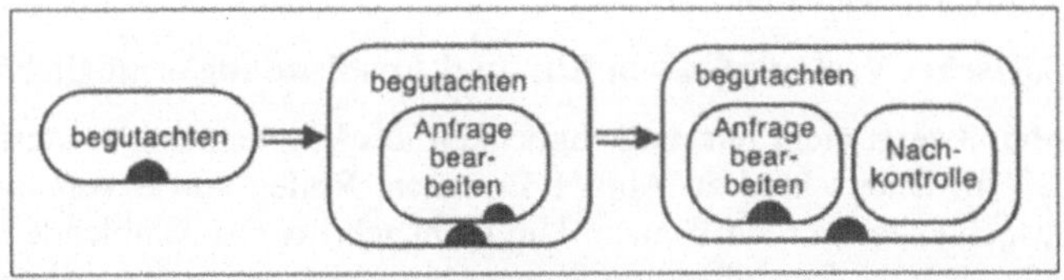

Abbildung 4a

Während Verständlichkeit auf die Auswahl des Symbolvorrates abzielt (s. ISO 9241, Teil 12), referiert Übersichtlichkeit auf Hervorhebungen sowie auf Formen der Gruppierung und Untergliederung mittels geeigneter Symbole und unter Nutzung des Gesetzes der Nähe. Deutlichkeit meint, daß Informationen möglichst direkt und ohne gemehrte Ableitungsschritte erschließbar sein müssen (s. auch Lohse 1995). Von zentraler Bedeutung ist nach Hoffmann (1998) die Reduktion von Komplexität, indem man z.B. pro Diagramm möglichst mit einer begrenzten Zahl von Elementen arbeitet. Da es schwerfällt, eine sinnvolle Obergrenze für diese Zahl im vorhinein anzugeben, empfehlen wir hier, Flexibilität durch Ein- und Ausblenden zu ermöglichen.

4 Schrittweises Einblenden von Teilen der Prozeßdarstellung

Wenn Betrachter eines Diagrammes Zusatzinformationen abrufen, so können diese auf dem Weg der Einbettung dargestellt werden, wie es in Abb. 3 für die Aktivität begutachten der Fall ist. Bei Abb. 3 sind alle Sub-Aktivitäten in einem Schritt eingeblendet. Dies ist nicht immer angemessen.

Verschiedene Strategien des schrittweisen Einblendens anbieten

Neben dem Einblenden aller Sub-Elemente in einem Schritt sollten auch andere Strategien mittels eines Präsentationstools selektierbar sein. Zum einen können schrittweise die Ele-

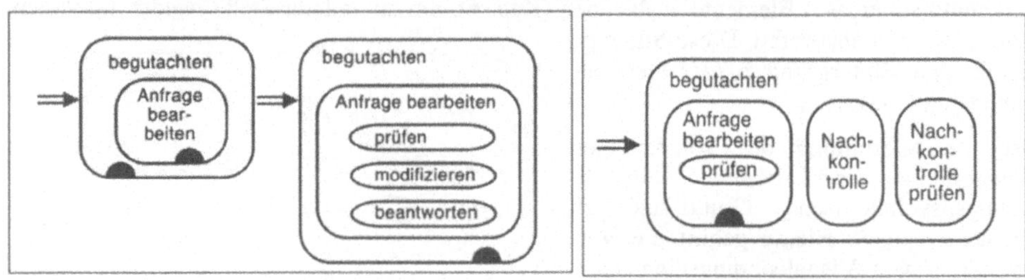

Abbildung 4b   Abbildung 4c

mente eingeblendet werden, die auf der gleichen Einbettungsebene liegen, wie bei Abb. 4a gezeigt. Der Pfeil mit zwei Strichen deutet jeweils einen Einblendungsschritt an. Zum anderen kann demgegenüber mit jedem Schritt die nächst tiefer gelegene Einbettungsebene eingeblendet werden, wie dies bei Abb. 4b erkennbar ist. Schwieriger liegt der Fall, wenn man beide Strategien mischen möchte. Aber auch dies kann unter semantischen oder pragmatischen Gesichtspunkten Sinn machen, wenn man z.B. wie in Abb. 4c alle Sub-Aktiviäten mit einer semantischen Gemeinsamkeit einblenden möchte, hier etwa alle Sub-Aktivitäten, die Überprüfungen beinhalten.

Logisches Verknüpfen von Ein- und Ausblendung ermöglichen

Abb. 3 zeigt nicht nur die Möglichkeit des Einblendens, sondern ist auch ein Beispiel dafür, daß im Unterschied zu Abb. 1 Entitäten, Rollen sowie vor- und nachgelagerte Aktivitäten ausgeblendet werden können. Um beim schrittweise Einblenden einen Informationsoverflow oder auch das Zerreißen von Sinnzusammenhängen zu vermeiden, kann es sinnvoll sein, Ein-

blendungen eines Elementes mit Ausblendungen anderer Elemente logisch zu verknüpfen. Abbildung 4d zeigt, daß mit dem Einblen-

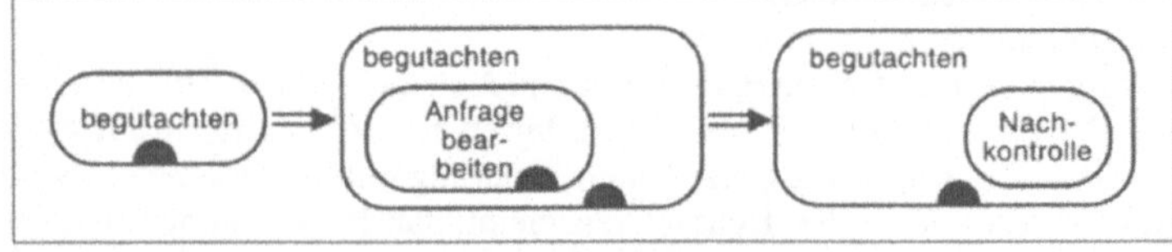

Abbildung 4d

den von Nachkontrolle die Sub-Aktivität Anfrage bearbeiten nicht mehr angezeigt wird. Es liegt also ein XOR-Verknüpfung vor. Genauso können auch UND-Verknüpfungen oder logische INKLUSION möglich sein. So könnte man festlegen, daß beim Ausblenden von Anfrage bearbeiten auch immer Nachkontrolle automatisch ausgeblendet wird, während Nachkontrolle selbst auch ohne Nebenwirkungen ausgeblendet werden kann. Es ist zu beachten, daß diese logischen Zusammenhänge des Visualisierungsgeschehens zwar mit den semantischen Zusammenhängen des Geschäftsprozeßmodells in Einklang stehen können, sich aber auch durchaus von ihnen lösen können. Abb. 4c, bei der logische Zwischenschritte (nämlich beantworten, siehe Abb. 3) ausgeblendet werden, ist ein Beispiel hierfür. Mit einer Präsentationskomponente sollte man solche Abhängigkeiten vorab definieren oder flexible erzeugen können.

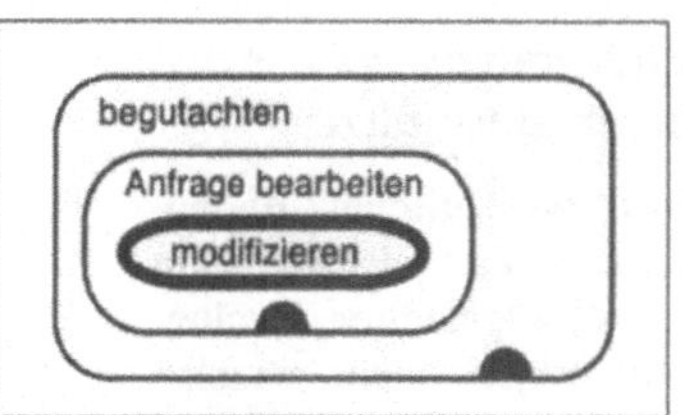

Abbildung 5a

Kontext muß darstellbar sein

Aus kommunikationstheoretischer Sicht (s. Herrmann, 1997) empfiehlt es sich, bei Bedarf den Kontext einzelner Sub-Elemente ebenfalls darzustellen. Abb. 5a stellt zum Beispiel die übergeordneten Elemente der Aktivität modifizieren dar. Somit wird sofort erkennbar, zu welchem übergeordneten Kontext modifizieren gehört, wodurch das Verständnis eines Diagrammes erleichtert wird. Beispielsweise kann man dann auch mittels Abbildung 1 die zugehörige Rolle nachvollziehen,

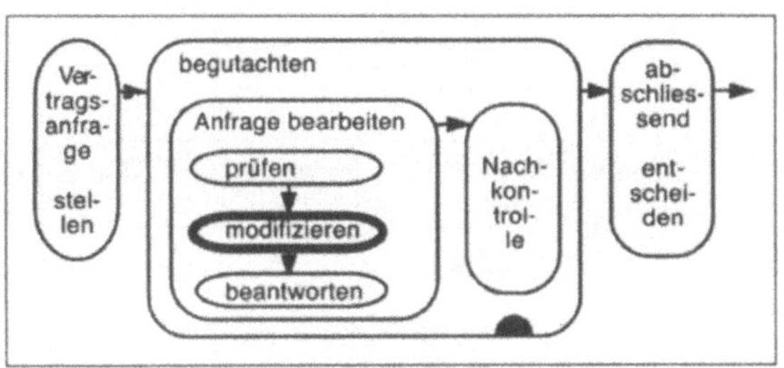

Abbildung 5b

die modifizieren ausführt. Eine andere Strategie zum Aufzeigen von Kontext kann darin bestehen, die vor- und nachgelagerten Elemente anzuzeigen, im Falle von modifizieren also die Aktivitäten prüfen und beantworten. Abb. 5b zeigt die Kombination beider Strategien: Es werden die Vorgänger und Nachfolger auf allen Einbettungsebenen gezeigt. Zur Kennzeichnung der Vorgänger- und Nachfolger werden Pfeile benutzt. Es ist zu beachten, daß Abb. 5b den Kontext von modifizieren darstellt und somit die Existenz von modifizieren als gegeben unterstellt wird, was aus der Sicht der Logik des Geschäftsprozesses nicht immer der Fall sein muß (vergl. Abb. 11).

Hinzunahme typ-fremder Elemente ist anzubieten

Schrittweises Einblenden muß sich nicht nur auf die Darstellung von Elementen des gleichen Typs beziehen. Im zweiten Schritt von Abb. 6 werden z.B. die Relationen und die Bedingung dargestellt, die den Ablauf und seine Verzweigung repräsentieren. Um eine Informationsüberladung zu vermeiden, sollte stets wählbar sein, ob man die Sub-Elemente mit den zwischen ihnen bestehenden Relationen einblendet oder ohne sie. Pfeile können auch genutzt werden, um das schrittweise Aufblenden zu unterstützen. Würde man z.B. den untersten Pfeil in Abb. 6 im Sinne einer Schaltfläche aktivieren, dann würde die nächste Aktivität, begutachten, (vergl. Abb. 1) angezeigt.

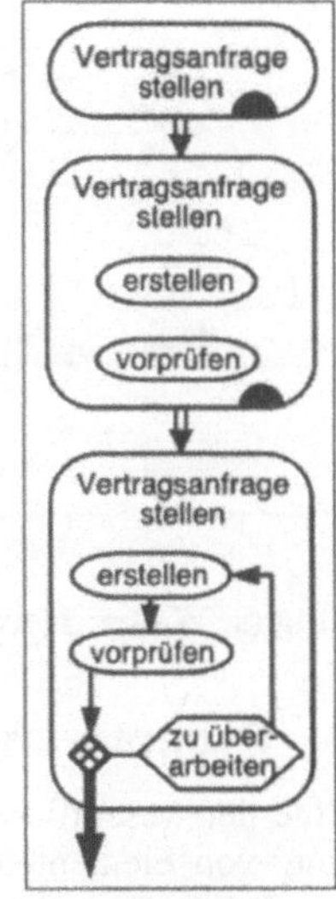

Abbildung 6

Einblendung von Attributen unterstützen

Von wesentlicher Bedeutung ist es, daß man zum Zwecke gezielter Evaluationsfragen auch einzelne Attribute einblenden kann. ESP in Abb. 7 steht für Entscheidungsspielraum.
Wir unterstellen hier, daß dieses Merkmal im Rahmen einer Aufgabenanalyse erhoben wurde. Es kann sinnvoll sein, dieses Attribut für einzelne Aktivitäten anzuzeigen. Ein entsprechender software-unterstützter Mechanismus muß darüber hinaus in der

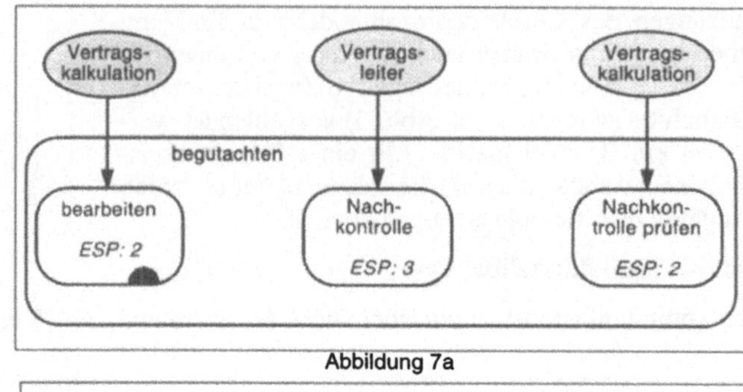

Abbildung 7a

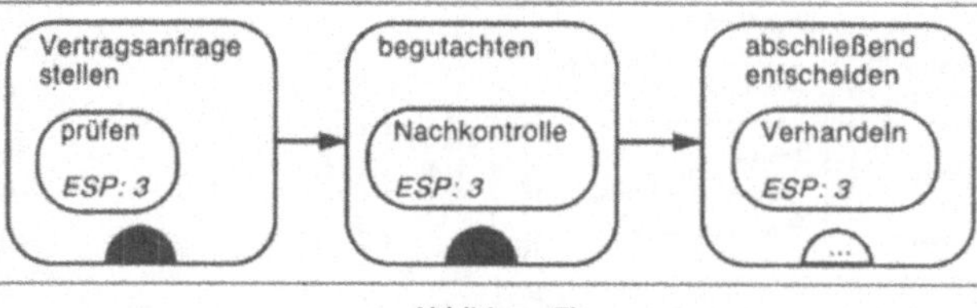

Abbildung 7b

Lage sein, alle Attribute der Art ESP in einem Schritt anzuzeigen, um etwa die Wertigkeiten verschiedener Tätigkeiten mit Hinblick auf die ausführenden Rollen vergleichbar zu machen (s. Abb. 7a). Darüber hinaus sollten auch alle Attribute mit der gleichen Merkmalsausprägung (z. B. ESP=3) in einem Schritt angezeigt werden können (s. Abb. 7b).

Eine besondere Herausforderung unter dem Aspekt des Einblendens stellt die Hinzunahme multimedialer Elemente dar, wie etwa Ton, Video-Animationen oder Screenshots. Dies bedarf einer gesonderten Vertiefung (s. Walter; Herrmann, 1998).

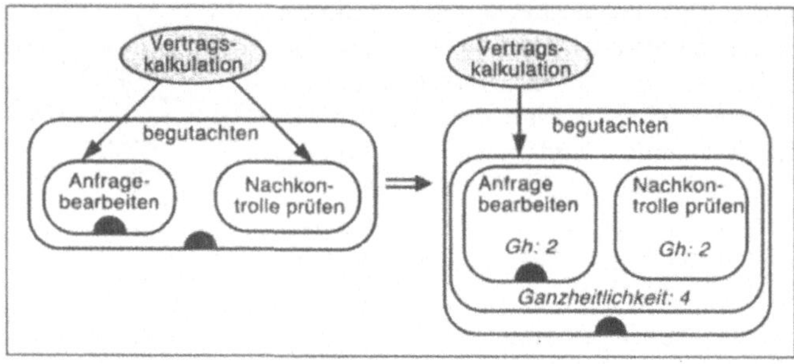

Abbildung 8

5 Gruppierung

Eine der wesentlichen Möglichkeiten zur Unterstützung der Wahrnehmung ist die Gruppierung von Elementen, etwa durch Umrahmungen, Nähe, oder Art der Anordnung. Modellierungsmethoden, die die Einbettung von Elementen erlauben, unterstützen auf semantischer Ebene die Bildung von Gruppen und damit auch die überschaubare Darstellung umfangreiche Zusammenhänge. Es müssen zu diesem Zweck Abstraktionen eingeführt werden wie z.B. in Abb. 1 (Vertragsanfrage stellen, etc.).

Die Einführung abstrakter Gruppierungselemente ist zu ermöglichen

Abb. 8 verdeutlicht den Vorteil von Gruppierung durch Einbettung. Es sollte möglich sein, unvollständig zu bleiben, indem man dem

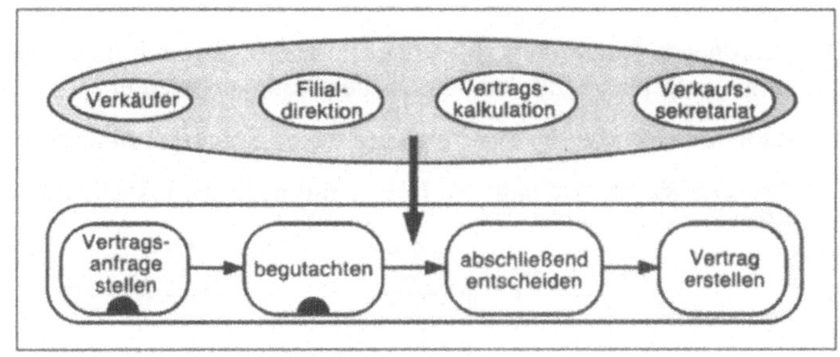

Abbildung 9

neu entstehenden, übergeordneten Element keinen Namen geben muß, weil dieser u.U. in der Erfahrungswelt der Betrachter keine Entsprechung hat. Zum einen erlaubt die Zusammenfassung eine angemessenere Bewertung von Aktivitäten, etwa hinsichtlich des Kriteriums *Ganzheitlichkeit* (GH). Die Ganzheitlichkeit der Aufgabe der Vertragskalkulation fällt bei einer Zusammenfassung höher aus als bei Betrachtung der einzelnen Aktivitäten. Zum anderen spart man durch die Zusammenfassung einen Pfeil, um darzustellen, daß die Rolle Vertragskalkulation beide Aktivitäten ausführt. Solche Einsparungen kann die Überschaubarkeit größerer Diagramme erhöhen und vor allem die Verwendung sich überschneidender Pfeile vermeiden.

Gruppierung und Unvollständigkeit kombinieren

Abb. 9 stellt eine weitergehende Methode dar. Einzelheiten, nämlich Relationen, werden zu Gunsten einer Gruppierung zusammengefaßt. Durch die Einführung zweier übergeordneter Elemente werden zum einen alle Rollen und zum anderen alle Aktivitäten zusammengefaßt. Der die Begrenzungslinien schneidende Pfeil ist ein Beispiel für mögliche Unvollständigkeit. Er sagt nur aus, daß einzelne Rollen einzelnen Aktivitäten zugeordnet sind und daß die diesbezüglichen Details in diesem Betrachtungskontext nicht von Interesse sind. Die Details können durch Aktivierung des als Schaltfläche dargestellten Pfeiles (s. Abb. 1) eingeblendet werden. Darüber hinaus bietet die Anordnung der Rollen und Aktivitäten schon Hinweise, wie die Zuordnung in etwa aussehen kann.

Umwandlung von Symbolen in Text unterstützen

Eine besondere Art der Gruppierung ist gegeben, wenn mehrere Elemente zu einem Text zusammengefaßt werden (s. Abb. 10). Diese Methode ist oftmals unerläßlich, um Überblicksdiagramme zu erzeugen.

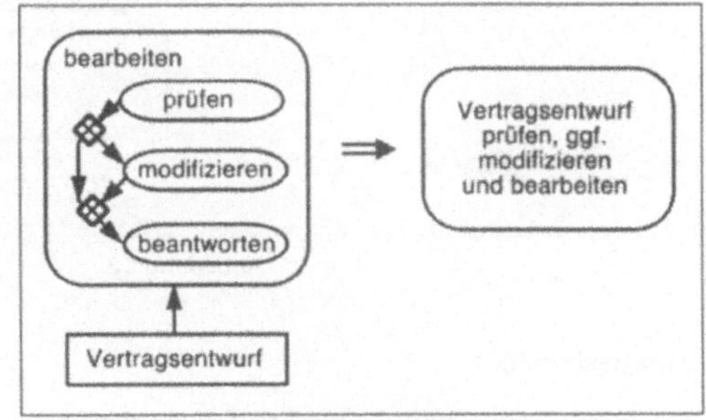

Abbildung 10

6 Ausblendung von Elementen

Ausblendung abstrakter Elemente ermöglichen

Abstraktionen, die Gruppen bilden, sind den Beschäftigten nicht unbedingt vertraut (s. Moody, 96). In der Praxis kann daher ein Darstellung wie bei Abb. 11 den Beschäftigten vertrauter erscheinen. Hier wurden bei der Detaillierung von Vertragsanfrage stellen und begutachten diese

übergeordneten Aktivitäten selbst nicht eingetragen und die Gruppierung nur durch die Anordnung, durch Nähe und den Verlauf der Pfeile angedeutet.

Verschiedene Strategien der Nutzung freier Flächen anbieten

Die Art und Weise, wie man ein Diagramm nach dem Ausblenden eines Elementes darstellt,
vermit-
telt dem
Be-
trachter
unter-
schiedli-
che Bot-
schaf-
ten. Bei
Über-
gang a)
in Abb.

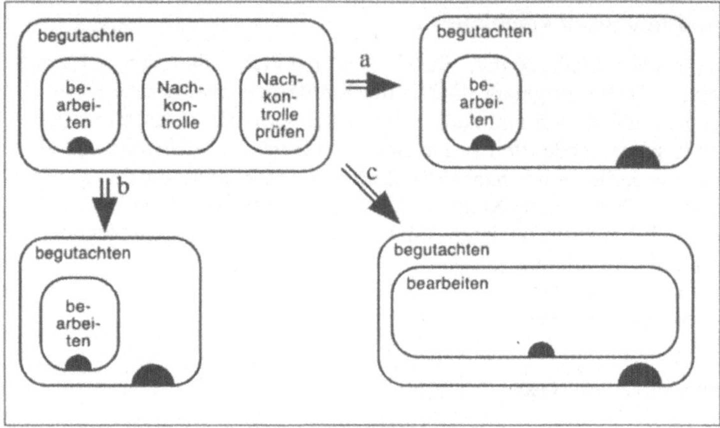

Abbildung 11

12 kann er aufgrund des freigelassenen Platzes leichter erkennen, daß bestimmte Einzelheiten
nicht dargestellt sind. Alternative b) spart Platz, um ggf. ein Diagramm, das begutachten als
ein Element um-
faßt, überschau-
barer zu machen.
Variante c) gibt
die Möglichkeit,
die Relevanz von
bearbeiten stärker
hervorzuheben
und   vermittelt
stärker den Ein-
druck, daß noch
weitere   Details
über diese Sub-
Aktivität   ab-
rufbar sind. Alle
drei Alternativen
können   sinnvoll
sein und sollten

Abbildung 12

selektierbar sein.

Relationen sind beim Ausblenden anzupassen

Abb. 13 befaßt sich mit dem Problem, wie man mit den Relationen zwischen dem ausgeblen-
deten Element und anderen Elementen verfährt. Nach dem ersten Transformationsschritt werden alle mit begutachten verbundene Rollen und Darstellungen der „Führt-aus-Relation" beibehalten, allerdings muß aufgrund der Reduktion der Größe des Elementes begutachten die
Ausrichtung eines Pfeiles verändert werden. Im nächsten Schritt könnte die redundante Wie-
derholung von Vertragskalkulation (VK) entfernt werden. Sie wurde nur eingeführt um eine
Überschneidung der „Führt-aus-Relation"-Pfeile" zu vermeiden (vergl. Abb. 1). Dem letzten
Schritt liegt dann die Strategie zu Grunde, nur noch solche Elemente zu zeigen, deren Rela-

tionen nicht ins Leere zeigen. Ein Präsentationswerkzeug muß alle drei Varianten unterstützen, einschließlich der Pfeile, die ins Leere weisen. Während man die Ausführung des ersten und letzten Schrittes u.U. vorprogrammieren kann, wird beim zweiten Schritt in der Regel die

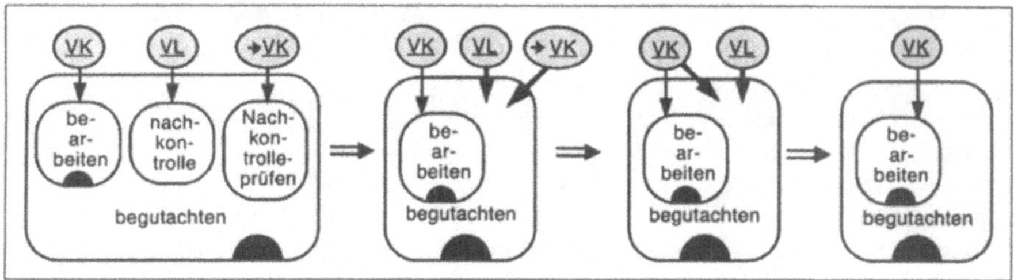

Abbildung 13

Hilfe eines menschliche Entscheiders erforderlich sein.

Relationen nach der Ausblendung optimiert darstellen

Weitere Arten der ästhetischen Optimierung nach dem Ausblenden eines Elementes (hier die Entität Benachrichtigung, vergl. Abb. 1) verdeutlicht Abb. 14. Durch die Verlängerung des Elementes Vertragsentwurf können die Pfeile direkter eingezeichnet werden, insbesondere unterbleibt die Verwendung eines Pfeiles, der nur über einen Umweg auf sein Ziel zeigt (vormals um Benachrichtigung herum, s. Abb. 1). Sol-

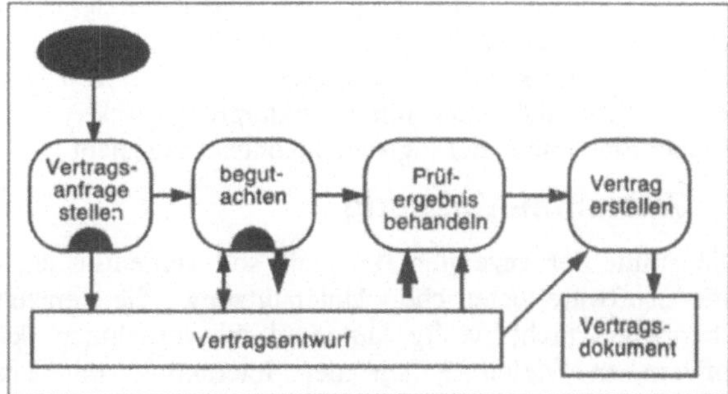

Abbildung 14

che Umwegpfeile reduzieren ebenfalls die Nachvollziehbarkeit. Sie wirken sich jedoch nicht so negativ aus, wie sich überschneidende Linien (s. Purchase, 1997). Abb. 14 verdeutlicht auch die Verwendung schwarzer (Schalt)-flächen, die man verwenden kann, um ausgeblendete Elemente wieder anzuzeigen, wie etwa ausgeblendete Rollen, der Name einer Rolle oder die Entität Benachrichtigung. Letztere ist über die fett dargestellten Pfeile erreichbar.

Gleichartige Elemente in einem Schritt entfernen

Neben dem Ausblenden einzelner Elemente sollten auch mehrere Elemente der gleichen Art in einem Schritt ausblendbar sein. Das gilt für Attribute oder auch für Kardinalitäten. Nach Moody (1996) sind Kardinalitäten bei ER-Diagrammen für Laien besonders schwer nach-

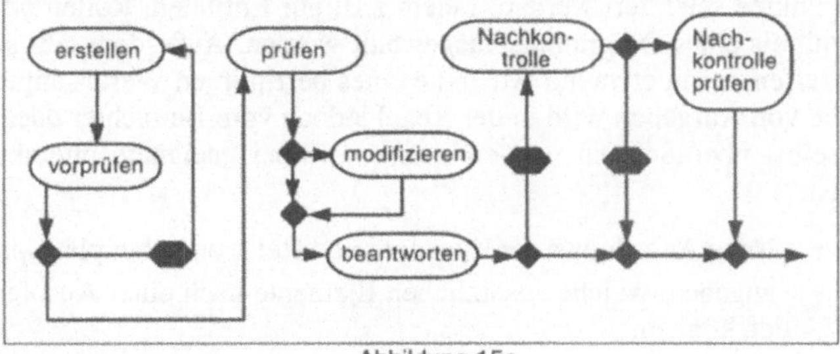

Abbildung 15a

vollziehbar. Ein besonde-
rer Fall ist die Ausblen-
dung logischer Konnekto-
ren, die bei Laien eben-
falls für Verwirrung sor-
gen. Abb. 15a wird aus
Abb. 11 erzeugt, indem
man die Spezifikation der
Konnektoren und Bedin-
gungen schwärzt (sie sind
dann ggf. unter Nutzung

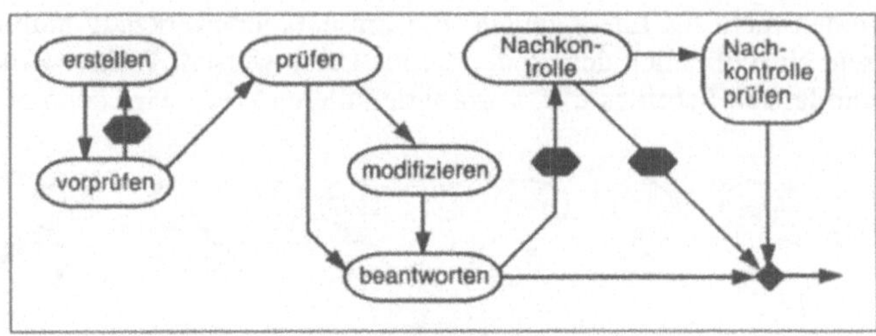

Abbildung 15b

der schwarzen Fläche einzeln wieder einblendbar). Wenn man den logischen Zusammenhang
zwischen Relationen gleichen Typs in einem bestimmten Darstellungskontext nicht näher
spezifiziert, dann kann man auch zwei oder mehrere dieser Relationen zu einem Element hin-
oder von einem Element wegführen, wie bei Abb. 15b.
Hierdurch wird der Fluß der Aktivitäten leichter
überschaubar. Dieselbe Abkürzungsmöglichkeit wurde auch
in Abb. 1 zur Vermeidung von Konnektoren bei den
Relationen zwischen Rollen und Aktivitäten sowie zwischen
Entitäten und Aktivitäten verwendet. Bei Abb. 15b besteht
der Nachteil, daß man keine Anknüpfungspunkte für das
erneute Einblenden der logischen Konnektoren sieht.

7 Zusammenfassung

Die Fülle der gezeigten Beispiele soll verdeutlichen, daß
eine software-technische Unterstützung die einzelnen
Übergänge nicht völlig automatisch vornehmen kann,
sondern in vielen Fällen der Interaktion mit einem
menschlichen Entscheider bedarf. Abb. 16 zeigt eine mögliche Form der interaktiven Realisierung. Sobald die
Schaltfläche für Einblendungen aktiviert wird, erscheint ein
Menü. Die wählbaren Optionen geben einen Teil der
einzelnen Möglichkeiten wieder, die oben diskutiert wurden.
Desweiteren kann man davon ausgehen, daß die Modelle in
Datenbanken abgespeichert sind und daß eine Selektion der

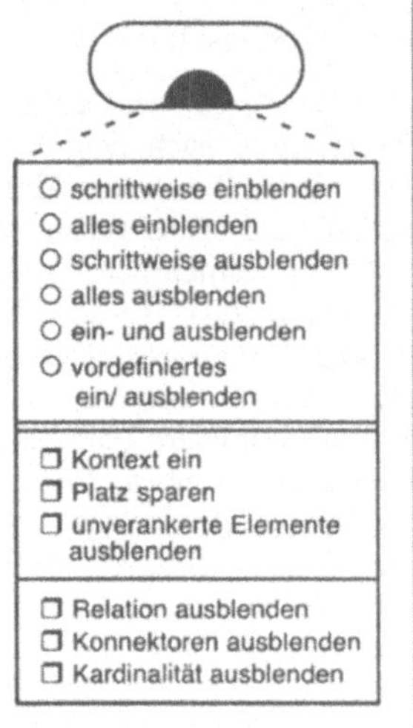

Abbildung 16

Elemente, die sich ein Betrachter anzeigen lassen möchte, über die Verwendung einer geeigneten Abfragesprache gesteuert werden kann. So kann dann nach syntaktischen Gesichtspunkten selektiert werden, indem z.B. nur Entitäten, Rollen oder Aktivitäten zur Darstellung
mittels eines Diagramms ausgewählt werden. Außerdem läßt sich eine semantische Auswahl
treffen, wenn etwa nur Attribute eines bestimmten Wertes angezeigt werden sollen. Eine Reihe von Aufgaben wird in der Regel jedoch vom Betrachter oder Präsentierer eines Diagramms
selbst übernommen werden müssen, wobei geeignete interaktive Unterstützung anzubieten
ist:

- Neue Anordnung der Elemente und Relationen bei platzsparenden Ausblendungen,
- angeben, welche zusätzlichen Elemente nach einer Ausblendung noch ausgeblendet werden können,

- Elemente so gestalten, etwa bzgl. ihrer Größe, daß Pfeile übersichtlich eingetragen werden können,
- Einführung von abstrakten Elementen zum Zwecke von Zusammenfassungen,
- Relationen in aktivierbare Schaltflächen zusammenfassen,
- Einsparung überflüssiger Pfeile

Die Ausgestaltung der Darstellung einzelner Diagramme ist mit Kommunikationssituationen vergleichbar: Jemand, der eine Diagramm präsentiert, muß entscheiden, welche Wahl der Ausdrucksmöglichkeiten für den Adressaten am geeignetsten ist. Unter diesem Aspekt sehen wir auch das Problem der empirischen Validierbarkeit: Die Frage, welche der oftmals möglichen Alternativen am angemessensten ist, läßt sich unseres Erachtens nicht mit kontrollierbaren Experimenten nachweisen, da die Angemessenheit in hohem Maße situationsabhängig ist. Der Versuch einer empirischen validierten Entscheidung wäre hier so ähnlich einzuschätzen, wie das Bemühen um eine gesicherte Auswahl von Paraphrasen (i.e. verschieden Möglichkeiten dasselbe auszudrücken) im Kommunikationsgeschehen.

Wir planen die Entwicklung eines Prototypen, um Erfahrung hinsichtlich der technischen Umsetzung der aufgezeigten Anforderungen und Möglichkeiten zu sammeln. Allerdings empfehlen wir aufgrund unserer Erfahrungen in den Fallstudien, die genannten Anforderungen bereits jetzt bei der Entwicklung von Editoren und Präsentationswerkzeugen im Bereich der Geschäftsprozeßmodellierung umzusetzen.

8 Literatur

Blythe, Jim; McGrath, Cathleen; Krackhardt, David (1995): The Effect of Graph Layout on Inference from Social Network Data. In: Brandenburg, F.J. (Ed.) (1995): Graph Drawing. Symposium on Graph Drawing, GD '95. Passau, Germany, Sept. 20-22, 1995. Proceedings: Berlin et al. Springer. S. 40-51.

Glaser, Wilhelm R. (1994): Menschliche Informationsverarbeitung. In: Eberleh, Edmund; Oberquelle, Horst; Oppermann, Reinhard (1994): Einführung in die Software-Ergonomie. Gestaltung graphisch-interaktiver Systeme: Prinzipien, Werkzeuge, Lösungen. 2., völlig neu bearbeitete Auflage. Berlin, New York. Walter de Gruyter. S. 7-47.

Herrmann, Th.; Scheer, A-W.; Weber, H. (1998):Verbesserung von Geschäftsprozessen mit flexiblen Workflow-Management- Systemen - Von der Erhebung zum Sollkonzept. Physica-Verlag, Heidelberg. S. 73 - 106.

Herrmann, Thomas (1997): Communicable Models for Cooperative Processes. In: Slavendy, G. (ed.): HCI International '97. Proc. of the 7th International Conference on Human-Computer Interaction,San Francisco. Amsterdam: Elsevier . S. 285 - 288.

Herrmann, Thomas; Hoffmann, Marcel; Loser, Kai-Uwe (1998): Sozio-orientierte und semi-strukturierte Modellierung mit SeeMe. (im Erscheinen).

Herrmann, Thomas; Loser, Kai-Uwe (1998): Vagueness in models of socio-technical systems. (eingereicht bei BIT).

Hoffmann, M. (1998): Mitarbeiter-orientierte Erhebung und Modellierung von Geschäftsprozessen bei der Einführung von Workflow-Management. Forschungsbericht Nr. 681. Dortmund: Universität (FB Informatik).

Moody, Daniel (1996): Graphical Entity Relationship Models: Towards a More User Understandable Representation of Data. In: Thalheim, Bernhard (Ed.) (1996): Conceptual Modeling, ER96. Proceedings of the 15th International Conference on Conceptual Modeling, Cottbus, Germany, October 1996. Berlin et al. Springer. S. 227-244.

Purchase, Helen (1997): Which Aestetic Has the Greatest Effect on Human Understanding? In: Dibattista, Giuseppe (Ed.) (1997): Graph Drawing. 5 th International Symposium on Graph Drawing, GD '97. Proceedings. Berlin et al.: Springer. S. 248-261.

Purchase, Helen C.; Cohen, Robert F.; Murray James (1995): Validating Graph Drawing Aestetics. In: Brandenburg, F.J. (Ed.) (1995): Graph Drawing. Symposium on Graph Drawing, GD '95. Passau, Germany, Sept. 20-22, 1995. Proceedings. Berlin et al. Springer. S. 435-446.

Metzger, W. (1953): Gesetze des Sehens. Frankfurt/M.: Kramer.

Rosemann, Michael (1996): Komplexitätsmangement in Prozeßmodellen. Methodenspezifische Gestaltungsemp-
    fehlungen für die Informationsmodellierung. Wiesbaden: Gabler.
Walter T., Herrmann, Th.(1998): The relevance of Showcases for the Participative Improvement of Business
    Processes and Workflow. In: PDC 98. New York: ACM (im Erscheinen).

Adresse des Autors

Prof. Dr.-Ing. Thomas Herrmann
Universität Dortmund (FB4, LS6, IuG)
44221 Dortmund
herrmann@iug.informatik.uni-dortmund

Neue Möglichkeiten der Analyse der Mensch-Computer-Interaktion zur Evaluation von computerunterstützten Gruppensitzungen

Torsten Holmer und Norbert Streitz

GMD- Forschungszentrum Informationstechnik GmbH

IPSI -Institut für Integrierte Publikations- und Informationssysteme, Darmstadt

Zusammenfassung

Bei der Evaluation von Software zur Computerunterstützung von Gruppenarbeit hat sich gezeigt, daß es einen Bedarf nach neuen Datenquellen gibt. Diese sollen insbesondere eine detailliertere Untersuchung des zeitlichen Verlaufs von Interaktions- und Kommunikationsverhalten in Gruppen erlauben als dies mit traditionellen Beobachtungsmethoden wie z.B. Video- und Beobachtungsprotokollen möglich ist. In diesem Beitrag stellen wir das Programm LOGAN vor, eine Logfile-Analyse von Interaktionen in Hypermedia-Dokumentstrukturen, die von einer Gruppe mit vernetzten Computern erzeugt und modifiziert werden. Über die detaillierten Prozessdaten hinaus bietet LOGAN neuen Aggregations- und Auswertungsverfahren an. Nach einer Darstellung der Prinzipien und Möglichkeiten von LOGAN illustrieren wir die Verwendung an Daten aus einem von uns durchgeführten Experiment zu computerunterstützten Gruppensitzungen.

Abstract

Since the evaluation of software for computer-supported cooperative work has to go beyond the standard measures and criteria, it became obvious that new types of data have to be collected. These data should enable a more detailed account of the complex interaction, communication and cooperation processes than it is possible with traditional methods as, e.g. video and observation protocols. In this paper, we present the program LOGAN, a logfile-based analysis of interactions in hypermedia information structures created and modified by a group of people using networked computers. Besides a detailed step-by-step process account, LOGAN provides new ways of aggregating and evaluating these data. The principles and possibilities of LOGAN are illustrated by examples with data collected in an experiment investigating computer-supported meetings of teams in an electronic meeting room.

1 Einleitung

Die zunehmende Vernetzung von Computern und die daraus resultierenden neuen Möglichkeiten zur Unterstützung kooperativen Arbeitens in Teams stellen neue Anforderungen an die Gestaltung der Mensch-Computer-Interaktion. Die ergonomische Gestaltung von Software für CSCW (Computer-Supported Cooperative Work) ist dabei durch entsprechende Evaluationsmöglichkeiten zu ergänzen. Die traditionellen Ansätze in der Software-Ergonomie sind für die Evaluation von CSCW-Systemen zu überdenken und auf die neue Situation anzupassen. Dabei sind drei Problemfelder zu betrachten: die Erweiterung von Single-User-Systemen auf Multi-User-Systeme, die Berücksichtigung synchronen Arbeitens mehrerer Personen und die räumliche Verteiltheit der Benutzer. Die daraus resultierenden Konsequenzen für die Evaluation von Systemen sind weder in der traditionellen Software-Ergonomie noch in der CSCW-Forschung bisher ausreichend reflektiert worden.

Im Bereich des Electronic Meeting Support (EMS), der Computerunterstützung von Gruppensitzungen, ist erst vor kurzem damit begonnen worden, Anwendungssoftware in experimentellen Situationen zu untersuchen. Dabei wird eine doppelte Zielsetzung verfolgt. Einerseits will man Einsichten erhalten, wie sich die Benutzung der Software auf die Problemlöseprozesse in Gruppen und auf die Kommunikationsstrukturen auswirkt. Andererseits will man Anforderungen für ein Re-Design der Software ableiten. Die erhobenen Daten werden meistens explorativ oder unter experimentellen Gesichtspunkten analysiert. Diese Untersuchungen zur Nutzung und Effektivität von EMS-Systemen bedürfen aussagekräftiger Daten, um die Effekte und Prozesse erklären zu können [1][2][3].

Das Hauptaugenmerk in diesen Untersuchungen liegt meistens auf den Arbeitsergebnissen in Form von elektronisch erzeugten Dokumenten und zum Teil auch auf den Gruppenprozessen, vorrangig auf den mündlichen Kommunikationsstrukturen. Die Produkte werden meist nur im Endstadium bewertet, der Verlauf der Erstellung wird selten untersucht. Der prozess-orientierte Anteil der Beobachtung ist für die Erklärung des Zustandekommens des Produktes jedoch unerlässlich und kann wichtige Einsichten vermitteln. Die meisten Untersuchungen vernachlässigen diese Prozessbeobachtung, wie schon öfter bemängelt wurde [4][5].

Die verwendeten Methoden sind meistens an psychologischen und soziologischen Verfahren orientiert: Fragebögen, Interviews, Beobachterprotokolle und Videoaufnahmen.

Die Auswertung prozessorientierter Beobachtungen ist schwierig und aufwendig. Der kontinuierliche Fluß von Informationen kann den Beobachter schnell überfordern. Die Auswertung von Videoaufnahmen kann sehr aufwendig werden und und oft beträgt der Aufwand für die Analyse einer Minute Videoprotokoll bis zu einer Stunde.

Ein weiteres Problem bei der Analyse computerunterstützter Sitzungen ist die Tatsache, daß nicht nur das Benutzerverhalten (verbal, non-verbal), sondern auch die Inhalte ihrer Bildschirme wichtige Informationsquellen darstellen. Das Produkt der Arbeit, das elektronische Dokument, steht für die Benutzer im Mittelpunkt. Es ist sowohl Ziel als auch Mittel der Kooperation. Damit ist der Prozess der Erstellung des Dokumentes ein zentraler Aspekt der Gruppenaktivitäten.

Will man nun die Entstehung eines elektronischen Dokumentes, das gleichzeitig auf mehreren Bildschirmen bearbeitet wird, mitverfolgen, so steht man vor einem großen Problem. Die Aufzeichnung aller Bildschirme per Video ist technisch sehr aufwendig. Man muß für jeden Bildschirm eine Videokamera vorsehen und diese so positionieren, daß der Benutzer die Sicht der Kamera zu keinem Zeitpunkt verdeckt. Gleichzeitig ist eine optimale Position der Kamera zu finden, so daß die Lesbarkeit der aufgezeichneten Bildschirme für die Auswertung gewährleistet ist. Neben der Aufzeichnung ist die Synchronisation und eine sehr aufwendige, da mehrere Interaktionsorte berücksichtigende, Video-Analyse durchzuführen.

Bei der kooperativen Erstellung und Bearbeitung eines vernetzten Hypermediadokumentes, wie in unseren Experimenten, ist die Situation noch schwieriger. Die Bildschirme können unterschiedliche Teile des Hypermediadokumentes anzeigen und die Benutzer können an verschiedenen Stellen verteilt arbeiten, so daß es auf der Dokumentebene zu Untergruppenbildungen kommen kann. Einige Gruppenmitglieder können ein Thema auch eine Zeit lang alleine bearbeiten [6].

Die Bildung und das Arbeiten in Untergruppen und als Einzelperson sowie das Zusammenführen zur Kooperation in der Gesamtgruppe sind Prozesse, die sich in dem Navigationsverhalten in der elektronischen Dokumentstruktur reflektieren. Für einen „externen" Beobachter im Besprechungszimmer ist es ziemlich unmöglich, dieses Navigationsverhalten und die inhaltsbezogenen Aktivitäten mehrerer Benutzer – wer arbeitet wann mit wem in welchem Dokumentteil – adäquat nachzuvollziehen.

Die Herausforderung besteht nun darin, eine Methode zu finden, die einerseits imstande ist, den gewünschten hohen Detaillierungsgrad zu gewährleisten. Andererseits müssen die dadurch anfallenden großen Datenmengen so verarbeitet werden können, daß die Daten sinnvoll aggregiert werden können. Erst dann ist man in der Lage, aufschlußreiche Aussagen über die Gruppenprozesse, Problemlösestrategien und Trends zu machen.

Da alle Interaktionen des Benutzers von dem verwendeten CSCW-System ausgeführt werden, ist es möglich, die Benutzeraktionen durch dieses Programm auch protokollieren und in ein sogenanntes Logfile schreiben zu lassen. Diese detaillierten Daten können dann durch geeignete Auswertungsprogramme analysiert werden - ohne den Umweg über menschliche Beobachter und den damit verbundenen Risiken von Beobachtungsfehlern.

Im weiteren Verlauf dieses Beitrags werden wir zunächst kurz über zwei Vorarbeiten berichten, die bereits wichtige Schritte in die angestrebte Richtung darstellten. Dabei handelt es sich um die Protokollierungssysteme VideoGrip und KOPROT, die wir bei unseren bisherigen Evaluationsuntersuchungen von DOLPHIN eingesetzt haben. Der Hauptteil wird dann LOGAN, einem Tool zur Logfile-Analyse und seinen Auswertungsmöglichkeiten gewidmet sein. Zum Schluß werden wir die Bedeutung dieser neuen Methode und zukünftige Anwendungsmöglichkeiten diskutieren.

1.1 Vorarbeiten

Im Laufe der Entwicklungsarbeiten an dem kooperativen Hypermedia-System DOLPHIN [7] haben wir mehrere eigene Werkzeuge entwickelt, die speziell auf die Datengewinnung bei computerunterstützten Sitzungen zugeschnitten sind. Für die Untersuchungsreihen von DOLPHIN 1 benutzten wir das Tool „VideoGrip" [8][6]. In der Untersuchung von DOLPHIN 2 [9] verwendeten wir zwei neue Tools: KOPROT und LOGAN. Nach einer kurzen Darstellung von VideoGrip und KOPROT berichten wir ausführlich über LOGAN.

1.1.1 VideoGrip

Ein erster Ansatz, Interaktionsdaten jenseits der traditionellen Videoaufnahme zu erhalten, war das System „Video Grip". Das Experiment [8], für das VideoGrip entwickelt wurde, diente der Untersuchung, welchen Einfluß die Bereitstellung von Hypermediafunktionalität auf den Gruppenprozess in einer Kleingruppensitzung hat. Während der Sitzungen wurde der Gruppenarbeitsraum in der Totale auf Videoband aufgenommen. Zusätzlich wurden alle 15 Sekunden Screenshots von den in diesem Experiment verwendeten drei Computern (zwei Workstations und eine interaktive elektronische Wandtafel) gemacht. Diese vier „Filme" wurden mittels VideoGrip nebeneinander auf einem Computerbildschirm angezeigt, so daß der Beobachter eine integrierte Sicht über die drei verschiedenen Bildschirme hatte und gleichzeitig auf dem Videobild die verbalen und nonverbalen Interaktionen der Gruppe im Zusammenhang sehen konnte. Ein wesentliches Ergebnis, das durch den Einsatz dieses Werkzeuge gefunden wurde, war, daß Gruppen, die Hypermedia-Strukturen verwenden, dazu tendierten, die Arbeit innerhalb einer Sitzung aufzuteilen und in parallelen Arbeitsmodi zu arbeiten [6]. In der Analyse der Videobänder wurde eine gegenläufige Abhängigkeit von Computerinteraktionen und mündlicher Kommunikation festgestellt. Eine ausführliche

Darstellung der Ergebnisse zur Wechselwirkung zwischen Arbeitsteilung und Hypermedia-funktionalität ist in [10] zu finden.

VideoGrip bot zwar die Möglichkeit, einen guten Überblick über die unterschiedlichen Interaktionen (mündlich und computerbasiert) innerhalb der Gruppe zu bekommen. Die Auswertung der Daten hatte jedoch noch Nachteile. Die Zeiten verbaler Interaktion und die Computerbenutzungszeiten mußten mit der Stoppuhr erfaßt werden. Durch die geringe Zeitauflösung der Screenshots (ein Bild alle 15 Sekunden) bestand z.B. die Gefahr, daß einige computerbasierte Interaktionen übersehen wurden, da die Zeitspanne zu groß war, in denen Veränderungen stattfinden konnten. Das erste Problem wird durch KOPROT gelöst, das zweite durch LOGAN.

1.1.2 Kooperations- und Kommunikationsprotokoll (KOPROT)

Die Untersuchungen mit der Nachfolgeversion DOLPHIN 2 fanden in einer veränderten Hardwareumgebung in dem Gruppenarbeitsraum statt. Diese bestand nun aus vier Arbeitsplatzrechnern, die in einen speziellen Tisch eingebaut waren, und einer interaktiven elektronischen Wandtafel [9]. Für dort relevanten Fragestellungen entwickelten wir ein Tool, das die Befunde der vorherigen Untersuchung bestätigen und neue Zusammenhänge aufzeigen sollte. KOPROT ist ein Programm, mit dem ein Protokollant schon während der Sitzung eine Analyse der mündlichen Sprechzeiten und der Kommunikations- und Kooperationsstrukturen - wer also wann mit wem zusammenarbeitet - vornehmen kann. Eine ausführliche Beschreibung findet sich in [9]. Es ist damit nicht nur möglich, die Dauer der einzelnen Sprechzeiten der Personen zu ermitteln, sondern es können auch die Anteile und der zeitliche Verlauf der unterschiedlichen Formen der Zusammenarbeit (Arbeit in der ganzen Gruppe, Parallelarbeit von Einzelnen oder Arbeit in Untergruppen) festgestellt werden. KOPROT erleichtert zwar die Protokollierung der Sprechzeiten und der mündlichen Kooperationsmuster. Die oben beschriebenen Interaktionen mit den elektronischen Dokumenten werden durch diese Daten aber nicht erfaßt.

2   Das Programm LOGAN – der LOGfile-Analyzer

Als weiteren Schritt zur Verfeinerung unseres Methodenrepertoires integrierten wir eine spezielle Funktion in DOLPHIN 2, mit der Logfiles der Benutzeraktionen erstellt werden können. Diese Daten werden dann mit dem von uns entwickelten Analyseprogramm verarbeitet. Dieses Programm trägt den Namen LOGAN – Logfile-Analyzer. LOGAN kann die von DOLPHIN erstellten Logfiles unter verschiedenen Gesichtspunkten analysieren und daraus Datensätze und Tabellen generieren, die mit entsprechenden Visualisierungstools weiterverarbeitet werden können.

2.1   Datenerhebung

In unseren Folgeexperimenten wurde die mit einer Logfile-Funktionalität versehene Version von DOLPHIN 2 benutzt. Für jeden Benutzer wurde damit während der Sitzung ein zeitlich geordnetes Protokoll der inhaltlichen Eingaben und der Systeminteraktionen erstellt. Für jede Aktion eines Benutzers wurde eine neue Zeile im Logfile angelegt, die u.a. folgende Angaben enthielt:

- Benutzername
- Kategorie der Aktion (Inhaltserzeugung: Zeichnen und Schreiben von Text, Strukturveränderung: Verschieben von Text im Knoten, Navigation: Betreten und Verlassen von Hypermedia-Knoten, Hyperstrukturerstellung: Erzeugen neuer Knoten)
- Zeitpunkt der Aktion
- Ort der Aktion, also des Hypermedia-Knotens, in dem die Aktion stattfand
- Namen anderer Benutzer, die sich in demselben Knoten befanden

Am Ende einer Sitzung war damit für jeden Benutzer eine komplette Aufzeichnung der Interaktionen mit dem DOLPHIN-System vorhanden. Die Einzelprotokolle wurden in ein Gruppen-Logfile integriert, das die zeitlich geordneten Interaktionen der gesamten Gruppe (in unserem Experiment vier Personen, die 4 + 1= 5 Computer benutzten) enthielt.

2.2   Auswertungsmethoden von LOGAN

Die Auszählung der Häufigkeit eines Aktionstyps über die Zeit summiert gibt zwar Auskunft über die Gewichtung der Aktivitäten (Textproduktion, Navigation, Dokumentstrukturierung), berücksichtigt jedoch nicht die zeitliche Verteilung. Neben der zeitlichen Perspektive ist außerdem die Betrachtung der Daten im Gruppenkontext notwendig, um die Strukturen der Zusammenarbeit aufdecken zu können. Wir stellen im folgenden drei Auswertungsmethoden von LOGAN vor: Gruppen- und Untergruppenaktivitätskurven zur zeitabhängigen Darstellung der Kooperationsmuster, „Awareness"-Kurven als Darstellung des Wissensstandes über Dokumentinhalte und Aktivitätstabellen als integrierte Übersichten von Navigations-, Inhaltserzeugungs- und Awareness-Informationen.

2.2.1 Gruppen- und Untergruppenaktivitätskurven

Wenn eine Gruppe zusammenarbeitet, lassen sich verschiedene Muster der Zusammenarbeit beobachten. Die Gruppe kann gemeinsam an einem Thema arbeiten oder sich in verschiedene Untergruppen aufteilen bis hin zur parallelen Einzelarbeit der Gruppenmitglieder. Die Beobachtung dieser Aktivitäten auf der verbalen Interaktionsebene ist einfach und mit KOPROT zu protokollieren (s. Kap. 1.1.2). Die Arbeitsteilung auf der Dokumentebene kann nur durch eine Auswertung der Navigationsaktionen (wer war wann in welchem Hypermedia-Knoten) und der Inhaltsaktionen (wer hat was und wo geschrieben) erfolgen. Das Ergebnis sind Zeitverlaufskurven, die angeben, wieviel Prozent des Dokumentes in welchem Zeitabschnitt durch welche Art der Zusammenarbeit erzeugt wurde.

2.2.2 Awareness und nicht gesehene Veränderungen

Wir waren insbesondere auch an der Frage interessiert, in welchem Ausmaß die Gruppenmitglieder die Aktivitäten der anderen mitverfolgen und wissen, welche Inhalte von ihnen erzeugt wurden. Wegen der von DOLPHIN bereitgestellten Möglichkeit der Parallelarbeit in verschiedenen Knoten kann es vorkommen, daß ein Mitglied der Gruppe Ideen in einem Bereich des Dokumentes erzeugt, den die anderen Mitglieder danach aber nicht mehr besucht haben, also den Inhalt nicht kennen. Um die daraus resultierenden Wissensverteilungen zu analysieren, können wir die Daten über die Navigationsaktionen (wer war wann wo) mit den Informationen über die Inhaltsaktionen (wer hat wann wo gearbeitet) kombinieren. Auf diese Weise können wir angeben, welche Information von einem bestimmten Mitglied der Gruppe wann gesehen wurde. Die Menge der gesehenen Informationen ist ein Maß für die „awareness", d.h. das bewußte („aware") Wissen um die Aktionen und Beiträge der anderen Gruppenmitglieder. Dies ist vergleichbar mit dem von Dourish und Bellotti [11] verwendeten „awareness"-Begriff, bzw. eine Form seiner Operationalisierung. Das Ausmaß der „awareness" kann für jede einzelne Person erhoben werden. Besonders interessant sind dabei die Veränderungen im Verlauf einer Sitzung. Zu diesem Zweck haben wir die Werte als Zeitverlaufskurve visualisiert. Damit können dann auch Wissenstrends (auf- oder absteigend) für einzelne Gruppenmitglieder dargestellt werden.

2.2.3 Aktivitätstabellen

Die Awareness-Kurven zeigen den zeitlichen Verlauf des Wissens über das Dokument. Es ist aber auch notwendig zu wissen, wo die Wissensdefizite über die Dokumenteninhalte am Ende der Sitzung sind. Für diesen Zweck kann LOGAN eine Aktivitätstabelle erstellen, die für jeden Knoten Auskunft gibt, ob und wie oft ein Benutzer einen Knoten besucht hat, ob und wieviele Inhaltsaktivitäten dort getätigt und wieviele der in diesem Knoten enthaltenen Informationen nicht gesehen wurden.

3   Ein Anwendungsbeispiel von LOGAN

In diesem Kapitel werden wir anhand von Daten einer zuvor am GMD-IPSI durchgeführten Untersuchung zeigen, wie die Resultate der Logfile-Analysen aussehen und welche Schlußfolgerungen man aus diesen Daten ziehen kann.

3.1   Das „Roomware"-Experiment

Um die Rolle verschiedener computerbasierter Informationssysteme in Gruppenarbeitsräumen zu klären, führten wir das „Roomware"-Experiment durch, in dem unterschiedliche Kombination von persönlichen und öffentlichen Informationssystemen in einem Raum untersucht wurden [9]. Unter „Roomware" verstehen wir allgemein die Integration von computerbasierten Informationssystemen mit Raumelementen [12]. Beispiele sind in Wände integrierte interaktive elektronische Wandtafeln, Tische oder Sessel mit integrierten Computer, etc. In dem Experiment, auf das wir hier Bezug nehmen, war diese Integration noch nicht so ausgeprägt wie in unserer neuen i-LAND Umgebung [12], aber z.B. durch vier in den Besprechungstisch eingelassene vernetzte Computer realisiert. Unsere Hypothese war, daß unterschiedliche Roomware-Konstellationen (nur öffentliche vs. nur persönliche vs. Kombination von öffentlichen und persönlichen Informationssystemen) die Prozesse computerunterstützten Problemlösens entscheidend beeinflussen würden.

Die Problemstellung für die Versuchspersonen in diesem Experiment bestand in der Entwicklung eines Programmschemas für einen neuen Fernsehsender. Diese Aufgabe sollte mit DOLPHIN 2 als computerbasiertes Brainstorming- und Ideenstrukturierungswerkzeug bearbeitet werden. Die Sitzung der „Programmkommission" in dem Experiment dauerte vier Stunden. Jede Versuchsgruppe bestand aus vier Mitgliedern, die eine bestimmte Rolle in der „Kommission" zugewiesen bekamen. Diese Rollen waren Leiter (L), Marketing (M), Redakteur 1 (R1) und Redakteur 2 (R2). Eine nähere Beschreibung des Experimentes findet sich in [9]. Die wichtigsten Ergebnisse dieser Studie waren, daß die Gruppen mit der Kombination aus öffentlichen (interaktive elektronische Wandtafel) und persönlichen Informationssystemen (im Tisch eingelassene Computer) mehr Ideen und qualitativ bessere

Ideen produzierten. Außerdem hatten diese eine effektivere Art entwickelt, zwischen
verschiedenen Kooperationsmodi zu wechseln, bzw. den Anteil an Einzel- und Untergruppen-
arbeit den Erfordernissen der Aufgabe optimal anzupassen. Diese Ergebnisse sind sehr
bedeutsam, könnten aber durch weitere Analysen noch detaillierter interpretiert werden.
Insbesondere der Verlauf der Kooperationsphasen in Zusammenhang mit der Awareness über
die Dokumentinhalte eröffnet neue Perspektiven für das Design computerbasierter
Unterstützung für Teamarbeit.

3.2 Ergebnisse der LOGAN-Auswertungen

Im diesem Abschnitt berichten wir am Beispiel einer ausgewählten Gruppe über mit LOGAN
durchgeführte Auswertungen, die über die bisherigen Ergebnisse [9] hinausgehen. Sie weisen
einen höheren Detaillierungsgrad auf und erlauben eine genauere Interpretation der Daten.

3.2.1 Gruppen- und Untergruppenaktivitäten

In der Abbildung 1 zeigen wir ein Beispiel für den zeitlichen Verlauf der Inhaltsproduktion
und welcher Anteil der Gesamtproduktion in welchem Kooperationsmodus erbracht wurde.

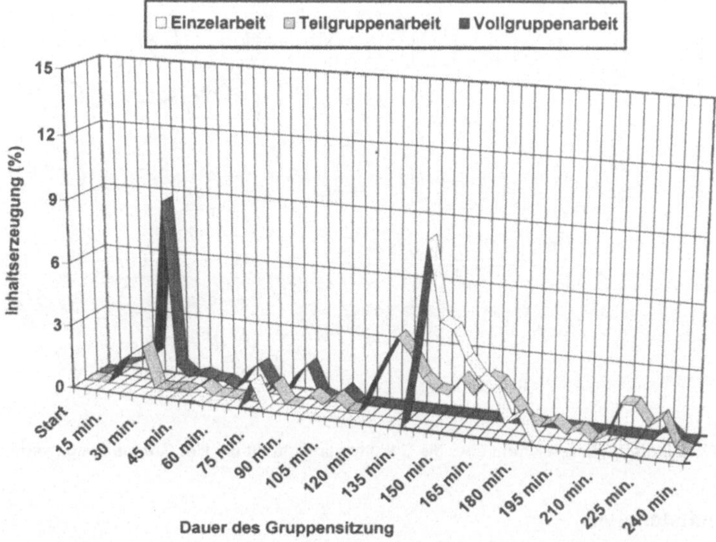

Abbildung 1: Zeitverlauf der Arbeitsanteile in Prozent der Inhaltsproduktion

Man erkennt deutlich eine Phase der Vollgruppenarbeit am Anfang, in der ca. 10% des Dokumentes erzeugt wurden (15. – 30. Minute). In dieser Phase wird gemeinsam diskutiert, wie die
Grobstruktur des Dokumentes aussehen soll. Es werden erste Ideen generiert und in einem
gemeinsam erstellten Knoten gruppiert. Nach der Mitte der Sitzung wird zunehmend in Teilgruppen und insbesondere auch getrennt individuell gearbeitet (120. – 185. Minute). In dieser
Phase beschließt die Gruppe, die Teilaufgaben in verschiedenen Hypermediaknoten parallel
zu bearbeiten. Am Ende steigt das Ausmaß der Teilgruppenarbeit noch einmal an.

3.2.2 Awareness und nicht gesehene Veränderungen

In Abbildung 2 haben wir für dieselbe Gruppe den zeitlichen Verlauf des individuellen Wissens der Gruppenmitglieder über die Dokumentinhalte dargestellt. Bis zur Mitte der Sitzung (120 Minuten) haben alle Mitglieder der Gruppe im Prinzip alle Informationen im Dokument gesehen. Erst mit Beginn der Teilgruppen- und Einzelarbeit sinkt das Ausmaß an Wissen über die aktuellen Informationen im Dokument. Nach 180 Minuten kann man ein Ansteigen der Wissenskurven von Leiter und Marketing beobachten. Dies erklärt sich daraus, daß Leiter und Marketing in der letzten Phase der Sitzung die von ihnen zuvor noch nicht gesehenen Knoten anschauen, während die Redakteure das anscheinend nicht tun. Die Gründe dafür sind uns nicht bekannt. Man kann aber vermuten, daß sie vielleicht gar nicht wußten, wo neue Informationen erzeugt wurden und daher auch nicht danach suchten.

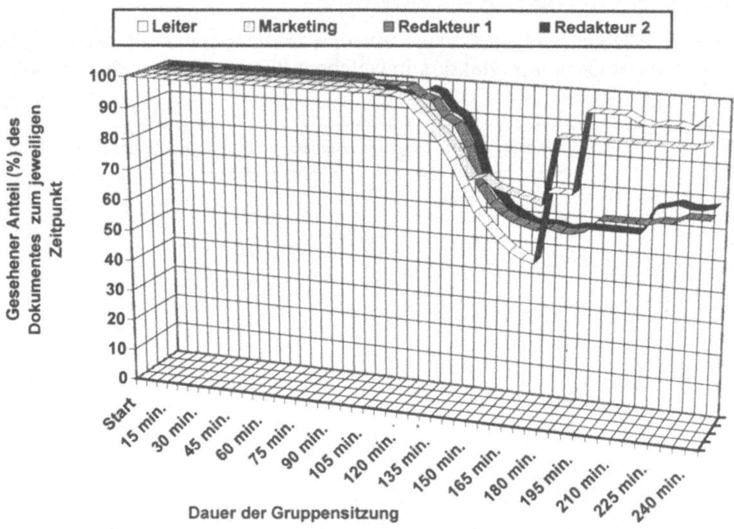

Abbildung 2: Wissensstand über die Dokumentinhalte für die vier Gruppenmitglieder

3.2.3 Aktivitätstabellen

Die Aktivitätstabelle gibt Aufschluß darüber, welche Knoten von einem Benutzer gesehen worden sind, in welchen er Inhalt produziert hat und wieviele Inhalte er in welchem Knoten nicht gesehen hat. Wir zeigen hier nur einen kleinen Teil einer kompletten Tabelle, die ca. 20-30 Knoten enthalten würde und wegen Platzmangel nicht vollständig dargestellt werden kann.

Knotenname	Besuche im Knoten				Produzierte Inhalte im Knoten				Nicht gesehene Informationen im Knoten			
	L	M	R1	R2	L	M	R1	R2	L	M	R1	R2
Brainstorming	20	24	31	30	373	326	313	31	0	0	0	0
Name und Logo	2	1	0	0	113	149	0	0	0	0	262	262
Qualität	3	0	4	3	7	0	26	26	0	59	0	0
Entwicklung	7	0	1	0	271	0	0	0	0	271	0	271
....												

Tabelle 1: Ausschnitt einer Aktivitätstabelle mit Anzahl der besuchten Knoten,
dort produzierten Inhalte und nicht gesehenen Informationen pro Person

Man kann in Tabelle 1 erkennen, daß Redakteur 1 (R1) und Redakteur 2 (R2) die Inhalte des
Knotens „Name und Logo" nicht gesehen haben, weil sie niemals diesen Knoten betreten
haben (Besuche im Knoten = 0). Diese Tabelle kann somit Auskunft geben, über welche
Bereiche und Themen ein gemeinsames Wissen vorhanden ist und welche Knoten bisher nur
von einer Teilgruppe gesehen wurden. Der Vergleich der Spalten mit den produzierten
Inhalten im Knoten läßt erkennen, welche inhaltlichen Schwerpunkte die Mitglieder der
Gruppe bearbeitet haben.

3.2.4 Diskussion der Ergebnisse

Die in unseren Untersuchungen zuvor gefundene Untergruppenbildung und Parallelarbeit
kann mit den Möglichkeiten der Analyse von Logfiles auf Gruppenebene wesentlich genauer
dokumentiert werden als dies durch die Beobachtung der verbalen Kommunikationsstrukturen möglich ist. Die für den Beobachter sonst kaum nachvollziehbaren Ereignisse im
Hypermedia-Dokument (wer mit wem in einem bestimmten Knoten zusammenarbeitet und
wer welche Informationen gesehen hat) können nun detailliert erhoben und analysiert werden.
Man kann damit erkennen, daß die Möglichkeit der Parallelarbeit, die ein gleichzeitiges
Arbeiten in Untergruppen ermöglicht und effizienzfördernd wirken kann, auch mögliche
Risiken beinhaltet. Es kann passieren, daß die Gruppenmitglieder unterschiedliches Wissen
über den Stand der gemeinsamen Arbeit haben. Wir werden diese Auswirkungen im Design
unserer Software reflektieren und Möglichkeiten vorsehen, daß die Benutzer sich über das
Ausmaß der nicht gesehenen Änderungen informieren können.

4 Schlußfolgerungen

Das Ausmaß und die Komplexität der Interaktionen innerhalb von verzweigten Hypermedia-
dokumentstrukturen sind mit bisherigen Methoden nicht angemessen zu erfassen. Wie wir
hier gezeigt haben, bietet die Erstellung und Analyse von Logfiles eine wertvolle Ergänzung
zu den traditionellen Beobachtungsansätzen. Die Genauigkeit und Objektivität der
Datenerfassung, die Unabhängigkeit von der Anzahl der Benutzer und der Länge der Beo-
bachtungszeit sowie die umfangreichen Auswertungsmöglichkeiten sind eine Bereicherung
im Methodeninventar der Software-Ergonomie für CSCW-Systeme. Durch die Möglichkeit,
bisher nicht direkt Beobachtbares nun erfassen zu können, ergeben sich neue Einsichten, aber
auch neue Fragestellungen, wie am Beispiel der „awareness" gezeigt wurde.

Der hier verfolgte Ansatz, durch Analyse der Systeminteraktionen zusätzliche Informationen über das Individual- und das Gruppenverhalten und den individuellen Wissensstand über einen gemeinsamen Arbeitsbereich zu gewinnen, ist nicht nur für die Evaluation von Systemen interessant. Eine Komponente dieser Art kann z.B. als Basis für ein adaptives System verwendet werden, das dem jeweiligen Benutzer selbständig mitteilt, in welchen Bereichen der Dokumentstruktur neue Informationen erzeugt, bzw. wo Veränderungen vorgenommen wurden. Der Benutzer weiß damit immer, was er (noch) nicht weiß.

Diese Funktionalität kann weiterhin dazu benutzt werden, dem Benutzer bei der laufenden Arbeit wichtige Informationen zu vermitteln. So wäre zum Beispiel eine Übersichtsanzeige während der Sitzungen vorstellbar, mit der das sich angesammelte Ausmaß an Unwissenheit angezeigt wird. Dies könnte mit Hinweisen verbunden werden, welche Dokumentbereiche von den jeweiligen Benutzern angesehen werden sollten. Diese Übersicht wäre sicher auch für den Leiter einer Arbeitsgruppe von Nutzen, der so besser dafür sorgen kann, daß alle Gruppenmitglieder auf dem aktuellen Wissensstand sind.

Die Möglichkeit der Analyse und Dokumentation, welche Informationen zu welcher Zeit und in welchem Modus der Kooperation erzeugt worden sind, kann der Gruppe wichtige Daten über die eigene Arbeitsweise liefern. So könnte man z.B. feststellen, ob bestimmte Ideen immer in bestimmten Gruppenkonstellationen auftreten. Diese Information könnte dazu benutzt werden, dieses Verhalten absichtlich zu fördern oder zu verändern.

Wir haben mit diesem Beitrag zeigen können, daß die Logfileanalyse sowohl eine geeignete Methode für die Evaluation von Systemen zur Computerunterstützung von Gruppenarbeit darstellt als auch neue Einsichten in Interaktionsprozesse von Gruppen gewonnen werden können.

5 Literatur

[1] J. Nunamaker et al.: Electronic meeting systems to support group work. Communications of the ACM, 34 (1991), 7, 40-61.

[2] J. Nunamaker, R. Briggs, D. Mittleman: Electronic Meeting Systems: Ten Years of Lessons Learned. In: D. Coleman, R. Khanna (Eds.), Groupware: Technology and Applications. Prentice-Hall Inc, 1995, 149-193.

[3] J. Olson, G.Olson, M. Storrosten, M. Carter: Groupwork close up: A comparison of the group design process with and without a simple group editor. In: T. Malone, N. Streitz (Eds.), Special Issue on CSCW of ACM Transactions on Information Systems. 11 (4),1996, 321-348.

[4] H. Lewe: Der Einfluß von Teamgröße und Computerunterstützung auf Sitzungen. In: U. Hasenkamp (Hg.). Einführung von CSCW-Systemen in Organisationen, Braunschweig, 1994: Vieweg, 147-166.

[5] A. Pinsonneault, K. Kraemer: The effects of electronic meetings on group processes and outcomes: An assessment of the empirical research. In: European Journal of Operational Research 46 (1990), 143-161.

[6] G. Mark, J. Haake, N. Streitz: Hypermedia Structures and the Division of Labor in Meeting Room Collaboration In: Proceedings of the ACM CSCW'96, Boston, MA., November 16-20, 1996, 170-179.

[7] N. Streitz, J. Geißler, J. Haake, J. Hol: DOLPHIN: Integrated meeting support cross Liveboards, local and remote desktop environments. In: Proceedings of the ACM CSCW'94, Chapel Hill, 1994, 345 - 358.

[8] G. Mark, J. Haake, N. Streitz: The use of hypermedia in group problem solving: An evaluation of the DOLPHIN electronic meeting room environment. In: Proceedings of the E-CSCW'95 Conference, Amsterdam, 1995: Kluwer Publishers, 197-213.

[9] N. Streitz, P.Rexroth, T. Holmer: Does "roomware" matter? Investigating the role of personal and public information devices and their combination in meeting room collaboration. In: Proceedings of the European Conference on Computer-Supported Cooperative Work (E-CSCW'97). Amsterdam, 1997: Kluwer Academic Publishers, 297 - 312.

[10] G. Mark, J. Haake, N. Streitz: Hypermedia Use in Group Work: Changing the Product, Process, and
Strategy. In: Computer Supported Cooperative Work: The Journal of Collaborative Computing 6 (1996),
327-368.

[11] P. Dourish; V. Bellotti: Awareness and coordination in shared workspaces. In: Proc. of ACM Conference
CSCW '92 (Toronto), 1992, 107-114.

[12] N. Streitz, J. Geißler, T. Holmer: Roomware for cooperative buildings: Integrated design of architectural
spaces and information spaces. In: N. Streitz, S. Konomi, H. Burkhardt (Eds.), Cooperative Buildings –
Integrating Information, Organization and Architecture. Proceedings of the First International Workshop on
Cooperative Buildings (CoBuild'98), Heidelberg 1998: Springer. 4-21.

Adressen der Autoren

Dipl. Psych. Torsten Holmer
GMD-IPSI
Institut für Integrierte Publikations-
und Informationssysteme
Dolivostr 15, 64293 Darmstadt
Email: holmer@darmstadt.gmd.de

Dr. Dr. Norbert Streitz
GMD-IPSI
Institut für Integrierte Publikations-
und Informationssysteme
Dolivostr 15, 64293 Darmstadt
Email: streitz@darmstadt.gmd.de

Gegenständliche Modellierung virtueller Informationswelten

Eva Hornecker und Kai Schäfer

Forschungszentrum Arbeit und Technik (artec), Universität Bremen

Zusammenfassung

In diesem Beitrag stellen wir ein Konzept greifbarer, gegenständlicher Benutzungsschnittstellen vor. Es werden Übergänge zwischen dem Modellieren im Realen und dem Erstellen virtueller Modelle geschaffen. Synchron zum gegenständlichen Modell entsteht ein virtuelles Abbild, das für Simulation, Animation und für Hilfen genutzt werden kann. Intuitive, spielerische und vorbegriffliche Herangehensweisen werden auf diese Weise mit abstraktem, analytischem Vorgehen verbunden. Es werden Anwendungsbeispiele gegeben und ein Projekt näher erläutert, in dem Förderbandanlagen synchron im Gegenständlichen und Virtuellen modelliert werden. Dabei wird das Programmieren durch Vormachen eingesetzt: Durch Bewegen von Gegenständen durch das Modell werden Regeln für Steuerungen von Förderbändern generiert. Anschließend stellen wir dar, wie gegenständliche Modellumgebungen, die durch ein synchronisiertes Computermodell unterstützt werden, die Zusammenarbeit fördern, indem sie auf natürliche Weise einen gemeinsamen Interaktionsraum herstellen, dessen zentrale Eigenschaften erläutert werden.

Abstract

In this text we present a concept of graspable computer-interfaces. It links modeling in physical reality with building virtual models. Synchronous to the graspable model a virtual twin-model is generated. The latter can be used for simulation, animation and giving help on the subject. Thus we connect intuitive, game-like and pre-conceptual approaches with abstract and analytic ways of thinking. We present some application areas, focusing on a project for the modeling of conveyor systems. We explain Programming by Demonstration: manually demonstrating the movement of objects through the scene we generate rules for programmable controls for the conveyor system. Then we focus on how working in computer supported concrete environments fosters group work and communication because it provides a shared workspace in a natural way.

1 Einleitung

Stellen Sie sich eine Fahrschule der Zukunft vor. Die Lernenden spielen zusammen Verkehrssituationen durch. Aber sie haben diese weder als Text noch als Zeichnung vor sich – sie spielen diese mit Hilfe von Spielzeugmodellen nach. Sie wählen eine Basisplatte mit dem Abbild einer Kreuzung, auf der einige Verkehrsschilder bereits plaziert sind, verändern diese evtl. durch Hinzufügen weiterer Schilder und setzen Modellautos, Fahrräder, Menschenfiguren auf Startpositionen am Rand der Platte. Jeder der Lernenden bewegt dann ein bis zwei Figuren über die Platte. Plötzlich leuchtet der Untergrund rund um zwei der Fahrzeugmodelle auf und aus einem Lautsprecher kommt eine Stimme: "Nicht eingehaltene Regel: 'Rechts hat Vorfahrt' sowie 'Fußgänger am Zebrastreifen hat Vorrecht'." Die Lernenden diskutieren die Situation, winken dann aber einen Fahrlehrer herbei. Da sie die gespielte Situation nicht mehr gänzlich aus dem Kopf rekonstruieren können, sehen sie sich gemeinsam die Aufzeichnung des Computers an, die in Form einer Animation gezeigt wird und analysieren diese.

Dieses bewußt spielerisch gehaltene Beispiel soll hier die Möglichkeiten aufzeigen, die unsere grifforientierte, gegenständliche Benutzungsschnittstelle eröffnet. Leitidee unserer Arbeitsgruppe ist es, Übergänge zwischen dem Modellieren im Realen und dem Erstellen virtueller

Modelle zu schaffen. Wie die Erfahrungen in mehreren Industriekooperationen zeigten, werden gegenständliche Modelle (z.B. von Fabrikanlagen und Hallenlayouts) nach wie vor aufgrund ihrer Anschaulichkeit und Kommunikationsförderlichkeit bevorzugt. Virtuelle Modelle im Computer sind jedoch für Animation, Simulation, Analyse und Archivierung notwendig. Daraus entstand das Ziel, die Modellarten zu koppeln und Übergänge zu schaffen.

Während der Modellierung im Realen wird synchron ein Modell im Rechner erstellt, vermittelt über moderne Interfacetechniken, die im folgenden noch erläutert werden. Das gegenständliche, greifbare Modell ist selber die Benutzungsschnittstelle. Die Modellierung mit realstofflichen Elementen und die technisch ungebrochene Kommunikation mit ihnen und über sie steht im Vordergrund [1] [2]. Der "Rechner im Rücken" soll unterstützendes Werkzeug und Medium für die Anwender sein. Bisher klafft eine Lücke zwischen den "Reality Techniken" Virtual Reality und Augmented Reality; der Bereich des Arbeitens mit den eigentlichen stofflichen Gegenständen wird unzureichend ausgefüllt. Da wir uns auf gegenständliche Modelle als Eingabemedium und Interaktionsraum konzentrieren und damit diese Lücke schließen, haben wir unser Konzept 1993 intern "Real Reality" genannt und den Begriff in [3] eingeführt.

Modellierung und Diskussion über Modelle sind für viele Lern- wie Arbeitsbereiche typische Situationen, in denen Real Reality Systeme eingesetzt werden können. Die gegenständliche Modellierung ermöglicht einen intuitiven Zugang, da sie an alltägliche Handlungsweisen anknüpft und nicht-sprachliche, vorbegriffliche Modellbildung erlaubt. Sie ist anschaulich und spricht das räumliche Denken an. Unsere Kopplung realer Modelle mit virtuellen verbindet das abstrahierende, distanzierende Vorgehen – das typischerweise von technischen Modellbildungssystemen unterstützt wird – mit dem intuitiven, spielerischen, sich einlassenden Vorgehen [1] und erleichtert die vielperspektivische, kreative Kommunikation. Real Reality Umgebungen ermöglichen und fördern so ein subjektivierendes Arbeitshandeln, wie es vom Arbeitswissenschaftler Fritz Böhle beschrieben wurde [4]. Unsere These besagt, daß ein hoher Grad an Stofflichkeit von Modellen positiven Einfluß auf Verständnis, Ideenreichtum und Kommunikation der Modellierenden hat.

2 Modellieren mit Real Reality

Wenn wir in einer Real Reality Umgebung modellieren, geschieht dieses nicht vor einem Computer, sondern mit alltäglichen Gegenständen oder mit Modellbausteinen. Lediglich zwei Voraussetzungen müssen für die stetige Aktualisierung des Computermodells erfüllt werden:

1. Die Geometrie und die Anfangsposition der Modellteile müssen im Computer als virtuelles Modell vorliegen.
2. Alle Modellierenden müssen bei ihrer Tätigkeit Datenhandschuhe tragen.

Nun können vorhandene Gegenstände beliebig bewegt werden, das Computermodell wird automatisch angepaßt. Zu jedem verwendeten Bauelement wird ein entsprechendes virtuelles Objekt erzeugt. Jeder Zustand des realen Modells hat ein kongruentes virtuelles Abbild im Rechner [2]. Die Paare aus virtuellem und realem Objekt heißen *Complex Objects* und können jeweils über unterschiedliche Eigenschaften verfügen.

Wichtig ist, daß während der Modellierung die Modellobjekte selbst die Informationsträger sind. Das virtuelle Computermodell dient lediglich zu Analyse- und Dokumentationszwecken und bleibt während der Modellierung im Hintergrund. Für die Grifferkennung sind nur die Außenkonturen der Gegenstände relevant, so daß auf eine exakte Geometrie verzichtet werden kann. Neben den in Abb. 1 abgebildeten Bauklötzen verwenden wir häufig Fischertechnik, so daß sowohl im Gegenständlichen wie im Virtuellen unterschiedliche Abstraktionsstufen möglich sind.

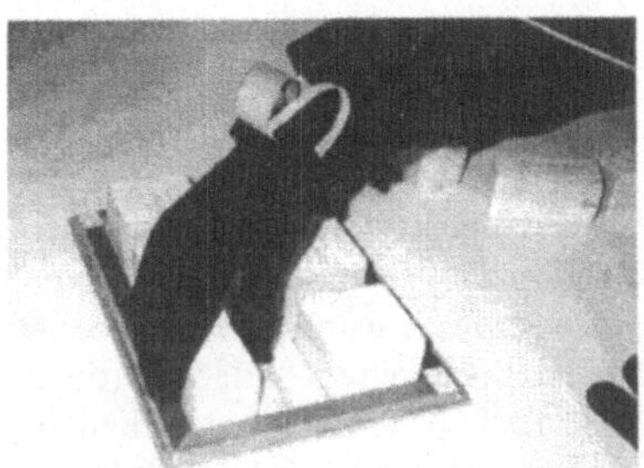

Abb. 1: Synchrone Modellierung im Realen und Virtuellen

Alle Veränderungen im Szenario gehen von der sensorisierten Hand aus. Mit Hilfe des Handschuhs werden Griffe erkannt. Die Position und Orientierung der Hand wird über ein elektromagnetisches Trackingsystem räumlich bestimmt. Entsprechen die Krümmungswerte am Handschuh einem Griffmuster, verknüpft die Real Object Manipulator Software (ROMAN) ein reales mit dem zugehörigen virtuellen Objekt und bewegt diese jetzt synchron, bis der Griff durch Verlassen des Griffmusters gelöst wird (Abb. 1) [3]. Die Architektur des ROMANs erlaubt die gleichzeitige Verwendung mehrerer Handschuhe.

Wenn wir mit Gegenständen handeln und modellieren, hat dieses für uns eine Bedeutung. Das ist für die Real Reality Software nicht der Fall. Diese behandelt alle Objekte gleich und verarbeitet lediglich die für die Handhabung relevanten Attribute Position, Orientierung und Geometrie. Aus diesem Grund läßt sich das Real Reality Konzept für jeden beliebigen Kontext universell einsetzen. Die unteren vier Ebenen in Abb. 2 stellen diesen Kern von Real Reality dar. Den Bezug zu den Anwendungs- und Modellkontexten stellen nachgeschaltete Softwarekomponenten durch die weitere Verarbeitung der Modelldaten her.

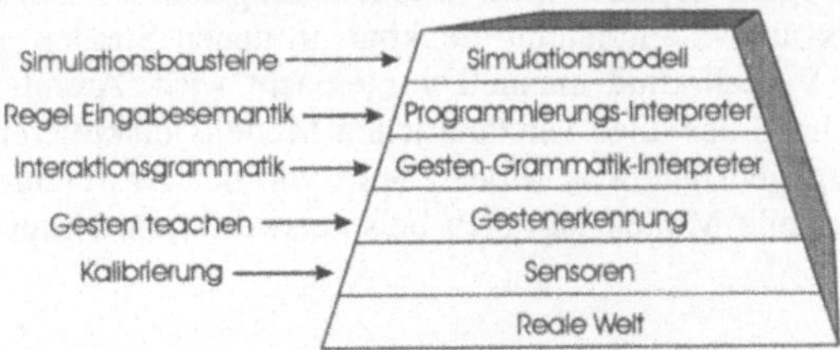

Abb. 2: Automatische Abstraktion der realen Welt

Der ROMAN verwaltet für die kontextspezifische Verarbeitung beliebige benutzerdefinierte Attribute, die mit den Objekten verknüpft sind. Die Modellierung erhält hierdurch eine semantische Bedeutung, die sich für festgelegte Anwendungsfälle durch spezifische Software automatisch verarbeiten läßt. Beim Modellieren werden nicht nur die aktuellen Positionen der

Objekte gespeichert, sondern auch die Bahnen, auf denen sie bewegt wurden. Dieses ermöglicht es, dynamisches Verhalten durch Vormachen am gegenständlichen Modell zu programmieren, es zu interpretieren und zu simulieren. Die durchgeführten Bewegungen und Aktionen werden dabei zu abstrakten Steuerprogrammen für Simulatoren und Maschinen verarbeitet. Über einen mehrstufigen Abstraktionsprozeß (Abb. 2, die beiden oberen Ebenen), können aus Modellen in der realen Welt simulierte und animierte Welten im Computer geschaffen werden.

3 Anwendungsbereiche

Der erste Anwendungsbereich, in dem Real Reality eingesetzt wurde, ist die Modellierung von Förderbandanlagen für die Produktionstechnik [8]. Das Verhalten der Anlage kann mittels manuellen *Programmierens durch Vormachen* spezifiziert und in einem Simulator evaluiert werden. Dies wird im folgenden genauer dargestellt. In einem anderen Projekt wird eine Lernumgebung für den Pneumatikunterricht an beruflichen Schulen entwickelt [6]. Parallel zum Aufbau einer Schaltung mit gegenständlichen Symbolen oder aus echten Pneumatikbauteilen entsteht simultan in einem Pneumatiksimulator ein Abbild der Schaltung, das simuliert und evaluiert werden kann. Der Simulator kann durch Gesten gesteuert werden und liefert z.B. Hilfe zu einem Bauelement oder startet und stoppt den Simulationslauf. Die Lernumgebung soll fließende Übergänge zwischen dem erfahrungsorientierten Lernen im Realen und dem Umgang mit Symbolen in Schaltplänen und Simulatoren ermöglichen und handlungsorientierten Unterricht unterstützen.

In einem laufenden Projekt wird die Freiformmodellierung im Sinne einer Skizzierungsmethode untersucht. Die ModelliererIn verformt mit ihren sensorisierten Händen eine Modelliermasse. Alle Bewegungen werden aufgezeichnet, um einen dreidimensionalen Freiformkörper zu generieren. Gesten und synchrone Spracheingaben spezifizieren das Endziel weiter.

Eine mögliche weitere Anwendung wäre eine kombiniert real-virtuelle Plantafel mit in Größe, Farbe und Form unterscheidbaren Zeit- und Auftragsbausteinen. Die reale Plantafel bietet aufgrund ihrer Größe und Physikalität einen besseren Überblick und einen intuitiven Umgang. Der virtuelle Teil kann zur Konsistenzprüfung und zur Dokumentation eingesetzt werden. In der Einleitung wurde bereits das Szenario eines Lernsystems für Fahrschulen dargestellt, in dem das gegenständliche Spiel von Verkehrssituationen vom Computer auf die Einhaltung der Verkehrsregeln geprüft wird. In einem umgekehrten Ansatz sind Real Reality Systeme zur Verkehrs- und Stadtplanung denkbar, in denen Straßen gegenständlich modelliert werden und der Verkehrsfluß manuell vorgemacht wird. Architekten, Stadtplaner und Anwohner können anhand des allen verständlichen Modells diskutieren und daran Änderungen vornehmen. Das gegenständliche Modell wird von der Anwendung in Planskizzen etc. umgewandelt; das virtuelle Modell läßt sich beispielsweise zur Simulation des Verkehrsablaufes nutzen.

4 Erstellen von Steuerprogrammen für industrielle Automatisierungseinrichtungen durch Vormachen

Vergleichende Studien zur Simulation und Modellierung von Hallenlayouts mit Förderbändern, Verzweigungen, Bearbeitungsstationen haben gezeigt, daß verschiedene Resultate von verschiedenen Personen mit verschiedenen Simulatoren erzielt wurden [7]. Zum einen lag das an Unklarheiten in der Aufgabenstellung, die von den Modellierenden aufgrund mangelnder

Anschaulichkeit nicht erkannt wurden und zum anderen an der unterschiedlichen Implementierung des Verhaltens der Simulationsbausteine, die zu unkritisch ins Modell eingesetzt wurden. Wenn das Verhalten der Anlage an einem Modell mit kleinen Paletten vorgemacht wird, ist nicht zu erwarten, daß solche Fehler unbemerkt bleiben. Im Projekt RUGAMS[1] wird deshalb ein solches Problem gegenständlich modelliert, um zu zeigen, daß Hallenlayout und Anlagensteuerung mit Real Reality geplant und simuliert werden können [8].

Bereits in der Entwurfsphase wird eine ereignisorientierte Computersimulation zukünftiger Fertigungsabläufe ermöglicht. Jeder Baustein ist hierfür mit einem charakteristischen Vorgabeverhalten ausgestattet. Die topologische Analyse des virtuellen Computermodells verknüpft die einzelnen Simulationsbausteine zu einem lauffähigen Gesamtmodell. In darauf folgenden Phasen kann das Verhalten des Modells durch Vormachen weiter beeinflußt werden (Abb. 3).

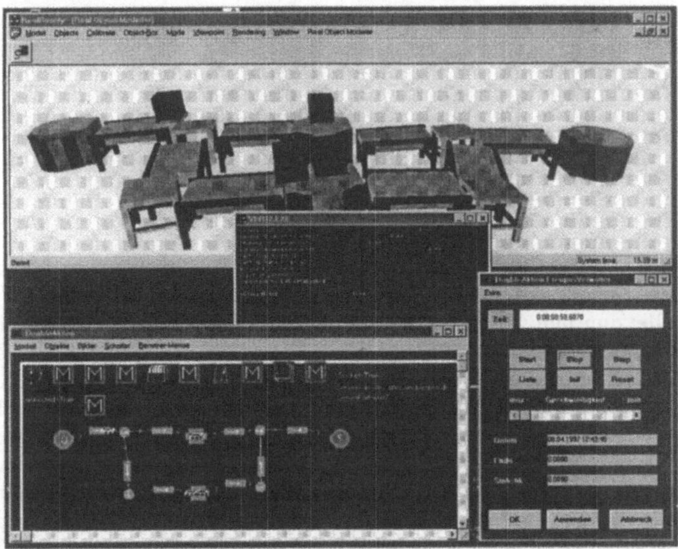

Abb. 3: Visualisierung von Modelldynamik am VR Modell

Beim Vormachen der Route von Werkstücken durch das System wird die Reihenfolge überfahrener Positionen aufgezeichnet. Diese *Sensepathes* werden zusammen mit der Topologie und Informationen über die Förderbausteine als *Kontextwissen* für eine Regelerkennung genutzt. Wird ein Verzweigungspunkt überfahren (Abb. 4 links), wird eine Entscheidungsregel generiert und als Attribut der Verzweigung gespeichert. Der Bearbeitungszustand des Werkstücks, der durch die Farbe des Klotzes repräsentiert ist, wird in die Verzweigungsregel einbezogen. Wenn das vorgegebene Verteilungsverhalten an der Verzweigung z.B. zufällig oder alternierend war, wird an der Verzweigung jetzt auf alle Paletten mit gleicher farblicher Kennzeichnung das vorgemachte Verhalten angewandt. Alle anderen Verzweigungsentscheidungen werden weiterhin nach dem Vorgabeverhalten des Förderbausteins behandelt [9].

---

1 Rechnergestützte Übergänge zwischen gegenständlichen und abstrakten Modellen produktionstechnischer Systeme (DFG)

Zusätzlich können Verzweigungen in Abhängigkeit von Maschinen- und Pufferbelegungen spezifiziert werden. Reale Anlagen sind hierfür mit Sensoren ausgestattet. Entscheidungsrelevante Sensorpositionen werden vor dem Überfahren des Verzweigungsknotens in ihrem aktuellen Belegungszustand mit Token gekennzeichnet (Abb. 4 rechts). Neben dem Zustand des Materialflußelements hängt die Verzweigung jetzt von der Verfügbarkeit einzelner Ressourcen ab. Werden verschiedene widersprüchliche Verzweigungsregeln für eine Situation vorgemacht, werden diese alternierend angewendet. Das vorgemachte Verhalten wird von einer wissensbasierten Regelanalyse verarbeitet, die das Verhalten in einer Programmiersprache für Anlagensteuerungen (AWL für SPS) ausgibt.

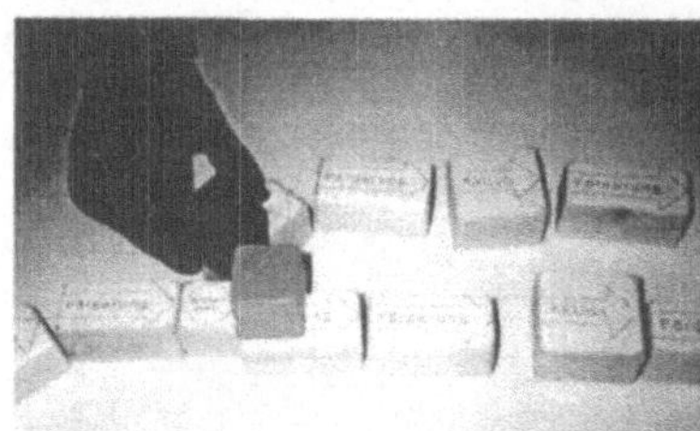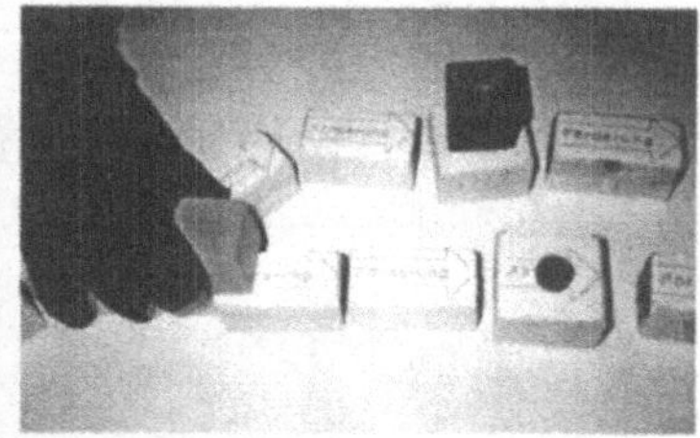

Abb. 4: Vormachen einer Verzweigungsregel

Aus der Modelltopologie, dem Vorgabeverhalten für die Bausteintypen und dem vorgemachten Verhaltensregeln läßt sich automatisch das Simulationsmodell aufbauen. Zur Zeit existiert eine Schnittstelle zum industriellen Materialflußsimulator SIMPLE++. Wenn dieses Modell in den Simulator geladen und ausgeführt wird, kann das von der geplanten Anlage zu erwartende Verhalten auf dem Bildschirm beobachtet und beurteilt werden. Da die Animation des Simulators auf dem Bidschirm sehr abstrakt und damit nicht für jeden verständlich ist (Abb. 3, links unten), werden die Veränderungen auf das dreidimensionale virtuelle Modell übertragen und dort visualisiert. Da der Wechsel vom Arbeitstisch zum Computer einen Medienbruch darstellt, projizieren wir die Bildschirmausgabe häufig mit einem Videoprojektor (Beamer) auf den Tisch und in das gegenständliche Modell hinein (Abb. 5). In diesem Sinne verwenden wir Augmented Reality Techniken, um die greifbare Repräsentation durch die dynamische Simulationsvisualisierung zu erweitern. Die große Ausgabefläche erleichtert die Diskussion in der Gruppe und erhöht die Beteiligung spürbar. Auf dieser Visualisierung aufbauend kann das Modell auf dem Tisch verändert und neu programmiert werden, bis es den Anforderungen der Akteure entspricht.

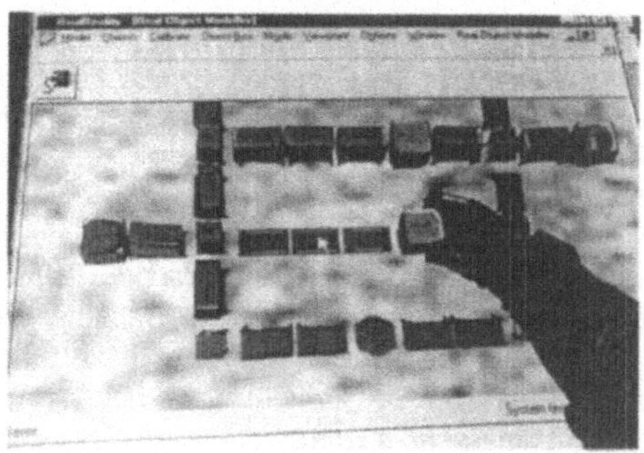

Abb. 5: Statische und dynamische Informationen werden ins Modell eingeblendet

Der besondere Vorzug des Programmierens durch Vormachen am gegenständlichen Modell ist die Intuitivität beim Umgang mit den Objekten und deren Verhalten, die es allen Beteiligten erlaubt, sich aktiv in den Planungsablauf einzubringen. Gute Modelle sind jene, die passende Ideen von allen Modellierenden berücksichtigen und die von allen Beteiligten verstanden und akzeptiert sind. Anders als bei klassischen Simulatoren ist eine kooperative Modellierung in der Gruppe möglich, die es unterschiedlichsten Qualifikationsstufen erlaubt, sich auszudrücken, aktiv zu werden und das Ergebnis zu verstehen.

5 Real Reality als kooperationsunterstützendes Arbeitsmedium

Die verbreiteten computergestützten Modellierungs- und Simulationswerkzeuge erfordern es, am Monitor zu arbeiten, auch wenn alle Interaktionspartner anwesend sind. Die Sicht auf den Bildschirm beschränkt die Gruppengröße; Tastatur und Maus restringieren die aktive Kontrolle. Real Reality ermöglicht dagegen die Arbeit in einem gemeinsamen Interaktionsraum und stellt dabei Computerunterstützung zur Verfügung, wie in Abb. 6 zu sehen.

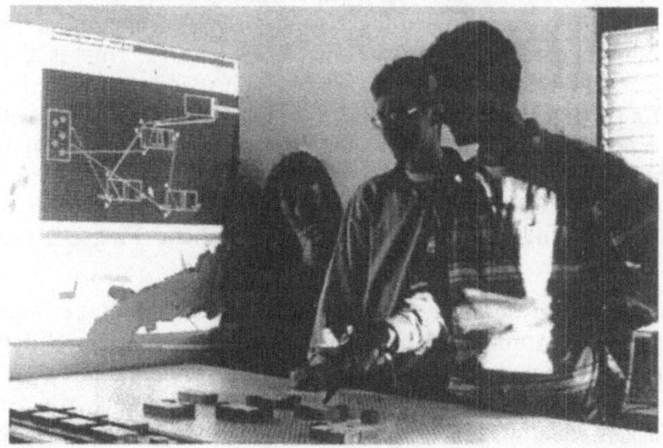

Abb. 6: Gemeinsame Arbeit am Modelliertisch, im Hintergrund die Projektion

Kooperatives Lernen und Arbeiten wird durch die Arbeit in einem solchen gemeinsam geteilten Interaktionsraum erleichtert. Die Schwierigkeiten, die durch das Fehlen sozialer Hinweisreize und Präsenz (nonverbale Kommunikation), fehlende Gruppenkoordination und "awareness" (Wahrnehmung der Handlungen und Intentionen der Interaktionspartner) entstehen, werden in vielen Studien zur computerunterstützten Kommunikation deutlich. Videokonferenzsysteme erhöhen den Grad der sozialen Präsenz, ermöglichen in der Regel jedoch nicht das gesamte Spektrum nonverbaler Kommunikation, da die Teilnehmer(bilder) in keinem räumlichen Bezug zueinander stehen, wodurch z.B. kein Blickkontakt entsteht. Als effektiver hat sich die Sichtbarkeit der gemeinsam bearbeiteten bzw. diskutierten Objekte erwiesen, die als gemeinsame Referenz den Konversationsinhalt koordinieren (video as data) [10]. Daher wird viel Aufwand in die Schaffung von "shared workspace" investiert; gemeinsame Arbeitsbereiche, in denen z.B. die Stelle, an der eine Kollegin gerade arbeitet, visuell markiert ist oder ihr Sichtbarkeitsbereich durch einen Kasten repräsentiert wird. [11]

Mehrere Studien zeigen die Vorteile von realen gemeinsamen Interaktionsräumen auf und beschreiben ihre Merkmale. Rauterberg [12] und Suzuki [13] argumentieren, daß durch einen solchen "shared social space" Kommunikation und gemeinsam geteiltes Verständnis gefördert werden. In ihm ist das gesamte Spektrum verbaler und nonverbaler Kommunikation offen. Seine Besonderheiten [14] bestehen darin, daß er drei wichtige Kommunikationsaspekte unterstützt: Sichtbarkeit, Hörbarkeit und soziale Nähe. Dies wird von Böhme [15, S. 177, 211] gestützt, der Kommunikation als Ausübung und Vermischung von Präsenz beschreibt. Dabei sei die leibliche Anwesenheit wichtig, denn der Mensch strahle eine spürbare Atmosphäre aus, die sich durch ihre Konkretheit und Lebendigkeit auszeichne.

Die Gegenständlichkeit – als Möglichkeit zum Greifen und Zeigen – erleichtert die Verständigung, insbesondere wenn noch keine gemeinsamen Begriffe existieren oder die Fachtermini fremd sind [13] [16]. Dies gilt sowohl für heterogene Entwurfsgruppen [17], wie für Lernende, die eine Fachsprache erst unvollkommen beherrschen. Die gegenständlichen Symbole sind für alle sichtbar und greifbar, können von Hand zu Hand weitergereicht werden. Dinge werden zu Werkzeugen des Denkens und Kommunizierens, wie D. Norman am Beispiel des Vorspielens eines Autounfalls mit zufällig ergriffenen Gegenständen erläutert: "We can make marks or symbols that represent something else and then do our reasoning by using these marks. (... They) help the mind keep track of complex events. (... This) is also a tool for social communication. (...) The tabletop becomes a shared workspace with shared representations of the event." [18]. Es gibt zudem Hinweise, daß die manuelle Aktivität zu mehr Reflektion vor der Ausführung einer Handlung führt, die Sichtbarkeit und (scheinbare) Endgültigkeit die Verbindlichkeit erhöht [19]. Daher interessiert uns, wie Menschen in ihren Denk- und Kommunikationsprozessen mit Gegenständen interagieren und wie die Gegenstände sie darin beeinflussen. Bisher scheinen nur vereinzelte Studien ähnliche Fragen zu untersuchen (z.B. [20]).

Der gemeinsame Interaktionsraum ist im Real Reality Ansatz durch die Gegenstände und den Modelltisch selbst gegeben und muß nicht erst durch elektronische Hilfsmittel geschaffen werden. Da mehrere Handschuhe gleichzeitig verwendet werden können, ist die parallele Arbeit mehrerer Personen am Modell möglich. Dies bezieht sich auch auf das Programmieren durch Vormachen, das ebenfalls ein kooperativer, diskursiver Prozeß sein kann. Mehrere Personen können alternierend oder sogar gleichzeitig Regeln vormachen, da alle Bewegungen der Objekte aufgezeichnet werden. Denkbar wären spezielle Regelgenerierungsmodi, in denen vorgemachte Regeln je nach Benutzer anders zugeordnet werden oder parallel vorgemachte Regeln gemeinsam interpretiert werden.

Eine offene Frage besteht darin, wie bei einer interaktiven Steuerung eines Anwendungsprogrammes (z.B. des Pneumatiksimulators) vom Arbeitstisch aus gleichzeitige oder interferierende Befehle an den Computer behandelt werden. Bei einer visuellen Ausgabe ergeben sich Fragen der Zeitlichkeit, der Anordnung, der Zuordnung, sowie Platzeinschränkungen. Je nach Gestensprache sind weiterhin widersprüchliche gleichzeitige Befehle möglich. Ein Sperren von Aktionsmöglichkeiten, wie in Computersystemen, ist im Realen schwer möglich. Für die weitere Arbeit am Real Reality-Konzept ergeben sich interessante konzeptionelle Fragen. Wie lassen sich die hochgradig parallelen, nur bedingt synchronisierten, zum Teil widersprüchlichen und sozial ausgehandelten Interaktionen adäquat im Anwendungsprogramm spiegeln und verarbeiten, ohne den Benutzern ein Verhaltenskorsett und inadäquate Kommunikationsmodelle aufzuerlegen? Inwieweit kann auf aufwendige Lösungen verzichtet werden, weil die Benutzer in der Lage sind, ihr Arbeiten und die Reihenfolge von Aktionen unter sich auszuhandeln und abzustimmen?

6 Zusammenfassung und Einordnung

Unsere Forschungsperspektiven beinhalten die Kombination mit Telekooperation, die Erweiterung des Programmierens durch Vormachen, die Ergänzung durch Visualisierungstechniken der Augmented Reality, Konzepte zur Synchronisierung paralleler Aktionen im Anwendungsprogramm, die Erkundung von Merkmalen geeigneter Anwendungsbereiche, Eingabesemantiken und Interaktionsgrammatiken für Gestensprachen. Schwierigkeiten bereitet die uns verfügbare Hardware, die keine optimale Grifferkennung und Positionsbestimmung ermöglicht. Das elektromagnetische Trackingsystem wird zudem durch metallische Objekte beeinflußt und läßt deren Verwendung nicht zu.

Unser Ansatz hat mit Augmented Reality das Ziel gemeinsam, den Menschen in seiner alltäglichen Umwelt integriert zu lassen und einen intuitiven Umgang mit Benutzungsschnittstellen zu ermöglichen (siehe z.B. [21]). Verschiedene Augmented Reality Projekte befassen sich mit kooperativen Arbeitsumgebungen für gleichzeitig anwesende Personen, häufig unter Einblendung virtueller Gegenstände in halbdurchlässige Head-Mounted-Displays (siehe z.B. [22]). Auch Konzepte wie die elektronischen WhiteBoards oder "Roomware" [23] integrieren aktives und reaktives Verhalten in Gegenstände bzw. ganze Räume, um kreative kooperative Arbeit zu unterstützen.

Unser Ansatz läßt sich jedoch eher mit den sog. "graspable user interfaces" vergleichen. Ein wichtiges Prinzip von Real Reality ist es, alle Modellobjekte greifbar zu machen, dreidimensional und haptisch erfaßbar, nicht nur die Interaktionswerkzeuge (etwa in Form greifbarer Zeiger und Manipulationswerkzeuge). Dies unterscheidet Real Reality von Projekten wie Build-It [12] oder "tangible bits" des MIT [24]. In der Verbindung von Greifbarkeit aller Modellobjekte und Simulationsvisualisierung kommt das "Illuminating Light" Projekt des MIT [25] unserem Vorgehen am nächsten. Von diesem und den Labilbaukästen wie Triangles [26] oder AlgoBlocks [13] unterscheidet sich Real Reality wiederum durch die Sensorisierung der Hand (anstatt die physikalischen Objekte zu sensorisieren). Gesten werden durch Vormachen definiert, sie wirken sich zunächst auf alle Dinge gleich aus. Sie können aber auch ohne Objektbezug interpretiert werden. Durch die Sensorisierung der Hand wird zudem die Auswahl von Objekten vereinfacht, da diese nicht sensorisiert werden müssen.

Wir danken unseren Kollegen bei artec, die am Entstehen des Konzeptes und der beschriebenen Gedanken maßgeblich mitgewirkt haben: Prof. Willi Bruns, Volker Brauer, Bernd Robben, Ingrid Rügge, Dieter Müller, Achim Heimbucher, sowie unseren studentischen Mitarbeitern.

4. Literatur

[1] W. Bruns: Sinnlichkeit in der Technikgestaltung und Technikhandhabung – Ein konstruktiver Ansatz. In: C. Schachtner (Hg.): Technik und Subjektivität. 1997: suhrkamp taschenbuch wissenschaft.

[2] F. W. Bruns: Zur Rückgewinnung von Sinnlichkeit - Eine neue Form des Umgangs mit Rechnern. CH, 1993: Technische Rundschau Heft 29/30, 85. Jahrgang.

[3] F. W. Bruns, V. Brauer: Bridging the Gap between Real and Virtual Modeling - A new Approach to Human-Computer-Interaction. IFIP-Workshop ”Virtual Prototyping”, Texas 1996 (Auch als artec Paper Nr. 46, Universität Bremen).

[4] F. Böhle, B. Milkau: Vom Handrad zum Bildschirm – eine Untersuchung zur sinnlichen Erfahrung im Arbeitsprozeß. 1988: Campus Verlag.

[5] K. Schäfer, V. Brauer, F. W. Bruns: A new Approach to Human-Computer Interaction; Synchronous Modelling in Real and Virtual Spaces. DIS ‘97. 1997: ACM.

[6] B. Robben, E. Hornecker: Gegenständliche Modelle mit dem Datenhandschuh begreifen – Eine Lernumgebung für den Technikunterricht. In: V. Claus (Hg.): Informatik und Ausbildung ‘98. 1998: Springer.

[7] J. Krauth: Comparison 2, Flexible Assembly System. EUROSIM: Simulation News 2, May 1992.

[8] V. Brauer, W. Bruns, K. Schäfer: Rechnergestützte Übergänge zwischen gegenständlichen und abstrakten Modellen produktionstechnischer Systeme, Erster und zweiter Zwischenbericht zum DFG Forschungsprojekt RUGAMS. Bremen, 1998: artec Arbeitspapier 56.

[9] K. Schäfer: Real Reality - Simulationsunterstützung durch gegenständliche Modelle. Tagungsband Simulation und Visualisierung ‘98. Magdeburg, 1998: SCS.

[10] S. Whittaker: Rethinking video as a technology for interpersonal communications: theory and design implications. In: International Journal of Human-Computer-Studies (1995) 42

[11] R.B. Smith: What You See Is What I Think You See. In: Conference on Computer Supported Collaborative Learning 1992, Vol 21 #3, 192: ACM.

[12] M. Rauterberg, P. Steiger. Pattern Recognition as a Key Technology for the Next Generation of User Interfaces. In: 1996 IEEE International Conference on Systems, Man and Cybernetics – Information Intelligence and Systems, Vol 4 of 4. 1996: IEEE.

[13] H. Suzuki, H. Kato. Interaction-Level Support for Collaborative Learning: AlgoBlock – An Open Programming Language. In: Conference on Computer Supported Cooperative Learning ’95. 1995.

[14] M. Rauterberg, M. Sperisen, M. Däwyler (1995): From Competition to Collaboration through a Shared Social Space. In: B. Blumental, J. Govnostaev, C. Unger. (eds.): East-West International Conference on Human Computer Interaction – EWHCI ’95 (Vol II), Moskau: International Centre for Scientific & Technical Information, 1995.

[15] G. Böhme: Einführung in die Philosophie – Weltweisheit Lebensform Wissenschaft. 2. Auflage. 1997: suhrkamp taschenbuch wissenschaft.

[16] E. Arias, H. Eden, G. Fischer. Enhancing Communication, Facilitating Shared Understanding, and Creating Better Artifacts by Integrating Physical and Computational Media for Design. In: DIS ’97. 1997: ACM.

[17] J. Scheel, K. Hacker, K Henning: Fabrikorganisation neu beGreifen. Köln , 1994: TÜV Rheinland.

[18] D. A. Norman: Things that Make Us Smart – Defending Human Attributes in the Age of the Machine. 1994: Addison Wesley.

[19] S. Whittaker, H. Schwarz: Back to the Future: Pen and Paper Technology Supports Complex Group Coordination. In: CHI ’95 Mosaic of Creativity 1995: ACM.

[20] I. Neilson, J. Lee: Conversations with graphics: implications for the design of natural language/graphics interfaces. In: International Journal of Human-Computer-Studies (1994) 40.

[21] P. Wellner, W. Mackay, R. Gold. Computer-Augmented Environments – Back to the Real World. In: Communications of the ACM, 36/7. 1993: ACM.

[22] Z. Szalavári, M. Gervautz. Interaktion mit virtuellen Informationen in realen Umgebungen - das "Personal Interaction Panel". In: I. Rügge, B. Robben, E. Hornecker, W. Bruns (Hg.): Arbeiten und Begreifen: Neue Mensch-Computer-Schnittstellen. 1998: Lit-Verlag.

[23] N. A. Streitz, J. Geißler, T. Holmer. Roomware for Cooperative Buildings: Integrated Design of Architectural Spaces and Information Spaces. In: N. A. Streitz, J. Konomi, H.-J. Burkhardt (Edt.): Cooperative Buildings. Proceedings of CoBuild '98. 1998: Springer.

[24] H. Ishii, B. Ullmer. Tangible Bits: Towards Seamless Interfaces between People, Bits and Atoms. In: CHI '97. 1997: ACM.

[25] J. Underkoffler, H. Ishii. Illuminating Light: An Optical Design Tool with a Luminous-Tangible Interface. In: CHI '98. 1998: ACM.

[26] M. G. Gorbet, M. Orth, H. Ishii. Triangles: Tangible Interface for Manipulation and Exploration of Digital Information Topography. In: CHI '98. 1998: ACM.

Adressen der Autoren

Eva Hornecker, Kai Schäfer
Universität Bremen
Forschungszentrum Arbeit und Technik (artec)
Enrique-Schmidt-Straße (SFG), Postfach 330440
D-28334 Bremen
Email: eva@artec.uni-bremen.de, schaefer@artec.uni-bremen.de

HyperCons: Eine Informationswelt für Bildungseinrichtungen

Peter Hubwieser, Johann Schlichter

Fakultät für Informatik, Technische Universität München

Zusammenfassung

In diesem Beitrag stellen wir eine Informationswelt für Schulen und andere Bildungseinrichtungen vor. Sie ist aus den Anforderungen der Praxis entstanden und soll als Grundlage für einen lokalen Informationsverbund zahlreicher Bildungsinstitutionen eingesetzt werden. Dabei steht vor allem die Unterstützung der Benutzer bei der Suche nach Informationen, bei der Wiederverwendung und beim Austausch von Dokumenten im Vordergrund. Die Basis des Systems stellt ein detailliertes Modell des Einsatzbereiches dar. Darauf kann eine Intelligenzkomponente aufsetzen, um dem Benutzer möglichst umfangreiche Unterstützung anbieten zu können.

Abstract

We intend to present an information system for schools and other educational institutions. The system will be the basement of a proposed local information network of many different educational institutions. With the construction of HyperCons we intend to provide an information system that supports teachers and students in searching, retrieving and publishing information from the Internet or from any local file system.

1 Einleitung

Auf den ersten Blick eröffnet der Anschluß von Schulen und anderen Bildungseinrichtungen an das Internet eine Reihe von faszinierenden Möglichkeiten. Zunächst denkt man natürlich an die Nutzung der schier unerschöpflichen Informationsvielfalt dieses Mediums zur Unterrichtsvorbereitung, Veranschaulichung von Lerninhalten oder Anfertigung von Referaten oder Abschlußarbeiten. Daneben bieten sich auch die Überwindung der Grenzen zwischen verschiedenen Bildungseinrichtungen durch Email-Diskussionen oder Austausch von Unterrichtsskizzen und Lehrmaterialien oder gar gemeinsame "Online"-Lehrveranstaltungen an. Ein weiterer wichtiger Einsatzbereich könnte die häusliche Versorgung längerfristig erkrankter, behinderter, wegen Schüleraustausch abwesender Schüler bzw. Lehrgangsteilnehmer mit wichtigen Lernunterlagen sein.

Allerdings hat sich in den ersten Jahren der schulischen Internetpraxis gezeigt, daß diese Anwendungen auf zum Teil unüberwindliche Schwierigkeiten unterschiedlicher Herkunft stoßen:

- Es wird immer schwieriger für Lehrer und Schüler, unter Zeitdruck im Internet Informationen aufzufinden. Dies hält viele potentielle Nutzer vom tatsächlichen Einsatz des Mediums ab.
- Falls man brauchbare Dokumente findet, muß man sie auf lokale Rechner übertragen, um ihre Persistenz zu sichern. Dadurch entstehen riesige Datenmengen auf den schuleigenen Festplatten, in denen man ein gespeichertes Dokument nach einiger Zeit nicht mehr wiederfindet. Dasselbe Schicksal erleiden die Datenspuren von Email-Partnerschaften.

- Die pädagogisch und didaktisch sehr fruchtbare Eigenproduktion von Dokumenten wird enorm angeregt, wenn die Schüler das Gefühl bekommen, daß andere Netzteilnehmer von ihren Produktionen erfahren und diese wiederum im Unterricht einsetzen. Leider werden derartige Neuproduktionen im allgemeinen kaum wahrgenommen.

Die Technische Universität München versucht derzeit, durch Konstruktion eines speziellen Informationssystems zur Lösung dieser Probleme beizutragen. Unser Ansatz ist die Erstellung einer speziell für den Bildungsbereich zugeschnittenen Informationsstruktur, die alle Informationen über brauchbare Dokumente in Abhängigkeit von deren Kontext mit möglichst hoher Intelligenz verwaltet und zusätzlich die Suche nach neuen Informationsquellen unterstützt (siehe dazu auch [1], [2], [3]). Dieses System soll vor allem dazu dienen, konstruktivistische Lernansätze wie *Cognitive Flexibility* [4] zu fördern, indem es einerseits die Konstruktion von Wissen durch Einordnung von Informationen in den jeweiligen Kontext fördert, andererseits zu demselben Thema viele verschiedene Sichtweisen und Informationen anbietet. Wir bezeichnen das System deshalb als „hypermediale, konstruktivistische Lehr- und Lernumgebung", kurz *HyperCons*.

2 Das Einsatzszenario

Im oberbayerischen Regionalzentrum Rosenheim begannen die ersten Internetaktivitäten der örtlichen Schulen bereits im Herbst 1995, lange vor der Initiative "Schulen ans Netz". Zu diesem Zeitpunkt starteten wir den Schulversuch "Internet an der Schule", unterstützt von der örtlichen Fachhochschule, dem Bayerischen Staatsministerium für Unterricht, Kultus, Wissenschaft und Kunst sowie dem Leibniz-Rechenzentrum. Der Einwahlknoten zum Anschluß der ersten 25 Schulen an das Internet konnte im November 1995 an der Fachhochschule in Betrieb gehen. Inzwischen hat sich das örtliche Bürgernetz als idealer Provider für die mittlerweile 44 Schulen etabliert. Davon sind 10 über das ISDN-Netz der Stadt Rosenheim völlig kostenfrei angeschlossen. Seit Ende Juni 1998 sind zwei Schulen zusätzlich über das Rosenheimer Breitbandkabelnetz mit dem Bürgernetz verbunden. Diese einmalige Konstellation ermöglicht Datentransportraten im Bereich von 4 MBit/s. Zwei weitere Schulen werden im Laufe des Jahres 1998 ebenfalls einen Kabelanschluß erhalten.

Ab Herbst 1998 soll nun auf der vorhandenen örtlichen Infrastruktur zusammen mit der Volkshochschule, dem Stadtarchiv, der Stadtbibliothek und weiteren Bildungseinrichtungen unter der Leitung das *Rosenheimer Bildungsnetz* (RoBiN) aufgebaut werden. HyperCons soll dabei das Herzstück dieses Informationsverbundes bilden.

Da die bisher beteiligten Lehrer sich mit zunehmender Häufigkeit über den großen Zeitaufwand beklagen, den die Suche nach brauchbaren Informationen im Internet verschlingt, können wir mit einer relativ großen Akzeptanz unseres Systems rechnen, falls es uns gelingt, diesen Zeitaufwand deutlich zu reduzieren.

3 Die Idee

Der Grundgedanke von HyperCons ist die Konstruktion einer dynamischen Informations-struktur speziell für den Bildungsbereich. Auf der Basis einer sorgfältigen Untersuchung der Einsatzdomäne wurde zunächst ein detailliertes Benutzermodell erstellt und in einer Datenbank implementiert [5]. Wir verwalten darin Informationen über Benutzer, deren Projekte und die Themenbereiche, an denen sie interessiert sein könnten (siehe Abb. 1).

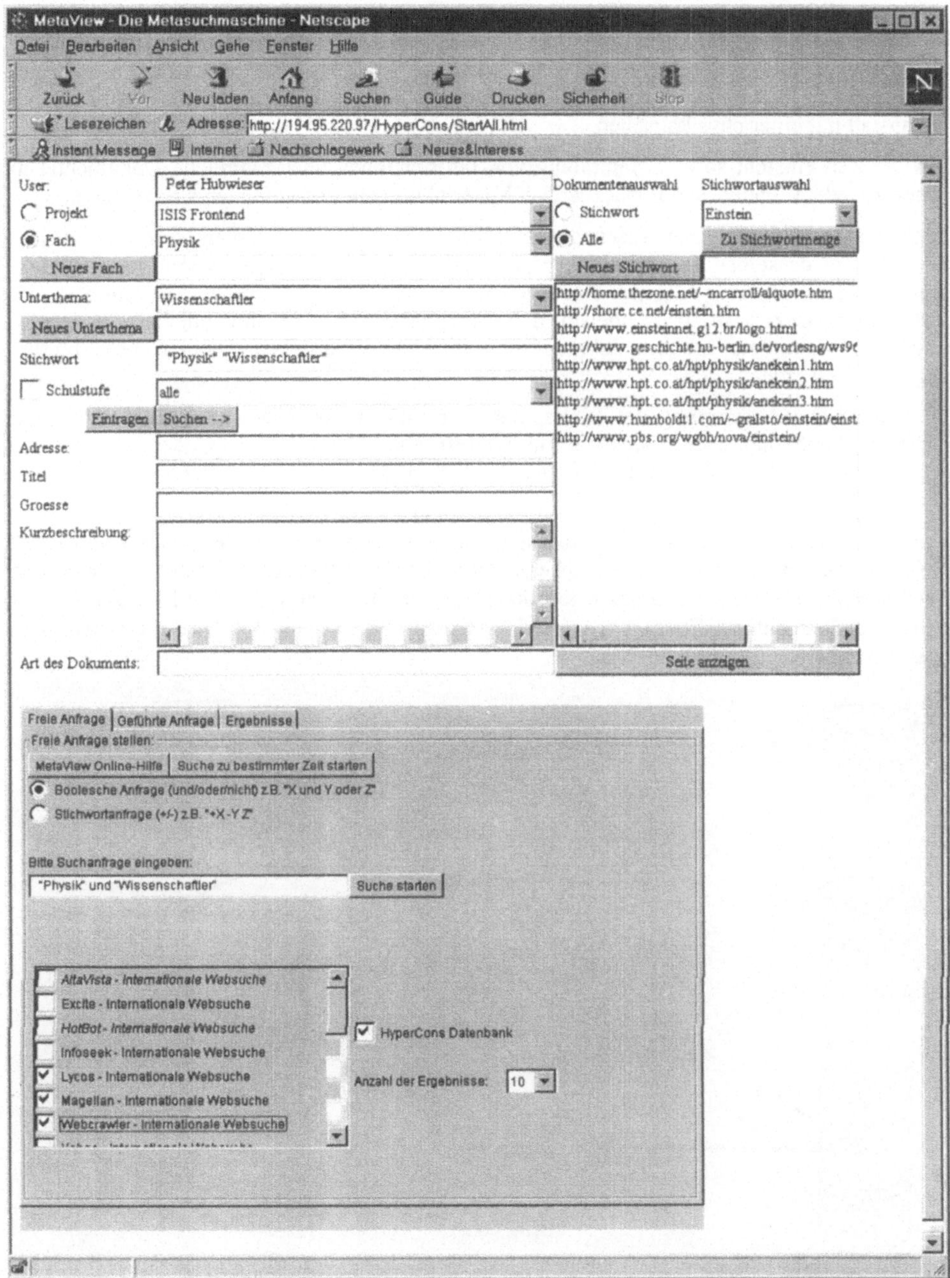

Abb. 1: HyperCons aus der Sicht des Benutzers

Falls ein Dokument ausgetauscht, veröffentlicht oder auch nur aufbewahrt werden soll, speichern wir alle wichtigen Informationen sowohl über seine physikalischen Eigenschaften als auch über seine Einordnung in den beschriebenen Kontext. Der Benutzer hat es dann im Gegensatz zu Dateisystemen oder URLs mit einer semantischen Ordnung der Dokumente zu tun anstatt mit einer physikalischen.

Mit der Zeit entsteht so eine Sammlung von Informationen über zahlreiche Dokumente, in denen zu jedem schulischen Einsatzbereich Material gesucht und gefunden werden kann.

Zusätzlich ermöglicht dieser Ansatz die Benachrichtigung von betroffenen Nutzern, etwa den Teilnehmern desselben Projektes, über neu veröffentlichte Dokumente aus ihren Interessenbereichen. Wenn genügend Eintragungen vorhanden sind, wäre es denkbar, die Informationsstruktur so abzuschließen, daß minderjährige Schüler nur innerhalb der eingetragenen Dokumente arbeiten können und so von gefährlichen Inhalten des Internet abgeschirmt werden.

Die statistische Auswertung der Eintragungen ermöglicht die Gewichtung der Bedeutung einzelner Subthemen oder Stichworte im Kontext eines Themenbereichs oder Projektes. So können wir dem Nutzer vor jeder Suche je nach Kontext eine Liste vermutlich interessanter Quellen anbieten.

Um die Aufnahme neuer Informationen anzuregen, unterstützen wir die Suche nach neuen Dokumenten, indem wir mit einfacher, einheitlicher Syntax eine Reihe von externen Suchmaschinen bedienen und deren Resultate geordnet und von jedem überflüssigen Ballast befreit zur Betrachtung, Eintragung in die Datenbank oder Aufnahme in ein Arbeitsdokument präsentieren.

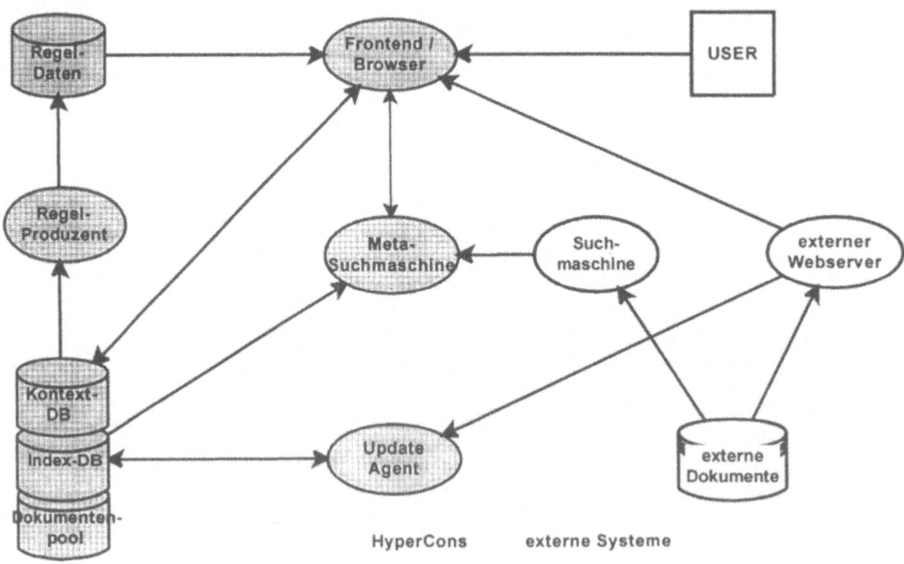

Abb. 2: Die Komponenten von HyperCons

4   Die Implementierung

4.1   Die Softwarearchitektur

Um das System möglichst universell nutzbar zu machen, sollten als Randbedingungen Unabhängigkeit von Hardware- und Betriebssystemplattform sowie das Entfallen jeglicher Installationssarbeiten vor Ort eingehalten werden. Dies führte zur Entscheidung für Java-Applets und Java-Applikationen, die untereinander, mit externen Suchmaschinen und einer zentralen relationalen Datenbank kommunizieren. Allerdings sind beide Bedingungen aufgrund der derzeit sehr mangelhaften Ausstattung der gängigen Web-Browser noch nicht erfüllt. Lediglich Netscape®-Browser auf MS-Windows® Systemen können mit einigen wenigen Installationsvorkehrungen so aufgerüstet werden, daß die Applets darauf laufen. Die Ursache dafür liegt einerseits in den hohen Ansprüchen an das Fenstersystem und andererseits in der nötigen Anpassung der Sicherheitsumgebung der Browser. Nun sollen die einzelnen Komponenten der Anlage (siehe

Abb. 2) kurz vorgestellt werden.

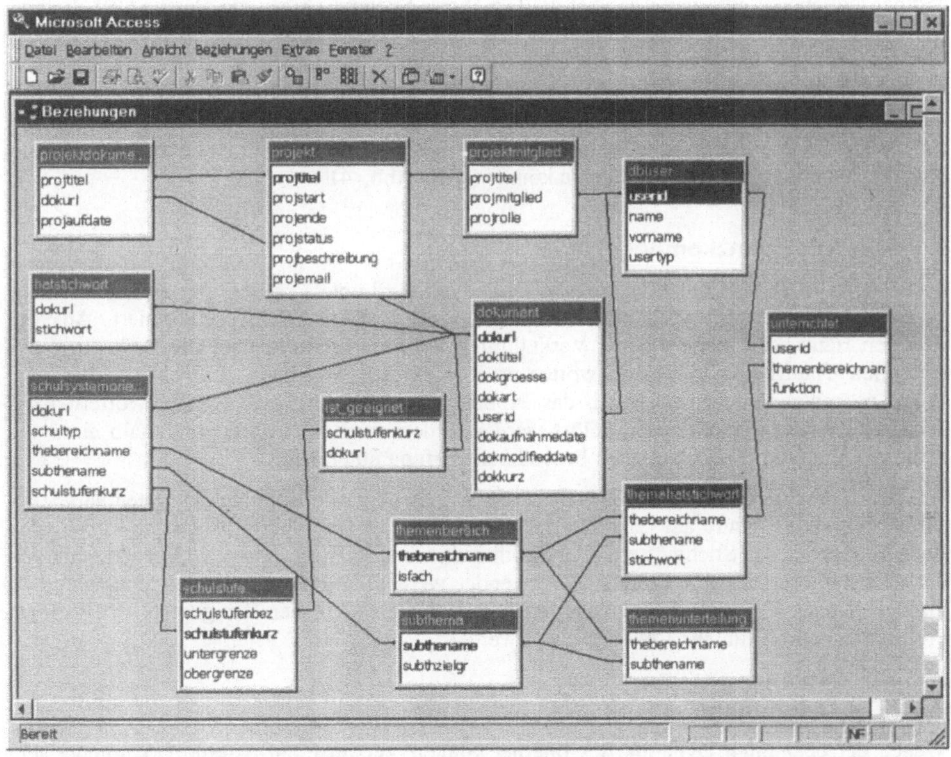

Abb. 3: Die Struktur der Datenbank

4.2. Die Datenbank

Das Entity-Relationship-Modell führt zu 24 Tabellen einer relationalen Datenbank. Im gegenwärtig laufenden Prototyp sind 14 davon in Betrieb (siehe

Abb. 3). Die Intelligenzkomponente wird zu weiteren Tabellen führen. Das Frontend-Applet zur Bedienung der Datenbank existiert derzeit nur als rudimentärer Prototyp [6]. Nach der Entwicklung der geplanten Intelligenzkomponente wird es deren Wissen nutzen, um dem Benutzer kontextabhängige Vorschläge für interessante Unterthemen, Stichworte oder Dokumente zu unterbreiten.

4.3   Die Suchmaschine

Die als Java-Applet konstruierte Suchmaschine *MetaView* [7] übernimmt einen Suchausdruck vom Benutzer oder vom Frontend und leitet diesen, falls gewünscht auch verzögert, zusammen mit einigen Verwaltungsinformationen an die vom Benutzer gewünschten externen Suchmaschinen weiter. Zusätzlich wird in der lokalen Datenbank gesucht. Alle Anfragen an Suchmaschinen laufen parallel als Java-Threads ab. Die Suchergebnisse werden nach den logischen Komponenten des Suchausdruckes und nach liefernden Suchmaschinen geordnet und bar jeglicher Werbung aufgelistet. Falls mehrere Suchmaschinen die gleiche URL liefern, werden derzeit alle diese Treffer angezeigt. Zukünftig könnte man solche Mehrfachtreffer als besonders wichtig kennzeichnen. Nach der Suche können die Treffer im Browser begutachtet werden. Falls sich ein Dokument als brauchbar erweist, kann es mit allen wichtigen Informationen in die Datenbank eingetragen werden. Zusätzlich kann man Dokumente markieren, um die Verweise anschließend in ein HTML-Arbeitsdokument aufzunehmen, mit dem man dann in den Unterricht gehen könnte (siehe Abb. 4).

4.4   Die Intelligenzkomponente

Durch jedes in HyperCons eingetragene Dokument wird eine Verknüpfung zwischen einem Themenbereich oder Projekt, einem Unterthema und einem Stichwort definiert. Aus der relativen Häufigkeit einer solchen Verknüpfung können Aussagen über die Bedeutung des speziellen Unterthemas oder Stichwortes im Kontext des jeweiligen Themenbereiches oder Projektes gewonnen werden. Auch das Problem der Eindeutigkeit von Stichwörtern kann damit, sozusagen demokratisch, gelöst werden, indem man Stichwörter unterhalb einer bestimmten relativen Häufigkeit einer besonderen Prüfung unterzieht.

Wir planen einen Agenten auf Java-Basis, der regelmäßig alle Einträge der Datenbank statistisch auswertet und auf der Basis fuzzy-logischer Schlüsse Aussagen über den Bedeutungsgrad von Stichwörtern oder Subthemen erzeugt. Diese Aussagen können dann von der nächsten Version des Frontends verwendet werden, um dem Benutzer Vorschläge zu machen. Dieser Teil von HyperCons befindet sich derzeit in der Designphase. Mit einem Prototyp kann im Frühjahr 1999 gerechnet werden.

4.5   Die Datenpflege

Wegen der bekannten Dynamik des Internet können Verweise auf externe Dokumente sehr schnell veralten. Die Lösung dieses Problems erfolgt durch einen Agenten, der diese Verweise pflegt und auf möglichst intelligente Weise den Verbleib verschollener Dokumente bestimmt.

4.6 Der laufende Prototyp

Alle bisher implementierten Komponenten von HyperCons sind im Rahmen eines Prototypen bereits lauffähig. Informationen zur Installation und Benutzung können unter der URL *http://www-schulen.informatik.tu-muenchen.de* über den Menüpunkt *HyperCons* abgerufen werden. Der Systemstart erfolgt ebenfalls von diesen Seiten aus. Derzeit kann sich jedermann als Gast einloggen. Die Aufnahme neuer Dokumente ist jedoch registrierten Nutzern vorbehalten.

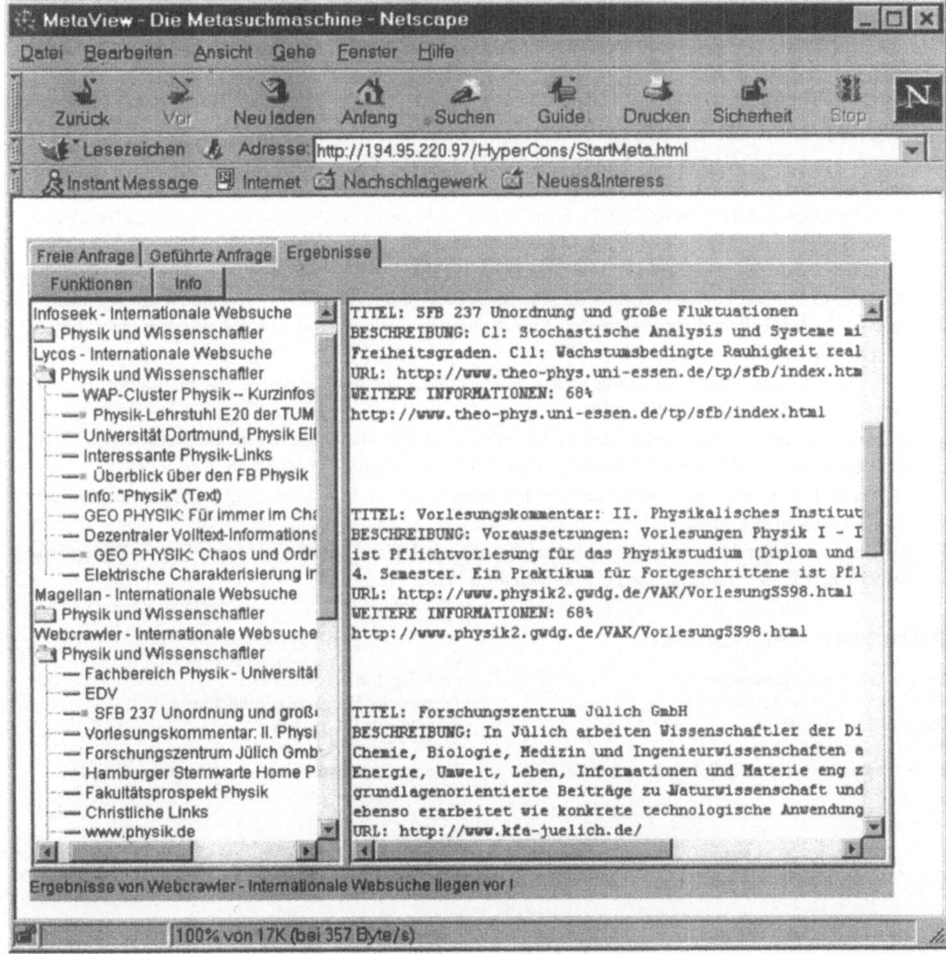

Abb. 4: Die Ergebnisse des Suchvorgangs

5 Ausblick

Aus technischer Sicht erwarten wir im Frühjahr 1999 den nächsten wesentlichen Entwicklungsschritt von HyperCons. Bis dahin werden einige Pflegearbeiten am System durchgeführt, um die Aufnahme erster Daten durch die beteiligten Schulen in Rosenheim voranzutreiben. Wir hoffen, im Sommer 1999 genügend Daten gesammelt zu haben, um die Funktion der Intelligenzkomponente beurteilen und anpassen zu können. Außerdem planen wir möglichst bald eine englische Version der Software, um mit ausländischen Schulen Dokumente austauschen zu können.

6 Literatur

[1] P. Hubwieser, J. Schlichter, B. Mair: HyperCons–eine hypermediale konstruktivistische Lernumgebung. In: C. Herzog C. (Hg.): 8.Arbeitstreffen der GI-Fachgruppe 1.1.5/7.0.1 "Intelligente Lehr-/ Lernsysteme", Duisburg 1997. Institut für Informatik der Technischen Universität München, TUM-I9736, August 1997. 13-23.

[2] P. Hubwieser Peter, J. Schlichter Johann, B. Mair: HyperCons – A Hypertext Learning Environment Based on Constructivism. In: J. Isaac, R. Dangwal, C. Chakraborty (Hg.): Proceedings of the International Conference On Cognitive Systems (ICCS'97). From Intelligent Systems to Cognitive User Interfaces for a Better Wired Society ? (Vol. II.) Allied Publishers Ltd., Delhi, 1997. 846-857.

[3] P. Hubwieser, J. Schlichter: HyperCons: Integrating and structuring school relevant information on the internet. In: L.C. Fulmer, P. Nolan, B.Z. Barta: The Integration of Information for Educational Management. Proc. of the 3rd IFIP working conference on information technology in educational mamangement, July 6-10, 1998 Damariscotta, ME (to appear).

[4] J. Gerstenmaier, H. Mandl: Wissenserwerb unter konstruktivistischer Perspektive. In: Zeitschrift für Pädagogik, 41. Jg., 1995, Heft 6, S. 867 – 885.

[5] B. Mair: Entwurf und Implementierung der serverbasierten Datenhaltung eines internetbasierten Schul-Informationsssystems. Diplomarbeit an der Fakultät für Informatik der TU München, Aug. 1997

[6] C. Weller: Intelligentes Frontend für ein Schulinformationssytem. Diplomarbeit an der Fakultät für Informatik der TU München, März 1998

[7] A. Konkow: Konstruktion einer Metasuchmaschine für ein Internet Schulinformationssystem. Diplomarbeit an der Fakultät für Informatik der TU München, Mai 1998.

Adressen der Autoren

Dr. rer. nat. Peter Hubwieser
Technische Universität München
Fakultät für Informatik
80290 München
Peter.Hubwieser@informatik.tu-muenchen.de

Prof. Dr. Johann Schlichter
Technische Universität München
Fakultät für Informatik
80290 München
Johann.Schlichter@informatik.tu-muenchen.de

A Classification of Evaluation Methods for Intelligent Tutoring Systems

Arif Iqbal*, Reinhard Oppermann**, Ashok Patel*** and Kinshuk**

*University College London UK, **GMD-FIT Germany, ***De Montfort University UK

Abstract

Evaluation of intelligent tutoring systems (ITS) is an important area of research in current educational practices. There are many evaluation methods available but the literature does not suggest any clear guidelines for an evaluator – normally an educator – which methods to use in particular contexts. This paper proposes a classification of evaluation methods to simplify the selection task. The classification is based on two primary questions relating to the target of evaluation and learning environment in which the evaluation would be pursued. The classification is hoped to help in improving quality of computer based education by providing a practical and to the point way of selecting the appropriate evaluation methods for intelligent tutoring systems.

Zusammenfassung

Die Evaluation intelligenter, tutorieller Systeme (ITS) ist ein wichtiges Forschungsfeld im Bereich computer-gestützten Lehrens und Lernens. Bei einer Vielzahl vorhandener Methoden finden sich in der Literatur aber keine klaren Richtlinien für einen Evaluator – üblicherweise ein Lehrender – dafür, welche Methoden in einem bestimmten Umfeld zu nutzen sind. Dieser Beitrag stellt eine Klassifikation der Evaluationsmethoden vor mit dem Ziel, deren Auswahl zu vereinfachen. Die Klassifikation basiert auf Angaben zu zwei grundlegenden Fragestellungen: Zweck und Ziel der Evaluation sowie den jeweiligen Rahmenbedingungen für das Lernen. Es besteht die Hoffnung, daß die Klassifikation hilft, die Qualität computer-gestützten Lehrens dadurch zu verbessern, daß ein praktikables und zielgerichtetes Verfahren zur Auswahl geeigneter Evaluationsmethoden für intelligente tutorielle Systeme bereitgestellt wird.

1   Introduction

Intelligent tutoring systems are increasingly being employed in education ([18], [21]) and as a consequence the need for careful and systematic scrutiny of these systems has become an important issue. The literature provides a number of evaluation methods that can be used to evaluate intelligent tutoring systems, but given the diversity of these methods, it is quite difficult for an evaluator, normally an educator, to judge which method is appropriate for a particular purpose. [39] argued that there are no guidelines available for use of evaluation methods and most of the existing evaluations are empirical endeavours. Although such empirical experiments are necessary for further research, they do not allow a generalisation of the approach, and may fail to consider all possible independent factors leading to inconclusive results and debatable conclusions.

This paper proposes a classification of evaluation methods based on two primary questions an evaluator needs to ask before conducting the evaluation:

i) What is being evaluated: the whole system or just a component?

ii) Is it possible to systematically manipulate variables in the evaluation, and how many users are available for the purpose of evaluation?

The evaluation methods are classified along two dimensions, each relating to one of the above questions. The first dimension focuses on the the *degree of evaluation* covered by the evaluation method. [17] pointed out that if a method solely concentrates on testing a component of a system, it can be considered suitable for *internal evaluation*. If the method evaluates whole system, it is suitable for *external evaluation*.

The second dimension is concerned with the local feasibility of using a particular evaluation method. It distinguishes between a method *establishing a causal association* from a controlled investigation i. e. *experimental research* and the *accumulation of a large amount of data* about a particular aspect of the system i. e. *exploratory research*. *Experimental research* requires experiments varying systematically the independent variable(s) while measuring the dependent variable(s) and ensuring random assignment of participants to conditions and require statistically significant groups. *Exploratory research* includes in-depth study of the system in a natural context using multiple sources of data, usually where sample size is small and the area is poorly understood.

The paper starts by discussing the great necessity for the evaluation of intelligent tutoring systems and then reviews a number of evaluation methods for intelligent tutoring systems suggested by various researchers. This is followed by the description of proposed classification of evaluation methods. The paper concludes by giving an example of the evaluation methods selection using the proposed classification.

2   Ensuring the effectiveness of ITSs

[23] pointed out that evaluation of ITSs has not been taken seriously. [20] noted that most ITS researchers have concerned themselves only with envisioning the potential of ITSs and investigating the implementation issues involved in constructing actual components and systems, and have paid little attention to the process of evaluation. [33] commented that despite the decade long existence of ITSs, the degree to which they have been successful is equivocal solely due to the lack of proper evaluation.

Though the evaluation of intelligent tutoring systems is a costly and time consuming affair, [17] argued that it pays off by helping to answer two questions that are central to cognitive science, artificial intelligence, and evaluation:

i) What is the educational impact of an ITS on students?

ii) What is the relationship between the architecture of an ITS and its behaviour?

Another advantage of evaluation is that it provides an opportunity to learn from the mistakes and is capable of improving the life-span of ITSs as well as their usability. [17] acknowledged that by evaluating PROUST, they learned a lot about how novices learn to program, how to teach programming, and how to build ITSs to actually do the teaching. [33] also expressed the same enthusiasm for evaluation by accepting that they found the results of careful experimental design always informative.

Furthermore, the benefit of carrying out ITS evaluation is to focus the attention away from short-term delivery and open up a dialogue about issues of appropriateness, usability and quality in system design. Evaluation can provide benefits of shared experiences and help in avoiding continuously reinventing the wheel. Researchers have felt the need to develop a more systematic approach to the evaluation of ITS which means ensuring availability of evaluation methods and adequate quality control in the evaluation process. [38] underlined this point by forecasting that *"in a future where [AI-ED systems] may be widely available to schools and training institutes, evaluations will shape what and how people learn, and what they become able to do"*.

Though many researchers have attempted to formalise the evaluation procedures of ITSs from different points of view, these formal methods vary from each other as much as the design and development methodologies of various ITSs vary. We take these formal evaluation methods as the basis of our research. The following section provides summary of these methods. The list of methods is not meant to be exhaustive, but covers a majority of methods discussed in the literature for the evaluation of ITSs.

3   Review of evaluation methods

The evaluation methods for ITSs suggested by various researchers provide a large variety of factors to be considered and the evaluation seems not to be a trivial task, as [20] pointed out: *"There are few agreed upon standards within the ITS community to guide investigators who wish to evaluate systems. However, other fields have developed evaluation methods which may be applicable to ITSs."* The methods discussed in this section are adapted by various ITS researchers from expert systems development, computer-based instruction, education, evaluation, computer science, engineering and psychology disciplines.

<u>Methods for experimental research</u>

The methods in this category require systematically varying the independent variable(s) while measuring dependent variable(s) and ensuring randomly assignment of participants to conditions and require statistically significant groups. As reasoned below, the first four methods are suitable for internal evaluation due to their focus on testing components of the systems rather than overall effectiveness. Methods 6 and 7 focus on overall effectiveness and therefore are good for external evaluation. Method 5 can be used for both component testing or overall effectiveness and hence is suitable for both internal and external evaluations.

1. Proof of correctness is a check whether the system fulfils the desired requirements or goals or whether there is a correspondence between its structure and behaviour and its specifications. The method evaluates the internal components of the systems in rather hypothesis testing fashion and therefore is suitable for internal and experimental nature of evaluation. [20] argued against this method in the case of ITSs due to the inherent AI nature of ITSs which deals with analytically intractable problems represented as incompletely specified functions [28].

2. Additive experimental design comparisons have been proposed to evaluate the impact of *large* ITS components that can be experimentally modified or withheld [24]. Thus they belong to experimental research and have limited suitability only for internal evaluations. These comparisons have been used for ITS component evaluation by [35] for competing tutoring approaches in Pixie tutor and by [3] for the effect of strategy workbooks in West tutor. This approach requires a large number of students and is not cost-effective if the ITS component is minor. The advantage of additive design includes possibility of individual aspects of the ITS to be manipulated in order to directly assess their impact and importance.

3. The diagnostic accuracy of an ITS can be estimated through procedures that address the quality of micro-theories used by the system. [16] contended that the method of diagnostic accuracy has gained more prominence for internal evaluation of ITSs because subsequent pedagogical interactions are dependent on the correct recognition and interpretation of student errors. The method provides a controlled examination of the student model across conditions and therefore belongs to experimental research. This method has been used to evaluate POSIT ITS by [25].

4. The immediate impact of **feedback/ instruction quality** on the student can be experimentally estimated through lag sequential analysis procedures which are suitable for internal evaluation. One way to measure this is to calculate a lag sequential probability, that is the ratio of actions in a specific category to the total number of actions that occur within a specific frame following a target action. The advantage of such a measure is that a calculation can be made for either a different level of detail in ITS feedback or even for different types of feedback. The ALM tutor ([26]) was evaluated for its feedback quality using lag sequential prabability.

5. In **sensitivity analysis**, an examination is made of how varying information to the component or system would result in diverse responses. This experimental approach is equally suitable for components and whole ITS evaluations and hence can be used for both internal and external evaluations. [20] argued that this method is relevant for ITSs since individualised instruction is a primary objective in intelligent tutoring.

6. The **experimental research** enables the researchers to obtain a relationship between a set of interventions and the outcome and as a consequence it is particularly suited to examining the effects of teaching. [20] pointed out a variety of experimental research designs such as single group designs, control group designs, and quasi-experimental designs. Experimental research has been used in ITS evaluation by [14] for Sherlock, [19] for VCR tutor and by [13] for Byzantium ITTs. Experimental research is more appropriate for external evaluations since it is likely to provide overall conclusions rather than acquisition of information [20].

7. [15] examined **product evaluation** as broad-based ITS evaluation studies to identify promising evaluation procedures that could justify continued ITS development. After analysing past ITS evaluation studies (for example, [3], [35]), three guidelines were specified:

i) the instructional effectiveness of ITS applications, human tutors, and traditional methods need to be compared on the basis of performance data;

ii) the instructional effectiveness of only extensive ITS applications should be evaluated; and

iii)large groups of subjects are required to precisely estimate ITS effectiveness.

Since the method manipulates three classic learning conditions, it belongs to experimental research.

Methods for exploratory research

Methods in this category involve an in-depth study of the system in a natural context using multiple sources of data, usually where sample size is small and the area is poorly understood. Similar to the reasoning given in experimental research methods section, methods 8 to 12 in this category are suitable for internal evaluation due to their focus on part evaluation rather than whole system evaluation. Methods 15 to 20 are suitable for external evaluation and methods 13 and 14 can be used for both.

8. An altogether different approach is to make use of an **expert inspection** to assess whether a program meets an explicit standard level of performance. This type of evaluation is often used in the development of knowledge-based systems and the most well-known example is the Turing test which compares human and computer behaviour [27]. Such inspection is useful for evaluating specific ITS components making the approach suitable for internal evaluations. This method can be classified as exploratory since no controlled hypothesis testing takes place.

9. The **level of agreement** between subject matter experts is related to knowledge base varification and can be estimated correlationally and is suitable for internal evaluations. This approach uses subject experts or manuals and correlational methods to estimate the consistency of knowledge and beliefs across experts. Therefore the method falls into exploratory research. The knowledge base of Media Selection Expert System (MSES) was verified using agreement of media specialists by [4].

10. Wizard of Oz experiments use a human to simulate the behaviour of a proposed system, so that one can test aspects of a program design before actually implementing it [36]. This method is more favourable for internal evaluation than external ones. [30] used this method for an intelligent help system of a text editor. The difficulty in predicting the way the human will behave makes it impossible to decide in advance, the way the parameters should be observed. This method therefore belongs to exploratory research.

11. Performance metrics are methods which use quantitative results to explore individual factors or features of an ITS and thus fall into the category of internal evaluations. These are important in exploratory settings when data pools are too small to allow statistically significant conclusions. But such methods of measuring system performance include false positive and false negative results. These methods encompass three sub-categories [22]: *diagnostic accuracy*, where measures of the success of ITS components are assessed by recording 'hit and miss rates' of students; *feature usage*, which assesses how users utilise various features of tutor; and *cognitive change*, which can describe differences across populations or time periods, as well as absolute metrics. [34] and [12] evaluated various ITS components using performance metrics methods.

12. [22] pointed to a category of **internal evaluation** methods which study the relationships between an ITS architecture and its behaviour, and often involve in-depth exploratory analysis of program traces and data structures. This method includes following sub-categories: *knowledge level analysis* to evaluate knowledge base, for example, [5] evaluated MYCIN to make NEOMYCIN for intelligent tutoring; *process analysis* for evaluation of algorithms, for example, [17] analysed the limitations of PROUST system; and *ablation & substitution experiments* ([6]) to observe system performance by replacing parts of the system with more primitive versions, for example, [7] analysed LISP tutor without its feedback capabilities.

13. A similar method is **criterion-based** evaluation, where the requirements and specifications are drawn up almost as a checklist which are then used in an exploratory way to compare the systems inadequacies. This method can be used to examine either the components of the ITSs or the whole system and hence is suitable for both internal and external evaluations. This method has been used by [9] for ITS evaluation.

14. The **pilot testing** method is used to check the system design for unanticipated outcomes of using the system with users. The method is in essence an exploratory tool for identifying problems in the system. Three types of pilot tests are identified ([10]) as providing opportunities to learn about different parts of the development process both at component and whole system level. *One-to-one* pilot tests are useful in the early stages of development. They have been used by [29] to evaluate 'Guidon-watch' knowledge-base interface. *Small group testing* is conducted later in the development with a representative group of students once the system has begun to stabilise and finally *field testing* takes place towards the end when completion is only a short period away.

15. Another approach to ITS evaluation is **certification**, based on methods currently applied in identifying competent human teachers. Such a procedure has the benefits of an authoritative endorsement of the correctness of the ITS and due to this reason it falls under exploratory research. The suitability of the approach for overall program makes it adequate for external evaluations. [20] favoured this method as a way of obtaining feedback on the strengths and weaknesses of ITSs in formative evaluation and of obtaining adequacy ratings in summative evaluation.

16. The opinions of experts or a large number of (potential) users of a system, known as **outside assessment**, can be significant if there is agreement. [22] argued that there are two types of such assessments: *on-site expert evaluation* and *panel of experts*. The former is the normal way of getting experts to observe and assess the behaviour of system, for example several teachers rating an ITS effectiveness with some students using it. The latter relies on a selected group of experts being questioned by convening them or through correspondence. Since the method operates on overall ITS and is based upon an exploration of users' opinions, it is suitable for exploratory external evaluations.

17. Existence proofs method bases its conclusions on the successful implementation of a system or application of a method to propose a new system architecture or design method. Since the method applies on whole system, it is suitable for external evaluations. This method is only powerful if the researcher describes the goals and assumptions of the study, discusses design trade-offs, and documents surprises and failures. It is useful in exploratory research in beginning to identify key issues and providing a baseline from which further studies can be designed [22]. This method has been used by [2] and [1] showed by different means that various ITSs were possible.

18. The intention of **observation and the qualitative classification of phenomena is to identify classes of phenomena**, patterns, and trends in the interaction of people with instructional systems. This data is gathered for exploratory research primarily through observation, anywhere from natural to contrived situations as part of external evaluations. [22] stated that *"The observation can be of novices (learners) or experts (teachers), and can be aimed at identifying behavioural phenomena or at inferring cognitive phenomena"*.

19. The approach of focusing on the collection of data by limiting or organising the responses of the subject under exploratory research is collectively known as **structured tasks and the quantitative classification of phenomena**. Here quantitative data is gathered in contrived or structured situations in external evaluations, for example by interviews, questionnaires and surveys. The benefit of this approach is that more voluminous and precise data can be obtained by using more structured tasks. The method has been used by [31] in a study of Anderson's LISP tutor.

20. Comparison studies are a general class of methods which note similarities and differences between the behaviour or design of an ITS with a standard or even another system. These methods are suitable for external evaluations and apply to overall system performance in exploratory research. [22] suggested three methods to achieve such comparisons: judging system performance against a well-known successful case (i.e. gold standard), for example, [9] evaluated Highway code tutor against [32]'s definition of "intelligence"; arguing at a theoretical level for the generality and extendibility of a new approach to other systems (i.e. theoretical corroboration), for example, [8] compared PRO-TEG system with six other ITSs using a three dimensional ITS classification system; and, simulating the behaviour of another system (i.e. empirical corroboration and duplication).

4 Proposed classification of evaluation methods

Given the variety of evaluation methods, it is difficult to decide which one is appropriate in a particular context. From an examination of the evaluation requirements two important questions emerge which confront the educator who is going to employ the ITS and hence evaluating it:

(i)   What is being evaluated - the whole system or simply a part of the system?

(ii)   Are there enough students and other suitable conditions available to conduct evaluation based on an experimental design?

These two questions can be used to classify various methods, so that a method could be differentiated from a number of others on a scale between external evaluation (considering the whole system) and internal evaluation (testing a component of the system). In addition, a method could be classified along a dimension consisting of exploratory research versus experimental research. In the case of exploratory research such methods should be applicable where samples are small and the area is poorly understood. Evaluation methods which are categorised as experimental are likely to involve manipulation of variables and require statistically significant groups. Based on the illustration of each method in the section 3 of this paper, figure 1 classifies the methods relative to these two dimensions.

The methods falling towards lower left part of the chart are suitable for evaluation of ITS components and do not require large sample sizes or rigorous statistical studies. Methods which require control group studies and statistical procedures for component evaluation are grouped towards the lower right part of the chart. Upper part of the chart shows methods which evaluate overall system performance. The methods at the right side require statistical studies whereas the methods at left side do not require such studies.

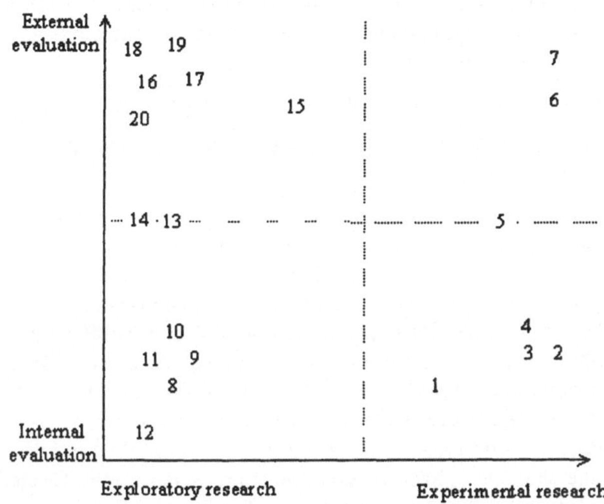

Key to evaluation methods:

1. Proof of Correctness
3. Diagnostic accuracy
5. Sensitivity Analysis
7. Product evaluation
9. Level of agreement
11. Performance metrics
13. Criterion-based
15. Certification
17. Existence proofs
19. Structured tasks & quantitative classification

2. Additive experimental design
4. Feedback/instruction quality
6. Experimental research
8. Expert knowledge
10. Wizard of Oz experiment
12. Internal evaluation
14. Pilot testing
16. Outside assessment
18. Observation & qualitative classification
20. Comparison studies

Figure 1: Classification chart of evaluation methods

The following section gives an example of using the classification chart for selection of methods appropriate to the local requirements and conditions at hand.

5   Selecting evaluation methods by classification chart – An example

An evaluation was carried out on a fully operational intelligent tutoring system [37], aimed to teach marginal costing subject in cost engineering/ accounting domain, to assess its value and effectiveness with target users (i.e. students). The evaluation was conducted at three universities in United Kingdom. The results of evaluation are compiled in [11].

The research design for evaluation was selected with reference to the aims of the study, that is firstly a comparison of how effective computer aided instruction was in relation to conventional teaching and secondly a determination of the features significant in assessing a teaching system. The classification chart was used to select the methods used for evaluation. At first the two evaluation requirements were examined:

(i)   What is being evaluated - the whole system or simply a part of the system?

(ii)   Are there enough students and other suitable conditions available to conduct evaluation based on an experimental design?

Since the study aimed to evaluate the whole system and not just a part of the system it was classified as an external evaluation. However, the other question provided a mixed reply. The comparison of instruction techniques demanded testing of hypothesis to evaluate the question: which type of instruction is "better". This led to the conclusion that an experimental design should be used. On the other hand the requirement to collect a multifaceted assessment from the users of the computer aided instruction suggested an exploratory evaluation. As a consequence two evaluation methods were chosen, one each from exploratory and experimental techniques but both suitable for external evaluation. A structured questionnaire was used simulating *Structured tasks and the quantitative classification of phenomena* (labelled as 17 in the chart) to collect information from users for multifaceted assessment and then a before-after two-group experiment was used for *Product Evaluation* (labelled as 12 in the chart) to determine the differences between two instruction methods.

6   Conclusion

The usefulness of the proposed classification of evaluation methods for ITSs is in the fact that it enables the evaluators to select the most appropriate evaluation methods for their investigation. The evaluators have to decide whether the investigation will primarily involve the whole system or just a few components. Then they have to judge whether it is possible to conduct the evaluation in an experimental style. At this stage, it should be clear from which region of the chart the method has to be picked. The final decision is dependent on practical factors faced by the evaluator and the quality of information being sought, and ultimately the chosen method is characteristic of those individual circumstances.

While the proposed classification provides a simple yet robust way to select evaluation methods, the classification requires a fully developed ITS available in hand. The ITS field in fact, has very few fully developed products in market. Most of the systems are still at their prototyping stages, confined in the research laboratories. To get most out of the research, future work is planned to add another dimension of formative vs. summative evaluation to the classification chart. This dimension will facilitate the educators to formatively evaluate the prototypes for adequacy to their curriculum and cohorts while keeping the classification open for summative evaluations of any fully developed products.

7   Acknowledgement

The research presented in this paper is partially supported by Science and Engineering
Research Council (UK) studentship awarded to the first author.

8   References

[1] Buchanan B.: Artificial intelligence as an experimental science. Stanford University Knowledge Systems Lab
    Technical Report KSL 87-03, 1987.

[2] Burke R. & Funaro G. M.: Case-based environments for learning. Working Notes of the AAAI Spring
    Symposium on Knowledge-Based Environments for Learning and Teaching, March, 1990, Stanford, CA, 63-
    67.

[3] Center for the Study of Evaluation: Intelligent computer aided instruction (ICAI): Formative evaluation of
    two systems (ARI RN 86-29). Alexandria, VA: U. S. Army Research Institute, 1986.

[4] Chao P. C. & Legree P. J.: The Media Selection Expert System knowledge base verification and performance
    evaluation analyses. ARI Technical Report, U. S. Army Research Institute, VA, 1991.

[5] Clancey W. J.: Tutoring rules for guiding a case-method dialogue. In: Sleeman D. & Brown J. S. (eds.)
    Intelligent Tutoring Systems, New York, 1988: Academic Press, 201-225.

[6] Cohen P. & Howe A.: How evaluation guides AI research. AI Magazine, Winter (1988).

[7] Corbett A. T. & Anderson J. R.: The effect of feedback control on learning to program with the LISP tutor.
    Proceedings of the Twelfth Annual Conference of the Cognitive Science Society, July, 1990, Cambridge,
    MA, 796-806.

[8] Dillenbourg P.: A model of knowledge acquisition by intelligent tutoring systems. Proceedings of ITS-88,
    1988, Montreal, Canada, 145-153.

[9] Ford L.: The appraisal of an ICAI system. In: Self J. (ed.) Artificial Intelligence and Human Learning:
    Intelligent Computer-aided Instruction, London, 1988: Chapman & Hall.

[10] Gagne R. M., Briggs L. J. & Wager W. W.: Principles of Instructional Design, New York, 1988: Holt,
    Rinehart and Winston.

[11] Iqbal A. M.: Evaluation of computer aided instruction: Assessing the value and effectiveness of operational
    systems. PhD thesis submitted for assessment, University of London, UK, 1997.

[12] Kimball R.: A self-improving tutor for symbolic integration. In: Sleeman D. & Brown J. S. (eds.) Intelligent
    Tutoring Systems, New York, 1988: Academic Press, 283-307.

[13] Kinshuk: Computer aided learning for entry level Accountancy students. PhD Thesis, De Montfort
    University, England, July, 1996.

[14] Lajoie S. P. & Lesgold A. M.: The SHERLOCK experience: An evaluation of a computer-based supported
    practice environment for electronics trouble-shooting training. Proceedings of the International Conference
    for Cognitive Science for the Development of Organizations, May 2-4, 1991, Montreal, 56-62.

[15] Legree P. J. & Gillis P. D.: Product effectiveness evaluation criteria for intelligent tutoring systems. Journal
    of Computer-Based Instruction, 18 (2) (1991), 57-62.

[16] Legree P. J., Gillis P. D. & Orey M. A.: The quantitative evaluation of intelligent tutoring system
    applications: Product and process criteria. Journal of Artificial Intelligence and Education, 4 (2/3) (1993),
    p209-226.

[17] Littman D. & Soloway E.: Evaluating ITSs: The cognitive science perspective. In: Polson M. C. &
    Richardson J. J. (eds.) Foundations of Intelligent Tutoring Systems, New Jersey, 1988: Lawrence Erlbaum
    Associates, 209-242.

[18] Major N., Ainsworth S. & Wood D.: REDEEM: Exploiting sumbiosis between psychology and authoring
    environments. International Journal of Artificial Intelligence in Education, to appear.

[19] Mark M. A. & Greer J. E.: The VCR Tutor: Evaluating instructional effectiveness. Proceedings of the 13th Annual Conference of the Cognitive Society, 1991, 564-569.

[20] Mark M. A. & Greer J. E.: Evaluation methodologies for intelligent tutoring systems. Journal of Artificial Intelligence and Education, 4 (2/3) (1993), 129-153.

[21] Mizoguchi R., Sinitsa K. & Ikeda M.: Task ontology design for intelligent educational/training systems. Position paper for ITS'96 Workshop on Architectures and Methods for Designing Cost-Effective and Reusable ITSs, June 10th 1996, Montreal.

[22] Murray T.: Formative qualitative evaluation for `exploratory' ITS research. Journal of Artificial Intelligence and Education, 4 (2/3) (1993), 179-207.

[23] Nwana H. S.: Mathematical Intelligent Learning Environments, Oxford, 1993: Intellect Books.

[24] O'Neil H. & Baker E.: Issues in intelligent computer aided instruction: Evaluation and measurement. In: Conoley J. C. (ed.) The Computer as Adjunct to the Decision Making Process, Hillsdale, NJ, 1987: Lea/Wiley.

[25] Orey M. A. & Burton J. K.: POSIT: Process oriented subtraction – interface for tutoring. Journal of Artificial Intelligence in Education, 1(2) (1989/90), 77-104.

[26] Orey M. A., Park J. S., Chanlin L. J., Jih H., Gillis P. D., Legree P. J. & Sanders M. G.: High bandwidth diagnosis within the framework of a microcomputer-based intelligent tutoring system. Journal of Artificial Intelligence in Education, 3(1) (1992), 63-80.

[27] Parry J. D. & Hofmeister A. M.: The development and validation of an expert system for special educators. Learning Disability Quarterly, 9(2) (1986), 124-132.

[28] Partridge D.: Artificial Intelligence: Applications in the future of software engineering, New York, 1986: Ellis Horwood.

[29] Richer M. H. & Clancey W. J.: Guidon-watch: A graphic interface for viewing a knowledge-based system. In: Lawler R. W. & Yazdani M. (eds.) Artificial Intelligence and Education: Volume one, Learning Environments and Tutoring Systems, Norwood, NJ, 1987: Ablex.

[30] Sandberg J., Breuker J. & Winkels R.: Research on HELP-Systems: Empirical study and model construction. Submitted for ECAI, Munchen, 1988.

[31] Schofield J. & Verban D.: Barriers and incentives to computer usage in teaching. LRDC Technical Report No. 1, 1988.

[32] Self J.: Intelligent computer assisted instruction. Unpublished paper presented at the ICAI Spring Seminar, 1985, Cambridge.

[33] Shute V. J. & Regian J. W.: Principles for evaluating intelligent tutoring systems. Journal of Artificial Intelligence and Education, 4 (2/3) (1993), 245-271.

[34] Sleeman D.: Assessing aspects of competence in basic algebra. In: Sleeman D. & Brown J. S. (eds.) Intelligent Tutoring Systems, New York, 1988: Academic Press, 185-199.

[35] Sleeman D., Kelly A. E., Martinak R., Ward R. D. & Moore J. L.: Studies in the diagnosis and remediation of high school algebra students. Cognitive Science, 13 (1989), 551-568.

[36] Twidale M.: Redressing the balance: The advantages of informal evaluation techniques for intelligent learning environments. Journal of Artificial Intelligence and Education, 4 (2/3) (1993), 155-178.

[37] Wilkinson-Riddle G. J. & Patel A.: "Human Tutor" Emulation in Teaching Numerical Business Subjects - a Software Breakthrough. Proceedings of the tenth International Conference on Technology and Evaluation, March 21-24, 1993, 174-176.

[38] Winne P. H.: A landscape of issues in evaluating adaptive learning systems. Journal of Artificial Intelligence and Education, 4 (4) (1993), 309-332.

[39] Wu A. K. W.: On the formal evaluation of learning systems. Lecture Notes in Computer Science, 1086 (1996), 324-332.

Authors' addresses

Mr. Arif Iqbal
42, New Rowley Road
Dudley, West Midlands DY2 8AS
United Kingdom
Tel/Fax: (+44) 1384 459 129

Prof. Dr. R. Oppermann
GMD-FIT, German National Research Center for
    Information Technology
Schloss Birlinghoven, 53754 St. Augustin
Germany
Email: oppermann@gmd.de

Mr. A Patel
CAL Research & Software Engineering Centre,
    Bosworth House, De Montfort University
The Gateway, Leicester LE1 9BH
United Kingdom
Email: apatel@dmu.ac.uk

Dr. Kinshuk
GMD-FIT, German National Research Center for
    Information Technology
Schloss Birlinghoven, 53754 St. Augustin
Germany
Email: kinshuk@ieee.org

Gemeinsame Anpassung von Einzelplatzanwendungen

Helge Kahler, Oliver Stiemerling, Volker Wulf

ProSEC – Institut für Informatik III, Universität Bonn

Jörg-Guido Hoepfner

SAP AG

Zusammenfassung

Einzelplatzanwendungen wie z.B. Textverarbeitungsprogramme werden oft kooperativ angepaßt. Schwierige Anpassungen werden dabei von erfahrenen Benutzern erstellt und dann an deren Kollegen verteilt. Trotzdem sehen Einzelplatzanwendungen keine Unterstützung dieser kooperativen Anpassungen vor. Um Gestaltungsanforderungen für ein solches Tool zu generieren, haben wir eine qualitative empirische Studie bezüglich der kooperativen Anpassungsgewohnheiten von MS Word Benutzern durchgeführt. Die Ergebnisse werden anhand von vier Szenarien dargestellt. Auf diesen Szenarien basierend haben wir ein Tool entwickelt, welches sowohl eine öffentliche und private Ablage für Anpassungen, als auch ein direktes Verschicken derselben unterstützt. Das Tool ist vollständig in die Arbeitsoberfläche des Textverarbeitungsprogrammes integriert. Ergebnisse des Usability Testings ergaben, daß die Benutzer unabhängig vom Grad ihrer Qualifikation in der Lage waren das Tool anzuwenden.

Abstract

Generic single user applications are often tailored cooperatively. Complex adaptations are carried out by power users and distributed to their colleagues. Nevertheless generic single user applications do not provide groupware support to share adaptations among its users. To analyze design requirements for such a tool we have carried out a qualitative empirical study on the cooperative tailoring habits of users of MS Word. The results are presented by means of four scenarios. Based on these scenarios, we have developed a tool which provides a public and a private repository for tailored artifacts as well as a mailing function. The tool is fully integrated into the user interface of the word processor. Results of a usability test indicate that users of different levels of qualification are able to handle the tool.

1 Einleitung

Anpaßbarkeit wird im Hinblick auf die Dynamik der Einsatzumgebungen von Software als deren Schlüsseleigenschaft angesehen. Da sich das Verhalten der Nutzer ändern kann, Aufgaben variieren und Organisationen sich weiterentwickeln können (Oppermann [1]) muß Software flexibel, sprich *anpaßbar* gestaltet werden. Dies gilt für Einzelplatz- wie auch für Groupware-Systeme, wobei letztere in der Regel einer stärkeren Dynamik unterliegen. Allerdings konzentriert sich die Softwareentwicklung bzgl. Anpaßbarkeit fast ausschließlich auf den einzelnen Endbenutzer, auch in Groupware-Systemen. Obwohl es viele Forschungsarbeiten und zahlreiche empirische Untersuchungen zu Anpaßbarkeit und deren Prozessen gibt, die ausdrücklich unter anderem auf einen kooperativen Anpassungsprozeß hinweisen, ist dies praktisch noch nicht umgesetzt worden, da die Beiträge innerhalb der Literatur in erster Linie von beobachtender Natur sind:

So beschreibt Mackay [2], wie Benutzer verschiedener Qualifikationsgrade Anpassungsdateien austauschen. Während es einen hohen Qualifikationsgrad erfordert, Anpassungen selber vorzunehmen, ist es dagegen recht leicht, diese in Form von Dateien von einem Kollegen zu kopieren und dann im eigenen System zu benutzen. Sie beschreibt dabei verschiedene „Muster der Verteilung" ("patterns of sharing") solcher Anpassungen in realen Anwendungsfeldern.

Nardi [3] stellt die Ergebnisse zweier Feldstudien dar, die sich mit kooperativer Anpassung von Spreadsheet- und CAD-Benutzern beschäftigen. Sie sieht kooperative Anpassung als eine natürliche Konsequenz der Arbeitsteilung und betont, daß dieser Aspekt der Anpassung bei der Gestaltung von Softwaresystemen berücksichtigt werden muß.

Oppermann [4] betont, daß Anpassung und Nutzung der EDV-Systeme nicht nur individuell, sondern vielfach in einem kooperativen Arbeitszusammenhang durchgeführt werden. Aus diesen kooperativen Erfordernissen, die schon im Designprozeß von Software vorzusehen und im System zu integrieren sind, leitet Oppermann ein weiteres Ziel für alle Adaptierungsmöglichkeiten ab: nämlich auf die kooperativen Nutzungsmöglichkeiten der Benutzer im Zuge der Softwarehandhabung und der Aufgabendurchführbarkeit Rücksicht zu nehmen und Werkzeuge bereitzustellen, die den kooperativen Charakter von Adaptionsprozessen zu unterstützen vermögen. Außerdem sollten individuelle und (teil-) gruppenbezogene Anpassungsstandards verwalt- und austauschbar sein.

Ebenso geht Paetau [5] bei Systemanpassungen von einem kooperativen Prozeß aus. Auch er fordert, daß Anpassungswerkzeuge dazu beitragen müssen, die Kooperation dieses Prozesses zu unterstützen.
Andere Arbeiten untersuchen kooperative Anpassung in organisatorischen Rahmen. Carter und Henderson [6] beispielsweise postulieren die Notwendigkeit einer „Kultur der Anpassung" ("tailoring culture") innerhalb einer Organisation. Da Anpassungen des technischen Systems auch die Art und Weise verändern, wie Individuen und Gruppen miteinander arbeiten, sollte eine Kultur geschaffen werden muß, in der technische und organisatorische Veränderungen etwas sind, an dem jeder teilhaben und dazu beitragen kann.

Paetau [7] diskutiert auch den negativen Effekt den unkoordinierte, individuelle Anpassungen auf eine Organisation haben. Beispielsweise sind höchst individualisierte Systeme schwer zu unterstützen/unterhalten, schwer von Vertretern (z.B. bei Krankheit) zu benutzen und für das Management nicht transparent. Des weiteren kann Individualisierung sogar einen Punkt erreichen, an dem Kooperation tatsächlich behindert wird, und zwar durch Inkompatibilitäten individueller Anteile bei gemeinsamer Arbeit (z.B. Formulare). In der Tat fanden Trigg und Bødker [8] eine aufkommende Systematisierung von kooperativen Anpassungsversuchen in einer Regierungsfiliale. In ihrer Studie untersuchten sie ebenfalls die Anpassung von Textverarbeitungen.

Einer der wenigen Beiträge, die nicht nur beobachten und analysieren, sondern auch kooperative Anpassung bei der Implementation von Software Systemen berücksichtigen, stammt von MacLean et al. [9]. Die Autoren beschreiben das „Buttons System", dessen wesentlicher Bestandteil Schaltflächenartige Objekte (*buttons*) sind. Diese Objekte sind so gestaltet, daß sie per E-Mail verschickt werden können. Folglich können erfahrenere Benutzer, die z.B. den LISP Code eines Button anpassen, diese Anpassungen mit ihren Kollegen teilen. Obwohl das „Buttons" System tatsächlich auch in den nicht-akademischen Teilen des Forschungsinstitutes genutzt wurde, war es auf die Xerox InterLISP Umgebung begrenzt und wurde darum nicht für Benutzer in anderen Organisationen zugänglich.

Zusammengefaßt ergibt sich, daß

- es sich schlüssig gezeigt hat, daß Anpassungen oft kooperativ vorgenommen werden,
- es fast keine Arbeiten gibt, die dieser Tatsache bei der Implementierung von Softwaresystemen Rechnung getragen haben, insbesondere in Anwendungsfeldern außerhalb von Forschungsinstituten.

Deshalb untersuchen wir in unserer Arbeit die Frage, wie kooperative Anpassung bei realen Anwendungen durch technische Mechanismen unterstützt werden kann. Wir haben Microsoft Word als ein Beispiel für ein weit verbreitetes und oft genutztes Produkt benutzt. Als ersten Schritt nahmen wir eine Feldstudie in vier verschieden Anwendungsbereichen vor, mit dem Ziel, mehr darüber zu erfahren, wie Gruppen von Benutzern in ihrer täglichen Arbeit insbesondere kooperativ anpassen und diese Anpassungsdaten benutzen. Das Ergebnis der Studie sind vier typische Szenarien kooperativer Anpassungen, die sich auf den Austausch von Dokumentvorlagen und Symbolleisten für das Textverarbeitung beziehen (Abschnitt 2). Ausgehend von software-ergonomisch allgemeingültigen Anforderungen aus der Literatur und den Ergebnissen diesen Szenarien haben wir einen Prototypen entwickelt, der komplett in die Textverarbeitung MS WORD integriert wurde und das kooperative Anpassen unterstützt. Die Implementierung des Prototypen I wird im Abschnitt 3 beschrieben. Die durch die Feldstudie gewonnenen Szenarien dienen außerdem als Basis für die Evaluierung des Prototypen, beschrieben in Abschnitt 4. Die Ergebnisse der Evaluierung werden momentan in Prototyp II implementiert. Der Beitrag schließt mit einem Ausblick auf zukünftige Arbeit.

2   Empirische Vorstudie

Obwohl es zahlreiche quantitative Studien zur Untersuchung von Anpassungsgewohnheiten in Textverarbeitungssystemen gibt (z.B. Page et al. [10], Mackay [11]), insbesondere bzgl. der Barrieren und Auslösern derselben, haben diese dabei auftretende kooperative Aspekte vernachlässigt. Um gerade in diesem Zusammenhang Gestaltungshinweise für den Prototyp zu gewinnen, führten wir eine qualitative Feldstudie mit Benutzern von MS Word durch. Insgesamt fanden 12 semistrukturierte Interviews mit Benutzern aus vier unterschiedlichen Anwendungsfeldern statt (Öffentliche Verwaltung, Privatfirmen, Forschungsinstitute und Heimbenutzer). Die Interviews bestanden aus drei Teilen. Zuerst wurden die Gesprächspartner nach ihrer Position und ihren Arbeitsaufgaben befragt. Im zweiten Teil wurde nach konkretem Wissen bzgl. Anpassungen und deren Durchführung in MS WORD befragt. Zuletzt wurde mittels Ausdrucken ein „technisches" Tool vorgestellt, das die Interviewten bewerten sollten und Ideen und Wünsche einbringen konnten.

Abhängig von ihrem Anwendungsfeld berichteten die Benutzer über Unterschiede im Umfang und der Art und Weise wie Anpassung als eine kooperative Aktivität gesehen wird. Um einen Eindruck dieser Vielfalt zu geben und die Motivation der in Abschnitt 3 vorgestellten Implementierung zu zeigen, werden wir vier typische Szenarien kooperativer Anpassung beschreiben, die wir aus den Interviews extrahiert haben.

2.1   Szenario I: Erfahrungsaustausch unter isolierten Heimbenutzern

Die Heimbenutzer unter den Interviewten waren zwei Studenten, die MS WORD neben ihrem normalen Schriftverkehr vor allem für Seminararbeiten benutzten. Die Vorgaben der Fakultäten für diese ca. 30 Seiten langen Arbeiten sind recht umfangreich, so daß die Studenten gezwungen sind, Formatvorlagen zu entwerfen, oder sich diese von Komilitonen zu besorgen.

Die Studenten berichteten über ziemlich wenige kooperative Anpassungsaktivitäten aufgrund ihrer individuellen Arbeitsaufgaben. Trotzdem beschrieben beide eine in etwa gleiche „Anpassungssituation". Gelegentlich, wenn einer der beiden Studenten andere Komilitonen am PC arbeiten sah die für ihn unbekannte Anpassungen vornahmen oder solche benutzten, fragte er nach, wie diese konstruiert wurden. Nachdem er eine Demonstration erhalten hatte, ging er nach Hause und versuchte auf seinem eigenen System diese zu wiederholen bzw. für seine Aufgaben in abgewandelter Form wieder zu verwenden.

2.2 Szenario II: Zentrale Ablage für standardisierte Formulare

Ferner interviewten wir zwei Systemadministratoren und zwei Forscher eines großen Informartik-Forschungsinstitutes in der Nähe von Bonn. Die Systemadministratoren waren für den Support der Unix und PC Umgebung in einer der Abteilungen des Forschungsinstitutes verantwortlich. Die Forscher waren Beschäftigte der selben Abteilung und arbeiteten in zwei verschiedenen Forschungsgruppen.

Die Interviewten berichteten über wenig kooperative Anpassungsaktivitäten ihrer Kollegen, da innerhalb dieser Forschungseinrichtung ein heterogenes Spektrum von Textverarbeitungen auf verschiedenen Plattformen genutzt wird und zudem die Aufgaben von individualisierter Natur sind. Außerdem sind die meisten Mitarbeiter sehr erfahren Systembenutzer, die alle selber Anpassungen vornehmen können, was auch ein Grund für die wenigen kooperativen Adaptierungen ist. Trotzdem benutzt die Organisation ein Intranet, um bestimmte Dokumentvorlagen in einer standardisierten Form bereitzustellen. Die Mitglieder der Organisation finden Dokumentvorlagen für verschiedene administrative Zwecke auf einem dieser Intranet-Server (z.B. Bestell- und Rechnungsformulare). Diese Vorlagen werden von einer zentralen Verwaltungsabteilung, die erst kürzlich ins Leben gerufen wurde, erstellt und auf den neusten Stand gebracht. Somit können alle Benutzer diese Formulare einfach kopieren und benutzen. Ideen für neue Formulare werden der Verwaltungsabteilung vorgeschlagen, die diese dann im Netz beteitstellt.

2.3 Szenario III: Kooperative Anpassung und organisationsweite Verteilung

Vier der Interviewten arbeiteten für die Vertretung eines norddeutschen Bundeslandes in Bonn. Ungefähr drei Jahre zuvor war die Organisation mit Arbeitsplatzrechnern ausgestattet worden. Zwei der Interviewten leiteten Abteilungen, die für die Vertretung der Interessen ihres Bundeslandes in Angelegenheiten der föderalen Gesetzgebung verantwortlich waren. Die zwei anderen arbeiteten in der Verwaltung der Landesvertretung. Einer von ihnen war für den Systemsupport für die anderen Benutzer verantwortlich.

Jeder der vier berichtete von einem recht intensiven Austausch von Anpassungen. Eine der Beschäftigten der Verwaltung erzählte, wie sie eine Dokumentvorlage zusammen mit einem Kollegen erstellt hatte. Jeder führte einen Teil der Arbeit aus. Dann speicherte sie ihren Teil auf Diskette und brachte diese zu ihrem Kollegen, der die Teile dann zusammenfügte.

In der Landesvertretung gibt es keine formale Prozedur dafür, wie mit häufig genutzten Vorlagen umzugehen ist. Einer der Beschäftigten berichtete, daß eine diesbezügliche Einigung oft eine schwierige Aufgabe darstellt. Momentan werden die Vorlagen ausgedruckt und von Beschäftigtem zu Beschäftigtem weitergereicht. Jeder kann dabei den Ausdruck kommentieren und Vorschläge einbringen. Der Interviewte, der letztendlich für die Erstellung der Dokumentvorlagen verantwortlich ist, ist oft durch widersprüchliche Anforderungen und Wünsche überfordert das endgültige Layout zu definieren. In den Fällen, in denen er nicht alle Anforderungen erfüllen kann, empfiehlt er seinen Kollegen sich selber individuelle Versionen der Vorlage zu erstellen. Folglich ist der Entstehungsprozeß von Dokumentvorlagen ziemlich unstrukturiert. Die Landesvertretung nutzt das vorhandene Groupwaresystem, dessen Funktionalität ein gemeinsamen Arbeitsbereich zum Austausch von Dokumenten vorsieht. Momentan wird allerdings ein spezieller Arbeitsbereich benutzt, der manuell erzeugt und gewartet wird, um neu erstellte Dokumentvorlagen auf diesem Wege in der gesamten Organisation zu veröffentlichen. Einfache Benutzer haben innerhalb dieses Arbeitsbereichs nur das Recht Dokumente zu lesen. Da einige der Beschäftigten nur über ein sehr rudimentären Computerwissen verfügen, hat nur der Systemadministrator das Recht, Vorlagen zu verändern oder hinzuzufügen. Demzufolge werden die Vorlagen eher als eine kollektive Ressource angesehen und weniger als ein Mittel, um die gemeinsame Verwaltungsstandards zu bestimmten.

2.4 Szenario IV: Verteilte Dokumentvorlagen und Benachrichtigung von Benutzern

Ein erfahrener Benutzer, der in der Marketingabteilung eines Fahrzeugherstellers arbeitete, beschrieb, wie er Dokumentenvorlagen als abteilungsweite Normen erstellt und verteilt hatte. Vorher hatte jeder in „seiner" Abteilung eigene Vorlagen erstellt. Er begann, das Layout der Dokumente zu standardisieren, indem er eine erste Version einer Vorlage entwarf, die dann mit seinen Kollegen inhaltlich abgestimmt wurde. Als letztes zeichnetet der Abteilungsleiter diesen neuen Standard gegen, bevor die Vorlage im LAN abgelegt wurde. Die meisten Kollegen hatten nur Leseberechtigung auf diese händisch eingerichtete und gepflegte Ablage, da aufgrund ihrer mangelnden Fähigkeiten zu befürchten war, daß versehentlich Veränderungen vornehmen würden (z.B. durch Überschreiben). Immer, wenn eine Dokumentenvorlage abgelegt wurde, informierte er anschließend seine Kollegen via Telefon.

2.5 Gestaltungsvoraussetzungen für Groupwareunterstützung

Am Ende der Interviews diskutierten wir mögliche Gestaltungsformen der Groupwareunterstützung für kooperative Anpassungsaktivitäten mit den Interviewten. Bezogen darauf, ob ein integriertes Tool zur Verteilung von Anpassungen wirklich gebraucht wird, waren die Reaktionen eher geteilt. Sie reichten von „Ausgezeichnet, auf so ein Tool warte ich schon lange!" (erfahrener Benutzer der Privatfirma), bis zu „Völliger Unsinn, das würde ich nie benutzen!" (Benutzer des Forschungsinstitutes). Insgesamt bewerteten die Interviewten ein solches Tool positiv.

Nach Evaluierung der Szenarien und Zusammenfassung der Resultate der abschließenden Diskussion mit den Interviewten, ergaben sich die folgenden Gestaltungsanforderungen:

- enge Integration in die Textverarbeitung, von wo aus Anpassungen auch generiert und geladen werden können,

- zusätzliche Adaptierungsmöglichkeiten, wie z.B. Symbolleisten zu speichern und zu laden oder Verknüpfung verschiedener Anpassungen ,

- ein zentrales Verwaltungstool mit:

  - einer privaten Ablage für die individuelle Anpassungen,

  - einer öffentlichen Ablage, um über gemeinsame Standards zu verfügen und diese zentral zu verwalten,

  - Postfächern für alle Benutzer, um Anpassungen direkt zu ausgesuchten Benutzern oder definierbaren Benutzergruppen zuschicken zu können,

  - einem Ereignisdienst, der die Benutzer über Anpassungsereignisse (z.B. Einstellung von neuen Anpassungen in den öffentlichen Bereich) automatisch benachrichtigt.

3 Implementierung

In diesem Kapitel beschreiben wir die Implementierung unseres ersten Prototypen basierend auf den Anforderungen aus der Analyse. Zuerst beschreiben wir die grundlegende Architektur des Systems. Dann stellen wir die verschiedenen Strategien zur Verteilung und für gemeinsamen Zugriff (Sharing) dar, die unser Tool anbietet. Anschließend diskutieren wir Fragen, die Privatheitsaspekte betreffen und Probleme des Findens oder Identifizierens von Anpassungen und unsere Implementierung eines Benachrichtigungsdienstes.

3.1 Grundlegende Architektur

Prototyp I wurde in VBA (Visual Basic for Applications) entwickelt, der Makro Sprache für Microsoft Anwendungen, die direkten Zugriff auf das Objektmodell der Anwendung erlaubt. Außerdem bietet sie Sprachelemente und Komponenten zur Gestaltung von grafischen Arbeitsoberflächen an.

Der Austausch von Symbolleisten und Dokumentvorlagen geht transparent (im technischen Sinne) für den Benutzer über das Betriebssystem vonstatten. Die eigentliche Funktionalität bleibt vollkommen auf der Seite des Clients. Abbildung 1 zeigt die Grundarchitektur des Systems. Dabei richtet sich das System beim ersten Benutzen eines neuen Benutzers für diesen selbständig ein (Postfach, private Ablage, Eintrag in die Benutzerliste).

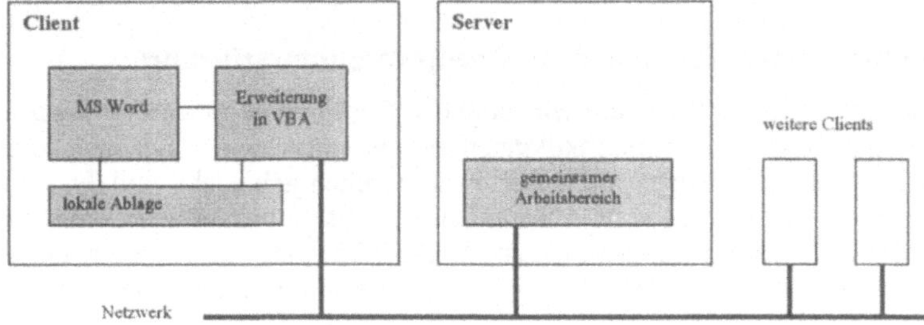

Abb. 1: Architektur des Prototypen

Die Erweiterungen sind in die MS Word Menüleiste integriert, um den Zugriff auf die gesamte Anpassungsfunktionalität für den Benutzer zu vereinfachen. Die Grundfunktionalität umfaßt das Laden und Speichern von Dokumentvorlagen und Symbolleisten. Es ist außerdem möglich, eine Dokumentvorlage und verschiedene Symbolleisten in einem Paket zu kombinieren, welches für bestimmte Aufgaben einer Textverarbeitung gedacht ist, z.B. die Gestaltung einer Webseite oder das Verfassen eines Textes mit vielen mathematischen Formeln.

3.2 Das Sharing von Dokumentvorlagen und Symbolleisten

Der Prototyp bietet sowohl einen Versende- als auch einen Zugriffsmodus zum gemeinsamen Nutzen von Anpassungen. Um zentral verwaltete Umgebungen zu unterstützen, können Anpassungen an Gruppen von Benutzern verschickt werden. Solch eine Operation könnte z.B. von Administratoren durchgeführt werden, die alle Word-Installationen mit dem neuen Firmenbriefkopf versehen wollen. Ebenso könnte ein Benutzer diese Operation nutzen, um z.B. eine bestimmten Vorlage an einen Kollegen zu mailen.

Auch ist es möglich, die Anpassung einfach in einem gemeinsamen Arbeitsbereich (s. Abb. 1) abzulegen. Wenn ein anderer Benutzer nun eine bestimmte Anpassung sucht, kann er/sie auf die gewünschten Vorlagen oder Symbolleisten in der gemeinsamen Ablage zugreifen.

Abbildung 2 zeigt den Anpassungsbrowser, der Versende- und Zugriffsfunktionalität für den Benutzer zur Verfügung stellt. Auf ihn kann einfach durch ein Menü in der Textverarbeitung zugegriffen werden.

Auf der linken Seite ist der Inhalt des gemeinsamen Arbeitsbereichs zu sehen, während die private, lokale Ablage in der Mitte zu finden ist. Die zwei Listen auf der rechten Seite zeigen die anderen Benutzer im System und die Benutzergruppen an. Auf dem Bildschirm kann der Benutzer Anpassungen auswählen und sie zwischen dem lokalen und dem öffentlichen Arbeitsbereich verschieben oder aber, wie oben beschrieben, an einzelne Benutzer oder Gruppen verschicken. Dazu können entsprechende Benutzergruppen definiert werden (unten rechts in Abb. 2).

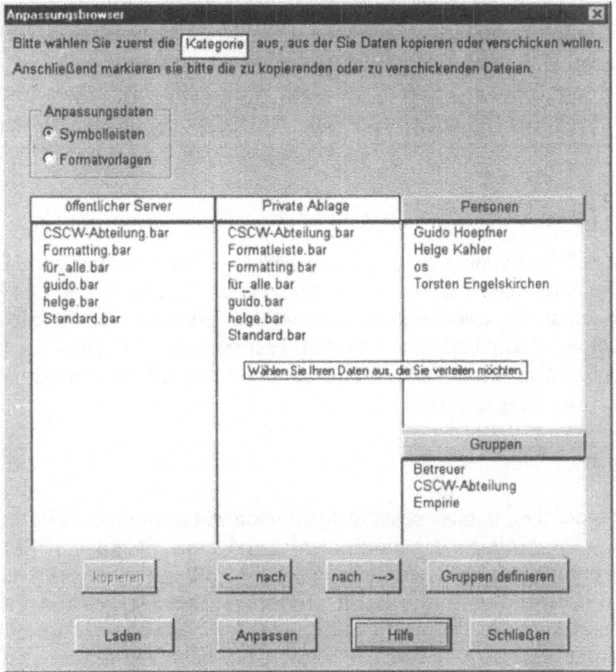

Abb. 2: Der Anpassungsbrowser zur gemeinsamen Nutzung von Anpassungen

3.3 Das Finden von Anpassungen in gemeinsamen Arbeitsbereichen

Die Benutzung des öffentlichen Servers für Anpassungen erfordert die Identifikation von relevanten Anpassungen in der, u.U. recht großen, gemeinsamen Ablage. Bis jetzt bietet der Prototyp hierzu drei Features an.

Erstens ist es möglich, eine Anpassung mit Hilfe einer *Textbeschreibung ihres Zwecks* zu kommentieren, z.B. eine Beschreibung der Umstände oder Aufgaben, für die sie nützlich sein könnte. Die Beschreibung wird bei der Verwendung des Browsers in beiden Ablagen angezeigt, sobald eine Anpassung ausgewählt wird. Weiterhin ist es möglich das Datum und die Größe der Anpassung zu identifizieren.

Zweitens haben wir für Symbolleisten einen Vorschaumodus implementiert, der die schnelle Einfügung (und Deaktivierung) von Symbolleisten auf den Bildschirm erlaubt. Der Benutzer kann dabei die Alternativen erkunden, ohne sie schon permanent übernehmen zu müssen.

Drittens kann der Benutzer beim Suchen in den privaten oder öffentlichen Ablagen zwischen Kategorien von Anpassungen wählen. Momentan unterscheiden wir nur zwischen Symbolleisten und Dokumentvorlagen, aber wir glauben, daß eine stärkere Differenzierung nützlich sein könnte. Dabei bietet es sich an, das Tool mit einer logischen Suchfunktion auszurüsten, die auf bestimmten Attributen der verschiedenen Anpassungen basiert.

3.4 Benachrichtigung von Benutzern beim Eingang von Anpassungen

Das Versenden von Anpassungsdaten macht es nötig, die Benutzer zu informieren, wenn ihnen neue Anpassungen zugemailt worden sind, damit sie über deren Annahme entscheiden können. Dazu haben wir einen Benachrichtigungsdienst implementiert, der die Benutzer sowohl beim Starten der Textverarbeitung als auch beim Aktivieren einer Anpassungsfunktion per Nachrichtenfensters informiert. Dieses Fenster stellt die Anpassung dar und fragt nach, ob der Benutzer sie entweder in seine private Ablage speichern oder sofort löschen möchte.

4 Usability Test

Die Evaluation des Prototypen I fand in zwei Sitzungen mit Teams von jeweils zwei Teilnehmern statt. Die Sitzungen bestanden aus drei Teilen. Zuerst wurde den Teilnehmern die Handhabung des Anpassungstools erklärt. Im zweiten Teil bekamen sie ein Blatt mit gemeinsam zu bearbeitenden Aufgaben, die von den Testpersonen auf zwei vernetzten Rechnern durchzuführen waren. Der dritte Teil bestand aus einer Anzahl von Fragen an die Teilnehmer zur Nutzung und Bewertung des Tools.

4.1 Aufgaben

Die erste Aufgabe der Teilnehmer bestand für Person A darin, eine Dokumentvorlage zu erstellen, dann eine Symbolleiste zu modifizieren und eine andere von Person B erhaltene Symbolleiste mit der Vorlage zu verknüpfen. Danach sollte all das von A als Dokumentvorlage verbunden mit einer Symbolleiste im privaten Ordner gespeichert und schließlich an Person B verschickt werden. Person B sollte eine Symbolleiste mit bestimmten Symbolen erstellen und dann zu Person A senden, die diese dann benutzen mußte. Die zweite Aufgabe erforderte von A, eine Gruppe zu definieren und dieser eine Dokumentvorlage zuzuschicken, dann eine Symbolleiste zu verändern und im privaten Ordner abzuspeichern und schließlich diese Symbolleiste im öffentlichen Ordner zugänglich zu machen. Bei dieser Aufgabe mußte Person B diese Symbolleiste vom öffentlichen in den privaten Ordner kopieren und sie dann mittels des Vorschaumodus laden. Die Aufgaben erforderten einiges an Koordination zwischen den Teilnehmern. Bei der Beobachtung dieser Koordination verwandten wir die Methode der *Constructive Interaction* (vgl. [12]), bei der einem Team eine Aufgabe gestellt wird und dessen (vornehmlich verbalen) Kommunikation zur Kooperation festgehalten werden. Dieser Methode haftet nicht die Künstlichkeit der Thinking-Aloud Methode an, bei der Teilnehmer gegenüber den neutralen Beobachtern äußern, was sie denken, während sie beispielsweise eine Software benutzen bzw. evaluieren (vergl. z.B. [13]).

4.2 Resultate

Neben Vorschlägen zur Verbesserung der Benutzungsschnittstelle erbrachte die Evaluation, daß alle Teilnehmer die Möglichkeit, Anpassungen zu speichern, verbinden und verteilen als sehr hilfreich bei ihrer Arbeit erachteten. Obwohl nicht alle Teilnehmer erfahrene Benutzer waren, waren sie alle in der Lage, die Anpassungs- und die Verteilungsfunktionalität nach nur 15-minütiger Einweisung zu benutzen. Die generelle Brauchbarkeit des Tools wurde als gut empfunden. Ein Teilnehmer sagte explizit, daß er Word in der Zukunft häufiger anpassen werde, da er nun wisse, wie dies zu bewerkstelligen sei und er sich nicht länger davor fürchte müsse, da der Vorgang und die Effekte von Anpassungen nun transparenter sei und man zusätzlich einfach Adaptierungen rückgängig machen könne. Das wurde vor allem dem Vorschaumodus zugeschrieben. Die beiden Teilnehmer, die Netzwerkadministrator und Power-User waren, äußerten, daß solch eine Verteilung von Anpassungen für ihre Organisationen sehr hilfreich sei. Die Diskussion, die den Aufgaben folgte, enthüllte, daß das konzeptionelle Modell, welches die Teilnehmer darüber hatten, wie die Verteilung der Dateien erfolgte, dem sehr nahe kam wie wir, die Designer, die Verteilung geplant und implementiert hatten. Dies ist insofern ein wichtiges Ergebnis als daß insbesondere in einer komplexeren Gruppenarbeitsumgebung eine falsche Wahrnehmung des zugrunde liegenden Modells, z.B. über das Konzept von Verweisen, zu einem ineffizienten Gebrauch führt oder die Akzeptanz eines Systems reduziert (vergl. [14]).

5 Schlußfolgerung

Obwohl die Tatsache, daß Anpassungsaktivitäten oft kooperativ ausgeführt werden, durchaus seit einiger Zeit wohlbekannt ist, besteht ein Mangel von Groupwareunterstützung für diesen Bereich. Im vorliegenden Beitrag wurde ein Tool vorgestellt, das den Austausch von Anpassungsdaten einer Textverarbeitung erleichtert. Auf unserer empirischen Untersuchung basierend nehmen wir an, daß solch ein Tool als ein Mittel dienen könnte, das Gruppen dazu ermutigt, Gruppenstandards zu diskutieren, z.B. für Briefvorlagen, die dann gemeinsam genutzt werden können. Die Systematisierung von Anpassungen (vgl. [8]), die aus einem gemeinsamen Anpassungsprozeß resultiert, kann dann zu gemeinsamen Normen und Konventionen beitragen, die für kooperatives Arbeiten erforderlich sind (vgl. [15, 16]).

Derzeit fehlt unserem Tool noch eine adäquate Evaluation. Dazu ist als nächster Schritt die Einführung bei mehreren Organisationen geplant, um dann noch konkretere Aussagen über den Gebrauch des Tools für echte Kooperation auch außerhalb künstlicher Testsituationen machen zu können. Die Ergebnisse einer solchen Studie werden nicht auf Anforderungen des technischen Designs begrenzt sein, sondern sollen auch organisatorische Vorschläge enthalten. Wir sind davon überzeugt, daß die Einrichtung eines *gardeners* (vergl. [3]) oder eines *translators* (vergl. [2]), d.h. eines lokalen Experten, der verantwortlich für die Koordination der Anpassungsaktivitäten ist, eine entscheidende Rolle bei Anpassungsmaßnahmen in Organisationen spielen wird.

Nach unserem bisherigen Kenntnisstand können Anpassungen am nützlichsten im organisatorischen Kontext ihrer Erstellung verwendet werden, wo sie diejenigen Aufgaben unterstützen, für die sie gemacht werden. Das vorgestellte Tool zum Austausch von Anpassungen ist in seiner jetzigen Form besonders hilfreich für kleine Arbeitsgruppen mit einem eher ähnlichen Arbeitskontext. Unsere zukünftige Arbeit wird außerdem die Frage der technischen und organisatorischen Skalierbarkeit eines solche Tools betreffen. Die Hypothese hierbei ist, daß das Modell von privaten und öffentlichen Räumen, als auch die Unterscheidung zwischen Benutzer und Ersteller der Anpassung auf mehr als zwei Ebenen erweitert werden müßte, wenn die

Gruppengröße eine bestimmtes Grenze überschreitet. Wie in gemeinsamen Arbeitsbereichen für generelle Zwecke wird hier ein fortgeschritteneres Modell der Zugangsberechtigung benötigt werden (vergl. [17]). Außerdem ist eine Historienaufzeichnung für Anpassungen angedacht, aus der hervorgehen würde, wer wann welche Veränderungen an Anpassungen vorgenommen hat.

Eine weitere Erweiterung des Tools soll erlauben, Anpassungen weltweit, also z.B. über das World Wide Web (WWW) zu verteilen. So könnte sogar daran gedacht werden, Teams global zu unterstützen oder weitreichend zugängliche Bibliotheken für Anpassungen einzurichten. Ob dies jedoch im Hinblick auf den dabei mangelnden Kontext von Aufgaben und Organisationsstrukturen sinnvoll ist, bleibt zu untersuchen.

6 Literatur

Oppermann, R. : Evaluation von adaptierbaren und adaptiven Leistungen im Tabellenkalkulationsprogramm EXCEL. Arbeitspapiere der GMD, Nr. 596, 1991.

W. Mackay: Patterns of Sharing Customizable Software. In: Proceedings of CSCW '90. 209 - 221.

B.M. Nardi: A Small Matter of Programming. MIT Press. Cambridge. Massachusetts, 1993.

Oppermann, R. : Möglichkeiten und Probleme individualisierter Systemnutzung. Arbeitspapiere der GMD, Nr. 539, 1991.

Paetau, M. : Kooperative Konfiguration – Ein Konzept zur Systemanpassung an die Dynamik kooperativer Arbeit. In: Proceedings CSCW '91, Bremen, Teubner, S. 137 – 151.

K. Carter und A. Henderson: Tailoring Culture. In: R. Hellman et al. (Hg.): Proceedings of 13th IRIS. Abo Akademic University. Reports on Computer Science & Mathematics. No. 107. 1990. 103 - 116.

M. Paetau: Kooperative Konfiguration - Ein Konzept zur Systemanpassung an die Dynamik kooperativer Arbeit. In: J. Friedrich, K.-H. Rödiger (Hg.): Computerunterstützte Gruppenarbeit (CSCW). Teubner. Stuttgart, 1991. 137 - 152.

R. Trigg und S. Bødker: From Implementation to Design: Tailoring and the Emergence of Systematization in CSCW. In: Proceedings of CSCW '94. ACM-Press. New York, 1994. 45 - 55.

MacLean, K. Carter, L. Lövstrand und T. Moran: User-Tailorable Systems: Pressing the Issues with Buttons. In: CHI '90 Proceedings. 175 - 182.

S. Page, T. Johnsgard, U. Albert und C. Allen: User Customization of a Word Processor. In: Proceedings of CHI '96. April 13.-18 1996. 340 - 346.

Mackay, W.E. : Triggers and barries to customizing software. Proceedings of CHI '91, ACM SGICHI, New Orleans, S. 153 – 160.

C.E. O'Malley, S.W. Draper und M.S. Riley: Constructive Interaction: A Method for Studying Human-Computer-Human Interaction. In: Proceedings of IFIP INTERACT '84: Human-Computer Interaction. 269 - 274.

J. Nielsen: Usability Engineering. Academic Press. Boston, 1993.

G. Mark und W. Prinz: The Establishment and Support of Conventions for Groupware. In: S. Howard, J. Hammond, G. Lindgaard (Hg.): Human Computer Interaction: INTERACT 97. Chapman & Hall. 1997. 413 - 420.

G. Mark: Merging Multiple Perspectives in Groupware Use: Intra- and Intergroup Conventions. In: Proceedings of International ACM SIGGROUP Conference on Supporting Group Work (GROUP 97). 19 – 28.

V. Wulf: Storing and Retrieving Documents in a Shared Workspace: Experiences from the Political Administration. In: S. Howard, J. Hammond, G. Lindgaard (Hg.): Human Computer Interaction: INTERACT 97. Chapman & Hall. 1997a. 469 - 476.

U. Pankoke und A. Syri: Collaborative Workspaces for Time deferred Electronic Cooperation. In: Proceedings of International ACM SIGGROUP Conference on Supporting Group Work (GROUP 97). 187 – 196

Adressen der Autoren

Helge Kahler
ProSEC
Institut für Informatik III
Universität Bonn
Römerstr. 164
53117 Bonn
Tel.: 0228 / 734299
Email: kahler@cs.uni-bonn.de

Volker Wulf
ProSEC
Institut für Informatik III
Universität Bonn
Römerstr. 164
53117 Bonn
Tel.: 0228 / 734276
Email: volker@cs.uni-bonn.de

Oliver Stiemerling
ProSEC
Institut für Informatik III
Universität Bonn
Römerstr. 164
53117 Bonn
Tel.: 0228 / 734503
Email: os@cs.uni-bonn.de

Jörg-Guido Hoepfner
Basis Vertrieb SAP
Ketscher Str. 20a
68782 Brühl
Tel: 06227-7-61070
Email: joerg-guido.hoepfner@sap-ag.de

Ein Werkzeug zur Moderationsunterstützung

Barbara Kleinen

Institut für Multimediale und Interaktive Systeme, Medizinische Universität zu Lübeck

Zusammenfassung

In diesem Beitrag werden „Ideenwerkstätten", ein Werkzeug zur Unterstützung von Moderatoren in Arbeitstreffen, vorgestellt. Neben Electronic Meeting Systems stellen sie eine neue Klasse von Meeting Support Systemen dar, da sie im Unterschied zu EMS allein durch die Moderatorin bedient werden. Ideenwerkstätten sind in Anlehnung an die Moderationsmethode entworfen worden und halten sich in ihrer minimal gehaltenen Gestaltung an diese Metapher. Es wird die Hypothese aufgestellt, daß mit Ideenwerkstätten eine neue Qualität der elektronischen Unterstützung von Arbeitstreffen erreicht werden kann. Als Abschluß wird ein Konzept für die Erweiterung von Ideenwerkstätten für den räumlich verteilten Einsatz vorgestellt.

1 Einführung

Im Verhältnis zu anderen Anwendungsbereichen und insbesondere der zunehmenden Verbreitung von CSCW allgemein wird in räumlich und zeitlich synchronen Arbeitstreffen bislang wenig Informationstechnologie eingesetzt. Verbreitet ist allein spezielle Hardware für Sitzungsräume, wie insbesondere Präsentationsperipherie. Das Spektrum an „Meetingware", an speziell für den Einsatz in Arbeitstreffen entwickelter Software, ist hingegen sehr eng. Dabei wären weitaus mehr Einsatzmöglichkeiten auch herkömmlicher Anwendungsprogramme in Sitzungen denkbar. In diesem Beitrag werden einige Probleme des Einsatzes von Computerunterstützung in Arbeitstreffen besprochen sowie ein neuartiges Werkzeug zur Moderationsunterstützung vorgestellt.

Im ersten Abschnitt wird auf „Electronic Meeting Systems" eingegangen, die bislang die in der Literatur am weitesten behandelte Klasse von Meetingware ist, und einige Gründe besprochen, warum sich dieser Typ von Meetingware noch nicht durchsetzen konnte.

Ausgehend von der These, daß der breite Einsatz von Informationstechnologie in Arbeitssitzungen auch an einer für diesen Einsatzbereich ungeeigneten Gestaltung scheitert, werden im zweiten Abschnitt einige spezielle Gestaltungsanforderungen für Software, die in Sitzungen eingesetzt werden soll, formuliert.

Diese Gestaltungsanforderungen wurden im Moderationsunterstützungswerkzeug „Ideenwerkstätten" umgesetzt, das im darauf folgenden Abschnitt vorgestellt wird. Dabei wird die mit diesem Werkzeug mögliche Moderationsunterstützung ausführlich beschrieben.

Im letzten Abschnitt wird schließlich das Konzept für die geplante Erweiterung von „Ideenwerkstätten" auf räumlich verteilte Konferenzen vorgestellt.

2 Electronic Meeting Systems

Kommerziell erhältliche Meetingware – auch Electronic Meeting Systems (EMS) genannt – geht von der theoretischen Grundlage aus, daß das Identifizieren von Problembereichen, das Sammeln und Bewerten von Ideen (Vorschlägen, dieses Problem anzugehen) und schließlich ein Konsens über das weitere Vorgehen Kernbereiche erfolgreicher Arbeitssitzungen sind. Demzufolge bieten diese Systeme Funktionen wie Planen der Tagesordnung, Sammeln von Ideen (Electronic Brainstorming) sowie Werkzeuge zum Kategorisieren und Abstimmen über diese Ideen. Sie sind konzipiert für den Einsatz in speziell eingerichteten Räumen, den „Electronic Meeting Rooms", die über einen vernetzten PC pro Teilnehmer und i.d.R. einen großen gemeinsamen Bildschirm verfügen. Obwohl derartige Systeme seit über zehn Jahren kommerziell erhältlich sind und Studien gezeigt haben, daß ihr Einsatz hohe Produktivitätsgewinne – sowohl Zeitersparnis als auch eine höhere Qualität der Ergebnisse – ermöglicht, sind sie bislang nur sehr wenig verbreitet [1],[2].

Während zunächst organisatorische Probleme für die schleppende Verbreitung verantwortlich gemacht wurden, wird inzwischen angenommen, daß eine fehlende emotionale Zufriedenheit nach Sitzungen mit EMS eine weitere Verbreitung erschwert. Obwohl EMS direkte Kommunikation zwischen den Gruppenmitgliedern nicht vollständig ersetzen, vermitteln und formalisieren sie doch einen erheblichen Teil der Kommunikation. Es kann angenommen werden, daß dieser Umstand an sich zu einer geringeren Zufriedenheit der Teilnehmer mit dem Sitzungverlauf führt.

So stellen Reinig u.a. fest, daß eine *Affective Reward,* eine „Affektive Belohnung", die sich in herkömmlichen Sitzungen durch emotionale Involviertheit und das Gefühl „gewonnen zu haben" ergibt, fehlt – unbeeinflußt von der Tatsache, daß EMS auch in der Wahrnehmung von Teilnehmern zu besseren Ergebnissen führen. Dementsprechend wurde versucht, die *Affective Reward* durch die Integration von künstlichen Konkurrenzsituationen zu einer fiktiven Vergleichsgruppe in das EMS zu erreichen [3]. Der Verlust der *Affective Reward* findet allerdings offensichtlich durch die Einschränkung und Formalisierung der direkten Kommunikation statt: „Table-pounding, arguments, and face-to-face confrontations in unsupported meetings may lead to physiological arousal. At the end of a meeting the participants may attribute that arousal to their satisfaction with the process and outcomes, resulting in high Affective Reward. GSS[1] eliminates many traditional meeting behaviors that induce arousal – shouting, pounding and heated debates – which could account for reduced Affective Reward in people whose normal meeting style is active and confrontational." [3, S. 3].

Es gibt weitere Hinweise darauf, daß die Einschränkung einer direkten Kommunikation zwischen den Teilnehmern Nachteile für die Arbeitszufriedenheit und Gruppendynamik hat. So werden etwa in der Arbeitspsychologie Kommunikationserfordernisse als Qualitätsmerkmal eines Arbeitsplatzes betrachtet [4]. Dies legt nahe, daß eine Einschränkung und Formalisierung der direkten Kommunikation durch EMS die Zufriedenheit generell sinken läßt.

Die Kommunikationstheorie von Watzlawick, Beavin und Jackson besagt u.a., daß jede zwischenmenschliche Kommunikation neben der Inhaltsebene eine Beziehungsebene hat, über die Beziehungen zwischen den Sprechenden definiert und gefestigt werden [5]. Diese Beziehungsebene fällt beim Einsatz eines Electronic Brainstorming Tools offensichtlich weg.

Diese Ergebnisse lassen es sinnvoll erscheinen, direkte Kommunikation zwischen Kooperationspartnern nicht zu ersetzen, sondern als solche zu unterstützen bzw. nach Unterstützungsstrategien zu suchen, welche die Kommunikation nicht einschränken oder formalisieren.

---

1 Group Support Systems

Damit könnten Vorteile der direkten Kommunikation mit den Vorteilen elektronischer Sitzungsunterstützung verbunden werden.

3 Gestaltungsanforderungen an Meeting Support Systems

Über Electronic Meeting Systems hinaus ist der Einsatz von Informationstechnik zur Unterstützung von Arbeitstreffen bislang wenig untersucht worden. Etwa wäre Zugriff auf das World-Wide-Web oder Unternehmensdatenbanken, speziellen Anwendungsprogrammen sowie eine Unterstützung der Protokollführung in vielen Sitzungen nützlich.

Da die hierzu notwendige Technik in den vielen Unternehmen vorhanden ist, stellt sich die Frage, warum Informationstechnologien nicht häufiger in Arbeitstreffen eingesetzt werden. Es ist zu vermuten, daß dies neben organisatorischen Barrieren daran liegt, daß Software für den Einsatz in Gruppensituationen nicht geeignet gestaltet ist. So ist zum einen die Berücksichtigung software-ergonomischer Kriterien für die Dialoggestaltung für Groupware von besonderer Bedeutung. Während eine umständliche, nicht an den konkreten Arbeitsvorhaben orientierte Bedienung am Einzelarbeitsplatz häufig hingenommen wird, ist sie in einem größerem Treffen aber inakzeptabel, da sie zuviel Zeit in Anspruch nimmt und den Diskussionsprozeß stört.

Aus der Forderung, schnell zu bedienen zu sein und den Gruppenprozeß und damit die direkte Kommunikation der Teilnehmer so wenig wie möglich zu beeinflussen, ergeben sich vier spezielle Gestaltungsanforderungen an Meeting Support Systeme, auf die im folgenden eingegangen wird.

3.1 Schnelle Erlernbarkeit und eine Übersichtliche Anzahl von Funktionen

Nunamaker fordert eine einfache Erlernbarkeit für Electronic Meeting Systems: ihre Funktion sollte in 30 Sekunden erklärt sein [1] (vgl. auch [6]). Diese Forderung ist auch auf Software zu übertragen, die nur von der Moderatorin oder dem Vortragendem bedient wird, da die Gruppenmitglieder versuchen, die Aktionen oder Bildschirmsymbole nachzuvollziehen und zu verstehen. Deshalb werden sie durch ein Interface, das eine Vielzahl an Funktionen bietet – seien diese nun sichtbar oder werden sie vom Moderator angewendet – eher von der eigentlichen Arbeit abgelenkt.[2]

3.2 Schnelle und direkte Bedienbarkeit

Eine begrenzte Anzahl von Funktionen erleichtert auch, diese Funktionen direkt – beispielsweise durch einen einzelnen Mausklick – zugänglich zu machen. Dadurch kann eine für Moderationen erforderliche Geschwindigkeit erreicht werden, insbesondere, wenn eine Diskussion mitprotokolliert werden soll. Eine Ablenkung der Gruppe durch technische Details des Werkzeuges wird weitestgehend vermieden.

---

2 Wenn eine komplexe Funktionalität unumgänglich ist, sollte der Bildschirm in einen Kontroll- und Sichtbereich aufgeteilt werden können, wovon nur der letztere Teil für alle sichtbar projiziert wird –wie es auch Wulf und Schinzel für Videokonferenzsysteme beim Teleteaching für sinnvoll erachten. [11]

3.3   Flexibilität bei einfacher Bedienbarkeit

Gleichzeitig muß ein Werkzeug flexibel sein, um sich dem Gruppenprozeß anzupassen und nicht umgekehrt. Nur so kann eine Störung oder Beeinflussung des Gruppenprozesses vermieden werden. Aber auch diese Flexibilität darf keine komplexe Bedienung erfordern, sondern sollte auf möglichst natürliche Weise in das Werkzeug integriert sein.

3.4   Hohe Ausfallsicherheit und schnelle Antwortzeiten

Ausfallsicherheit und schnelle Anwortzeiten sollten zwar selbstverständlich sein, diesbezügliche Mängel werden jedoch bei Groupware besonders schnell zur Nichtakzeptanz führen, da – insbesondere bei der Unterstützung von Arbeitstreffen – immer eine größere Anzahl von Benutzern unmittelbar betroffen ist.[3]

4   Ideenwerkstätten: Ein Werkzeug zur Moderationsunterstützung

4.1   Ziel

Bei der Entwicklung von Ideenwerkstätten stand, wie bei EMS auch, die Absicht im Vordergrund, das Sammeln und Bewerten von Ideen in einer Gruppe zu unterstützen. Im Unterschied zu EMS sollte mit Ideenwerkstätten dabei aber in erster Linie ein Unterstützungswerkzeug für die Moderatorin bzw. ein Werkzeug, das allein vom Moderator bedient wird, entwickelt werden. Leitgedanke ihrer Gestaltung war, Ideen als „unfertige Gedanken" in einer flexiblen Weise zu visualisieren. Dabei sollte sowohl die Möglichkeit zur Änderung und (Um)Strukturierung gegeben sein als auch der Ideencharakter, das Unfertige der eingegebenen Ideen durch die Gestaltung deutlich werden. Als Entwurfsparadigma und Oberflächenmetapher wurde daher das Material der weithin bekannten Moderations- oder auch Metaplan-Methode – farbige Karten auf Packpapier – herangezogen.

4.2   Soft- und Hardwarevoraussetzungen

Derzeit sind die Ideenwerkstätten ein Teil des Mehrbenutzer-Hypertextautorensystems „HyperCom", das prototypisch in der Forschung Systemtechnik der Daimler-Benz AG unter Leitung von Alexander Mankowsky in Berlin Moabit entwickelt wird. Derzeit werden sie am Institut für Multimediale und Interaktive Systeme nach Java portiert, um plattformübergreifend einsetzbar zu sein und die Konferenzprotokolle als Ideenwerkstätten unabhängig von HyperCom verfügbar zu machen.

Zum Einsatz der Moderationsunterstützung ist ein Beamer mit einer Auflösung von 800x600 sowie ein PC notwendig. Experimente mit einem interaktivem Whiteboard stehen noch aus, es ist zu vermuten, daß es sich insbesondere für die Ideenstrukturierung eignet. Der technische Aufwand bleibt damit erheblich geringer als bspw. für einen voll ausgestatteten Electronic Meeting Room mit einem vernetzen Rechner für jeden Teilnehmer und ist in vielen Institutionen bereits vorhanden.

---

3 Z.B. wird die Zeit, die das Laden einer Webseite in Anspruch nimmt, am Einzelarbeitsplatz in Kauf genommen. In einer – insbesondere größeren – Gruppe ist diese Wartezeit aber zu lang. Dies führt dazu, daß Webpräsentationen i.d.R. durch vorherigen Download der Seiten vorbereitet werden müssen, und die spontane Einbeziehung einer Webseite in einen Vortrag oder eine Argumentation nicht praktikabel ist.

4.3   Funktionsbeschreibung

Ausgangspunkt für die Gestaltung der Ideenwerkstätten war die Zielsetzung, daß sie sowohl von Einzelpersonen am Rechner als auch von einer Moderatorin in einer Gruppe zum Sammeln und Strukturieren von Ideen verwendet werden können. Oberste Priorität lag daher auf dem Entwurf eines Interfaces, das die wesentlichen Funktionen für die Ideensammlung – das Eingeben und Strukturieren – derart verfügbar macht, daß sie die Gedanken des Einzelnen bzw. den Diskussionsprozeß in der Gruppe möglichst nicht beeinflussen oder stören.

Ideenwerkstätten bieten die Möglichkeit, beschriftete Karten an eine Packpapierwand zu heften. Diese Karten können später beliebig verschoben, mit Linien oder Pfeilen verbunden, in Text, Farbe und Form verändert sowie gelöscht werden. Soll ein auf einer Karte notiertes Stichwort weiter ausgeführt werden, läßt sich mit Hilfe des ,Create-Buttons' eine neue Ideenwerkstatt erzeugen, die automatisch mit dieser Karte verknüpft wird⁵ und ihren Text als Titel erhält. Ein Rücksprung zur erzeugenden Ideenwerkstatt ist über das Stellwand-Icon (in Abb.1 und 2 links oben) möglich.

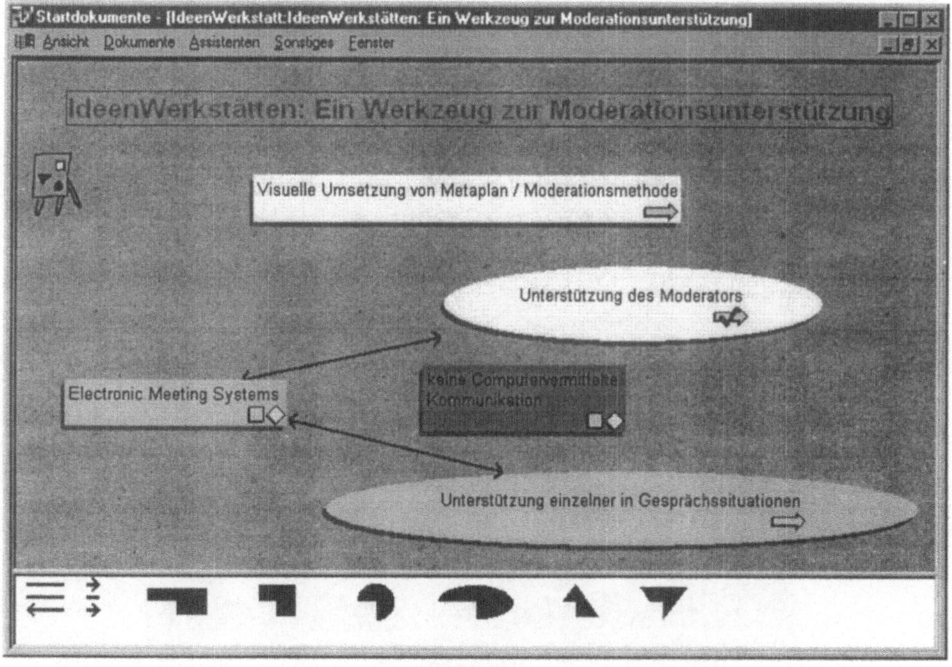

Abbildung 1.

---

4 Der türkisfarbene quadratische Create- sowie der gelbe rautenförmige Connect-Button auf jeder Karte (vgl. Abb. 1), sind ein Standard in HyperCom. Bei einer von HyperCom unabhängigen Reimplementierung wird hier ein intuitiveres Symbol zu finden sein.
5 Wenn eine Karte einen Link zu einer anderen Ideenwerkstatt repräsentiert, werden die Create-/Connect-Buttons durch einen Pfeil ersetzt. (vgl. Abb. 1).

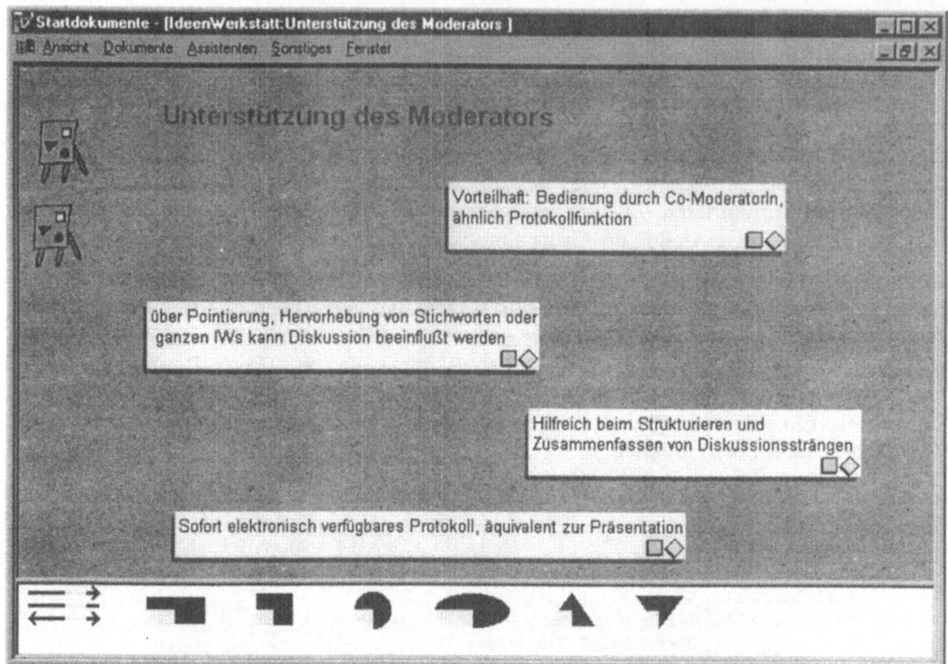

Abbildung 2.

Die Oberfläche ist auf wesentliche Funktionen beschränkt. Es wurde Wert darauf gelegt, daß die wesentlichen Funktionen wie Einfügen neuer Karten und Verbindungen, sowie deren Verschieben und Verändern mit dem geringem Aufwand möglich sind. Dies wurde durch die mehrfarbigen Knöpfe am unterem Bildschirmrand erreicht, über die eine Karte in gewünschter Form und Farbe direkt angewählt werden kann. So entfällt z.B. die Auswahl der Farbe über ein Farbauswahlmenü. Die sich als Kehrseite der direkten Erreichbarkeit ergebenen Beschränkungen – hier die Beschränkung auf lediglich vier Farben – wurde dabei bewußt in Kauf genommen.

Um ein einfaches Umstrukturieren der eingegebenen Ideen zu gewährleisten, sind alle Veränderungen und Anpassungen mit geringem Aufwand möglich. So kann z.B. die Farbe und Form einer Karte geändert werden, indem die entsprechende Form aus der Auswahlleiste ausgewählt wird und über eine bestehende Karte gelegt wird. Linien können auf dieselbe Weise eingefügt und verändert werden. Selbstverständlich werden die Linien beim Verschieben der Karten mitgeführt.

In einer HyperCom-Netzansicht kann eine Übersicht über alle erzeugten Ideenwerkstätten mit ihren Verknüpfungen angezeigt werden.

5 Moderation mit Ideenwerkstätten

Ideenwerkstätten können in Arbeitstreffen als flexibles Hilfsmittel für den Moderator eingesetzt werden. Sie lassen sich sowohl als Protokoll- und Visualisierungshilfsmittel in einer ansonsten unstrukturierten Diskussion, als auch als universelles Hilfsmittel für unterschiedliche Moderationsmethoden verwenden.

Während die Ideenwerkstätten vom Erscheinungsbild her an der Moderationsmethode orientiert sind (vgl. dazu [8]), wurde bewußt darauf verzichtet, starre Formen der Methode vorzugeben. So gibt es beispielsweise keine Richtlinien dafür, welche Bedeutung spezielle Farben und Formen haben. Diese kann von der Moderatorin frei bestimmt werden. Bei der Gestaltung der Ideenwerkstätten wurde hingegen Wert gelegt auf eine kreative Ästhetik, durch die der Benutzer angeregt werden soll, mit dem Werkzeug zu spielen und neue Anwendungsmöglichkeiten zu erfinden:

„Was uns faszinierte, waren: das braune Packpapier, die bunten Kärtchen, daß jeder dabei zu Wort kommt, daß Aussagen geclustert bzw. schwarz umrandet werden können und daß sich schließlich mit roten Punkten Prioritäten festlegen lassen." [7, S. 147].

Im Sinne diesen Zitats sind Ideenwerkstätten mehr „Material" als Methode.

So eignen sie sich zunächst für alle Visualisierungen und Präsentationen in einer Moderation, die der Gruppe helfen, Ideen und Themen einzuordnen, Ziele klarzustellen oder die Agenda vor Augen zu haben. Ideenwerkstätten ermöglichen Präsentationen mit erheblich geringerem Aufwand, da das Handling der Packpapierwände und Karten, wie Befestigung, Sortieren usw. entfällt. Ein Nachteil ist jedoch, daß im Vergleich zu Packpapierwänden der Platz sehr begrenzt ist und somit eine gleichzeitige Visualisierung mehrerer „Packpapierwände" schwierig ist.[6]

Zur Unterstützung von Präsentationen werden bereits besuchte Ideenwerkstätten „abgehakt" (vgl. Abb. 1, Karte rechts oben). So fällt es leicht, auch ein komplexes Netz von Ideenwerkstätten vorzustellen. Ein wesentlicher Unterschied zu anderen Präsentationshilfsmitteln ist, daß nahtlos zur Protokollierung gewechselt werden kann – die Diskussion wird so direkt an Ort und Stelle aufgenommen und mit den „Vortragsfolien" verknüpft. Dies ist bei anderen Präsentationsprogrammen aufgrund einer zu komplexen Bedienung kaum möglich. Ideenwerkstätten bieten weiter einen Vorteil bei der Anordnung der „Folien", die nicht

---

6 Dieses Problem könnte durch die Verwendung modularer Videowände, wie sie z.B. als "Overview mX" von Seufert angeboten werden, bei entsprechendem finanziellem und räumlichem Aufwand behoben werden.

notwendigerweise sequentiell ist. So können Rück- und Vorgriffe auf andere „Folien" durch
Verknüpfungen vorbereitet werden.

Wie bereits erwähnt, legen Ideenwerkstätten weder ein spezielles Vorgehen bei der Moderation noch spezielle Methoden fest. So lassen sich alle bei Klebert u.a. beschriebenen Schritte
der Moderationsmethode, die keine direkten Aktionen der Gruppenteilnehmer mit dem
Präsentationsmaterial erfordern, mit Hilfe von Ideenwerkstätten durchführen. Moderations-
schritte, die ein gleichzeitiges Schreiben der Gruppenteilnehmer erfordern, wie insbesondere
die „Karten-Frage'" können modifiziert durchgeführt werden, indem die Karten auf Zuruf von
der Moderatorin ausgefüllt werden.[8] Diese Modifikation verändert die Methode allerdings
erheblich: zum einem fällt die Anonymität weg, die gegeben ist, wenn die Teilnehmerinnen
die Karten selbst ausfüllen. Zum anderem werden die Teilnehmer durch die Ideen der anderen
beeinflußt, was beides kreative Ideen hemmen kann. Andererseits fällt der hohe
organisatorische Aufwand, der sonst mit der „Karten-Frage" verbunden ist, weg, wodurch sie
häufiger und einfacher eingesetzt werden kann.

Unsere Erfahrungen mit Ideenwerkstätten haben gezeigt, daß eine „sanfte Moderation", die
Schwerpunkte auf das Protokollieren, Strukturieren und Spiegeln der Diskussion legt, auch
von Gruppen, die „Gruppenmethoden" eher skeptisch gegenüberstehen, als besondere Form
der Protokollführung gern angenommen wird. Es kann sinnvoll sein, die eigentliche
Moderation und „das Protokoll" bzw. die Bedienung der Ideenwerkstätten auf zwei
Moderatoren aufzuteilen. Durch das Protokollieren wird kein unmittelbarer Einfluß auf den
Diskussionsprozeß genommen. Die Teilnehmer sprechen frei und unmoderiert; ihre
Äußerungen werden für alle sichtbar protokolliert. Es kann aber davon ausgegangen werden,
daß die Wahl der notierten Stichworte, Auswahl der zugehörigen Karte sowie deren
Positionierung am Bildschirm einen mittelbaren Einfluß auf den Diskussionsprozeß hat, auch
ohne eine vorherige explizite Einigung über die Bedeutung dieser Variablen.

Durch den Einsatz von Ideenwerkstätten wird die Möglichkeit und Notwendigkeit zur direkten Kommunikation nicht verändert. Ob und wie die Kommunikation strukturiert bzw. forma-
lisiert wird, hängt allein von der Strategie und den Methoden des Moderators ab.

Ideenwerkstätten unterstützten den Moderator, den Gruppenfokus auf ein bestimmtes Thema
zu lenken sowie den Diskussionsablauf zu strukturieren: Eine sprunghafte oder
unstrukturierte Diskussion kann durch die schnellen Sprungmöglichkeiten zwischen den
Ideenwerkstätten strukturiert protokolliert werden. Dies macht der Gruppe die Struktur
deutlich, und es fällt leichter, aus dem Zusammenhang gerissene Äußerungen (wie sie z.B.
auch durch eine Rednerliste entstehen) im Kontext des vorher zu diesem Thema Gesagtem zu
sehen.

Daher dient das Geschriebene der Moderatorin gleichzeitig als Feedback für die Sprechenden:
Welche Stichworte werden . in die Ideenwerkstatt aufgenommen, wie werden diese
eingeordnet, werden überhaupt neue Stichworte notiert oder stehen sie bereits in einer Ideenwerkstatt?

Die Netzübersicht über die während eines Treffens erzeugten Ideenwerkstätten ermöglicht
einen Rückblick über die bisherige Diskussion und erleichtert eine Entscheidung über das
weitere Vorgehen.

---

7 Teilnehmerinnen schreiben ihre Äußerungen auf Karten, die dann an der Präsentationswand sortiert werden.

6. Räumlich verteilter Einsatz von Ideenwerkstätten

Mit der Portierung zu Java sollen Ideenwerkstätten um die Möglichkeit eines räumlich verteilten Einsatzes bei zeitlicher Synchronität erweitert werden.

Dafür sind – neben der eigentlichen technischen Realisierung – drei wesentliche Erweiterungen notwendig, die auch die Gestaltung der Ideenwerkstätten berühren. Zunächst muß eine Integration von Audio- und Videokonferenzsystemen realisiert werden. Die Ideenwerkstätten selbst müssen um einen Zugriffskontrollmechanismus sowie Group-Awareness-Support erweitert werden. Das Konzept hierzu soll im folgendem kurz erläutert werden.

Workspace Awareness – das „Gewahr-Sein" der Aktivitäten anderer Gruppenmitglieder im gemeinsamen Arbeitsbereich – sowie deren technische Unterstützung und Einfluß auf die Benutzbarkeit von Groupware ist seit längerem Thema der CSCW-Forschung, und Forschungsergebnisse bestätigen, daß ein Awareness-Support die Benutzbarkeit von Mehrbenutzersystemen erhöht.[9,10]. Für die Weiterentwicklung der Ideenwerkstätten ist daher ein Awareness-Support geplant.

Kernstück des Awareness-Support wird eine Visualisierung aktiver Teilnehmer durch Fotos oder Videobilder sein. Dabei bietet es sich an, an dieser Stelle die Integration des Verbindungsaufbau- und Sitzungskontrollmechanismus des Videokonferenzsystems vorzunehmen.

Für ein strenges WYSYWIS' – beide Gruppen bzw. alle Teilnehmer sehen dieselbe Ideenwerkstatt im identischen Zustand – werden die Teilnehmer unterhalb der Ideenwerkstatt gezeigt, gemeinsam mit einer ihnen zugeordneten Farbe. Diese Farbe wird zur Markierung und Blockierung einer Karte verwendet, der von der jeweiligen Person bearbeitet wird.

Für entspanntes (*relaxed*) WYSIWIS, in dem unterschiedliche Ideenwerkstätten in einem Netz bearbeitet werden können, werden alle im Workspace aktiven Teilnehmerinnen angezeigt. Zusätzlich soll über einen Graph-Radar-View, wie bei Gutwin beschrieben [9,10], die Position aller Teilnehmer im Netz der Ideenwerkstätten visualisiert werden. Ein derartiges entspanntes WYSIWIS wird aber eher in zeitlich asynchronen Treffen von Bedeutung sein.

Die Zugriffskontrolle wird über ein Blockieren aktuell bearbeiteter Karten und Verbindungen realisiert werden. Eine einzelne Karte/Verbindung erscheint als geeignete Granularität für das Blockieren, da sie klein genug ist, um beim Bearbeiten nicht zu lange blockiert zu sein. Andererseits ist sie auch groß genug, um keine Konflikte mit anderen Aktionen entstehen zu lassen. Darüber hinaus entspricht diese Granularität der Metapher des gemeinsamen Arbeitens an der Packpapierwand, bei dem eine Person i.d.R. auch jeweils eine Karte verändern kann, die solange für alle anderen unerreichbar ist.

Durch diese Erweiterungen werden einige interessante Anwendungsbereiche in räumlich und/oder zeitlich verteilten Gruppenkonstellationen erschlossen. So können z.B. zwei räumlich getrennte Gruppen zusammenarbeiten, einzelne Teilnehmer von außen an einer moderierten Sitzung teilnehmen, oder auch eine vollständig räumlich verteilte Sitzung mit Moderationshilfsmitteln unterstützt werden.

Ein wesentlicher Teil der Forschungsarbeit wird sich dabei auf eine benutzungsgerechte Integration eines Videokonferenzsystems konzentrieren.

---

8 WYSIWIS – What you see is what I see

7. Zusammenfassung und Ausblick

Ideenwerkstätten haben sich in unserem Einsatz als nützliches Hilfsmittel für die Moderation von Arbeitstreffen erwiesen. Sie lassen sich flexibel in einem breiten Spektrum von Situationen verwenden. Der Einsatz der Ideenwerkstätten wurde i.a. wohlwollend aufgenommen, wobei die sofort elektronisch und in ansprechender Form verfügbaren Sitzungsergebnisse für die Teilnehmer eine sehr wichtige Rolle spielen.

Die geplante Erweiterung für den räumlich verteilten Einsatz wird insbesondere für Videokonferenzen interessant werden. Ideenwerkstätten werden es ermöglichen, zahlreiche Moderationsmethoden auch für Telekonferenzen einzusetzen.

Mit der Java-Version der Ideenwerkstätten ist am Institut für Multimediale und Interaktive Systeme eine genauere Untersuchung der Benutzbarkeit in Gruppensituationen sowie der Auswirkungen ihres Einsatzes auf Produktivität, Gruppenverhalten und Zufriedenheit geplant.

8 Literatur

[1] Jr. Nunamaker et al.: *Electronic Meeting Systems: Ten Years of Lessons Learned.* In: D. Coleman, David; R. Khanna (ed.): *Groupware. Technology and Applications.* Upper Saddle River: Prentice-Hall, 1995, S. 146-193.

[2] J. Schiestl: Groupware zur Unterstützung von verteilten Kommunikationsprozessen in Entwicklungsprojekten aus der Sicht des Projektmanagements. Dissertation, Universität Innsbruck, Januar 1995. http://www.informatik.unibw-muenchen.de/inst5/diss/disdeck.html

[3] Reinig, B. et al.: *Beyond Productivity: Group Support Systems and Affective Reward* JMIS, In Press. Erhältlich über http://www.cmi.arizona.edu/users/bbriggs/

[4] E. Ulich: *Arbeitspsychologie.* Zürich: vdf Hochschulverl. AG an der ETH Zürich; Stuttgart; Schäffer-Poeschel, 1994.

[5] P. Watzlawick, J.H. Beavin, D.D. Jackson: *Menschliche Kommunikation. Formen, Störungen, Paradoxien.* Bern: Huber, 1990

[6] F. Niederman, C.M. Beise, PM. Beranek: *Facilitation Issues in Distributed Group Support Systems .*In: SIGCPR '93. Proceedings of the 1993 conference on Computer personnel research, S. 299-312.

[7] L. Dirr: *Die Angst des Tormanns vorm Elfmeter. Widerstände in großen Organisationen gegen die moderatorische Arbeitskultur.* In: J. Freimuth, F. Straub: *Demokratisierung von Organisationen. Philosopie, Ursprünge und Perspektiven der Metaplan®-Idee.* Wiesbaden: Gabler, 1996.

[8] K. Klebert, E. Schrader, W. G. Straub: *KurzModeration.* Hamburg: Windmühle, 1997.

[9] C. Gutwin: *Workspace Awareness Research* Beschreibung des Forschungsbereiches an der University of Calgary: http://www.cpsc.ucalgary.ca/projects/grouplab/people/carl/research/awareness.html

[10] C. Gutwin, S. Greenberg: *Effects of Awareness Support on Groupware Usability.* Proceedings of the ACM Conference on Computer Human Interaction CHI '98, Los Angeles, 1998, P. 511-518.

[11] V. Wulf, B. Schinzel: Erfahrungsbericht zur Televorlesung und Teleübung „Informatik und Gesellschaft". In: Ergonomie & Informatik, Mitteilungen des Fachausschusses 2.3 „Ergonomie in der Informatik", Nr. 32, Februar 1998.

Adresse der Autorin

Dipl. Inf. Barbara Kleinen
Institut für Multimediale und Interaktive Systeme
Medizinische Universität zu Lübeck
Seelandstr. 1a
23569 Lübeck
Email: kleinen@informatik.mu-luebeck.de

Designleitlinien und Bewertungskriterien für die Strukturgeometrie technischer Informationswelten

Falk Mletzko

FG Mensch-Maschine-Systeme, Universität –Gesamthochschule- Kassel

Zusammenfassung

Für das Design komplexer grafischer Informationswelten existiert ein Überschuß an Allgemeinplätzen („Aufgabenangemessenheit") und Restriktionen („nicht mehr als 5 Farben je Display!"). Zugleich gibt es einen Mangel an konkreten ergonomischen Leitlinien, die die Designerkreativität lenken, statt sie einzuengen.

Ziel dieses Beitrages ist es, solche handhabbaren Leitlinien vor den Augen des Lesers zu entwickeln. Ausgehend von Ziel und Aufgabe des Prozessdisplays in der industriellen Mensch-Maschine-Interaktion wird dazu am konkreten Designbeispiel auf geometrische Stolpersteine aufmerksam gemacht und das an der menschlichen Wahrnehmung orientierte verbesserte Design vorgestellt. Die gefundenen ergonomischen Leitlinien und Bewertungskriterien werden anschließend zusammengefaßt.

Abstract

There's a surplus of commonplaces („suitability for task") and restrictions („not more than 5 colours per display!") for the design of complex graphical information worlds. At the same time there's a lack of concrete ergonomic guidelines to direct the designers creativity instead of restricting it.

The objective of this contribution is to develop such applyable guidelines in front of the reader. Starting with objective and task of the process display in the industrial human-machine-intercation there will be called attention to geometric stumbling blocks on a concrete design example and presented the corrected design oriented on the human perception. Finally the found ergonomic guidelines and criterias of valuation will be summarized.

1   Einleitung

Ein in der industriellen Prozessleittechnik seit jeher verfolgtes Ziel ist die Maximierung des Automatisierungsgrades technischer Anlagen und Prozesse. Die mit geringstem Aufwand zu automatisierenden Funktionen und Aufgaben wurden und werden dem Automatisierungssystem übertragen, während der Mensch den verbleibenden „Rest" zu lösen hat [1]. Dies ist keine von den Fähigkeiten und Schwächen des Menschen, sondern denen der aktuell verfügbaren Automatisierungs- und Regelungstechnik ausgehende Aufgabenteilung.

Die Funktion des Menschen in der Prozeßüberwachung besteht zum Großteil im Erkennen und Beseitigen gefährlicher Prozeßzustände. Da die Prozessüberwachung ausschliesslich bei der Führung dynamischer Systeme eingesetzt wird, spielt die Rechtzeitigkeit der menschlichen Reaktion eine zentrale Rolle. Die Anforderungen an die Prozeßbediener, eine Handlung rechtzeitig auszuführen, werden von der Dynamik des Prozeßgeschehens diktiert.

Die Qualität der zugrundeliegenden Handlungsentscheidung ergibt sich aus dem Verhältnis von objektiven Aufgabenanforderungen einerseits und den in der vorgegebenen Zeitspanne subjektiv aktivierbaren Arbeitspotentialen des Prozeßbedieners andererseits.

Entsteht eine Differenz zwischen entscheidungsrelevanten und tatsächlich entscheidungswirksamen Arbeitspotentialen, so ist eine unangemessene, u.U. sogar kontraproduktive Handlungsentscheidung zu erwarten. Um dies zu vermeiden und den Prozeßbediener in seiner Arbeit zu entlasten, muß das handlungsunterstützende Informationssystem die schnellstmögliche Aktivierung von Arbeitspotentialen unterstützen.

2 Motivation und Forderung: Effektives Problemlösen

Typisches Merkmal moderner Prozeßführung ist die räumliche Trennung von Mensch und Prozeß und die damit verbundene Beschränkung auf visuelle und eingeschränkt auditive Reize als Informationsträger. Die grafisch zu übertragenden Informationen können zu einer der zwei wesentlichen Klassen von Informationen gehören, die als Entscheidungsgrundlage dienen können. Dies sind Informationen über   a) *Zustände*

und   b) *Strukturen.*

Zustandsinformationen liefern Aussagen über äußerlich beobachtbares Verhalten ohne Kenntnis innerer Strukturen oder Abhängigkeiten. Ziel ist das Finden eines mental abgespeicherten Musters zur Ableitung vorgefertigter Handlungsschemata. Dies ist Oberflächenwissen und schnell aktivierbar.

Sollte für eine aktuelle Kombination von Symptomen kein Muster existieren, so muß zusätzlich Strukturwissen hergezogen oder erzeugt werden, falls noch nicht vorhanden. Dies ist Tiefenwissen und nur durch bewußtes Nachdenken aktivierbar.

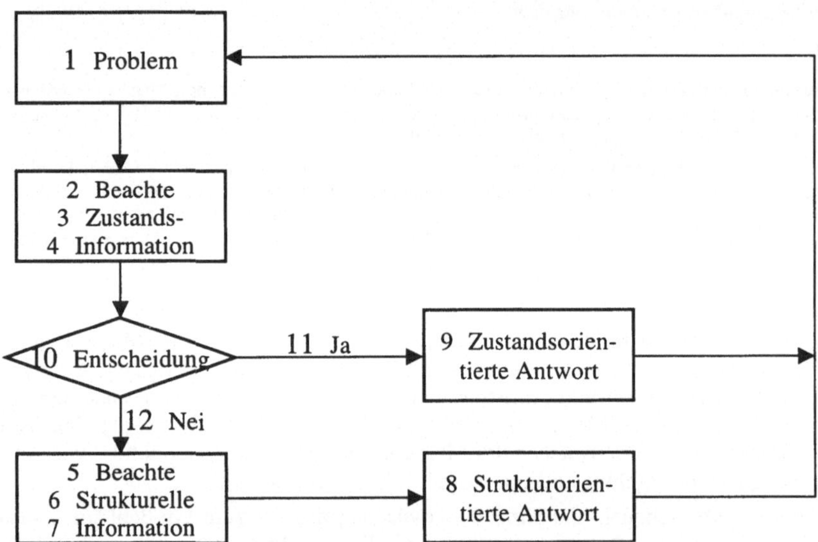

Bild 1: Modell menschlichen Problemlösens [4].

Für diese beiden unterschiedlichen Vorgehensweisen bei der Nutzung ein und derselben vorhandenen Information zur Entscheidungsfindung wurden die Begriffe der symptomatischen (Zustands-/Symptomorientierten) bzw. der topografischen (strukturorientierten) Suche eingeführt [2][3]. Wie im Bild 1 dargestellt, versucht der Mensch zunächst mittels der kognitiv weniger anspruchsvollen und schnellen symptomatischen Suche zum Ziel zu kommen („Weg des geringsten Widerstandes"), ehe er bei dessen Nichterreichen mit der mühevollen und langsamen topografischen Suche beginnt.

Wenn also das Ziel komplexer industrieller Informationssysteme in der schnellen Aktivierung von Arbeitspotentialen besteht, sollte die ressourcenschonende Strategie einer schnellen symptomatischen Suche und zustandsorientierten Antwort des Menschen unterstützt werden.

Aber wie ?

3 Bedingung: Effektives Wahrnehmen

Aus Sicht des Menschen gibt es mehrere Faktoren, die den Schwierigkeitsgrad einer Problemstellung und damit ihre kognitiven Anforderungen bestimmen. Dies sind ihre Komplexität; ihre Vernetztheit als Menge und Stärke von Verknüpfungen zwischen Komponenten und Zuständen; ihre (subjektiv empfundene) Transparenz; und schliesslich , ihre Dynamik [5].

Da der Mensch in der modernen Prozeßleittechnik immer weniger Kontakt zum realen Prozeß hat, nimmt er die Realität nicht mehr wahr wie sie ist, sondern *wie sie dargestellt wird.*

Somit wird die grafische Darstellung selbst zum Prozeß, und die Art und Weise ihrer Gestaltung bestimmt wesentlich Wahrnehmung und mentales Modell auf Seiten des Anlagenfahrers. Die Prozeßvisualisierung beeinflußt folglich durch die gewählte Präsentationstechnik direkt die kognitive Methode ihrer Verarbeitung [6]. Durch den erwähnten schwindenden Kontakt und den damit abnehmenden Zwang zur genauen Widerspiegelung geografischer Verhältnisse in Prozessbildern ergeben sich neue Chancen für die Anwendung ergonomischer Grundsätze.

Die am Anfang der menschlichen Informationsverarbeitung stehende *Wahrnehmung* erfolgt im Falle grafischer Informationsübertragung statisch mittels Formen, Farben und Intensitäten und dynamisch durch Formänderungen, Bewegungen oder Blinken. Der vorliegende Artikel beschäftigt sich explizit mit Formen, genauer Strukturen als komplexen Formen, da durch Farbeinsatz allzu schnell Mängel bei der Formgebung überdeckt werden können. Ein Farbkonzept für die Prozeßführung findet sich in [7].

Ergonomische Grundzüge für die Strukturierung eines Prozessabbildes ergeben sich aus einigen im Unterbewußtsein ablaufenden Strategien zur visuellen Wahrnehmung. Diese werden in der Gestaltpsychologie beschrieben [8] und erfassen visuelle Integrationseffekte, die den Betrachter mehrere grafische Objekte als zusammengehörig wahrnehmen lassen, nämlich:

- Nähe   räumlich / zeitlich nahe beieinanderliegende Objekte
- Ähnlichkeit   ähnliche Form von Objekten
- Geschlossenheit   Objekte bilden gemeinsam eine Figur, z.B. Dreieck
- Gute Fortführung   Mensch gruppiert Objekte richtungsabhängig verschieden stark
- Gemeinsames Schicksal Objekte als Einheit bei gemeinsamer Änderung.

Da diese Prinzipien im Unterbewußtsein verankert und durch täglichen Gebrauch gut trainiert sind, sind sie extrem schnell, können quasiparallel ablaufen und belasten die mentalen Ressourcen kaum, die dann uneingeschränkt für komplexere Interpretationsprozesse zur Verfügung stehen.

Auf der Grundlage weiterer theoretischer Grundlagen, wie Woods ‚visual momentum‘ [9] und Gibsons ‚ecological optics‘ [10], entwickelten Anfang der 90er Jahre Benneth und Flach die ‚Emergent Features‘ [11] und Vicente das ‚Ecological Interface Design‘ EID, das die Prinzipien der automatisierten Wahrnehmung mittels dynamischer Geometrierelationen nutzt [12]. Diese Methoden beruhen auf einer Nutzung der angesprochenen Prinzipien zur schnellen Interpretation *dynamischer* Prozess*parameter.*

Demgegenüber ist die Anwendung dieser Prinzipien auf die *statische* Struktur technischer Informationswelten wie etwa die topologischer Fließbilder nach wie vor mangelhaft. Das liegt an fehlenden Kriterien für eine grafische Umsetzung dieser Prinzipien und für eine anschließende Bewertung der entstandenen Strukturen. Im speziellen Fall kommt noch die starr an der Komponentengeografie ausgerichteten Interpretation des Begriffes Topologie hinzu.

Dies führt zu einer weltweit in Tausenden Prozeßvisualisierungen beobachtbaren maßstäblichen Wiedergabe nicht nur der Form einzelner Systemkomponenten, sondern auch der Strukturgeometrie der Gesamtanlage. Diese geometrischen Relationen aber sind als Folge technologischer, finanzieller, platzökonomischer und zufälliger Faktoren entstanden.

Aus der praktisch maßstabsgetreuen Abbildung örtlicher Gegebenheiten im zweidimensionalen Prozeßbild ergeben sich mitunter sehr gedrängte Darstellungen, die keiner ergonomischen Kritik standhalten.

Desweiteren scheint ein ungeschriebenes Gesetz zu existieren, welches leere Flächen auf dem Bildschirm *verbietet*. Dabei wird völlig außer acht gelassen, daß sich benachbarte Bildschirmregionen in ihrer Wahrnehmung durch den Menschen gegenseitig beeinflussen.

Wenn sie durch ein Museum gehen, und die Wände hängen voller Bilder, werden Sie möglicherweise Schwierigkeiten haben, sich auf eines von ihnen vollständig zu konzentrieren. Denn nicht nur die Wände sind voller Bilder, sondern mittels der peripheren Wahrnehmung auch Ihre „kognitive Warteschlange" der anderen in Ihr Bewußtsein drängenden Objekte.

Ganz anders stellt sich die Situation in einem Raum dar, in dem nur ein einziges Bild hängt: es wird *automatisch* ihre gesamte Aufmerksamkeit auf sich ziehen – so wie dieser isoliert stehende Satz.

Eine leere Umgebung verbessert die Wahrnehmung eines informationsreichen Zentrums.

Geschickt eingesetzt, müssen leere Flächen auf dem Bildschirm also nicht unbedingt auch ungenutzte Flächen sein. So wie der Hintergrund eines Prozeßbildes dessen Wahrnehmung beeinflußt, tut es auch dessen „Nebengrund".

Ein weiterer nicht zu übersehender, aber oft ignorierter Aspekt sind kulturelle Einflüsse. Fast ausnahmslos alle Prozeßbilder folgen dem Technologieablauf von links nach rechts. Dies ist kein Zufall, sondern Folge der unserem Kulturkreis eigenen Schriftausrichtung von links nach rechts. Das ist ein Beispiel, wie die Art der Visualisierung (Schrift) unsere Art der Wahrnehmung beeinflußt. Kaum ein Entwickler denkt beim Bilddesign bewußt daran, er tut es einfach: die Dinge entwickeln sich für uns völlig natürlich von links nach rechts, wie auch oft von oben nach unten (z.B. Trenddiagramme, Ablaufpläne; s.a. Bild 1). Jede dem widersprechende Darstellung würde als unnatürlich empfunden und läßt uns stocken, da dies nicht *erwartungskonform* wäre [13].

Wenn zwei Sachen durch Abhängigkeit miteinander verknüpft sind, steht die Ursache zumeist links, die Wirkung folgt rechts, auch wenn keinerlei Schriftzusammenhang diese visuelle Reihenfolge erzwingt. Viele der in Deutschland empfangbaren Fernsehsender haben ihr Logo (links) oben: *weil Du uns eingeschaltet hast, bekommst Du das Bild (rechts) unten zu sehen.*

Für die Wahrnehmung folgt, daß ein Objekt um so mehr als Ursache betrachtet wird, je weiter links (und oben) es erscheint, und um so mehr als Folge, je weiter rechts (und unten). Da sich der Zusammenhang zwischen mehreren Vorgängen erfahrungsgemäß oft durch ihre zeitliche

Reihenfolge als Ursache und Wirkung erschließt, folgt bei einer klar links-rechts differenzierten Darstellung die Wahrnehmung auch zeitlich dem logischen Gang der Ereignisse.

4 Ergonomische Prozessbildkritik am Beispiel

Bild 2 zeigt die topologische Prozeßvisualisierung einer Destillationskolonne, wie sie in der
petrolchemischen Industrie Anwendung findet [14]. Bei dem betrachteten technologischen
Prozeß handelt es sich um die thermische, also physikalische Trennung eines Flüssigkeits-
gemisches durch die Ausnutzung unterschiedlicher Siedetemperaturen.

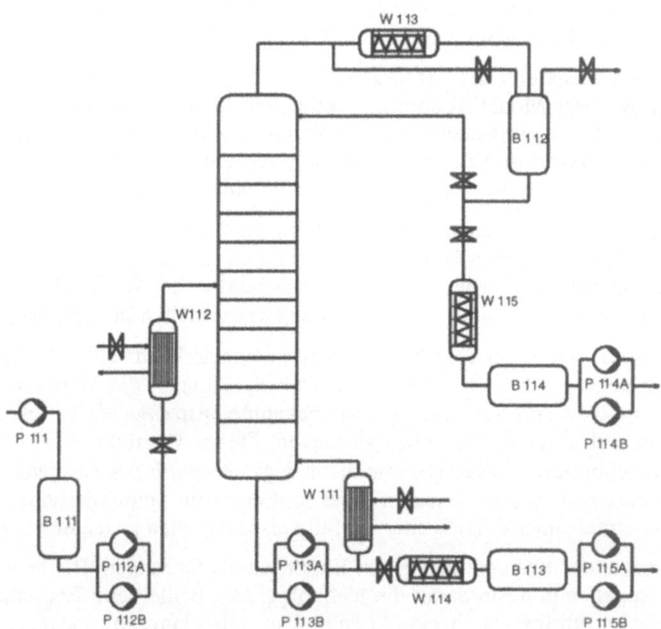

Bild 2: Topologisches Prozessbild einer Destillationskolonne (nach [15])

Im Inneren der im Zentrum des Bildes dargestellten Destillationskolonne wird eine Temperatur aufrechterhalten, die zwischen der Siedetemperatur des leichtersiedenden Stoffes und
der des schwerersiedenden liegt. Dadurch verdampft ersterer und steigt zum Kopf der Kolonne, wohingegen sich letzterer am Boden der Kolonne als Sumpfprodukt sammelt. Zur Steigerung der Trenngüte wird ein Teil der eben separierten Anteile der Kolonne erneut zugeführt.

Betrachten wir nun als erstes detaillierter das Kolonnen-Teilsystem ‚Produktzuführung'.

Möglicherweise haben Sie eben intuitiv zuerst auf den linken Bildteil geschaut, obwohl sie
das Bild vorher noch nie gesehen haben. Dazu mußte hier das Wort „Produktzuführung" statt
des in der Technik üblichen „*Ausgang*sproduktes" verwendet werden, um die Assoziation mit
‚Ausgang' zu vermeiden. Das ist übrigens unter ergonomischem Gesichtspunkt ein extrem
gutes Beispiel für ein extrem schlechtes Mapping zwischen Inhalt (*Input*) und Form („*Ausgang*sprodukt") und kann getrost als sprachlich-ergonomischer SuperGAU betrachtet werden.

Formassoziation und Inhalt sind hier durch den theoretisch wie praktisch größtmöglichen Abstand in einem System getrennt, obwohl das eigentliche Ziel Deckungsgleichheit wäre.

Also Input. Die Pumpe P111 pumpt das Flüssigkeitsgemisch in den Vorratsbehälter B111, aus diesem fördern es die Pumpen P112A/B durch das nachfolgende Ventil und den Erwärmer W112 in die Kolonne. Die Darstellung dieses Teilsystems („Gemischzulauf.") verstößt gleich gegen mehrere ergonomischen Prinzipien und behindert dadurch die Wahrnehmung.

Folgt man dem technologischen Ablauf im Prozeßbild, muß man visuell vom Teilsystemeingang bis zu dessen Ausgang viermal „um die Ecke wahrnehmen", genauergesagt 4 mal 90°, macht insgesamt 360°, und das in einem wirklich simplen Prozeßabschnitt. Das verstößt gegen das Prinzip der ‚Guten Fortführung', da der natürliche, geradlinige Fluß der Wahrnehmung praktisch nach jeder Komponente unterbrochen wird.

Da aber wie oben ausgeführt die Art und Weise der grafischen Darstellung wesentlich Wahrnehmung und mentales Modell bestimmt, ergibt sich die Notwendigkeit, nach jeder Komponente nicht nur wahrnehmungstechnisch, sondern auch mental ‚neu anzusetzen'. Daraus folgt ein Geschwindigkeitsverlust bei der Wahrnehmung, und durch die fragmentierte Wahrnehmung auch ein fragmentiertes Prozeßmodell, was gefährliche Konsequenzen haben kann.

Sicher ist es ein kritischer Zustand, wenn die Pumpen P112A oder B gegen das *geschlossene* folgende Ventil laufen. Optisch ist dies aber nicht wahrnehmbar, sondern erschließt sich erst mühsam durch bewußte kognitive Überlegung. Das liegt an der Winkeldistanz zwischen Pumpen und Ventil von 90° *und* an der Nichtbeachtung des visuellen Integrationsprinzips ‚Nähe'.

Ein weiterer Aspekt wurde im Text bereits durch geeignete Wortwahl unterhalb der bewußten Wahrnehmung angesprochen. Es hieß oben, daß die Pumpen das Gemisch in die Kolonne *fördern*. Das assoziiert eine aus dem Bergbau bekannte *anstrengende* Tätigkeit, etwas aus der Tiefe gegen die Schwerkraft nach oben zu hieven. Dieses Wortbild entspricht der Symbolik des obigen Prozeßbildes: der Teilsystemeinlauf liegt *unterhalb* des Ausgangs, noch tiefer die Pumpen. Das mag real so sein, assoziiert aber statt eines für kontinuierlichen Kolonnebetrieb unerläßlichen permanenten Zulaufs eher ein fallweises Arbeiten gegen einen Widerstand.

Nun zu den rechts der Kolonne gelegenen Endproduktabführungen. Bevor wir näher auf die Technologie eingehen und Sie damit die technologische Brille beim Betrachten übergestülpt bekommen, die im übrigen auch eine kognitive ist (also langsam und ressourcenverschlingend), testen Sie die ergonomische Qualität dieses Prozeßbildes durch schnellstmögliche Produktstromverfolgung und Zuordnung der 2 (!) Endproduktausgänge.

Ok, die halbe Stunde ist um, fahren wir fort.

Die Notwendigkeit, bei der eben vorgeschlagenen Aufgabe überhaupt nachdenken zu müssen, liegt -wieder- an der Nichtbeachtung des Visualisierungsprinzips ‚Nähe'. Dieses kann nicht nur objektgruppierend, sondern auch -separierend wirken. Die Ausgänge der beiden Endprodukt-Vorratsbehälter B114 und B113 sind viel zu nahe, verglichen mit der Distanz zwischen dem Behälterausgang B114 und dem nur gelegentlich aktiven Bypass oberhalb von B112.

Weiterer Schwachpunkt ist die unförmige Figur, die den Kopfprodukt*kreis*lauf darstellt. Statt eines schnell wahrnehmbaren Rechteckes bläst das vorliegende Gebilde die Winkelsumme der Verbindungslinien von 270° auf 450° auf. Außerdem wird durch die in der Senkrechten erfolgende Gemischteilung und den Zulauf von RECHTS aus B112 die Gleichwertigkeit beider Fließalternativen nicht transparent. Grund ist, daß Symmetrie besser horizontal wahrnehmbar ist. Vom Behälterausgang gesehen, müssen schließlich beide Teilströme einen Umweg von je 270° überwinden, ehe sie in ihrer beabsichtigten Richtung weiterfliessen können.

Zum Sumpfabschnitt. Auch dieser ist ein dankbares Objekt für ergonomische Betrachtungen. Offensichtlich handelt es sich technologisch bei den beiden unmittelbar nach den Zwillingspumpen 113A/B entstehenden Teilströmen um gleichrangige Wege, denn sowohl die im Rücklauf erfolgende Kolonnenbeheizung als auch die Produktabführung erfolgen kontinuierlich. Optisch gibt es aber einen waagerechten Hauptstrang der Produktabführung, während der Rücklauf durch seinen 90°-Versatz nach oben optisch zum Nebenstrang degradiert wird (wie der Bypass im Kopfabschnitt, wo das gewollt ist). Weiterhin ragt der Erhitzer W111 durch die gedrängte Enge in den „Wahrnehmungsraum" der Produktabführung hinein und behindert die ablenkungsfreie Verfolgung der Produktabführung. Insbesondere erschwert er die akkurate Wahrnehmung, daß die Produktstromaufteilung dadurch erfolgt, daß die Pumpen das Sumpfprodukt gegen das mehr oder weniger geschlossene Ventil vor Kühler W114 pressen.

Daneben wird durch die rechtsseitige Lage der Heizung W111 zur Destillationskolonne (Wirkungsseite statt Ursachenseite) ihre essentielle Bedeutung für den Kolonnenbetrieb ‚unwahrnehmbar'. Die aktuelle Variante suggeriert, daß sie ausschließlich zur Aufrechterhaltung der Temperatur der Produktrückführung dient und nicht zur Beheizung der gesamten Kolonne. *Diese* Aufgabe scheint, wenn überhaupt wahrnehmbar, der im Gemischzulaufstrang liegenden Heizung W112 zuzufallen, also den tatsächlichen Gegebenheiten diametral entgegengesetzt!

5 Ergonomisch optimiertes Prozessbild

Der vorige Abschnitt zeigte, wie durch eine techniklastige Visualisierung aus dem Bauch heraus und Nichtbeachtung relativ simpler ergonomischer Stolpersteine eine beliebig komplexe, unlesbare Prozeßdarstellung entsteht (da auch geschulte Operateure, und gerade sie, schon vorab ihre subjektive technologische Brille aufhaben, läßt sich *dieser* Effekt übrigens *nicht* durch partizipatives Design verhindern, das andere Stärken hat [16]).

In der Praxis wird nun gern versucht, solche Darstellungen durch massiven Farbeinsatz wieder transparent zu gestalten. Diese Farbe kann dann folgerichtig nicht mehr für die Übermittlung anderer wichtiger Informationen verwendet werden, ohne den Betrachter zu verwirren.

Bild 3 zeigt einen Vorschlag zur Verbesserung durch *Beachtung ergonomischer Geometriekriterien ohne Farbeinsatz, allein durch wahrnehmungsorientierte Strukturoptimierung.*

Auf die einzelnen ergonomischen Besonderheiten soll hier nach der ausführlichen Betrachtung der Schwächen im ersten Bild etwas kürzer eingegangen werden. Auch soll nach der erfolgten Sensibilisierung für spezifische wahrnehmungstechnische Abläufe eine Bewertung, ob eine Verbesserung erreicht wurde, dem geneigten, also kritischen Leser überlassen bleiben.

Die horizontale Streckung unterstützt die klare wahrnehmungstechnische und damit mentale Unterscheidung der vier Untersysteme Gemischzulauf, Kolonne, Kopfbereich und Sumpfbereich und verbessert die Wahrnehmung ihrer jeweiligen inneren funktionellen Zusammenhänge. Dadurch wird der aus der Systemtheorie bekannte Ansatz der Lösung komplexer Probleme durch deren Aufspaltung in kleinere Teilprobleme optisch unterstützt. Als meßbares Er-

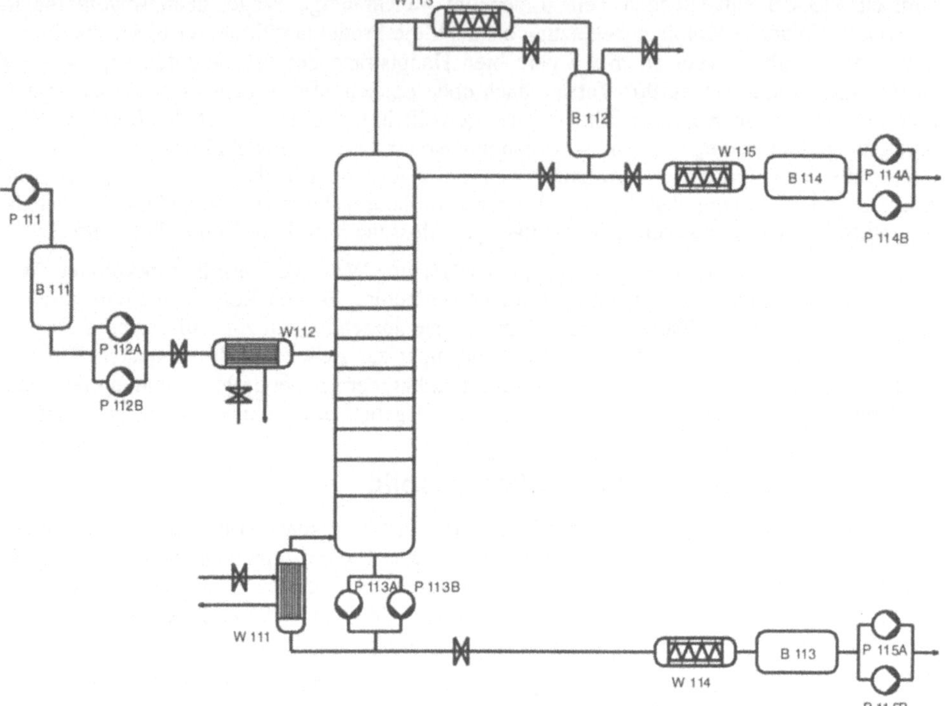

gebnis entfallen darstellungsweit 450° oder ein Drittel unnötiger Verbindungslinien-Winkel.

Bild 3: Ergonomisch optimiertes Topologisches Prozessbild

Die Visualisierung des Zulaufs als Gefällstrecke symbolisiert den kontinuierlichen Gemischzulauf zur Kolonne. Sowohl im Kopf- als auch Sumpfbereich soll die *horizontale* Teilung der Stoffströme die Gleichrangigkeit der jeweiligen Fließrichtungen wahrnehmbar machen, die anschließend nach links (technologisch zurück) oder rechts (technologisch voran) weiterfliessen. Die beiden Kreisläufe sind im Kopfbereich durch Herstellung einer geschlossen Figur bzw. im Sumpfbereich durch klare räumliche Abgrenzung zum korrespondierenden Produktablauf transparenter geworden. Die linksseitige, enge Position der Heizung W111 zur Kolonne unterstreicht ihre essentielle Bedeutung für den Kolonnenbetrieb.

Übrigens, die schnelle Wahrnehmung der beiden Ausgangsproduktströme sollte nun auch kein großes Problem mehr darstellen. Neben der verbesserten *Guten Fortführung* wurde sowohl durch optisches Herunterklappen des Kühlers W116 die *Ähnlichkeit* der beiden Hauptausgangsströme erhöht als auch durch größere horizontale Distanz vom Bypaßausgang zur Kopfproduktentnahme deren eindeutig größere Zugehörigkeit zur Wirkungsseite betont.

Was die Beschriftung in beiden Bildern angeht, habe ich Sie einfach mal mit den Widrigkeiten einer techniklastigen Prozessvisualisierung alleingelassen. Die Chance, hier *irgendeine* Namenslogik zu finden, die die Zusammenhänge transparenter macht als die existierende, ist kaum zu verfehlen. Oder hätten Sie vorab vermutet, daß die Komponenten W114, B113 und P115 in genau dieser Reihenfolge unmittelbar aufeinanderfolgen?

6 Visualisierungsleitlinen und Bewertungskriterien

Ausgangspunkt sind die bereits zitierten gestaltpsychologischen Gestaltungsgrundsätze Nähe, Ähnlichkeit, Geschlossenheit und Gute Fortführung. Für ihre Umsetzung gilt der Leitsatz:

Visualisierungsziel ist die gute Ausnutzung der mentalen Ressourcen, nicht die der Bildschirmfläche!

- Die Visualisierung eines komplexen Sachverhaltes muß grafisch so strukturiert sein, daß seine Zergliederung in abgeschlossene Teilaspekte augenfällig ist. Optische Nähe drückt Zusammengehörigkeit aus, optische Distanz impliziert logischen oder zeitlichen Abstand.

- Diese Zergliederung als auch die integrative Wahrnehmung der Teilaspekte verbessern sich durch Abbau optischer Zerklüftung mittels augenfälliger optischer Logik (*eine* Hauptentwicklungsrichtung/geometrische Figur) als Ausdruck innerer Zusammenhänge.

- Die Lage der optischen Hauptentwicklungsrichtung ist kulturell bestimmt; Ursachen sollten in unserem Kulturkreis möglichst weit links (und oben) erscheinen, Wirkung(en) dagegen relativ davon rechts (und unten).

- Die konsequente Einhaltung der Links-Rechts-Regel simplifiziert die schnelle Zuordnung von Ursache und Wirkung.

- Kreisläufe sollten in einer harmonischen, möglichst einfachen und in sich geschlossen wahrnehmbaren geometrischen Figur (Rechteck o.ä.) dargestellt werden.

- Für die Darstellung direkter Zusammenhänge, symbolisiert z.B. durch Verbindungslinien, sollte ein möglichst kurzer, kreuzungs- und kurvenarmer Weg gewählt werden unter Beachtung jeweiliger Designabreden, wie etwa keine schrägen Verbindungslinien usw.

- Die menschliche Wahrnehmung leitet aus der Schwerkraft im Realen eine ‚optische Gravitation' ab. Gleichwertige Wahrnehmung zweier Alternativen entsteht bei gleichem optischem Gefälle bezüglich dieser Gravitation (z.B. keines bei horizontaler Aufspaltung).

- Bedingt durch diese optische Gravitation, symbolisieren fallende Flüsse Leichtgängigkeit, z.B. für kontinuierliche Vorgänge, wohingegen steigende Flüsse eher eine anstrengende Tätigkeit assoziieren (sporadischer Betrieb, aufwendige Methode o.ä.)

- Bei ungleichwertigen Alternativen sollte sich die Nebenalternative optisch aus der als Hauptstrang visualisierten Hauptalternative abspalten, z.B. im 90°-Versatz. Auch eine gegenläufige Aufspaltung in der Vertikalen kann diesen Wahrnehmungseffekt auslösen.

- Logisch getrennte Teilaspekte sollten einen optischen Mindestabstand zueinander wahren.

- Eine leere Umgebung verbessert die Wahrnehmung eines informationsreichen Zentrums.

Über die jeweilige Anwendung muß fallweise nach Zweckmäßigkeit entschieden werden. Es läßt sich aber folgendes allgemeingültiges „*Formel-1-Prinzip der Visualisierung*" postulieren:

Kurvenarme und kreuzungsfreie Geraden sowie breite leere Seitenstreifen erhöhen die Wahrnehmungsgeschwindigkeit; unnötige Schikanen sind zu vermeiden !

7 Zusammenfassung

Universelle ergonomische Designleitlinien und Bewertungskriterien wurden abgeleitet.

An dem allgemein bekannten Beispiel topologische Prozessvisualisierung wurde auf ergonomische Probleme und Folgen ihrer Nichtbeachtung aufmerksam gemacht. An einer verbesserten Visualisierungvariante wurde demonstriert, wie sich durch Anwendung einfachster ergonomischer Prinzipien Bildtransparenz und Wahrnehmungsgeschwindigkeit steigern lassen. Dazu mußte die bisherige Lesart der Topologie als maßstäblicher geografischer Abbildung der realen 3D-Welt im 2D-Bild aufgegeben werden. Diese neue Sichtweise ist erst möglich, seit mit dem weltweiten Trend zur Fernwartung, auch und zunehmend via Internet, eine noch stärkere Entfremdung des Menschen vom überwachten Prozeß als bisher stattfindet. Als Folge davon tritt ein Abgleich zwischen Prozeßbild und realer Prozeßgeometrie in den Hintergrund, und es kommt zur Verwischung der Grenze zwischen realer und virtueller Welt.

8 Literatur

[1] G. Johannsen: Mensch-Maschine-Systeme. Springer, Berlin 1993

[2] J. Rasmussen: Models of Mental Strategies in Process Plant Diagnosis. In: J. Rasmussen, W.B. Rouse (Eds.): Human Detection and Diagnosis of System Failures. Plenum Press, New York 1981

[3] J. Rasmussen: Information Processing and Human-Machine Interaction.North Holland, New York 1986

[4] W.B. Rouse: Models of Human Problem Solving: Detection, Diagnosis, and Compensation for System Failures. In: Automatica – Special Issue on Control Frontiers in KB and MMS 1983, Vol. 19

[5] R.H. Kluwe: Problemlösen, Entscheiden und Denkfehler. In: C. Graf Hoyos, B. Zimolong (Hrsg.): Ingenieurpsychologie. Göttingen, Hogrefe 1990

[6] K.R. Hammond et al.: Direct Comparison of the Efficacy of Intuitive and Analytical Cognition in Expert Judgement. In: IEEE Trans. Systems, Man, and Cybernetics 1987, Vol. SMC-17

[7] H.-J. Charwat: Farbkonzept für die Prozeßführung mit Bildschirmen. In: atp, Oldenbourg, 1996, Nr. 5-7

[8] J. Wandmacher: Software-Ergonomie. de Gruyter Berlin New York 1993

[9] D.D. Woods: Visual Momentum: A concept to improve the cognitive coupling of person and computer. In: International Journal of Man-Machine Studies, 1984, No 21

[10] J.J. Gibson: The Ecological Approach to Visual Perception. Hillsdale (NJ) 1986

[11] K.B. Bennet & J.M. Flach: Graphical Displays: Implications for Divided Attention, Focused Attention, and Problem Solving. In: Human Factors 1992, Vol. 34, No. 5

[12] K.J. Vicente & J.Rasmussen: Ecological interface design: Theoretical Foundations. IEEE Transactions on Systems, Man, and Cybernetics 1992, Vol. 22

[13] Deutsches Institut für Normung: DIN EN ISO 9241: Ergonomische Anforderungen für Bürotätigkeiten mit Bildschirmgeräten. Teil 10: Grundsätze der Dialoggestaltung. Beuth 1996

[14] Gilles et al.: Ein Trainingssimulator zur Ausbildung von Betriebspersonal in der Chemischen Industrie. In: Automatisierungstechnische Praxis atp, Oldenbourg 1990, Nr. 7

[15] D. Settgast & F. Wiese: Analyse von Prozeßführungsproblemen bei einer Destillationskolonne. Universität Kassel 1993, FG Mench-Maschine-Systeme

[16] S. Ali, J. Heuer & M. Hollender: Partizipative Erstellung von Bedienoberflächen für Prozeßleitsysteme durch interaktive Gestaltung und Bewertung. Universität Kassel 1997, IMAT-MMS-Bericht 17

Diese Arbeit wurde ermöglicht durch ein von der DFG gefördertes Forschungsprojekt.

Adresse des Autors

Dipl.-Ing. Falk Mletzko, Systemtechnik und Mensch-Maschine-Systeme, Universität Kassel - FB 15, 34109 Kassel, eMail: fm@imat.maschinenbau.uni-kassel.de

Gestaltung von Software Agenten aus der Sicht des Benutzers

Jörg Nissler, Volker Thoma

Marktstrategieteam Interaktive Produkte, Fraunhofer IAO

Zusammenfassung

Software Agenten erledigen für ihre Besitzer die Informationssuche und andere Aktionen in elektronischen Diensten (z.B. im Internet). Die Bereitstellung interaktiver Dienste - wie z.B. Online-Shopping - über den Fernseher bietet die Möglichkeit, ein breites Kundenspektrum zu erreichen. Die Gestaltung der Benutzungsschnittstelle erfordert daher allerdings, daß sie von einem sehr heterogenem Anwenderkreis bedient werden kann und stellt hohe ergonomische Anforderungen an ein benutzerzentriertes Design. Zum einen wird der private Nutzer ohne Vorkenntnisse vor zunehmend komplexere Aufgaben gestellt, zum anderen müssen bei der Bedienung über den TV die eingeschränkten Möglichkeiten der direkten Manipulation berücksichtigt werden müssen [1]. Am Fraunhofer IAO wurde daher untersucht, wie potentielle Einsatzszenarien und grafischen Umsetzungsmöglichkeiten für Agenten im Bereich Online-Dienste über ITV aussehen können, um dem privaten Nutzer einfache Bedienkonzepte bereitzustellen. Aufgaben, Aussehen und Interaktionsmöglichkeiten von (anthropomorphen) Agenten aus der Sicht der Benutzer wurden in einem Focus Group Workshop mit potentiellen Anwendern erhoben. Die Ergebnisse wurden in einem Benutzertest validiert, in denen Anwender mit potentiellen Agenten in Form von anthropomorphen Avataren und virtuellen Help-Roboter in 3D Multi User Welten interagierten. Die Ergebnisse zeigen, daß herkömmliche anthropomorphe Agentenkonzepte erheblicher Schwierigkeiten bei der Interaktion und Bedienung bereiten. einer eher skeptischen Beurteilung eines zu hohen Ausmaßes an Intelligenz und Autonomie von Agenten eine breite Akzeptanzbasis bei den Benutzern für antropomorphen digitale Helfer vorhanden ist.

Die durch den Workshop und den Benutzertests gewonnenen Aspekte und Probleme hinsichtlich der gewünschten Dialogform, der Interaktion, dem Grad der anthropomorphen, realitätsnahen Gestaltung und der Intelligenz von Software-Agenten, waren Grundlage für eine internationale Benutzerbefragung sowie einer bevölkerungsrepräsentativen Fragebogenaktion in Deutschland.

1 Agenten für Online-Dienste im Fernsehen

1.1 Einleitung

„Ein Agent ist ein Stück Software, das Dinge für einen tut, die man vermutlich auch selber tun könnte, wenn man die Zeit dafür hätte" (Selker, IBM). Die von Ted Selker aufgestellte Agentendefinition erweckt große Erwartungen. Doch was muß passieren, daß diese Definition tatsächlich auch ihre Richtigkeit erhält. Welche Dinge sind das, die man erledigen lassen möchte; wie kann erreicht werden, daß ein einfacher Zugriff auch tatsächlich zur Zeitersparnis führt; und wie kann auch der private Nutzer von diesen Möglichkeiten profitieren?

Wenn man die rasante technische Entwicklungen der multimedialen und breitbandigen Datenübertragung und die zunehmende Verbreitung von Online-Diensten betrachtet, wird deutlich, daß elektronische Serviceangebote zunehmend auch den privaten Nutzer erreichen. Das Internet wächst unaufhaltsam; Experten prognostizieren, daß es in nahezu allen Bereichen des privaten Lebens Einzug halten wird und neue Dienste, vor allem im Bereich Infotainment, Edutainment und Electronic-Commerce, ein großes marktwirtschaftliches Potential besitzen

[2]. Die Akzeptanz, insbesondere beim privaten Nutzer, ist allerdings noch sehr stark von den tatsächlichen Möglichkeiten und der Bedienbarkeit, dem User Interface der Anwendung abhängig. Ein potentieller Ansatz Online-Dienste dem privaten Konsumenten zur Verfügung zu stellen, ist der Zugang über den Fernseher. Das Ziel interaktive Dienste über den Fernseher bereitzustellen, stellt allerdings hohe Herausforderungen an die Benutzungsschnittstelle. Mehrere Faktoren beeinflussen eine erfolgreiche Realisierung dieser Art des Zugangs zu Online-Diensten auf dem Fernseher. Aufgrund der, durch technische Restriktionen resultierenden, eingeschränkten technischen Visualisierungsmöglichkeiten [3] - wie die Verwendung großer Schriftgrößen (aufgrund geringer Auflösung) und der spezifischen Nutzungssituation (Abstand Benutzer UI, Verwendung von Fernbedienungen) - ergeben sich beträchtliche Restriktionen hinsichtlich Informationsdarstellung und Navigation in interaktiven Diensten. Eine Möglichkeit zur Vereinfachung der Nutzung komplexer Online-Dienste sind Agentenkonzepte in Form von persönlichen Assistenten. Diese Software-Agenten können z.T. autonom handeln und die Benutzer bei der Navigation und Interaktion sowie dem Abruf neuer Informationen oder auch beim Einkauf im Netz unterstützen [4,5,6]. Auch wenn die Verwendung von Agenten dem Nutzer nicht notwendigerweise direkt transparent gemacht werden muß, geschweige denn eine menschliche oder anthropomorphe Visualisierung unbedingt erforderlich ist, so liegt es doch nahe Agenten als mehr oder weniger menschliche Figur zu visualisieren. Dies scheint vor allem im privaten Bereich zuzutreffen, wo Aspekte wie intuitive Bedienbarkeit, Spaß und Unterhaltung eine noch wichtigere Rolle spielen als beispielsweise bei „Office" Anwendungen. Eine erste Auswahl denkbarer Darstellungsformen ist in Abbildung 1 skizziert.

Abb. 1: Mögliche Darstellung anthropomorpher Agenten in Online-Shopping Diensten

1.2 Hintergrund der Studie

Innerhalb des EU Projektes OCEANS (AC323, Open Communication Environment for Agent based Networked Services) werden netzwerkunterstützte Agentenkonzepte entwickelt, die dem privaten Nutzer kundenorientierte Anwendungen elektronischer Dienste über das Interaktive Fernsehen ermöglichen. Dabei wird in einem ersten Schritt untersucht, was künftige Benutzer beim Einsatz von Agenten im Bereich interaktiver Dienste im Fernsehen erwarten. Innerhalb eines iterativen Entwicklungsprozesses wurde daher zur Anforderungsanalyse ein ganztägiger Workshop durchgeführt, der unter Zuhilfenahme von Brainstroming-Methoden und Szenario Techniken erste Antworten zu potentiellen Einsatz- und Visualisierungsmöglichkeiten von Agenten erarbeitete. Die daraus gewonnenen Erkenntnisse dienten neben der Gewinnung von ersten Gestaltungsempfehlungen zur Konkretisierung und Ableitung von Fragestellungen für eine internationale Benutzerbefragung in fünf europäischen Ländern. Außerdem werden ausgewählte Fragestellungen in einer bevölkerungsrepräsentativen Fragebogenaktion in Deutschland evaluiert.

Als ein potentielles Anwendungsfeld für den Einsatz von visualisierten Software Agenten können 3D-Welten für virtuelle Communities, Malls und Online-Shops betrachtet werden. Daher wurden erste Ergebnisse des Workshops, die von dem Wunsch der Benutzer nach einer mehr oder weniger stark personifizierten und anthropomorphen Darstellung von Agenten ausgeht [7,8,9], innerhalb eines Benutzertest auf ihre Realisierbarkeit hin überprüft. Dabei wurde eine Evaluation von Interaktions- und Dialogformen von Benutzern mit Avataren in virtuellen 3D Welten durchgeführt. Als Avatare bezeichnet man mehr oder weniger abstrakte 3D-Figuren, die den User oder autonome 3D-Roboter in virtuellen Welten visualisieren und repräsentieren. Es sollte bei diesem Test weniger die Frage im Vordergrund stehen, ob ein Agent überhaupt visuell in Erscheinung treten soll oder nicht, sondern wie die Interaktion ausgeprägt sein kann.

2   Workshop

Zur Gewinnung von Gestaltungsrichtlinien für Agenten sollten neben den Erkenntnissen aus der klassischen Software-Ergonomie [1,14], aus Studien [15], sowie einer umfangreichen Literaturrecherche [10,11,12,13] potentielle Benutzer in einem Workshop beteiligt werden.. Als Ergebnis der Focus Group Diskussionen wurde eine Liste von Fragestellungen und Parametern erarbeitet, die für die weitere Entwicklung von Case Szenarien, User Interfaces und Dialogformen im Bereich agentenbasierter Dienste verwendet wird. Dazu wurde ein eintägiger Workshop mit 20 potentiellen Benutzern am Fraunhofer IAO durchgeführt. Die Teilnehmergruppe zeichnete sich durch eine große Heterogenität bezüglich Alter, Geschlecht, Berufsstand und PC-/Interneterfahrung aus. Es nahmen 12 männliche und 8 weibliche Teilnehmer im Alter von 15-78 Jahren teil. Um eine einheitliche Basis (Common-Understanding) für die Diskussion zu schaffen, wurde eine kurze Einführung in die technischen Grundvoraussetzungen und generelle Problematik gegeben. Wichtig war, ein möglichst umfassendes Bild über Potentiale der Agententechnologie zu vermitteln, ohne dabei die Kreativität und Unvoreingenommenheit der Teilnehmer einzuschränken.

2.1   Methoden des Workshops

Das Schwerpunkt des Workshops war die Erstellung und die Sammlung von Ideen, Meinungen und Einschätzungen von potentiellen Benutzern zum Thema Agenten im weiteren Sinne, sowie in Bezug auf die gestellten Themenfelder und Fragestellungen. Es wurde die Brainwriting Methode („6-3-5") angewendet und mit Hilfe der Metaplan-Technik eine Kategorisierung und Bewertung vorgenommen. Moderatoren des Fraunhofer IAO dienten lediglich zur Unterstützung der Prozesse und der Zusammenfassung der Ergebnisse.

2.2   Themenfelder des Workshops

Als potentielle Themenfelder wurden für den Einsatz von Agenten in ITV Online-Diensten vier Bereiche vorab ausgewählt:

1. Einkaufen direkt am Fernseher, Online Shopping via TV (Bummeln in virtuellen Läden, Produkte auswählen, anschauen, vergleichen, handeln und kaufen)
2. Persönliche Zeitung aus dem Internet, Personal Newspaper (Zusammenstellen von Informationen nach individuellen Interessensgebieten, Nachrichten auswählen nach persönlichen Schwerpunkten)
3. Persönliche Fernsehzeitung, Personal Electronic Program Guide/EPG (Auswahl von Sendungen aus wachsendem TV-Angebot, persönliches Fernsehprogramm auswählen, Fernsehzeitung über TV abrufbar, Zusatzinformationen, Querverweise)
4. Austauschbörse von privaten Dienstleistungen, Swap Shop (intelligentes „Schwarzes Brett", Suche und Biete, Dienstleistungen, Spiele, Partner)

2.3 Fragestellungen des Workshops

Auf folgende Fragestellungen wurden die Diskussionen innerhalb des Workshops fokussiert, mit den jeweiligen thematischen Schwerpunkten als gemeinsame Grundlage:

- Welche Aufgaben können von Agenten innerhalb des gestellten Anwendungsfeldes sinnvoll übernommen werden? (Was erwartet der Benutzer Zuhause von Agenten? Welche Anforderungen hat der Benutzer an die Agenten?)
- Wie „clever" soll ein Agent sein? (Wieviel Eigeninitiative soll er mitbringen? Soll der Agent lernfähig sein? Wie selbständig soll er sein? Welche Verantwortung würden Sie einem Agenten übertragen?
- Wie müssen Agenten gestaltet sein, um den Erwartungen des privaten Nutzers zu entsprechen? (Wie sollte sich der Agenten präsentieren? Soll der Agent überhaupt in Erscheinung treten? Welche Art der Darstellung für die Kommunikation mit Agenten sollte gewählt werden?)
- Wie präsentiert sich das Ergebnis der Dienste, die durch Agenten erbracht wurden? (Äußere Form der gefundenen Information? Welche Darbietungsformen der Ergebnisse sind denkbar/wünschenswert?

2.4 Ergebnisse des Workshops

Die Diskussionsgruppen sollten sich auf gewisse Aussagen hinsichtlich der allgemeinen Gestaltungskriterien in ihren Themenfeldern einigen. Die wesentlichen Ergebnisse sind unten aufgeführt. Dabei wurde noch keine Interpretation vorgenommen:

Potentielle Aufgaben von Agenten

- Erweiterte Suchfunktionalitäten aller Art werden gewünscht (Hersteller-, Produktsuche, Produktvergleich).
- Bereitstellung von Backgroundinformationen zu gesuchten Produkten/Themen sinnvoll.
- Querverweise aller Art zu verwandten Themen sollten vom Agenten aufbereitet werden.
- Agent sollte eine Vorauswahl von gefundenen Einträgen und Produkten unter Berücksichtigung des Benutzerprofiles treffen.
- Agent sollte Ranking der gefundenen Einträge und Produkte selbständig vornehmen.
- Ein virtueller Assistent als ständiger Begleiter (permanente Verfügbarkeit) ist gewünscht.
- Transaktionen mit limitierbarer Budgetierung sollte vom Agenten übernommen werden.

- Umfangreiche Speicher- und Archivierungsfunktionen für gefundenen Einträge und Produktinformationen erscheint den Teilnehmern wünschenswert.
- Jederzeit soll der Benutzer einen Statusbericht über die Aktivitäten des Benutzers erhalten oder ein Zwischenergebnisse abrufen können.
- Die Möglichkeit Ergebnisse auszudrucken sollte integriert werden.
- Flexible, vom Nutzer individuell einzustellende Wiederholfrequenzen und Intervalle der Aufgabenabarbeitung sind gewünscht.
- Eine „Kindersicherung", um bestimmten Nutzergruppen eingeschränkten Zugang zu Agentenfunktionen zu bieten, sollte implementiert sein (z.B. für Transaktionen).
- Der Agent sollte in der Lage sein, Werbung aus gesuchten oder gefundenen Informationen und Angeboten herauszufiltern (?!).

Gewünschter Grad an Autonomie und Intelligenz von Agenten

- Eigenständiges Handeln und Ausführen von Transaktionen ist von den meisten Teilnehmern gewünscht, allerdings scheint uneinheitliche Übereinstimmung bezüglich dem Grad der Autonomie des Agenten.
- Der Agent sollte selbständig und „intelligent" mit dem Benutzer interagieren und bei Bedarf aktiv nach fragen (pro-aktiv).
- Die Autonomie von Agenten sollte je nach individuellen Benutzervorstellungen und den zu bearbeiteten Aufgaben angepaßt sein.
- Der Sicherheitsaspekt von Agenten wird sowohl als wichtig als auch für problematisch erachtet.
- Die Idee des „Feilschens" schien den Teilnehmern interessant, allerdings auch problematisch zu sein (so kann z.B. der Agent den Preis drücken, allerdings aber auch über den Tisch gezogen werden).
- Die Idee eines Kurz- und Langzeitgedächtnisses, in Bezug auf das vom Agenten anzuwendende Benutzerprofiles, wurde als mögliche Lösung für den optimalen Grad der Autonomie und Intelligenz des Agenten vorgeschlagen:
  - Kurzzeitgedächtnis z.B. für die Suche nach aktuellen Produkten; strikte Beachtung des User Profiles beeinträchtigt unter Umständen Kreativität und Innovation des Agenten.
  - Langzeitgedächtnis liefert Ergebnisse in Abhängigkeit zum User Profile.
- Der Agent sollte nicht intelligenter sein als sein Besitzer - zumindest sollte er es dem Benutzer nicht zeigen (→ „Charming Agent").
- Die Unterstützung des Agenten sollte nach Vorstellung der meisten Benutzer in jedem Fall Zeitersparnis bringen und nicht zusätzlich Zeit kosten (Agent-Using = Time Saving).

Visualisierung, Präsentation und Interaktion mit Agenten

- Die einfache, intuitive Bedienung von Agenten muß im Vordergrund stehen.
- Der Agent sollte in Bezug auf Präsentation und Interaktion sehr benutzerfreundlich sein.
- Ein schneller Zugang zu Agenten muß gewährleistet sein.
- Es wird ein persönlicher Assistent gewünscht, mit unterschiedlichen Ausprägungen und Erscheinungsformen in Abhängigkeit von:
  - Individuellen Vorlieben und Vorstellungen
  - Status des Agenten bzw. der Aufgabenerfüllung

- Die Möglichkeit aus einem Baukasten seinen persönlichen Agenten zu selektieren oder gar selbst zu entwerfen wird begrüßt.
- Wiedererkennbarkeit des Agenten ist erwünscht, auch bei flexiblem Erscheinungsbild.
- Trend zu anthropomorphen Visualisierung von Agenten ist erkennbar (menschliches oder menschenähnliches Aussehen – z.B. auch Comicfiguren, Gesichtszüge, Augen, Mund).
- Trend zu anthropomorphen Verhaltensweisen von Agenten ist erkennbar (menschliches oder menschenähnliches Verhalten, Gesten, Emotionen, Sprache).
- Agent sollte „charmant" sein und sich stets diskret und zurückhaltend verhalten.

Visualisierung von gelieferten Daten (Einträge, Infos, Produkte, Querverweise)

- Der Agent sollte in der Lage sein, eine Vorselektion von Einträgen zu machen in Abhängigkeit von folgenden Parametern wie: Quelle und Herkunft des Eintrages; Trefferqualität; Wichtigkeit; Kategorie.
- Als konkrete Darstellungsform und Visualisierungsmetapher für die Präsentation von Ergebnissen wurden u.a. folgende Vorschläge genannt: Litfaßsäule (parallel in 2 Teams, sowohl für die Verwendung im Bereich News on Demand als auch für das persönliche Fernsehprogramm vorgeschlagen); Buchmetapher; Schwarzes Brett; Flip-Chart; Dart-Scheibe (Qualität des Ergebnisses im Zentrum); TV-orientierte Metaphern (Filmrolle).
- Der optimierte und angepaßter Einsatz von Multimedia (Sound/Film/Animation) wird gewünscht.
- Die Möglichkeit einer automatisierten Erstellung von Videoclips über gefundene Ergebnisse wurde vorgeschlagen (z.B. in Märchen- oder Nachrichtenform).
- Unterschiedliche Formen und Ebenen des Detailierungsgrades der präsentierten Ergebnisse (Vorschau, Zwischenergebnisse, Details) scheinen sinnvoll.

Der Workshop ergab eine Reihe von wichtigen Anregungen und Einschätzungen für den Einsatz und Interaktion mit Agenten. So war ein Trend zu persönlichen Assistenten mit permanenter Verfügbarkeit und konkreter Erscheinungsform zu erkennen. Dabei wurde bei der Frage nach dem Grad der realistischen/anthropomorphen Darstellung ein mittleres Maß gewünscht. Die Figur sollte glaubwürdig sein und Emotionen vermitteln können, allerdings nicht zu realistisch sein. Eine zu menschenähnliche Darstellung verunsichert die potentiellen Nutzer.

3 Benutzertest Interaktion mit anthropomorphen virtuellen Wesen

Um zusätzlich zu den qualitativen Ergebnissen des Workshops eine genauere Bewertung möglicher Interaktionsformen und den damit zusammenhängenden Schwierigkeiten, sowie der Akzeptanz bezüglich dem Umgang mit personifizierten Agenten bei potentiellen Kunden zu erhalten, wurden Benutzertests durchgeführt. Diese fanden in schon bestehenden Online Diensten in Form von 3D Multi User Welten (auch Multi User Dungeons - MUDs – genannt) statt. In ihnen können sich die Benutzer als virtuelle Verkörperungen (Avatare) bewegen und andere User (bzw. deren Avatare) treffen und kommunizieren. Diese Welten enthalten auch – meist anthropomorph dargestellte, autonome, mehr oder weniger intelligente - Robot Avatare, mit denen der Benutzer interagieren kann. Sie können dem Benutzer helfen, Orientierung bieten und Informationen bereitstellen. Diese Robot Avatare wurden zur Simulation von Software Agenten herangezogen. Die Versuche wurden in einer ersten Phase auf einer PC-

Plattform durchgeführt. In späteren Benutzertests wird die TV-Plattform mit ihren entsprechenden Restriktionen anvisiert.

Zehn Benutzer mit akademischen und nichtakademischen Hintergrund nahmen an den Benutzertests teil. Insgesamt hatten die Teilnehmer eher wenig Erfahrung mit Online-Diensten und dem Internet, um die Zielgruppe der potentiellen „neuen" Kunden von ITV Diensten zu repräsentieren . Dabei handelte es sich um die Multi User Welten "Active Worlds" [17] und "Blaxxun Community Server Ccpro 3.01"[16], die in Abbildung 2 dargestellt sind. Die Teilnehmer erhielten keine Einweisung, konnten aber während den Tests die Hilfefunktionen bzw. einen Ausdruck von Hilfethemen verwenden. Es wurden ein schriftliches Protokoll sowie Videoaufzeichnungen angefertigt. Ein Mitarbeiter der Evaluatoren steuerte als „Wizard" vom Regieraum des Usability Labors aus – unsichtbar für den Benutzer - einen Robot Avatar, der die Versuchsperson online unterstützen sollte. Die Teilnehmer bearbeiteten zwölf Navigations- und Interaktionsaufgaben, so z.B. Umhergehen in der 3D Welt, Wechseln der Welt, „chatten" mit anderen Avataren, Auslösen von Gesten, und so fort. Nach der Bearbeitung der Aufgaben mußten die Teilnehmer Fragebögen zur Benutzungsfreundlichkeit der Oberflächen sowie Präferenzbewertung ausfüllen sowie dem Evaluator einige zusätzliche Fragen in einem Interview beantworten.

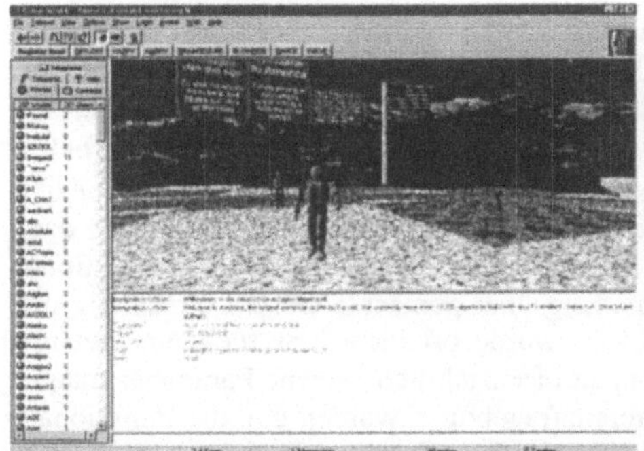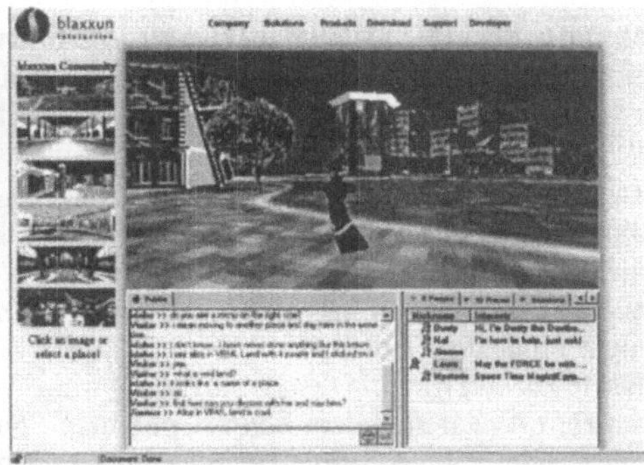

Abb.2: Die Multi-User Welten ActiveWorlds (links) und Blaxxun Ccpro 3.01(rechts). Die Oberflächen enthalten jeweils ein VRML Fenster mit den entsprechenden Welten in der Mitte des Screens. Das Chatprotokoll ist unterhalb des VRML Fensters. Bei Blaxxun kann man hier zusätzlich Gesten auswählen, die ein Avatar macht. Über „Tab Controls" können Anwender Welten wechseln und zahlreiche Einstellungen machen (links bei Active World, bzw. rechts unten bei Blaxxun). Active World bietet eine spezielle Menüleiste mit – je nach Welt verfügbaren –Gesten (oben).

Ergebnisse

Die Navigations- und Chatfunktionalitäten stellten die Benutzer vor große Probleme. So fand nur ein Benutzer die Eingabezeile für Text ohne Hilfe, da sie unauffällig wie eine Statusleiste war und kein Prompt enthielt. Statt dessen klickten die Benutzer auf Avatare, um mit ihnen zu interagieren oder versuchten Text in die virtuellen Werbetafeln innerhalb des VRML Fensters einzugeben. Alle Teilnehmer hatten Probleme damit, zwischen dem 3D VRML Fenster und der 2D Oberfläche mit den Menüs und Listenfeldern hin und her zu wechseln. So mußte man in der Blaxxun-Welt Avatarsymbole in einer Liste links unten anklicken, um sich zu einem anderen Avatar teleportieren (automatisches transportieren, „beamen") zu lassen, oder man konnte ihn zu einem privaten Chat einzuladen. In Active Worlds hingegen mußte man Soft-

Buttons in der Menüleiste klicken, um mit der Maus in der 3D Welt navigieren zu können oder Gesten auszulösen. Auch die 3D Welten konnte man in beiden Anwendungen nur wechseln, in dem man sie in einer Liste außerhalb des 3D Fensters auswählte (Abb. 3, links).

Acht von zehn Benutzer waren irritiert durch die Darstellung von Text über den Avataren. Da sich die Texte überlappten, wenn Avatare hintereinander stehen, waren sie oft unleserlich. Zudem waren Avatare nicht zu sehen, wenn sie weit entfernt oder von einem Objekt verdeckt waren – aber der dazugehörige „Chat"-Text schon, was zu weiteren Verwirrungen führte.

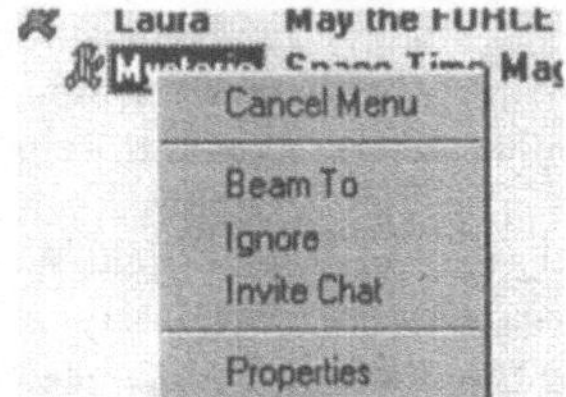

Abb. 3: Benutzer müssen ein Avatarsymbol, in einem Kontextmenü auswählen, um mit ihm/ihr z.B. privat zu "chatten" (links, Blaxxun). In Active World (rechts) verwirrten zu viele Texte über den Avataren die Benutzer.

Mehrere Benutzer nahmen nicht wahr, wenn Avatare mit ihnen redeten. Der Chat-Text, der in einer Chat Protokoll Box erschien, wurde oft übersehen, weil natürlich dauernd andere „Besucher" in der Welt etwas „chatten". Hilfreicher war hier die „private chat" Funktion von Blaxxun, mit der man sich gezielt mit einem bestimmten anderen Avatar/User unterhalten kann und wobei nur diese Texte in einem eigenen Chat-Protokoll angezeigt werden. Allerdings wurde diese Funktion nur von wenigen Benutzern von selbst gefunden und auch dann nicht ohne Probleme genutzt. Viele Benutzer waren sich unsicher, ob ausgelöste Aktionen auch Konsequenzen haben. So war die Rückmeldung über gemachte Gesten oft dürftig und nicht eindeutig. Auch das Wechseln von Avataren wurde oft nicht klar rückgemeldet und führte zur Verwirrung der Benutzer. Es war oft nicht offensichtlich, welche Parameter man an anderen Avataren ändern konnte. In einem Präferenzfragebogen wurden u.a. die Funktionalitäten, die die Avatare betreffen, von den Benutzern beurteilt. Während die Chat-Funktionalitäten als gleich gut bzw. schlecht eingeschätzt wurden, gab es bei den übrigen Bewertungen z.T. deutliche Präferenzen. So wurde die Menüdarstellung und das Auslösen von Gesten mittels Push-Buttons bei Active World der Drop-Down Menü Auswahl von Blaxxun vorgezogen. Dies lag auch an den zu langen Bezeichnungen für Gesten bzw. Emotionen in der Blaxxun Anwendung. Die Auswahl und Anzeige von Avataren wurde hingegen bei der Blaxxun Version bevorzugt, die ein reichhaltigeres und differenzierteres Angebot (Fantasiegestalten, anthropomorphe Darstellungen) von Figuren aufwies.

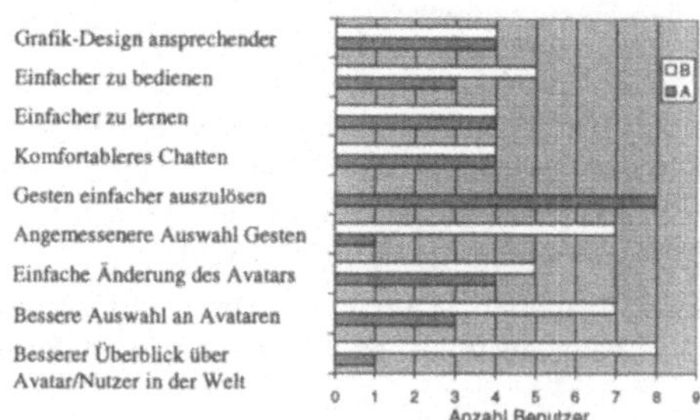

Abb 4: Anzahl der Nennungen in den Präferenzfragebögen.

Indifferente Urteile sind nicht dargestellt. In den Interviews äußerten sich alle Teilnehmer positiv über Robot Avatare, die ihnen gegebenenfalls in der Welt Tips und Hilfestellungen – z.B. zur Navigation - geben können. Dies sollte in Form von kurzen Dialogen geschehen und sich deutlich von der bisherigen Hilfefunktion unterscheiden. Sehr positiv fanden die Benutzer auch die Möglichkeit, sich zu einem anderen Avatar teleportieren zu können und private Chats zu führen.

4 Zusammenfassung der Ergebnisse

Für potentielle Kunden scheint der Einsatz von Agenten in Online-Diensten insgesamt sinnvoll und wünschenswert. So war die Akzeptanz relativ hoch besonders durch die Idee, einen persönlichen Helfer an der Seite zu haben. Bezüglich den potentiellen Aufgaben und dem Grad an Autonomie gibt es allerdings sehr unterschiedliche Einschätzungen und Vorstellungen. So gab es Begeisterung aber auch Unbehagen gegenüber sehr selbständigen Agenten. Etwa die Hälfte der Teilnehmer war sehr offen für neue Möglichkeiten und Potentiale der Agenten und wäre bereit, verantwortliche Aufgaben an den Agenten zu delegieren. Innerhalb der Gruppe des Online-Shopping wurde dagegen die Frage des Handelns und Feilschens sehr viel kontroverser diskutiert und einige Teilnehmer hatten eine sehr ablehnende Haltung gegenüber zu viel Autonomie von Agenten bei sensiblen Aktivitäten. So wurde die Transaktionsfunktionalität der Agenten als problematisch betrachtet, bezüglich der Autonomie des Agenten sowie der Intelligenz und Sicherheit von Transaktions- und Handels-Algorithmen. Dabei ergaben sich im Bereich der Shopping Agenten zwangsläufig folgende Fragen: Kann mein Agent feilschen? Was passiert wenn mein Agent schlecht im Verhandeln ist und von geschulten „Verkäufer-Agenten" übers Ohr gehauen wird? Wie kann ich mich gegen elektronische Betrüger schützen?

In weniger sensiblen Bereichen - wie der reinen Informationsbeschaffung, der Produktauswahl und dem Vergleich von Produkten und Angeboten- ist dagegen bei fast allen Beteiligten eine sehr große Akzeptanz bezüglich virtuellen Helfern zu erkennen.

Bei den Anforderungen an die visuelle Gestaltung und das generelle Erscheinungsbild des Agenten gab es einen klaren Trend zu sichtbaren Darstellung von Agenten und anthropomorphen oder zumindest lebendig wirkenden Ausprägungen.

Als ein potentielles Einsatzszenario, in Bezug auf die Interaktion mit Agenten, kann bei der Interaktion mit Avataren in virtuellen Welten festgestellt werden, daß hohe Anforderungen an einfache Dialog- und Interaktionsgestaltung gestellt werden. Außerdem müssen zusätzliche Navigations- und Orientierungshilfen zur Verfügung gestellt werden. Drei Hauptgestaltungsrichtlinien lassen sich aus den Benutzertests mit Avataren als potentielle Agenten in virtuellen Welten ableiten:

- Vermeiden von „Metapherbrüchen": das ständigeWechseln von einer herkömmlichen grafischen Oberfläche in eine „virtuelle Welt" führte zu einem eher geringen Gefühl von „Immersion" (Gefühl der Präsenz), dafür aber zu viel Verwirrung bei der Bedienung von Chat, Navigation, und Einstellungen der besuchten Welt.
- Direkte Manipulation in virtuellen Welten: Avatare und Objekte sollten durch direktes Anklicken manipulierbar sein, anstelle der verwendung von PullDown Menüs und Tab Controls.
- Verbesserte Rückmeldung von Aktionen, Positionen, und Chateingaben.

Metaphern wie Teleportation erhalten den Eindruck einer virtuellen Welt und erlauben trotzdem schnellen Zugang zu anderen Avataren und Funktionen. Benutzer wollen immer einen Überblick über die Welt, die sich darin befindlichen Avatare und Objekte haben. Die Anzeige von Namen über Avataren ist sinnvoll, nicht jedoch die konstante Anzeige von „Chat" Text in VRML Fenstern für alle sichtbaren Avatare (wie in Active World). Benutzer sollten Avatare und Objekte in der 3D Welt direkt anklicken und manipulieren können. Die übrige 2D Navigationsoberfläche sollte möglichst zurücktreten und wenig Interaktionselemente enthalten, um somit den Metaphernbruch zu reduzieren. Benutzer wollen soviel Feedback wie möglich. Verschiedene Sichten wie Vogelperspektive oder Überblicke sollten schnell aufgerufen werden können.

5 Diskussion

Das Einsatzpotential agentenunterstützter Online-Systeme ist sicher enorm. Allerdings stellen die zu übernehmende Aufgaben und das in die Agenten gesetzte Vertrauen weitere Fragen und werfen neue Diskussionspunkte auf. So herrschen stark unterschiedliche Einschätzung bezüglich der Autonomie von Agenten und den dadurch möglichen zu delegierenden Aufgaben. Um aussagekräftige Ergebnisse zu den innerhalb des Workshop gestellten Fragen zu bekommen, werden daher spezifischere Case-Szenarien entwickelt und evaluiert. Konkrete Aufgaben müssen definiert und von Benutzern bewertet werden. Weiterhin gibt es betreffend den Darstellungsformen und Möglichkeiten der Visualisierung noch große Herausforderungen bezüglich der technischen Realisierbarkeit der Gestaltung und der Interaktion mit Agenten. Zusammenfassend können als Ergebnisse des Workhops daher für die weitere Entwicklung zum Einsatz von Agenten in Online-Diensten des interaktiven Fernsehens, folgende Fragestellungen definiert werden:

- Welche spezifischen Aufgaben sollen von Agenten übernommen werden? Was sind die wirklich interessanten Aufgabenstellung, wo macht der Einsatz des Agenten tatsächlich Sinn? Welche Produktorders und Transaktionen sollen von Agenten getätigt werden, welche nicht (Lebensmittel, Konsumgüter, Autos, Dienstleistungen)?
- In welchen Phasen des Shoppings können Agenten sinnvoll eingesetzt werden [5]? Kann ein Agent unterstützen bei: z.B. der Erzeugung von Bedürfnissen, Produktsuche- und Vergleich, Lieferanten-/Händlerauswahl, Verhandlung, Transaktion, Auslieferung, Produktbewertung?

- Welchen Grad der Intelligenz und Autonomie soll ein Agent denn tatsächlich beinhalten? Welche Aufgaben würde der Benutzer tatsächlich delegieren welche nicht? Wie können Agenten und ihre Tätigkeiten überprüft werden?

- Wenn man von einer visuellen/anthropomorphen Darstellung des Agenten ausgeht, welche Ausprägungen soll der Agent haben? Wird eine Comicfigur mit den wesentlichen menschlichen Merkmalen gegenüber einer stark anthropomorphen Darstellungsform bevorzugt?

- Welche Formen der Interaktion unter Berücksichtigung der ITV-spezifischen Merkmale und Restriktionen sind geeignet? Ist die direkte Manipulation geeignet für den Dialog mit Agenten? Wie ist die Möglichkeit der Sprachein- und Sprachausgabe zu bewerten und wie kann sie realisiert werde? Wie kann sich der Agent sonst noch bemerkbar machen? Wie kann ich den Status des Agenten anzeigen? Welche Form der visualisierung von Ergebnissen ist gewünscht und realistisch?

- Bei welchen Parametern ist Customising des Agenten und des UI wichtig, wie sollte es von Benutzern vorgenommen werden?

- Wer bezahlt die zusätzlichen Möglichkeiten und Dienstleistungen der Agenten?

Diese Fragestellungen werden in internationalen Befragungen und einer nationalen bevölkerungsrepräsentativen Umfrage im weiteren Verlauf dieser Arbeit verfolgt und ausgewertet.

6   Literatur

[1] Shneiderman B., Maes P., Direct Manipulation vs. Interface Agents, Excerpts from debates at IUI 97 and CHI 97, ACM Interactions, November + December 1997, Volume IV.6, 1997, 42-61.

[2] Hitzges, Arno; EMNID-Institut (Hrsg.): media vision trend: Akzeptanz, Stand der Technik und Perspektiven ausgewählter Anwendungen – Dokumentation der Ergebnisse, Stuttgart,1997. Fraunhofer IRB Verlag

[3] Karepin, Rolf: "Das Zusammenwachsen von Fernseher und Computer liegt noch in weiter Ferne." Computer Zeitung, Nr. 50 / 11.12.97, p. 6

[4] Liebermann H., Autonomous Interface Agents, Proceedings of the 1997 Conference of Human Factors in Computing Systems, (CHI97), Atlanta, 1997, ACM Press.

[5] Guttman H.R., Moukas G., Maes P.: Agents as Mediators in Electronic Commerce, Electronic Markets, International Jounal of electronic markets, University of St.Gallen Switzerland, Vol.8, No.1, 1998, v/dfl

[6] Maes, P. Agents that Reduce Work and Information Overload, Communications of the ACM, 37, 7, 31-40.

[7] Lester, J.C. et al.: The Persona Effect: Affective impact of Animated Pedagogical Agents, Proceedings of the 1997 Conference on Human Factors in Computing Systems (CHI97), 1997

[8] King W.J., Ohya J.: The Representation of Agents: Anthropomophism, Agency and Intelligence. Proceedings of the 1996 Conference on Human Factors in Computing Systems (CHI96), 1996

[9] Laurel, B. Interface Agents: Metaphors with Character. In B. Laurel (Ed.). The Art of Human-Computer Interface Design, 1990, Addison-Wesley Publishing Company.

[10] Maes, P., Tomoka, K.: The Effects of Personification of Agents, Proceedings of the HCI'96 London, 1996 Springer-Verlag

[11] Oren, T., et. al. Guides: Characterizing the Interface. In B. Laurel (Ed.). The Art of Human-Computer Interface Design, 1990, Addison-Wesley Publishing Company.

[12] Naas, C., et. al. Computers are Social Actors, Proceedings of the 1994 Conference on Human Factors in Computing Systems (CHI94), New York: ACM Press, 72-77.

[13] Tomoka, K. Survey on people's impression on different faces as a pilot study, MIT, 1996 http://tomoko.www.media.mit.edu/people/tomoko/

[14] ISO 9241: Ergonomic requirements for office work with visual display terminals (VDTs)

[15] Beu, A.: Final Online Styleguide, MUSIST Project AC010 D20, AC010/GSM/WP1/DR/P/20/b1, 1998

[16] Blaxxun: http://www.blaxxun.com/

[17] Active Worlds: http://www.activeworlds.com/

Adressen der Autoren

Dipl.-Ing. Jörg Nissler
Fraunhofer IAO
MT Interaktive Produkte
Nobelstr. 12
70569 Stuttgart
Joerg.Nissler@iao.fhg.de

Dipl. Psych. Volker Thoma
Fraunhofer IAO
MT Interaktive Produkte
Nobelstr. 12
70569 Stuttgart
Volker.Thoma@iao.fhg.de

Online-Dienste für Alle:
Usability und Design einer PC/ITV Oberfläche

Jörg Nissler, Volker Thoma, Sylvia Manz

Marktstrategieteam Interaktive Produkte, Fraunhofer IAO

Zusammenfassung

Durch die zunehmende Konvergenz von elektronischen Medien (Internet, TV, Telekommunikation) eröffnen sich neue Potentiale, privaten Anwendern interaktive Online-Dienste zugänglich zu machen. Den Benutzern wird die Wahl gelassen, wo und wie sie elektronische Dienste nutzen – eine wichtige Voraussetzung, um neue Zielgruppen als Kunden zu gewinnen. Am Fraunhofer IAO wurde der Zugang zu einem Internetdienst auf zwei Plattformen (PC/ITV) untersucht. Dazu wurde ein einheitliches User Interface Konzept für einen Informationsdienst mit zwei Ausprägungen für PC und TV entwickelt und in iterativen Benutzertests evaluiert. Das Design der Oberfläche erwies sich durch die Bereitstellung unterschiedlicher Navigationselemente auf beiden Plattformen auch für Internet-Neulinge als benutzerfreundlich. PC-erfahrene Benutzer bevorzugten überwiegend die Ansteuerung der Oberfläche per Maus aufgrund der direkten Manipulation. Bei der Bedienung der ITV-Plattform wird die herkömmliche Tastenfernbedienung einer der PC-Maus ähnlichen Fernbedienung vorgezogen.

Die vorliegende Studie wurde unter anderem im Rahmen des EU Projektes MUSIST (AC010) durchgeführt und baut auf Erfahrungen und Erkenntnissen im Bereich der Gestaltung von ITV-Oberflächen auf. Als Basis für die Entwicklung eines Prototypen wurden die Inhalte und Anforderungen des „InfoCity"-Dienstes NRW als einen potentiellen Service-Provider für plattformübergreifender Dienste dieser Art ausgewählt. Es wurde ein einheitliches User Interface (UI)-Konzept für beide Plattformen (TV/PC) entwickelt, das dieselben Dienste und Funktionalitäten zur Verfügung stellt, auf dem gleichen „Look-and-Feel" und CI basiert, jedoch plattformabhängig angepaßt ist und unterschiedlich angesteuert wird (Maus/Fernbedienung).

1   Einleitung

Das Internet wächst rapide und Experten prognostizieren, daß es in nahezu allen Bereichen des privaten Lebens Einzug halten wird. Neue Dienste im Bereich Infotainment, Edutainment und Electronic-Commerce besitzen daher ein großes marktwirtschaftliches Potential [1]. Noch ist allerdings unklar, wie der private Nutzer Zugang zu diesen Diensten bekommen wird [2]. So konkurrieren Unterhaltungsindustrie, Telekommunikationsdienstleister und Computerindustrie um die Zugangsvoraussetzungen, wobei es besonders um Fragen hinsichtlich der technischen Umsetzung und der Basisplattform für zukünftige Online-Dienste geht. Die technologischen Lösungsansätze stehen also gegenwärtig im Vordergrund – und nicht die Anwender, die letztendlich über die Akzeptanz und den Erfolg dieser neuen Dienste entscheiden werden [3]. Dabei wird vergessen, daß z.B. der PC seinen kommerziellen Durchbruch erst nach einer technisch wenig aufwendigen, dafür gezielt anwenderorientierten Neuerung hatte – der Erfindung der grafischen Benutzeroberfläche im Xerox PARC. Das interaktive Fernsehen könnte ein ähnlicher Sprung nach vorn werden: Auch PC-unerfahrene Benutzer können nun das Internet und andere elektronische Dienste mit einem gewohnten Gerät nutzen – den Fernseher mit Fernbedienung. Gleichzeitig wollen auch erfahrene Internetnutzer sich nicht unbedingt extra einen PC für Zuhause besorgen, wenn dieser am Arbeitsplatz mit Internetanschluß vorhanden ist. Trotzdem aber will man daheim beispielweise auch im Internet Informationen über Freizeit oder Fahrpläne einholen können, was mit ITV bequem möglich ist während man fernsieht. Die Anwender verlangen jedoch gleichartige Navigations- und

Bedienkonzepte bei Online Diensten – trotz unterschiedlicher Plattformen und Eingabegeräte (Maus, Fernbedienung). Kein Benutzer will umlernen, wenn er denselben Dienst morgens im Büro und abends vorm Fernseher benutzt.

Daher wurde für die im Folgenden beschriebene Entwicklung einer Internet-Oberfäche ein benutzerzentrierter Ansatz verfolgt. Es wurde untersucht, wie ein ergonomisches und einheitliches User Interface gestaltet werden muß, das den Anwendern überläßt, welche Plattform – PC oder interaktiver Fernseher – sie als Zugang zu neuen Internet- und Online-Diensten nutzen. Als möglichst realistische Basis für die Umsetzung des einheitlichen UI Konzepts dienten die Inhalte des Service-Providers „InfoCity" in Nordrhein-Westfalen, und als Vorlage des graphischen Look&Feel wurde das vorhandene CI berücksichtigt. Als potentielle gemeinsame Zielplattform für einen vergleichbaren Dienst werden hier private Haushalte über Kabel (Coax) an ein Glasfaser-Backbone angeschlossen und erhalten somit mittels Kabelmodem breitbandigen Zugang zu multimedialen interaktiven Anwendungen über PC oder TV via Set-Top-Box [4]. Die Gestaltung der Oberflächen und die Benutzertests wurden innerhalb des EU-Projects MUSIST (ACTS AC010, Multimedia USer Interface for Interactive Services and TV) durchgeführt [5].

1.1 Fragestellung

Ziel der Studie war die Entwicklung eines plattformübergreifenden GUI Designs und die nachfolgende Evaluation mit Benutzern. Die Basis für die Benutzertests war ein Prototyp des „Info-City" Dienstes, der mit unterschiedlichen Werkzeugen (Macromedia Director, html) auf den Plattformen realisiert wurde. Im Folgenden sind die Fragestellungen aufgeführt, die anhand von Benutzertests in zwei Stufen untersucht und in einen iterativen Designprozeß integriert wurden.

- Wie muß eine einheitliche Oberfläche für zwei Plattformen (PC/TV) mit unterschiedlichen Interaktionskonzepten und Eingabemöglichkeiten gestaltet werden?
- Wie muß ein einheitliches Navigationskonzept für ITV und PC aussehen, damit es gängigen Web Usability Kriterien entspricht [6][7]?
- Kommen die Nutzer mit der hierarchischen Informationsstruktur (5 Ebenen) zurecht? Diese ist vor allem im Hinblick auf die eingeschränkten Darstellungsmöglichkeiten (große Schrift, geringe Auflösung) von ITV-Anwendungen notwendig.
- Wie ist die Akzeptanz der beiden Systeme im Vergleich, besonders hinsichtlich der Eingabegeräte (Maus, Fernbedienung)?
- Wie ist Akzeptanz interaktiver Dienste im Fernsehen generell?

Als Designphilosophie für ein benutzerfreundliches Navigationskonzept wurde die mehrfache Verwendung unterschiedlicher Navigationselemente (Thumbnails, Historienliste, Navigationsliste, Linkmenü) angestrebt, um Bedürfnissen unterschiedlicher Benutzergruppen Rechnung zu tragen. Daher wurden auch allgemeine Designprinzipien der Informationsdarstellung und Navigation im Internet untersucht, insbesondere die Gestaltung von Navigationstools und Links [8][9][10].

2 Design einer PC/ITV Web-Oberfläche

Die Parameter für die Entwicklung und Evaluation des Navigationkonzepts und das graphischen Design wurden durch die technischen Möglichkeiten der Plattformen, das CI des Service-Providers und den Umfang der zu integrierenden Inhalte eingeschränkt. Besondere Aufmerksamkeit wurde auf die Anforderungen gelegt, die durch die Ansteuerung über den Fernseher (TV-Bildschirm, Fernbedienung) resultieren [11].

2.1 Technische Realisierung

Die PC-Version basiert auf einer HTML 4.0 Anwendung und wird mit dem Netscape Browser dargestellt. Als Plattform dient ein Standard Pentium PC. Die Funktionsleiste des Browsers wird beim Öffnen der Anwendung ausgeblendet, da alle Funktionen die zur Navigation in den angebotenen Diensten nötig sind über die Anwendung angesteuert werden können. Bei der TV-Variante kam das u.a. innerhalb MUSIST entwickelte Terminal Xelos@Media des Projektpartners Loewe zum Einsatz [3].

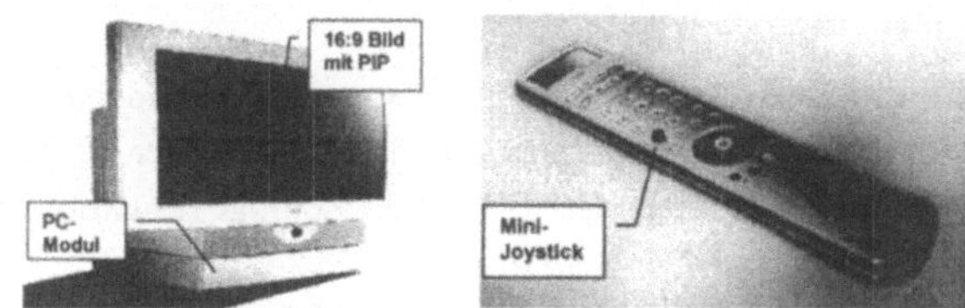

Abb. 1: Loewe Terminal und Fernbedienung mit Mini-Joystick

Hier dient ein 16:9 TV mit erhöhten Bildfrequenz und integriertem Pentium PC zur Übertragung, Darstellung und Interaktion von multimedialen Daten auf dem Bildschirm. Durch die auf 100Hz (60Hz progressiv) erhöhte Bildwiederholfrequenz und verbesserte Bildschärfe können die Vorteile des Fernsehers (Helligkeit und Brillanz) mit denen des PC-Monitors (Bildschärfe und –auflösung) kombiniert werden [12]. So wird eine Auflösung von bis zu 856*480 pkt erreicht (SVGA in vertikaler Ausrichtung). Als Auflösung für die Darstellung der TV-Variante wurde VGA zugrunde gelegt, um die Bild in Bild Funktion (PIP) des Loewe Terminals zu nutzen. Durch einfachen Tastendruck kann der Bildschirm geteilt werden, um parallel zwei Kanäle zu verfolgen oder, wie im konkreten Testszenario, die multimediale Anwendung sowie ein kleines Fernsehbild am Bildrand zu betrachten. Diese Funktionalität sollte von den Benutzern ebenfalls bewertet werden. In Abbildung 1 ist das Terminal mit integriertem PC, sowie die dazugehörige Fernbedienung mit Mini-Joystick zur direkten Manipulation auf dem Bildschirm in der TV Anwendungen dargestellt.

2.2 Bedienung und Ansteuerung

Die PC-Variante läßt sich auf die bekannte Art und Weise der direkten Manipulation und Interaktion mit der Maus steuern, allerdings ohne die Funktionen der Browserleiste. Die Oberfläche der TV-Variante ist dagegen durch eine Fernbedienung zu steuern, die vier Richtunsgtasten (links-rechts-oben-unten), eine Eingabetaste und eine Zurücktaste besitzt. Mit den Richtungstasten wird ein sensitiver Bereich des Bildschirms angewählt, ein sogenannter „Fokus". Der Fokus wird nur auf diskrete Felder (aktive Links) innerhalb der Anwendung bewegt („focus jump"), es gibt also keine kontinuierliche Zeigerbewegung wie mit der Maus. Zusätzlich zu der konventionellen Lösung testeten die Benutzer eine Fernbedienung mit Mini-Joystick. Dieses Eingabegerät ist durch die Kombination mit zwei auf der Rückseite angeordneten Tasten einer PC-Maus ähnlich (Abb. 1).

2.3   Navigations- und Gestaltungskonzept

Aus den Erfahrungen und Ergebnissen des Projekt MUSIST und weiteren Entwicklungen [13] im Bereich UI für ITV Anwendungen wurde ein Navigationskonzept erstellt. Nach Erkenntnissen aus einer Untersuchung von Metaphern in ITV-Anwendungen [14], erfolgt ein schneller und einfacher Zugang zu den obersten Navigationsebenen von hierarchischen Informationsräume zunächst über graphische Menüs. Erst auf einer tieferen Anwendungsebene sollten gegebenenfalls komplexere Metaphern, wie eine virtuelle Stadtrundfahrt für Neuanwender, eingesetzt werden. Neben diesen und anderen Studien wurden Regeln und Grundsätze von Web Style Guides berücksichtigt [8] [9] [10], mit besonderem Augenmerk auf Usability-Kriterien und optimaler Navigation. Die Entwicklung des Navigationskonzepts, des „Look & Feels" sowie die prototypische Implementierung der Anwendung basierten auf folgenden Einschränkungen:

- indirekte, bidirektionale Ansteuerung mit ´focus-jump´ auf der TV-Plattform,
- direkte Manipulation, direkte Ansteuerung mit Maus auf der PC-Plattform,
- klare, rasterförmige Struktur der Bereiche, Zoning (Funktionsbereich, Navigationsbereich und Inhalts- bzw. Menübereich),
- orthogonale Anordnung der Bereiche untereinander,
- orthogonale Anordnung der Elemente und Controls innerhalb eines Bereichen,
- seitenorientierter Aufbau (Blättern anstatt Scrollen)
- Integration des bestehenden Corporate-Design des Service-Providers,
- optimale Orientierungs- und Navigationsmöglichkeiten,
- Berücksichtigung unterschiedlicher Benutzerprofile (Erstbenutzer, ´Power-User´),
- Berücksichtigung der technischen Restriktionen bei Darstellung, Aktualisierung und Pflege der Seiten durch großflächigen Farbbereichen mit Graphiken und Text,
- keine transparente Flächen (für die Integration von Video ungeeignet).

Abb. 2:   Funktiosnbereiche der Oberfläche und Navigationstools am Besipiel der PC-Version

Die Umsetzung der oben genannten Voraussetzungen führte zu einem grafischen Konzept und einer prototypisch realisierten Anwendung mit drei Funktionsbereichen (Abb. 2). Der linke Bereich, die Funktionsleiste, bietet Zugang zu den Grundfunktionen und Hauptdiensten der Anwendung. Die Funktionsleiste bleibt immer sichtbar und enthält das „Hauptmenü", „Wir über uns", „Kommunikation" (Mail, SMS Telefonbuch etc.), „Internet", „Einstellungen" und „Hilfe" als Optionen. Diese Kapitel werden als Textlinks dargestellt, unterstützt durch ein grafisches Icon mit farblicher Kodierung. Die Icons dienen der besseren Wiedererkennung. Außerdem ist für den Internetzugang und die Integration weiterer Dienste vorgesehen diesen Bereich teilweise nach links auszublenden und bis auf die Breite der Icons zu reduzieren. Hier wurde den Erwartungen an ein flexibles Design Rechnung getragen, das den erfahrenen Benutzern die Vergrößerung der eigentlichen Arbeitsfläche ermöglicht. So wird beim Browsen in externen Diensten oder dem Internet eine optimale Platzausnutzung im rechten Content-Bereich gewährleistet ohne dabei ganz auf die Grundfunktionen zu verzichten.

Die Funktionsleiste der PC-Variante ist außerdem mit einer Shortcutleiste ausgestattet. Hier wird zusätzlich zu den beschriebenen Hauptbereichen die erste Ebene des Dienste-Menüs dargestellt und mit dem entsprechenden Link belegt. Die TV-Variante muß aus Platzgründen auf diese zusätzliche Vorschau und Auswahlmöglichkeit verzichten (Abb. 3, rechts).

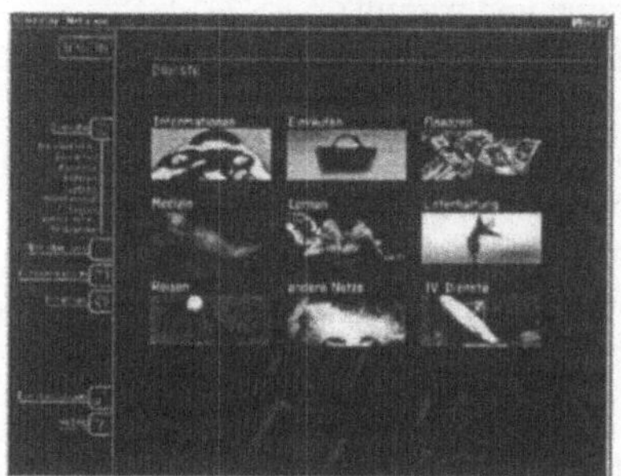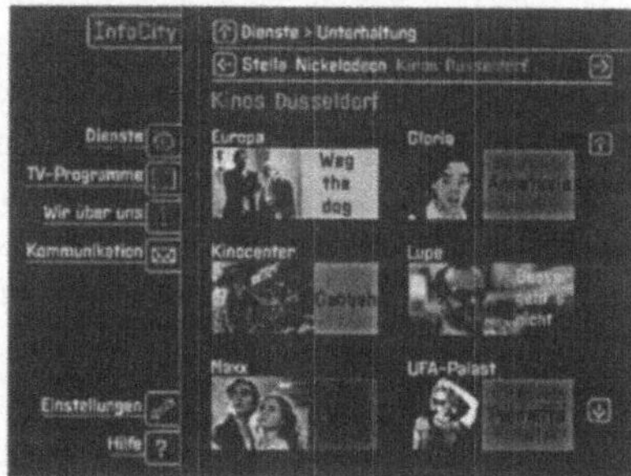

Abb. 3: Beispiel Oberflächen der PC-Variante (links) und der TV-Variante (rechts)

Die Navigationsleiste im oberen Bereich zeigt in der ersten Zeile eine interaktive Historie-Liste. In der zweiter Zeile werden die parallel zur aktuellen Auswahl aktivierbaren Menüeinträge der aktuellen Ebene angezeigt. Die Pfeile auf dieser Ebene ermöglichen die direkte Anwahl dieser parallelen Bereiche. Bei der TV-Variante kommt diesen Pfeilen eine besondere Bedeutung zu, da sie zum einen als Anker- und Navigationspunkte des Fokus dienen und der Benutzer durch Bestätigung mit „OK" zu dem benachbarten Menüeintrag wechselt (z.B. von Kino „MAXX" zu „AMBO"). Der Menü- und Content-Bereich beinhaltet entweder ein grafisches Menü mit „Thumbnails" (Kategorie-Links) sowie den eigentlichen Informationen (Content) bzw. externen Anwendungen in der untersten Navigationsebene. Die PC-Variante bietet hier Platz für 9 „Thumbnails", wohingegen die TV-Variante aufgrund reduzierter Darstellungsmöglichkeiten auf sechs Einträge beschränkt ist. Die PC-Oberfläche unterscheidet sich außerdem von der TV-Oberfläche durch die Shortcutleiste im linken Funktionsbereich, die in einer Gliederung die erste Menü-Ebene der Dienste-Seite anzeigt. Diese Leiste bleibt nach einmaligem Öffnen der Dienste-Seite konstant im linken Bereich stehen. Die Optionen der Shortcutleiste sind dann von jeder tieferen Ebene aus direkt anwählbar und aktiv. Durch einen Zurück-Pfeil am unteren Rand des Computer-Screens kann der Benutzer wie bei einem gängigen Internetbrowser die zuletzt aktivierte Seite öffnen. Die Zurück-Funktion wird am Fernseher durch eine „Zurück"-Taste auf der Fernbedienung realisiert, mit der man in die nächsthöhere Ebene navigiert.

3 Benutzertests

Die Usability-Tests wurden durchgeführt, um Design- und Navigationskonzepte sowie die grafische Gestaltung zu evaluieren, und um die Benutzerfreundlichkeit der Oberfläche hinsichtlich einer Optimierung zu überprüfen. Folgende Fragestellungen wurden untersucht:

1. Design- und Navigationskonzept:
   - Wie werden die Navigations-Tools genutzt (Shortcuts, Navigationsleiste)?
   - Wie gut können die Benutzer im hierarchischen Informationsraum navigieren?
2. Vergleich der Oberflächen:
   - Wie benutzerfreundlich ist die Navigation auf den jeweiligen Plattformen?
   - Auf welcher Oberfläche werden Informationen schneller gefunden?
   - Welche der beiden Versionen wird bevorzugt und warum?
   - Wie kommen die Benutzer mit der Fernbedienung im Vergleich zur Maus zurecht?
3. Akzeptanz ITV:
   - Welche Art der Fernbedienung (Tasten oder Mini-Joystick) wird bevorzugt?
   - Wie wird die Fernbedienung mit Mini-Joystick beurteilt?
   - Wird die „Bild-im-Bild" Funktion als störend empfunden?
   - Generelle Akzeptanz eines ITV Internetservice?

3.1 Test 1

In einem ersten Test mit 10 Benutzern im Alter von 20 bis 41 Jahren wurde das Design und Navigationskonzept einer Usability-Evaluation unterzogen. Darüber hinaus sollte konkret geklärt werden, ob der PC-Version eine Shortcutliste hinzugefügt werden kann, um sogenannten „Power Usern" einen schnellen Zugriff zu Inhalten zu erlauben. Zu diesem Zweck sollten die Benutzer Kategorien und Inhalte im Angebot „Dienste" suchen. Jede der vier Aufgaben war mit maximal 3 Klicks zu bewältigen. Die Aufgaben der beiden Versionen waren sehr ähnlich und erforderten die gleiche Anzahl an Bedienschritten. Jedem Benutzer wurden zwei Versionen angeboten: eine Anwendung mit einer Shortcutleiste für die Kategorie „Dienste" sowie eine ansonsten identische Version ohne Shortcutleiste. Eine Hälfte der Benutzer begann mit der ersten, die andere Hälfte mit der zweiten Version. Nach den Tests mußten Fragebögen zur Präferenz ausgefüllt werden. Alle Benutzer hatte Erfahrungen mit dem PC und dem Internet.

Die Ergebnisse des Tests zeigen, daß die Benutzer sehr gut mit der Oberfläche navigieren

konnten. Alle Benutzer fanden ohne eine Einweisung oder Hilfe die gesuchten Dienste und Angebote. In den Interviews und bei den Fragen zur Präferenz bevorzugten die Benutzer insgesamt die Version mit der Shortcutleiste (siehe Abbildung 4).

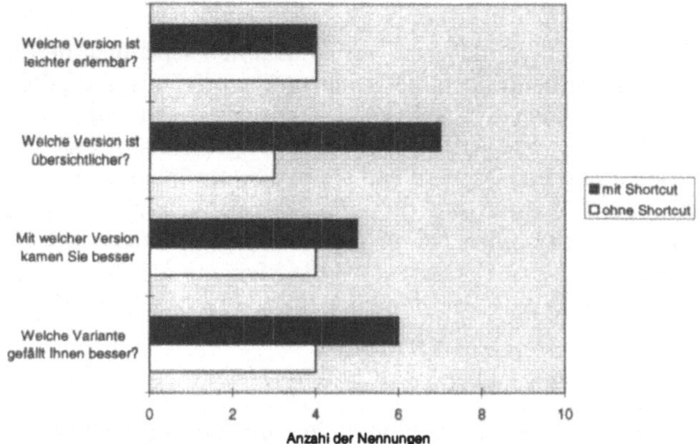

Abb. 4: Anzahl der Nennungen zur Präferenz der Versionen (mit/ohne Shortcuts)

Es wurde eine 2-faktorielle Varianzanalyse über die Anzahl von benötigten Bedienschritten gerechnet mit den Faktoren Reihenfolge der dargebotenen Version (Version mit Shorcutleiste zuerst oder zuletzt bearbeitet) und Navigation (mit Shortcutleiste oder ohne). Es gab einen statistisch signifikanten Effekt sowohl bei dem Faktor Reihenfolge [$F(1,8) = 5,714$; $p = 0,044$] als auch bei dem Faktor Navigation [$F(1,8) = 6,452$; $p = 0,035$], aber keine Interaktion beider Faktoren [$F(1,8) = 2,323$; $p = 0,166$]. Das heißt, Benutzer, die mit der Version mit Shortcutleiste begannen, benötigten insgesamt weniger Schritte als die Vergleichsgruppe. Unabhängig von der Reihenfolge der Darbietung ist der Unterschied zwischen den Versionen hinsichtlich der Anzahl der Klicks bis zur Lösung statistisch signifikant: die Benutzer sind mit der Shortcutleiste deutlich schneller (gemessen in Bedienschritten). Die Shortcuts helfen also den optimalen Navigationsweg zu finden. Dieses Ergebnis ist bedeutsam, da man in der Version ohne Shortcuts mittels der Navigationsleiste ebenso schnell (mit derselben Anzahl von Klicks) zum Ziel kommen konnte. Alle Navigationselemente wurden von den Benutzern verwendet, am häufigsten davon die grafischen Menüelemente sowie die Shortcutleiste. Die „Zurück"-Funktion und „Navigationsleiste" wurden sehr selten benutzt, obwohl sie in 2 Aufgaben der kürzeste Weg zur Informationsfindung waren. Aufgrund der Ergebnisse des Vortests wurden die Navigationselemente mit der Shortcutleiste und das grafische Layout beibehalten, da sie keine Usability Probleme verursachten. Im Haupttest sollte der Nutzen von multiplen Navigationselementen in tieferen Informationsebenen untersucht werden.

4 Haupttest

Die Navigationsoberfläche wurde von 13 Benutzern (sechs Frauen und sieben Männer) im Alter von 17 bis 69 Jahren auf beiden Plattformen (TV/PC) getestet. Da die Zielgruppe Personen mit unterschiedlicher Computer- und Interneterfahrung umfaßt, wurden auch Versuchsteilnehmer eingeladen, die den Computer bzw. das Internet selten (ein bzw. vier Teilnehmer) oder nie (drei bzw. vier Teilnehmer) nutzen. Um mögliche Lerneffekte zu kontrollieren, die bei der sequentiellen Anwendung beider Versionen auftreten können, begannen sechs Benutzer mit der Computerversion, die anderen sieben mit der TV-Version. Den Teilnehmern wurden zur PC-und zur TV-Variante jeweils sieben Aufgaben vorgelegt, in denen sie im Infocity-Angebot nach bestimmten Informationen suchen sollten. Die Navigationswege der Teilneh-

mer zur Lösung der Aufgaben wurden von den Evaluatoren schriftlich und mit Video protokolliert. Nach jeder Version beurteilten die Benutzer Usability-Eigenschaften der Oberfläche
auf siebenstufigen Ratingskalen. Nach Ende der Aufgabenbearbeitung zur Fernsehversion
mußten die Benutzer eine zweite Art der Fernbedienung (Mini-Joystick, siehe Abb. 1) für
interaktive TV-Geräte bewerten. Dazu sollten sie explorativ Inhalte auf der Oberfläche auswählen. Nach dem Tests wurden die die Versuchspersonen mit Fragebögen und halbstrukturierten Interviews hinsichtlich der Akzeptanz und Benutzbarkeit befragt. Schließlich
wurden die Akzeptanz der beiden Versionen und die Meinung der Benutzer zu verschiedenen
Aspekten der TV-Variante (Bild-in-Bild-Funktion, Mini-Joystick, Design) erfaßt.

4.1 Navigationsverhalten der Benutzer

Alle Aufgaben konnten von allen Teilnehmern ohne direkte Hilfe gelöst werden. Insgesamt
konnten die Benutzer auch ohne Vorkenntnisse in einer Informationstiefe von 5 Ebenen er
folgreich navigieren. Inhaltliche Information auf Webseiten muß also nicht unbedingt wie
oftmals gefordert mit 3 Klicks erreichbar sein [9].

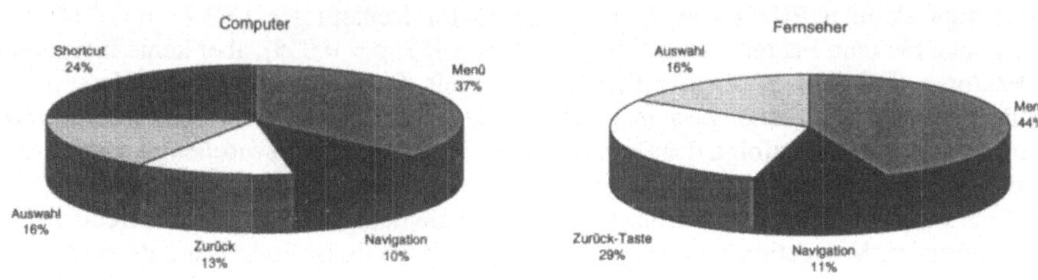

Abb. 5: Nutzung der verschiedenen Navigationselemente bei der PC- (links) und TV-Version (rechts).

Manche Navigationsschritte konnten nur mit bestimmten Navigationselementen durchgeführt
werden, so mußte z.B. zu Beginn eines Tests „Dienste" im Auswahlbereich gewählt werden,
bevor die Navigationsleiste und das grafische Menü erschien. Es wurde nur diejenigen Bedienschritte miteinander verglichen, bei denen die Benutzer eine Wahl zwischen den Bedienelementen Navigationsleiste, Zurückfunktion oder –taste, Shortcutleiste (nur PC) und
grafisches Menüs hatten (siehe auch Abbildung 5).

Auswahlmenü:
Von den Links im Auswahlmenü (16% der genutzten Auswahlaktionen bei beiden Versionen)
wurden am häufigsten „Dienste" ausgewählt, um nach einer Kategorie bzw. einem Inhalt von
der obersten Kategorie aus zu suchen. Hier war die Suchstrategie also „top-down" – die Benutzer gehen zurück zum gewohnten Ausgangspunkt und suchen von dort nach Inhalten.

Grafisches Menü:
Die Elemente des grafischen Menüs wurden am häufigsten benutzt (37% der Auswahlaktionen bei PC, 44% bei TV). Dies ist nicht überraschend, da ihre Größe und Plazierung zu

einer Auswahl aufforderten. Überdies waren die Optionen mit grafischen Elementen leichter zu merken, da das jeweilige Bild und die Position zusätzliche Gedächtnisstützen darstellen.

Shortcutleiste:
Insgesamt fielen 24% der Auswahlaktionen auf der PC-Version auf die Shortcutliste, was die Ergebnisse des Vortests bestätigt. Zu Beginn einer Testsitzung nutzten die Teilnehmer allerdings vorwiegend die grafischen Menüfelder in der Mitte des Bildschirms. Nur drei von 10 Benutzern wählten anfangs die Links in der Shortcutleiste. Im Verlauf der Testsitzung nahm die Nutzung der Shortcuts zu: in der siebten und letzten Aufgabe verwendeten bereits 8 Teilnehmer diese Navigationsmöglichkeit. Die Benutzer ließen sich aufgrund ihres Navigationsverhalten in zwei Gruppen einteilen: Auf der einen Seite gab es die „Power-User", die Shortcuts so oft wie möglich nutzten, weil die Navigation damit schneller geht. Die anderen Benutzer kehrten bei einer neuen Aufgabenstellung lieber noch einmal in die frühere Ausgangsposition zurück und rollten die Suche wieder von „oben" auf. Dies rechtfertigt den Einsatz einer eigentlich redundanten Menüführung: Die Auswahl über Bildmenü ist die offensichtlichste und einfachste für Benutzer, die zum ersten Mal mit dem System umgehen. Die Shortcutleiste ist sinnvoll, weil ein großer Teil der Benutzer nach einiger Zeit sehr schnell auf schon bekannte Inhalte zugreifen will.

Navigationsleiste:
Wie im Vortest wurde die Navigationsleiste bei beiden Versionen eher selten benutzt (10% bzw. 11% der Auswahlaktionen). Bei Aufgabe 4, bei welcher der Gebrauch der Navigationsleiste am schnellsten zur Lösung führt, wurde sie von keinem Versuchsteilnehmer benutzt, die den Versuch mit dem Computer begonnen haben. Dieses Ergebnis steht Studien bzw. Style Guides gegenüber, die hierarchische Angaben und „Historien" empfehlen [15][10]. Viele Benutzer gaben bei der PC-Version an, daß sie die Navigationsleiste nicht genutzt hätten, da sie etwas unauffällig sei. Allerdings war die Nutzung auch bei der TV-Version nicht höher, obwohl hier die Navigationsleiste im Verhältnis zu den anderen Bereichen mehr Platz bekam und somit deutlicher auffiel (siehe Abb. 3). Der Grund für die eher geringe Nutzung der Leiste ist wohl, daß diese auf dem Eingangsscreen noch nicht als Navigationsmöglichkeit erscheint, weil noch keine tieferen Ebenen durchschritten wurden. Viele Teilnehmer beachteten die Navigationsleiste auch in Folge nicht. Vier Versuchspersonen hielten die Navigationsleiste für redundant aber nicht störend, die anderen waren der Meinung, daß man sich erst daran gewöhnen müsse, dann aber zu tieferen Ebenen schneller gelangen könne.

Die häufige Nutzung des Auswahlmenüs bzw. der Shortcutleiste und die eher geringe Verwendung der Navigationsleiste widerspricht z.T. bisherigen Web-Designstudien, die eine Plazierung von Navigationstools eher am Seitenanfang als am Seitenrand empfehlen [8].

Zurück-Funktion:
Die Zurück-Funktion wurde in der PC-Version deutlich seltener benutzt (13% der Auswahlaktionen), als in der TV-Version (29%); so z.B. bei Aufgabe 2 und 3 nur zweimal, obwohl dies der schnellste Weg gewesen wäre. Dies liegt zum einen daran, daß die übrigen Navigationselemente genügend Möglichkeiten bieten, zu der gewünschten Option zu gelangen, zum anderen war die „Zurück"-Taste der TV-Version auffälliger und bequemer zu bedienen. Insgesamt war die „Zurück"-Funktion ein wichtiges Hilfsmittel für die Benutzer, was eine Darstellung – entgegen einigen Web Design Guides [10] - auf einer Webseite rechtfertigt.

Insgesamt benötigten die Benutzer der TV-Version mehr Bedienschritte als bei der PC-Version. Zum teststatistischen Vergleich der beiden Versionen wurde eine 2-faktorielle Varianzanalyse über die Anzahl der Fehler pro Benutzer mit den Faktoren Reihenfolge der dargebotenen Version (TV oder PC zuerst oder zuletzt bearbeitet) und Art der Plattform/Oberfläche (TV/PC) berechnet. Als Fehler wurden diejenigen Bedienschritte gewertet, die über der für eine Aufgabe mindestens benötigten Anzahl von Auswahlaktionen liegen. Es wurden Fehler statt Anzahl der Bedienschritte als Variable verwendet, weil durch die Verwendung der Shortcuts einige Aufgaben in der PC-Version mit weniger Aktionen gelöst werden konnten. Es gab keinen statistisch signifikanten Effekt sowohl beim Faktor Reihenfolge [F (1,11) = 0,541; p = 0,478] als auch beim Vergleich der beiden Versionen [F (1,11) = 2,368; p = 0,152]. Folglich gab es hinsichtlich der Anzahl der Bedienschritte keinen spezifischen Lerneffekt, es war unerheblich, ob die Benutzer mit der TV-Version oder der PC-Version begannen. In der TV-Version fanden die Teilnehmer die kürzeren Navigationswege weniger gut als in der PC-Version, die sich nur durch die Shortcutleiste und der größeren Menge an abgebildeter Information unterschied. Allerdings ist dieser Unterschied statistisch nicht bedeutsam. Im Vergleich zu den Ergebnissen des Vortests heißt dies, daß Vorteile durch bestimmte Navigationselemente (Shortcuts) mit steigender Komplexität der Navigationsaufgaben und einer höheren Anzahl von Informationsebenen aufgehoben werden. Dies spricht für die Verwendung von mehreren sich ergänzenden Navigationselementen.

4.2   Mausbedienung, Tastenfernbedienung, Mini-Joystick

Zwölf von dreizehn Benutzern zogen die Maus der Tastenfernbedienung vor, weil die gewünschten Optionen direkt anwählbar seien, während man mit der Fernbedienung auf vordefinierte Bereiche festgelegt ist und evtl. Umwege gehen muß, bevor man die gewünschte Option auswählen konnte. Die Auswahl mußte außerdem mit einer gesonderten Taste („OK") erfolgen. Den Mini-Joystick fanden 10 Benutzer deutlich schwerer zu bedienen als die Tastenfernbedienung. Vor allem bei kleinen Feldern war die Ansteuerung mit dem Joystick schwierig. Nur eine Versuchsperson gab dem Joystick den Vorzug. Zwei Benutzer bemerkten, daß ein Joystick bei einer stern- oder kreisförmigen Suchoberfläche geeignet wäre, bei einer „geometrischen" (orthogonalen) Oberfläche wie bei der Infocity Oberfläche seien die Tastenfernbedienung komfortabler. Einige Benutzer schlugen vor, das Gerät mit einer Art Infrarot-Mouse statt mit der Tastenfernbedienung auszustatten.

4.3   Bewertung der Benutzerfreundlichkeit

Der Fragebogen, der nach der Bearbeitung jeder Version auszufüllen war, erfaßte folgende Kriterien, die für die Evaluation der Oberfläche relevant schienen: Einfachheit der Bedienung, Selbstbeschreibungsfähigkeit, Kontrollierbarkeit, Erwartungskonformität, Erlernbarkeit, allgemeine Akzeptanz, ansprechende Gestaltung [7]. Beide Oberflächen wurden grundsätzlich sehr positiv beurteilt. In Abbildung 6 sind die Mittelwerte jeweils für die PC- und die TV-Version graphisch dargestellt.

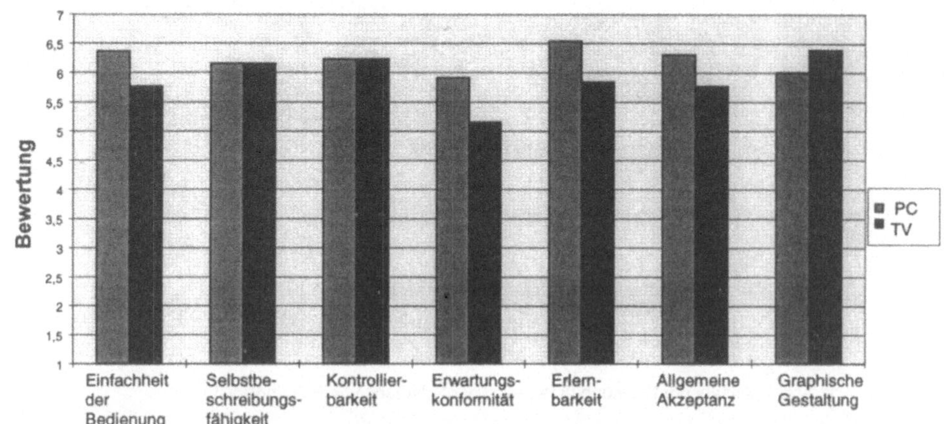

Abb. 6: Mittelwerte der Bewertungen (Skala von 1 –7) für PC- und TV-Version.
Je höher der Wert um so positiver ist die Bewertung.

Die PC-Version schnitt bei vier Fragen besser ab als die TV-Version: Einfachheit der Bedienung, Erwartungskonformität, Erlernbarkeit, und der allgemeinen Akzeptanz. Die TV-Version erreichte lediglich bei der grafischen Gestaltung eine etwas höhere Bewertung. Dies lag wohl am harmonischeren Größenverhältnis zwischen Menübereich (Inhalt) und der Navigations- und Auswahlleiste. Insgesamt sind die Unterschiede in den Beurteilungen der beiden Versionen jedoch gering.

Die Bewertungen auf den Fragebögen spiegelten den Umgang der Benutzer mit dem Programm wider: Selbst jene Teilnehmer, die angaben, nie einen Computer oder das Internet zu benutzen, fanden sich bemerkenswert schnell in dem Programm zurecht. Die Lösungen für die Aufgaben wurden immer gefunden und kein Benutzer mußte abbrechen oder benötigte explizite Hilfe. Alle Benutzer gaben an, daß sie sich im Umgang mit dem Programm nicht unsicher gefühlt haben. Sie wußten generell, in welcher Ebene sie sich befanden und konnten durch das Zurückkehren in höhere Ebenen falsche Schritte korrigieren und die Suche neu starten. Die vielen verschiedenen Möglichkeiten der Navigation haben - laut Aussagen der Teilnehmer - keine Verwirrung hervorgerufen.

4.4 Vergleich der beiden Versionen, Präferenzen der Benutzer

Abschließend wurden die Teilnehmer zur Präferenz hinsichtlich der Plattform befragt (siehe Abb. 7).

Die Erlernbarkeit des Umgangs wurde auf beiden Systemen gleich gut eingeschätzt, was den Vergleich der Usability-Ratings bestätigte. Überraschend ist, daß 12 der 13 Benutzer angaben, mit der PC-Version besser zurechtgekommen zu sein als mit der TV-Variante, obwohl beide Versionen in den Usability Ratings etwa gleich gut abschnitten und auch die Anzahl der Bedienschritte sich nicht gravierend unterschied. Die Erklärung ist hierfür wohl bei der Hardware-Schnittstelle zu suchen: 12 Personen fanden die Mausbedienung mit der Möglichkeit der direkten Anwahl von Links angenehmer als die Tastenfernbedienung, obwohl die Maus für einige Benutzer neu war.

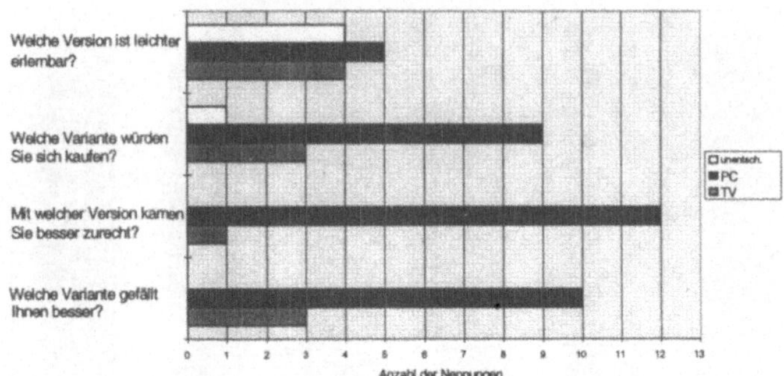

Abb 7: Anzahl der Nennung hinsichtlich der Präferenz von PC- und TV-Version.

Alle Benutzer, die der TV-Version in einer der restlichen Fragen (Kaufpräferenz, allgemeine Akzeptanz) den Vorzug gaben, waren Personen, die einen PC selten oder nie verwenden. Sie bevorzugen die TV-Version, weil sie mit dem Fernseher als Heimgerät vertraut sind. Für die PC-erfahrenen Benutzer ist die Bedienung mit der Maus komfortabler. Vor allem diejenigen, die schon viel Erfahrung mit einem PC haben und selbst einen besitzen, ziehen es vor, die Infocity-Dienste auch auf ihrer gewohnten PC-Oberfläche zu nutzen.

4.5 Akzeptanz eines Informationsdienstes via ITV

Sechs Benutzer begrüßten die Bild-in-Bild-Funktion bei der Fernsehversion. Sie stellte für sie eine Möglichkeit dar, das Fernsehprogramm nebenbei mitzuverfolgen, wenn man gerade den Informationsdienst benutzt. Die anderen Teilnehmer fühlten sich durch das Bild z.T. gestört. Die Musik, die im Versuch nebenbei lief, schien die Lösung der Aufgaben nicht zu beeinträchtigen.

Alle Benutzer äußerten sich positiv zu der Möglichkeit, den Fernseher als Zugang zu Online-Diensten nutzen zu können. Das Angebot ist für die Teilnehmer vor allem zum Internetsurfen attraktiv, sowie zum schnellen Abruf von Lokalinformationen wie z.B. dem Kinoprogramm.

5 Zusammenfassung und Diskussion

Die Verwendung von „multiplen Navigationselementen" scheint in dem Fall einer Cross-Plattform Oberfläche eine effiziente „Design for all" Strategie. Die getesteten Oberflächen erwiesen sich als benutzerfreundlich und einfach zu bedienen. Bildmenüs, Shortcut Links und „Zurück" Funktionen sind die am häufigsten genutzten Navigationselemente. Die Navigationsleiste (Historie) wurden eher kaum genutzt. Die Benutzer konnten in der beschriebenen Oberfläche auch ohne Vorkenntnisse in einer Informationstiefe von 5 Ebenen erfolgreich navigieren. Wichtig war dabei ein schneller Zugang zu den obersten Navigationsebenen.

Individuell sehr unterschiedlich ist die Bewertung der „Bild-im-Bild" Funktion des interaktiven Fernsehers. Ihre Akzeptanz und der Einfluß bei der Informationssuche ist noch unklar. Verglichen mit den bislang verfügbaren Möglichkeiten der ITV-Steuerung (z.B. Mini-Joystick) ist die herkömmliche Tastenfernbedienung mit „Windrose" und „Focus-Jump" eine akzeptable Alternative, zumindest bei einer orthogonal aufgebauten Oberfläche.

Usability hinsichtlich Informationssuche allein ist noch kein Kriterium für Akzeptanz von Internetdiensten im ITV. Die meisten Benutzer bevorzugten hinsichtlich der Bedienbarkeit eine PC-Plattform, was besonders an der vergleichsweise umständlichen Fernbedienung liegt. Dagegen würden PC-unerfahrene Benutzer beim Kauf einen Fernseher als Zugang zu elektronischen Diensten bevorzugen. Erfahrene PC-Nutzer werden sich sicherlich nicht von den erweiterten Möglichkeiten des PCs verabschieden, aber in Zukunft auch evtl. vorhandene ITV-Zugänge nutzen wollen. Dies unterstreicht die Bedeutung von benutzerzentrierten Entwicklungen von Online-Diensten sowie eines plattformübergreifenden Designs, um sowohl neue Benutzergruppen als auch erfahrene Anwender zu erreichen.

6 Literatur

[1] Hitzges, Arno; EMNID-Institut (Hrsg.): media vision trend: Akzeptanz, Stand der Technik und Perspektiven ausgewählter Anwendungen – Dokumentation der Ergebnisse, Stuttgart,1997. Fraunhofer IRB Verlag

[2] Karepin, Rolf: "Das Zusammenwachsen von Fernseher und Computer liegt noch in weiter Ferne." Computer Zeitung, Nr. 50 / 11.12.97, p. 6

[3] Computer Zeitung: "Loewe und Philips verheiraten die Glotze für viel Geld mit dem Multimedia-PC". Computer Zeitung, Nr. 37, 11.09.97, p. 13

[4] Eckhard E.: Das größte Multimediaprojekt entsteht. Funkschau, Fachzeitschrift für elektronische Kommunikation, 1997, Heft 18, 28-33.

[5] S.Antoniazzi, J.Buschmann, E.Marcozzi, A.Profumo, "Internet and New Interactive Multimedia Services: Integration Opportunities", European Conference on Multimedia Applications, Services and Techniques (ECMAST '96), ACTS Session, Louvain-la-Neuve, Belgium, 1996.

[6] Burmester, Michael (Hrsg.): FACE ESPRIT Project 6994 Guidelines and Rules for Design of User Interfaces for Electronic Home Devices. Stuttgart, 1997, Fraunhofer IRB Verlag.

[7] ISO 9241: Ergonomic requirements for office work with visual display terminals (VDTs)

[8] Spool, J.M., Scalon, T., Schroeder, W. et.al. (1997). Web Site Usability: A Designer´s Guide. North Andover, Massachusetts: User Interface Engineering, 1997.

[9] Apple Web Design Guide [http://applenet.apple.com/hi/web/intro.html]

[10] Sun: Guide to Web Style [http://www.sun.com/styleguide]

[11] Beu, A.: Final Online Styleguide, MUSIST Project AC010 D20, AC010/GSM/WP1/DR/P/20/b1, 1998

[12] Weickert, B.: Integrated Consumer Terminal for TV-i Deliverable D21, MUSIST Project (AC010) 1998

[13] Burmester, Michael; Koller, Franz: "Metaphern für interaktives Fernsehen: eine Fallstudie mit Endbenutzern". In: R. Liskowsky, B.M. Velichkovsky, W. Wünschmann (Hrsg.): Software-Ergonomie '97 - Usability Engineering: Integration von Mensch-Computer-Interaktion und Software-Entwicklung, Berichte des German Chapter of the ACM 49, Stuttgart, 1997, Teubner, pp. 93-110

[14] Burmester, Michael; Koller, Franz; König, Thilo: "Benutzerorientiertes Design – PREMIERE Benutzungsoberfläche für interaktives Pay-TV". In: R. Liskowsky, B.M. Velichkovsky, W. Wünschmann (Hrsg.): Software-Ergonomie '97 - Usability Engineering: Integration von Mensch-Computer-Interaktion und Software-Entwicklung, Berichte des German Chapter of the ACM 49, Stuttgart: Teubner, 1997, pp. 111-122

[15] Tauscher, L. Greenberg, S., 1997, „How People Revisit Web Pages: empirical Findings and Implications for the Design of History Systems". Spec. Issue Intern. Journal of Human-Computer Studies, 47 (1), p.97-137.

Adressen der Autoren

Dipl.-Ing. Jörg Nissler
Fraunhofer IAO
MT Interaktive Produkte
Nobelstr. 12
70569 Stuttgart
Joerg.Nissler@iao.fhg.de

Dipl. Psych. Volker Thoma
Fraunhofer IAO
MT Interaktive Produkte
Nobelstr. 12
70569 Stuttgart
Volker.Thoma@iao.fhg.de

Sylvia Manz
Fraunhofer IAO
MT Interaktive Produkte
Nobelstr. 12
70569 Stuttgart
Sylvia.Manz@iao.fhg.de

Supporting Learner in Exploratory Learning Process in an Interactive Simulation based Learning System

Reinhard Oppermann*, Kinshuk*, Akihiro Kashihara**, Rossen Rashev* and Helmut Simm*

*GMD-FIT, Germany   **I.S.I.R., Osaka University, Japan

Abstract

Learning-by-exploration is an effective way in the learning process, provided the educational system is capable of estimating the contextual requirements of the learners and of supporting them by providing facilities to keep the cognitive load at optimum level. Exploration Space Control methodology allows the systems to provide intelligent assistance to the learners by effectively presenting the information space and tools to explore that space in a way suitable for learners' current competence level. This paper describes the implementation of Exploration Space Control methodology in InterSim system, which aims to facilitate learning structure and functionality of organs and to improve appropriate skills in diagnosing and treating the related diseases.

Zusammenfassung

Lernen mittels Exploration ist eine effektive Form des Lernens vorausgesetzt, das Lehrsystem ist in der Lage, die Bedürfnisse des Lerners situationsgerecht einzuschätzen und ihnen durch Vorgabe eines adäquaten Niveaus cognitiver Belastung optimal zu entsprechen. Die Exploration Space Control Methodik ermöglicht es Systemen, den Lernenden intelligente Unterstützung zu bieten mittels angemessener Darstellung des Informationsraumes und mittels Werkzeugen, die eine dem aktuellen Wissenstand angepaßte Exploration dieses Informationsraumes erlauben. Dieser Beitrag beschreibt die Realisierung der Exploration Space Control Methodik im InterSim System, dessen Ziel es ist, das Lernen von Anatomie und Physiologie von Organen zu erleichtern und angemessene Fähigkeiten zur Diagnose und Behandlung ihrer Krankheiten zu entwickeln.

1 Introduction

The learning-by-exploration is an effective technique for learning [5], in particular for task oriented disciplines such as computer science and medicine. This technique not only provides skills of the domain but also the understanding of the embedded concepts ([2], [4],[7]). However, there are various considerations an educational system needs to take care to provide effective learning with the help of learning-by-exploration.

One primary way to support learning-by-exploration is to provide various techniques which enable the learners to explore domain concepts/knowledge to acquire cognitive and procedural skills. Current research on such techniques in educational systems include multi-media, hypermedia, simulation, demonstration and virtual reality, and the list of items is growing very fast due to extensive research going on in this field.

However, it is not an easy task for the learners to effectively explore the domain concepts/knowledge by themselves and develop adequate skills in domain related tasks. The extent and amount of complexity inherited in the techniques, provided by the educational system for exploration purposes, needs to depend on the learners' current level of competence and their current capacity to cope with cognitive load arising from such explorations. The systems should, therefore, be equipped with capabilities to assist the user in various tasks and to vary the extent and amount of complexity of exploration techniques. Typically, the system would need to limit the learning space (called exploration space) for the novice learner and would remove the restrictions gradually as the learner progresses in the learning process.

Current research on educational systems has proposed various intelligent assistant methods to support context based navigation, presentation tailoring etc. but these methods are not always sharable and reusable among researchers and developers as they do not correspond to a generalized protocol. [7] and [8] proposed a methodology, called Exploration Space Control (ESC for short), to generalize current intelligent methods of supporting exploration. This paper describes the use of ESC in the implementation of an interactive simulation based learning system, called InterSim.

In the following sections, we first give a summary of the ESC methodology. A detailed description of the methodology is given in [7] and [8]. Then we describe the use of ESC methodology in the implementation of the InterSim system, which aims to facilitate learning of structure and functionality of human ear and the acquisition of appropriate skills in diagnosing and treating the diseases. The paper finally concludes with identifying some research aspects which need further investigation.

2 Exploration Space Control

In most hypermedia systems, and simulation-based learning and training systems, learners learn a domain by accessing various information resources such as hypertext, demonstrations, simulations, and so on. In this sense, the exploration activity can be defined as manipulating these information resources to comprehend the information and to acquire domain concepts/knowledge.

Ideally, the learners should be free to explore since finding out domain concepts/knowledge by themselves would enhance their learning ([2], [5]). However, learners may not know what to and how to explore. They may also make excessive mental efforts to search and integrate the information from different information resources, which itself may cause cognitive overload [6]. The exploration space, in addition, may be quite wide so they may lose their ways.

Exploration Space Control (ESC) is a methodology encompassing various educational tasks which make it easier for the learner to explore domain material, for example, providing navigation in learners exploration paths ([1]) and information tailoring ([9]) to make the search and domain knowledge comprehension easier for the learner, restricting simulation parameters to make interpretation of the behavior of simulated results easier ([3]) and so on.

ESC methodology facilitates proper learning environment for all types of learners by controlling the extent of exploration space according not only to the domain complexity but also to the learners' competence, understanding levels, experiences, characteristics, etc. According to this methodology, the exploration space is restricted either in terms of the exploration tools and information to be presented or as recommending few choices according to the learners' perception and understanding level. This restriction/recommendation intends to reduce the learners' cognitive load. The purposes of ESC are as follows:

To facilitate active learning. This approach is suitable for the learners who have higher learning competence. The active learning is provided by reducing cognitive load as less as possible. The restrictions/recommendations are imposed only to protect the learners from cognitive overload.

To facilitate step-by-step learning. This approach is suitable for the learners who have lower learning competence. The step-by-step learning is provided by reducing cognitive load as much as possible. This approach gradually induces the learners to make exploration efforts.

The above mentioned purposes cover a whole spectrum of learners and the "active learning" and "step-by-step learning" are two extreme approaches covering that spectrum. The combination of these two approaches in varying quantity facilitates adequate learning for whole learner spectrum [6].

ESC is implemented at various levels of controls in the form of restrictions, warnings and suggestions imposed on learners, according to learner models and domain complexity. These control levels are as follows:

- **Embedding information.** This facilitates the creation of information space and involves scaffolding.
- **Limiting information resources.** Two kinds of controls are used to limit information resources:
    - Limiting the number of information resources to be presented to a learner at a particular moment.
    - Presenting information resources appropriate for looking into current domain material by a learner at a particular moment.
- **Limiting exploration paths.** For example, this can be done by restricting navigational paths in hypertext or by controlling various parameters in simulation environment. Two kinds of controls are used in the level:
    - Limiting the number of feasible exploration paths to be looked into.
    - Limiting the exploration paths which are unrelated to the current domain material.
- **Limiting information to be presented.** There are again two methods to provide such control:
    - Limiting the amount of information.
    - Adapting the contents of information to each learner.

Current technologies can be used to provide different kinds of controls as shown in Table 1.

Current Technologies	Control Levels
Scaffolding	Embedding information
Navigation	Limiting information resources & exploration paths
Problem ordering (Courseware)	Limiting exploration paths
Information tailoring	Limiting presented information
Simulation setting	Limiting exploration paths & presented information

Table 1: Relationships between current technologies and ESC.

3 Designing the educational systems

There are three main steps in designing the educational systems on the basis of ESC.

1) Learning goals which learners are expected to accomplish should be identified.
2) Scaffolding methods should be selected in the form of various information resources, taking into account the amount and contents of the information, and by considering various exploration operations to be used in and between each information resource. Examples of exploration operations include *Select* (selecting an information resource), *Trace* (tracing a sequential path within an exploration environment), *Interpret* (interpreting the results of previous actions, particularly useful in simulation environments) and *Apply* (to execute an action, such as application of changed parameters in a simulation).
3) Deciding which control level should be applied to which information resource.
   a) Deciding the purpose of ESC (support for active and/or step-by-step learning) so as to decide the guidelines for controlling exploration space.
   b) Deciding control levels (table 1) to restrict the exploration operations in the information resources.
   c) Deciding the application of controls according to learner models and domain models. Various factors which need to be considered in learner models are as follows:

   - Preferences   - Knowledge Levels   - Experiences
   - Competence   - Exploration Process   - Cognitive Load

   Possible factors to be considered in domain models include:
   - Knowledge type (Procedural/Declarative)   - Granularity   - Depth (Deep/Shallow).

4 Implementation of ESC in the InterSim project

The ESC methodology is being applied in the InterSim project for the system in the domain of ear. The objective of the system is to facilitate the learning of structural and functional aspects of healthy ear and related diseases, and to provide skills in the diagnosis and treatment of the diseases.

The design and development of the InterSim system on the basis of ESC includes the provision of several kinds of information resources for exploring ear and the provision of user interface to allow learners to make exploration operations. The system is aimed to facilitate active learning rather than step-by-step learning process. Figure 1 shows a partial view of conceptual layer of the learning material [8].

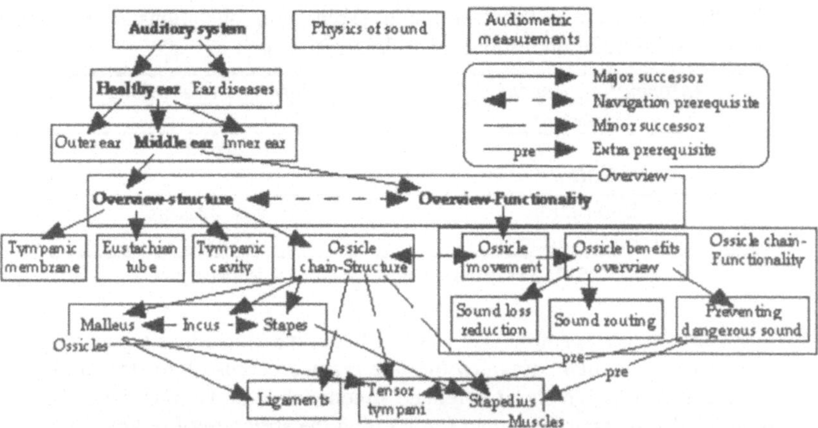

Figure 1: Partial conceptual layer of learning material for ear domain.

Various steps in the design of InterSim system according to ESC are as follows.

(1) Determination of learning goals.
  There are two main goals in the InterSim system:
  * Understanding the structure and functionality of ear.
  * Acquiring appropriate skills in diagnosing and treating the related diseases.

(2) Scaffolding.
  (a) Selecting and developing information resources:
   I. For the first learning goal, the suitable information resources are Hypertext, Demonstration, Simulation and Problem Ordering.
   * **Hypertext** gives learners the descriptions of structure, behaviour, and functionality of ear.
   * **Demonstration** facility shows learners the behaviour of normal ear.
   * **Simulation** facility enables learners to experiment on functionality of ear and to ·receive the simulated behaviour.
   * **Problem Ordering** facility provides learners with a number of problems sequenced according to learners' competence level.
   II. For the second learning goal, the suitable information resources are Simulations, Design and Problem Ordering.
   * **Simulation** facility provides an exploration environment for diagnosing and treating the diseases.
   * **Design** facility enables learners to introduce faults in the ear in order to deepen their understanding about interrelationships of various factors.
   * **Problem Ordering** facility enables learners to solve (diagnosis and treatment) problems in order.
  (b) Second step is to decide the control levels to restrict exploration operations in the information resources. For this purpose, we first consider which exploration operations are useful for the InterSim system:

I. For the first learning goal, suitable exploration operations within the above-mentioned information resources are shown in table 2.

Information Resources	*Exploration Operations*
Hypertext	Select & Trace
Demonstration	Interpret
Simulation	Apply & Interpret
Problem Ordering	Trace

Table 2: Suitable exploration operations for first learning goal.

The learners, for example, can select a hyperlink (including a part of image map) among many available to know the fine grained details of an structural part of ear, or can trace through the interrelated functionality of various parts. For example, figure 2 shows a partial screen for learning the structure of ossicle chain in middle ear. Each ossicle in the image is mouse sensitive. Taking mouse over an ossicle reveals the name of the ossicle, single clicking gives a short description and double click takes the learner to detailed explanation view of that ossicle.

Similarly, the activities shown in a demonstration need to be interpreted such as a muscle reflex needs to be related to the changes in various graphs. The learners need to apply various parameters in simulations and need to interpret the results. Within problem ordering, the learners need to trace through various parameters in order to solve the problems.

Since the learners can select one or more information resources and can integrate them for better understanding of the subject matter, the suitable exploration operations among above mentioned information resources are: Select, and Integrate. For example, the learners can select a static picture and a virtual reality scenario of tympanic cavity so as to have an spatial understanding of the cavity while still relating one part of cavity to another.

II. For the second learning goal, suitable exploration operations within the information resources are shown in table 3.

Information Resources	*Exploration Operations*
Simulation	Apply & Interpret
Design	Select, Apply, & Interpret
Problem Ordering	Trace

Table 3: Suitable exploration operations for first learning goal.

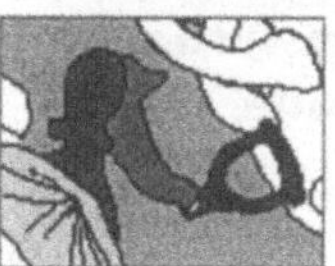

The three ossicles of the middle ear are connected to each other as depicted in the picture constituting the ossicle chain.

The joint between Malleus and Incus is assumed to be fixed (a small elastic movement up to 5° is possible) whereas the joint between Incus and Stapes is rather flexible.

Lateral the ossicle chain is connected to the Tympanic membrane by the handle of the Malleus while the medial endpoint is the base of the Stapes which is inserted into the Oval window.

Malleus and Incus are flexibly suspended from ligaments within the middle ear cavity. Malleus and Stapes are also connected with one muscle each.

The Ossicle chain forms a channel for sound energy transfer from Tympanic membrane to Oval window.

Figure 2: Partial screen for learning structure of ossicle chain.

For example, the learners can apply various causes for initiation of diseases and can interpret the results to see how one disease can advance. The learners can also select various parameters, change their values and apply them to a healthy ear to simulate a faulty ear and can interpret the results. Problem ordering requires the tracing of various relevant parameters in order to find the correct sequencing towards the solution.

For the second learning goal, the learners can not only select and integrate the information resources but also apply the results of one information resource to another. Therefore the suitable exploration operations among the information resources are: Select, Apply, Integrate. For example, the symptoms generated after selection and application of various treatment methods could then be integrated to another set of treatment methods to see the combined effect of one set on another.

(3) Deciding Control Levels.
    (a) The first step is to decide the purpose of ESC. In the case of InterSim system, the purpose is to provide active learning support to the learners.

    (b) The second step is to decide control levels (as shown in table 1). It is envisaged that all control levels are not suitable for both learning goals of the InterSim system. For first learning goal, the selected control levels are:

- **Limiting information resources**
- **Limiting exploration paths**
- **Limiting presented information**

For second learning goal, the selected control levels are:

- **Embedding information**
- **Limiting information resources**
- **Limiting exploration paths**

In InterSim system, these controls levels are implemented by activating, inactivating and recommending the choices in various controls in the interface, which allow the learners to select, trace, apply and so on while exploring. An important example in InterSim Ear system for control levels is the concept of "main path" and "excursions". While interacting with the system, the learner would explore along a main learning path within one knowledge module, and the system would provide intelligent support and guidance accordingly. Whenever the learner needs to deviate from that path to some loosely related unit of knowledge in some other domain module, the system would allow such excursion, but with limiting information resources, and the information presented in such excursions would also be limited to make it relevant with main learning path. The exploration paths in an excursion would also be limited for the sake of not letting the learner lost in hyperspace of excursions. For example, a learner, exploring the structure of middle ear, would be able to get an excursion in physics of sound to understand how the sound is travelled through the mechanical linkage of ossicles, but the information presented in the physics of sound would be tailored for better understanding of sound travel in ossicles. On the other hand, the learner, whose main learning path is physics of sound would be able to get the knowledge presentation in more depth.

(c) The third step is to decide how to control various information resources. As suggested in section 3, there are two ways to decide the application of controls: according to learner models and according to domain complexity. In InterSim system, it is assumed that the learner should access the complex domain material only when the competence level of the learner allows for this. Of course, learner can explicitly access any domain material, but in such cases, the system would not be able to recommend any preferred path to the learner. With this assumption, the implementation of ESC in InterSim system employs only learner models.

Following are the examples of each control level selected for two learning goals of the InterSim system.

First learning goal: Understanding the structure and functionality of ear

Example 1: Limiting information resources for understanding the structure and functionality of the ear.
- Restriction methods: Restricting the representation of various domain material in terms of complexity of representation (for example, static pictures Vs virtual reality scenarios)
- Exploration Experience: Low, Middle, High
- Cognitive load: High, Middle, Low
- The degree of limitation of multiple information resources for same domain material:
  - Strong when experience is low
  - Weak when experience is middle
  - No limitation when experience is high
- The degree of limitation of number of complex information resources:

- Strong when cognitive load is high   - Weak when cognitive load is middle
- No limitation when cognitive load is low

Example 2: Limiting exploration paths for understanding the structure and functionality of the ear.
- Restriction methods: Restricting/ recommending buttons, combo box choices, anchors/ links to be used in exploring Hypertext to limit Select & Trace operations.
- Exploration Competence: Low, Middle, High
- Knowledge Levels: Low, Middle, High
- The degree of limitation of feasible paths:
    - Strong when competence is low   - Weak when competence is middle
    - No limitation when competence is high
- The degree of limitation of unrelated paths:
    - Strong when knowledge level is low   - Weak when knowledge level is middle
    - No limitation when knowledge level is high

Example 3: Limiting presented information for understanding the structure and functionality of the ear.
- Restriction methods: Restricting the presented information to the learner
- Exploration preferences: Low, Middle, High
- Knowledge levels: Low, Middle, High
- Cognitive load: High, Middle, Low
- The degree of limitation of type of presented information:
    - Strong when preference is low   - Weak when preference is middle
    - No limitation when preference is high
- The degree of limitation of richness of presented information:
    - Strong when knowledge level is low   - Weak when knowledge level is middle
    - No limitation when knowledge level is high
- The degree of limitation of amount of simultaneously presented information:
    - Strong when cognitive load is high   - Weak when cognitive load is middle
    - No limitation when cognitive load is low

Second learning goal: Acquiring appropriate skills in diagnosing and treating the related diseases

Example 1: Embedding information for acquiring skills in diagnosing and treating the diseases.
- Restriction methods: Providing scaffolding so as to decrease domain complexity with regard to learner models (for example, first allowing the learner to semi-explore the disease development process in an animation wizard; then adding simulation capabilities to allow the full exploration; then adding extra simulation capabilities for diagnosis)
- Exploration Competence: Low, Middle, High
- The degree of scaffolding:
    - Strong when competence is low   - Weak when competence is middle
    - No limitation when competence is high

Example 2: Limiting information resources for acquiring skills in diagnosing and treating the diseases.

- Restriction methods: Restricting the representation of various domain material in terms of complexity of representation (for example, first providing exploration of anatomic details of the diseased ear, and then gradually providing various relevant graphs and diagrams used in the diagnosis of the disease. Figure 3 shows an example user interface of a disease simulation in the system. The learners can explore ear with various multimedia objects (such as image maps), experiment on the ear by manipulating parameters, and interpret the results.)
- Exploration Experience: Low, Middle, High
- Cognitive load: High, Middle, Low
- The degree of limitation of multiple information resources for same domain material:
    - Strong when experience is low   - Weak when experience is middle
    - No limitation when experience is high
- The degree of limitation of number of complex information resources:
    - Strong when cognitive load is high   - Weak when cognitive load is middle
    -     No limitation when cognitive load is low

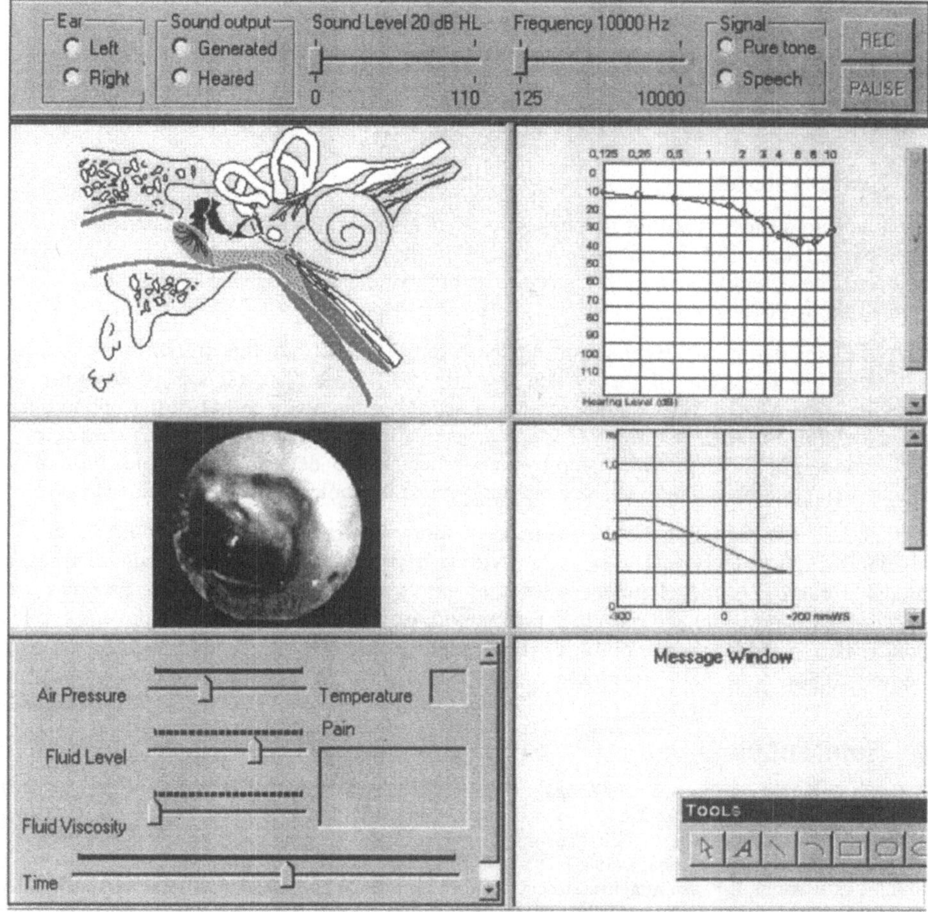

Figure 3: User Interface of a disease simulation in InterSim System.

Example 3: Limiting exploration paths for acquiring skills in diagnosing and treating the diseases.

- Restriction methods: Restricting various simulation operations to suit the learner models (for example, allowing the learner to explore the anatomy part of figure 3 to see the effects on the graph, and once the learner is capable of handling, providing the exploration activities in the graphs to see the effects in the anatomy part, and in other graphs.)
- Exploration Competence: Low, Middle, High
- Knowledge Levels: Low, Middle, High
- The degree of limitation of feasible paths:
    - Strong when competence is low
    - Weak when competence is middle
    - No limitation when competence is high

- The degree of limitation of unrelated paths:
- Strong when knowledge level is low   - Weak when knowledge level is middle
- No limitation when knowledge level is high

5 Conclusions

The InterSim system is still at its prototyping stage and no formal user evaluations have yet been carried out. But formative evaluation of the system by medical practitioners has validated the effectiveness of the system in both domain knowledge and cognitive skills acquisition by the learners.

The implementation of Exploration Space Control (ESC) in the InterSim system has facilitated the adaptivity of the system towards the learner. The success in designing the InterSim system using ESC methodology has validated the applicability of the methodology in successful implementations of adaptive educational systems supporting learning-by-exploration. The ESC provides step-by-step procedure to determine the objectives of the educational systems and help in selecting the suitable technologies to be used in such systems.

Further research regarding ESC is to make it more generalised so that it could be used in developing any educational systems providing learning-by-exploration. This requires an intensive survey of development efforts of the systems available in the literature and identifying the potential and pitfalls in the methodologies by which they were developed. This would also provide an appraisal of ESC methodology as it stands in front of other methodologies.

6 References

[1] Boyle, C. & Encarnacion, A.O.: An Adaptive Hypertext Reading System. User Modeling and User-Adapted Interaction, 3(3) (1993), 193-208.

[2] Carroll, J., Mack, R., Lewis, C., Grischkowsky N., Robertson S.: Exploring exploring a word processor. Human-Computer Interaction, 1 (1985), 283-307.

[3] Eliot, C. & Woolf, B.P.: An Adaptive Student Centered Curriculum for an Intelligent Training System. User Modeling and User-Adapted Interaction, 5(1) (1995), 67-85.

[4] InterSim Project HomePage: http://zeus.gmd.de/projects/intersim.html, 1996.

[5] Kamouri, A., Kamouri, J., & Smith, K.: Training by Exploration: Facilitating the Transfer of Procedural Knowledge Through Analogical Reasoning. Man-Machine Studies, 24 (1986), 171-192.

[6] Kashihara, A., Hirashima, T. & Toyoda, J.: A cognitive load application in tutoring. User Modeling and User-Adapted Interaction, 4(4) (1995), 279-303.

[7] Kashihara,A., Kinshuk, Oppermann,R., Rashev,R., Simm,H.: An Exploration Space Control as Intelligent Assistance in Enabling Systems. Proc. of ICCE 97, 1997, 114-121.

[8] Kashihara A., Kinshuk, Oppermann R., Rashev R., Simm H. & Toyoda J.: Towards Sharable Intelligent Assistance for Learning-by-Exploration. SIG Technical Notes in Japanese Society for Artificial Intelligence, 1998, SIG-IES-9703, JSAI, Shizuoka, Japan, 1-6.

[9] Kobsa, A., Mueller, D., & Nill, A.: KN-AHS: An Adaptive Hypertext Client of the User Modeling System BGP-MS. Proc. of User Modeling 94, Hyannis, Mass., 1994, 99-105.

Authors' addresses

Prof. Dr. R. Oppermann, Dr. Kinshuk, Mr. R. Rashev & Mr. H. Simm
GMD-FIT, German National Research Center for Information Technology
Schloss Birlinghoven, 53754 St. Augustin
Germany
Email: {oppermann, kinshuk, rashev, simm}@gmd.de

Dr. A. Kashihara
The Institute of Scientific and Industrial Research
I. S. I. R., Osaka University
8-1, Mihogaoka, Ibaraki, Osaka, 567
Japan
Email: kasihara@ai.sanken.osaka-u.ac.jp

Adaptive Information for Nomadic Activities.
A process oriented approach

Reinhard Oppermann and Marcus Specht
GMD FIT.HCI

Abstract

The paper describes the idea and the structure of a mobile information system under development in the HIPS[1] project that allows for supporting nomadic activities. Nomadic activities are a widespread class of human activities. Many activities are distributed in time, space and social groups. In contrast to print media till now electronic information and communication media did not support nomadic activities. The worldwide Web, wired and wireless connectivity to networks, and small mobile devices now allow for supporting nomadic activities by information and communication technology to an extensive extent. The paper describes ideas for the application of technological potentials for art excursions as an attractive class of nomadic activities. For the fit of user interests and needs adaptation facilities are described, to reflect the current user's local position (contextualisation) and interaction history (individualisation).

Zusammenfassung

Der Beitrag beschreibt die Idee und die Struktur eines mobilen Informationssystems, das im Projekt HIPS entwickelt wird und die Unterstützung nomadischer Aktivitäten erlaubt. Nomadische Aktivitäten sind eine weit verbreitete Klasse menschlicher Aktivitäten. Viele Aktivitäten erfolgen zeitlich, örtlich oder sozial verteilt. Im Gegensatz zu Druckerzeugnissen haben bisherige elektronische Informations- und Kommunikationsmedien keine nomadische Aktivitäten unterstützt. Weltweite Netze, drahtgebundene und drahtlose Verbindungen zu Netzen und kleine mobile Geräte erlauben mittlerweile auch die Unterstützung nomadischer Aktivitäten durch Informations- und Kommunikationstechnik. In diesem Beitrag werden Ideen für die Anwendung technologischer Potentiale für Kunstexkursionen, eine attraktive Klasse von nomadischen Aktivitäten. Für die Berücksichtigung von Benutzerinteressen und –bedürfnissen werden Anpassungsmöglichkeiten beschrieben, die die aktuelle örtliche Psotion (Kontextualisierung) und die Interaktionshistorie (Individualisierung) berücksichtigen.

Keywords: *Nomadic Activities, Mobile Computing, Contextualisation, Adaptive System, Museum Information System*

1 Introduction

A challenging feature of current IT development is mobile information technology [1]. Till now people using information technology were fixed to their hardware station (desktop). Currently two technical developments contribute to overcome this restriction. There are small computers like laptops and now even palmtops that allow for mobility in the physical space not being cut off from information technology. And there are net facilities that allow for accessing any information from any point in the physical space—given the user and the information is connected to a network. These developments support nomadic users using mobile information technology.

Mobile information technology may be curse or boon. It may fix the user to a technical system reducing or substituting the concentration on the real environment. Mobile information

---

1 The project Hyperinteraction within Physical spaces (HIPS) is an EU-supported LTR project in ESPRIT I3. The partners of the consortium are University of Siena (co-ordinating partner), University of Edinburgh, University College Dublin, ITC, SINTEF and GMD, CB&J, and Alcatel.

technology may so be a distraction and may reinforce one's obsession. But mobile information technology may also be a chance. It may support ongoing activities from the beginning to
the end. Traditional media like files, literature, memo-books are limited in accessibility. On
tour the user can only read those office files or private books that he or she has put into the
suitcase before leaving for a journey. The fit of the selection to the actual needs is typically
very limited.

Many activities are distributed over time, over space and between actors. Nomadic media allow for accessing information that are needed in any period of the activity at any place and
together with any partner involved. The actor is not released to plan his/her activity but he or
she is not required to plan the access to all information and all co-actors needed in several
phases beforehand.

Another challenging feature of current research is the adaptation of the selection and presentation of information according to the user's needs [2] [3]. The user uses different devices
with specific characteristics and restrictions both for information access (bandwidth) and information presentation (size and resolution). For mobile information technology the particular
challenge for adaptivity is the support of users at different locations of the user's interaction
with the system. Accordingly nomadic adaptive systems require both a user model where the
information according to user needs, knowledge, and preferences is evaluated and a usage
model where the information according to the current connectivity and hardware and software
is held up-to-date. To achieve this, mobile information technology can be combined with
technologies to identify the users' working environment and his or her position in the physical
space [4]. Infrared or General Positioning Systems (GPS) allow for contextualising the device
of the user in the sense mentioned above. The position and the movement in the physical
space can be observed and evaluated for required information [5].

In this paper we want to discuss demands and possibilities for an adaptive information system
with devices for nomadic activities distributed in time, in space, and between people. To illustrate the application of a mobile adaptive information system art excursions are selected including the preparation, the execution, and the evaluation of a museum visit. The navigation
in the physical and the navigation in the information space are used as indicators for the
user's interests, preferences, and the knowledge acquired so far.

2   Adaptive Information Support for Nomadic Activities

2.1   Activities in distributed time, space and social setting

Most human activities are not fixed to a particular point in time and space. Many activities
evolve and by their nature they are distributed in time and space. They may be executed at
different places and the execution may be interrupted by other activities. Activities often involve several people. Activities are planned by and for several individuals or groups. They are
more than individual short term acts at one place. Extensions of activities can thus be described in three dimensions: extension in time, extension in space, and extension in communities.

- *Extension of activities in time*
  Traditional IT-based systems support human activities by providing massive power to
  store and retrieve information. There has been effort to reflect the goals and the context of
  the user by adaptive features or agents to select a relevant subset of information for a given task for a given user to overcome the information overload or misfit. Machine memory is supposed to reinforce human memory. Memory power can be used to refer to information across time. Human capacity is powerful for the information processing in the pre-

sent; human capacity is selective and unreliable for retrieving information from the past and storing information for the future [6] [7]. For nomadic activities mobile information technology can support full information access at any time, i.e., from the past and for the future, with information retrieval specific for a given activity phase.

- *Extension of activities in space*
Modern network-systems support human activities by providing access to information resources at different places. They reinforce the mobility. The user can be mobile and can have access to information wherever he or she is, or the information can be mobile and the user at a constant place gets access to the information. The requirements for the information content and the information presentation vary for different places. Hardware and output features of information systems have been adapted to local constraints of usage long before laptops and palmtops complement office workstations. For nomadic activities information technology needs to support multiple device information access at any place and position aware (contextualised) information retrieval.

- *Extension of activities in communities*
Group support systems have been developed to contribute to the communication and co-operation of users. They reinforce the communication, co-operation or competition. Adequate technical support of human communication, co-operation and competition must reflect human behaviour to be empathic, role taking, and ambiguous. Reliable access to a communication and a communicator in any time and at any place has to be combined with the need of people to follow their own tasks and preferences not affected by obtrusive communications. For nomadic activities information technology needs to support access to communication tools at any time and place and social constellation aware communication exchange.
Integration of the extension of activities in time, space and communities
For an information and communication system to support human activities it is central to consider an integration of the requirements of the described features. For an appropriate support it is not enough to provide any information at any places in any time for anybody. The specific tasks and goals of an individual and his or her communicators, co-operators or competitors at specific points of time and space have to be taken into account. The adaptation has to consider the demands from the phase of the activity.

2.2 Adaptivity for nomadic activities in the physical space

Mobile adaptive information systems should support human activities considering time, location and social constellation in the physical space and the user's navigation in the information space. Nomadic activities–by its definition–take place at different places. Different devices support access to information technology at different places. In the office the user typically uses a well-equipped multimedia PC. At home the user has possibly low connection to the Internet via modem. During a journey the user is equipped with a laptop or a palmtop with low connectivity and display size. The system can be adapted to the specific conditions of the given place. It can prepare the connection to the Internet, adapt the composition of textual, graphical and video elements and adjust the presentation to the environmental conditions. Activities at different places allow for different interaction with an information system. In the office or at home the information system is the central focus of the user's attention. He or she is sitting in front of the system, selecting, reading and writing information presented on the screen. Outdoor the user interacts with objects within a physical space and the attention is directed towards the physical and social environment. The presentation of information by an information system should reflect the absorbed (mostly visual) sense of the user. Instead of visually displaying information audio output via headphones may be more suitable for a user while additional visual cues may be designed to complement the understanding of or the navigation through the audio information.

Audio presentations allow for viewing the physical environment supported synchronously by complementary information. Different modalities of perception have advantages and disadvantages for the recipient. Audio presentation is weak in providing control for the user. Control only allows for start, pause, move back or forward, and stop. Selection of parts of information and information about the kind and the length of the information is not supported. Compared to visual information audio output has to be received by the user sequentially. The speaker or more generally the audio composer defines the sequence of the information. The user can't scan audio information and can't address interesting parts selectively. The combination of visual and audio presentation offers the chance to integrate descriptions and explanations, but it also implies difficulties. Given the identification of an ideal sequence of perception and an ideal match between visual and audio information the synchronisation of visual and audio presentation requires the evaluation of the recipient's visual perception followed by synchronised audio output. One option is that the recipient controls the presentation by his or her eyes. In this case the visual perception of the recipient has to be identified by an eye tracking system that is at present only applicable for experimental environments. The alternative option is to guide the recipient by audio presentation followed by synchronised visual perception. In this case the recipient is fixed to the guidance of the audio presentation. To compensate some of these shortcomings the user could be given a visual display at least of the length and some characteristics of the kind of audio output. The length of an audio message could be displayed numerically and analogously. The characteristics of information could be displayed by a sequence of keywords (like vita, composition, geographical outline) or by a sequence of icons (like heads, sketch, maps). Elements of the sequence could be selectable for the user.

Adaptivity for nomadic activities should take device and bandwidth constraints into account and adapt to environmental and activity constraints.

2.3   Adaptivity for nomadic activities in the information space

Traditional adaptivity approaches also apply for information processing supporting nomadic activities. For the selection and presentation of information during nomadic activities additional indicators for the user's needs can be obtained from the context of use during the nomadic activity. Preparing an activity needs information different from the execution of the activity and the evaluation of an excursion needs yet another set of information or editing demands of information. During the execution of an activity the physical environment provides objects the user can refer to. During the preparation or evaluation phase of the activity the physical objects probably need to be visualised by a graphical representation. Following the user's activity process the user model of the system can evaluate the history of navigation both in the physical space and in the information space. The model can describe the information used by the user and the places visited by the user.

The adaptation of the selection and presentation of the information to the user supports the combination of new and familiar information for the user. For information behaviour in general and for learning behaviour in particular it is preferable in terms of effectiveness, efficiency and satisfaction for the person to learn new elements embedded into a familiar frame. The connection between already learned and still to be learned elements is essential for the learning success. Repetition is a reinforcing factor for learning. Combining new and familiar items can augment not only learning results. Also the enjoyment of attraction environments can be increased when the person gets new aspects of an already known topic. The actual and the perceived level of difficulty, i.e., the (preferred) relationship of new and old elements is a central issue in pedagogic. Self-assessment of discrepancy between own abilities and knowledge and the demand level of tasks control the achievement motivation of learners. Such factors are highly personality specific. They may depend on the learning history of the person. Success-oriented learners may be more open to new information than failure-experienced learners. One person feels uncomfortable and lost in information space when many new items are presented; others may enjoy serendipity in an innovative environment that provides stimulation and surprise [8, 109] [9, 55f]. Exploration as active and self-directed learning [10] [11] can be supported by an Exploration Space Control as proposed by [12] to prevent people to get lost in the information space but still providing a rich learning environment. What is true for learning also holds for more general experience. Understanding and enjoying reality most probably is effectual if new elements can be based on already known ones. This makes it yet more demanding and attractive to adapt the information to the person's individual knowledge in the given domain.

3 Information system for art excursions: A scenario

3.1 Description of the scenario and its aim

This chapter illustrates the application of a mobile adaptive information system for art excursions in a museum. A visitor is assumed to prepare the excursion to a museum at home by consulting the system via internet reading descriptions and recommendations or viewing representations of paintings. The user's information selection is the initial basis for the user model developed for the adaptive behaviour of the system. Being on site the visitor enters the museum and gets a palmtop device with the mobile system and contextualised information about the exhibit he or she is currently in front of. The visitor can walk through the exhibition following own ideas while the system supports the exploration by contextualised (adapted to the current position) and individualised (adapted to the interest and knowledge) descriptions and comments. He or she can also select a guided tour prepared by the curator of the exhibition or composed by the system based on an inference of interest and knowledge developed in the user model or the user can define a tour specified by explicit selections from the exhibition list.. Before, during, and after the visit in a museum the user can enter and exchange own notes, comments, and communications about his or her impressions for own or others' benefit. The evaluation of the selected exhibits and the selected information about the exhibits allows the system to update the user model continuously.

3.2 Empirical basis of the scenario

The scenario is based on observations, interviews and questionnaires performed in 5 museums in 4 European cities in the consortium of the HIPS project [13]. The design of the prototype of the system has been stimulated by consulting relevant literature about art and art reception of people being interested in art but are no experts or artists themselves. The empirical findings lead to several scenarios written by partners of the project consortium that are to guide the design, implementation and evaluation of the system. In this paper we don't describe the scenarios in detail but we only give the essentials for the understanding of the approach and of the presentation of information and the interaction of the user for the specific environment of art exhibitions.

3.3   Content of the scenario reflecting the demands of different phases

The art domain is attractive for mobile information technology first because art excursions require nomadic activities from visitors. Second, art can best be enjoyed when the art recipient can combine the impression of an art exhibition or an art event with meta-information synchronically explaining aspects of the art not being presented by the artwork itself.

Art excursions are typically planned at home reading books or guides. In future more then at present excursions will also take advantage of information systems given that more material about art and excursions is electronically available and mobile information technology more widespread used. During different phases several kinds of equipment can be used. At home an internet-connected multimedia PC provides information to be consulted, articles and notes prepared and to-do lists created for the journey. A visit of a city with several museums may be prepared by browsing through a list of contents of the museums, reading some descriptions of exhibited art works or artists. In this phase and this environment, i.e., at home without having the artwork in front of one's eyes, the user, for instance, is happy with a graphical representation of the artwork together with its explanation.

During the journey the system should offer seamless access to the previously collected information, for example, on a palmtop via a wireless local network. The presentation can take into account the already provided information and the current position of the user identified by infrared (indoor) or GPS (outdoors). In front of an artwork a high-resolution representation of the work is dispensable. If at all the user needs a silhouette of a painting and opportunities to address areas of the painting he or she wants to ask questions about.

For a system that users can use at several places the input and output of the system have to support immersive interaction. The user can concentrate on his or her activity in the physical and social space and is supported by an additional information layer of a virtual space with complementary information. For the viewer of a painting, for example, several options of verbal comments about the artist, the art period, the art style, the art technique etc., are provided to increase the intellectual benefit of the viewer without reducing his or her visual gusto. Roughly on site, in more detail before or after the visit the user can explore the painting style by overlaying the painting with typical outlines specific for the given or for a contrastive style. Virtual copies of a painting allow for exploring and manipulating its composition and presentation characteristics including explanatory support by remote experts via the net to increase the benefit of the viewer.

The following figure shows the structure of the usage in front of an example fresco (the *Guidoriccio da Fogliano* by Simone Martini).

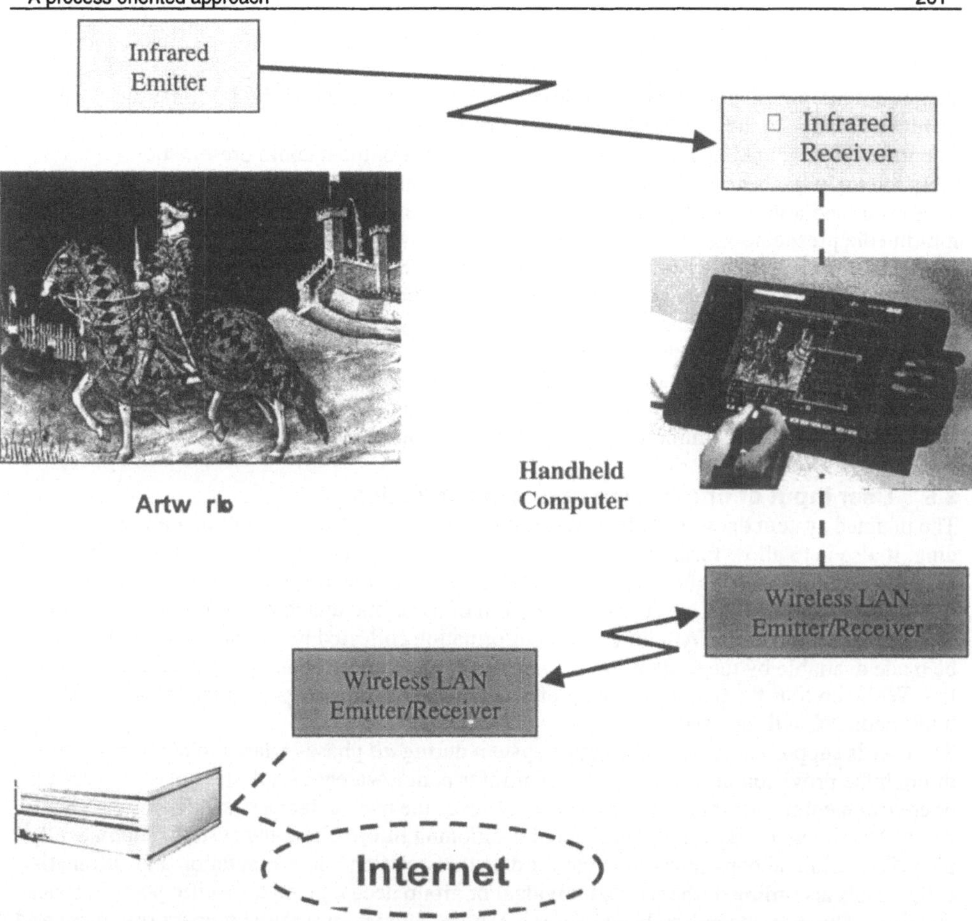

Figure 1: Structure of the system with an exhibit example

3.4 Main modality of system output in the scenario

What is further to be adapted to the user is the modality of the output of the system. During the visit of a museum the user wants to focus the attention on the exhibits, i.e., he or she wants to see the *Guidoriccio da Fogliano* in the example fresco in figure 1. Traditional print media like small labels attached to the exhibits, leaflets, and guidebooks compete with the art attraction itself for the attention of the visitor. The visitor has to switch from viewing the painting to viewing the description of the painting. Visitors of museums often spend plenty of time reading the small labels attached to the displayed items—sometimes as much time as they view the artwork itself. Many visitors consume exhibitions only by passing by the labels of the exhibits one after the other because it is of central interest for visitors to know the artist and the given title explaining the subject of the artwork. In the questionnaire-based study [13] we have found that visitors with more intensive art interest use to vary the distance and the perspective with respect to the artworks. During this navigation in front of the item none of the positions allows for reading the labels so that yet another position is required for reading the short artist and artwork information. More extensive written information requires changing if not the physical position but at least the focus of attention in the information space: from the artwork to the art description in a guide book.

The visual channel being absorbed by the physical environment audio presentations can suitably intensify the benefit of a visit. Experience with multimedia shows that the combination of several media into one device, i.e., the computer, is not enough to increase the benefit from multimedia presentations [14, 7] [15, 181]. Different modalities (sensory channels like eyes or ears) and different coding (like text, picture, table) have to be carefully composed on the media to suite the needs of the recipient. In museums audio guides like a Walkman, a Discman or a human guide are esteemed for their complementary modality. Systems have already been developed to take the position of the user into account when playing audio comments from a Discman [5]. But the presentation offered by these media is a standard one in terms of the visitor's knowledge, interest or time. More tailored audio presentation combined with visual silhouettes and textual summaries can be a complementary attraction.

3.5 User input of information and communication

The planned system does not only provide information output in different modalities and coding. It also is to allow for user entry and editing. Based on media available on site, like local guides, brochures, catalogues, etc., additional information can be collected during the execution phase of a journey, for example in the form of notes and attachments to previously collected information items. After the trip all information collected before and during the trip can be made available by the system for the visitor's evaluation using updating and editing facilities. While on tour the palmtop is the preferred device. At home again an internet-connected multimedia PC will be used.

The user is supported by the information system during *all* phases related to his or her journey through the provision of access to a familiar information warehouse, despite of the diversity of environmental surroundings. The system relieves the user as far as possible from the burden of this diversity. By employing global positioning information, the system automatically identifies technical constraints relevant for data transmission and presentation and automatically adapts according to the user's individual or group needs, given a specific point in space and time. The user can define hotspots for his or her journey during the preparatory phase and can count by predefined hotspots on the system's notice at the relevant situation during the execution indicated by specific space and time parameters.

Some art excursions are performed individually but most visitors come with friends or their family. In order to support social activities, the system has to support the access of several users with defined access rights to information resources and supports the exchange of information during the progress of activities wherever they take place. Information to be exchanged may be art oriented hints, questions or recommendations to other visitors but communications may also be of pragmatical intention to make appointments for a coffee break or the like. Restricted communication may be provided for a dedicated community (group, family) or public communication can show information and interpretation of individual visitors complementing official visitor guidance given by the museum curator.

The scenario integrates mobile appliances with communication and geographical positioning technologies in order to allow for interactions between geographically dispersed locations based on a multi-modal mobile access to distributed information resources, adaptive and adaptable information processing and presentation features. Intuitive interfaces on the background of user's experiences and environmental conditions reflect the user's needs.

4 Conclusions

The scenario stimulates ideas for future systems. Mobile adaptive information systems can be designed to be location aware for the current position of the user and to reflect both the user's navigation history in the physical space and his or her interaction history in the information space. Such systems can enrich the benefit of a nomadic user when developing activities over time and space and in communication with other users. The cut off from information and communication while moving in the physical space can be reduced, the selection of information can be adapted to the needs of the user at particular time and space and the presentation can be optimised due to the not-engaged modality of the user's perception.

The ideas described in this paper will be implemented and evaluated empirically[2]. Massive prototyping in the consortium will allow to follow different approaches in the design of the system and to evaluate the alternatives at different levels. The first level of evaluation will be the interaction level. How can users use the system, what techniques and metaphors are more and what alternatives less appropriate for the user to interact with the system?

The second level of evaluation will be the user's benefit of the system. Is the tour more attractive and more successful for the user? Gets the user a more continuous and richer support for the process of the activity and a better understanding of and a more positive feeling about the activity domain, e.g., the art exhibition?

The third level of evaluation addresses the implications and consequences of the system from a more general point of view. Is a user with a mobile information system a more autonomous user who can concentrate on the items of interest in the physical space of attraction? Is the information system the active guide the user becoming the receptive client? Determines massively available information the kind of reception and interpretation of the physical space of attraction? Enhances the information about the physical space of attraction the cognitive perception of the world reducing affective immersion?

The described questions will be evaluated empirically in trials based on mock-ups and prototypes of different scenarios. A first prototype has been developed by GMD for the castle of Birlinghoven and its art collection. The system can be used on-line to explore ideas of the information presentation or to simulate a preparation of a visit. To start the system use the link from the HIPS-homepage of GMD: http://zeus.gmd.de/ projects/hips.html. The prototype will be evaluated with real visitors of the art exhibition.

5 References

[1] L. Kleinrock: Nomadicity: Anytime, Anywhere In: A Disconnected World, Invited paper, Mobile Networks and Applications, Vol. 1, No. 4, January 1997, pp. 351-357.

[2] A. Kobsa, A. Nill, J. Fink: Hypertext and Hypermedia Clients of the User Modeling System BGP-MS. In: M. Maybury, ed.: Intelligent Multimedia Information Retrieval. Boston, MA, 1997: MIT Press, 339 – 356.

[3] A. Kobsa, W. Pohl: The User Modeling Shell System BGP-MS. User Modeling and User-Adaptive Interaction 4 (1995), 2, 59-106.

[4] G. D. Abowd et al.: Context-awareness in wearable and ubiquitous computing. 1st International Symposium on Wearable Computers, 1997. Proceedings of ISWC'97, October 13-14, 1997 in Cambridge, MA, USA. (http://www.cc.gatech.edu/fce/pubs/iswc97/wear-poster.html)

[5] E. Not et al.: Person-oriented guided visits in a physical museum. ICHIM'97, Paris, September 1997.

[6] J.R. Anderson: Language, Memory, and Thought. Hillsdale, N.J., 1976: Lawrence Erlbaum Associates.

[7] N.C. Waugh, D.A. Norman: Primary Memory. Psychological Review 72, 1965, 89-104.

[8] H. Schaumburg, L.J. Issing: Lernen mit Hypermedia: Verloren im Hyperraum? HMD 190 (1996), 108 - 121.

---

2 A prototype will be prepared for an open day of GMD at the end of September 1998.

[9] R. Schulmeister: Grundlagen hypermedialer Lernsysteme. Theorie - Didaktik - Design. Bonn, 1996: Addison-Wesley.

[10] H. Paul: Exploratives Agieren. Ein Beitrag zur ergonomischen Gestaltung interaktiver Systeme. Frankfurt am Main, 1995: Peter Lang Verlag.

[11] D. Mahling, B. Sorrows, I. Skogseid: A Collaborative Environment for SemiStructured Medical Problem Based Learning. CSCL '95 Proceedings 1995. (http://www-cscl95.indiana.edu/cscl95/mahling.html)

[12] A. Kashihara et al.: An Exploration Space Control as Intelligent Assistance in Enabling Systems. International Conference on Computers in Education Proceedings (1997), AACE, VA, pp. 114-121.

[13] M. Specht: Report on Visitor Questionnaire Data, HIPS internal report. Sankt Augustin, 1998: GMD.

[14] P. Klimsa: Multimedia aus psychologischer und didaktischer Sicht. In: L. Issing, P. Klimsa (Hrsg.): Information und Lernen mit Multimedia. Weinheim, 1995: Psychologie Verlags Union, 7 - 24.

[15] P. Reimann, T. Schult: Schneller schlauer. Bildung im Multimedia-Zeitalter. c't 9/1996, 178 - 186.

Authors´ addresses

Prof. Dr. phil. Reinhard Oppermann
GMD FIT
D-53754 Sankt Augustin
Email: Reinhard.Oppermann@gmd.de

Dr. rer.nat. des. Marcus Specht
GMD FIT
D-53754 Sankt Augustin
Email: Marcus.Specht@gmd.de

Soziale und kognitive Orientierung in einer computergestützten kooperativen Lernumgebung

Hans-Rüdiger Pfister, Martin Wessner und Jennifer Beck-Wilson

Institut für Integrierte Publikations- und Informationssysteme (IPSI),

GMD - Forschungszentrum Informationstechnik GmbH

Zusammenfassung

Es wird ein Ansatz zum computergestützten kooperativen Lernen (CSCL) verteilter Gruppen vorgestellt. Das Problem der sozialen und kognitiven Orientierung bei örtlich verteilten Lernern wird durch eine Benutzungsschnittstelle, die auf der Kombination der Metapher virtueller Räume mit der visuellen Repräsentation sozial konstruierten Wissens beruht, angegangen. Virtuelle Räume strukturieren die Lernumgebung in Analogie zu Räumen der wirklichen Welt; sie bilden Grenzen zwischen sozial und thematisch getrennten Bereichen und stellen spezifische Funktionalitäten je nach Typ des virtuellen Raums zur Verfügung. Eine Visualisierung des in kooperativen Lernprozessen gemeinsam konstruierten Wissens durch sogenannte Lernnetze soll die kognitive Orientierung erleichtern. Virtuelle Räume und Lernnetze zur Repräsentation gemeinsamen Wissens bilden die Kernkomponenten eines in Entwicklung befindlichen Prototypen.

Abstract

An approach for computer supported cooperative learning (CSCL) of distributed groups is presented. The problem of social and cognitive orientation of geographically distributed learners is approached by a user interface, which combines the metaphor of virtual rooms and the visual representation of socially constructed knowledge. Virtual rooms structure the learning environment analogous to rooms in the real world; they provide boundaries between socially and thematically separate regions and provide specific functionalities depending on the type of room. A visualisation of the shared knowledge space constructed during cooperative learning proces-ses through so called learning nets is designed to enhance cognitive orientation. Virtual rooms and knowledge nets for representing common knowledge are the core components of a prototype currently being implemented.

1 Einleitung

Computergestützte kooperative Lernprozesse gewinnen zunehmend an Bedeutung, nicht nur an Schulen und Hochschulen, sondern vor allem innerhalb von Unternehmen sowie in der außerbetrieblichen Weiterbildung; unter der Bezeichnung CSCL (computer supported cooperative learning) wird hier bereits von einem neuen Paradigma gesprochen [1]. Unter kooperativem Lernen verstehen wir einen Prozeß, in dem mehrere Personen in einer lernerzentrierten Lernumgebung mit dem Ziel kommunizieren und kooperieren, gemeinsam ein Wissensdefizit abzubauen.

Hinsichtlich der Ortsdimension können zwei Ausprägungen der Computerunterstützung kooperativen Lernens unterschieden werden: Während sich bei *lokalem CSCL* die Mitglieder der Lerngruppe (z.B. Lerner, Trainer, Tutoren) am selben Ort befinden, bezieht sich *verteiltes CSCL* auf die Unterstützung auf mehrere Standorte verteilter Lerngruppen. Die Unterstützung kooperativen Lernens kann hierbei auf verschiedene Arten erfolgen: ein CSCL-System kann z.B. der Beschaffung relevanter Informationen dienen, geeignete Kommunikationskanäle und gemeinsame Arbeitsbereiche zur Verfügung stellen, sowie Werkzeuge zur kooperativen Repräsentation, Konstruktion und Veränderung von Wissen anbieten.

Insbesondere von verteiltem CSCL werden zahlreiche Vorteile erwartet, etwa ein erhebliches Potential zur Kostenreduktion (z.B. durch die Einsparung von Reisekosten) oder die erhöhte Flexibilität in der Form der Nutzung (z.B. orts- und zeitunabhängige Lernformen, Lernen "nach Feierabend" oder unterwegs). Neben diesen Vorteilen erwachsen aus verteilten kooperativen Lerngruppen aber auch neue Problemfelder [2], die vor allem aus der Reduktion von Informationskanälen im Vergleich zu face-to-face Situationen resultieren.

Wir konzentrieren uns im folgenden auf die Betrachtung örtlich verteilter Lerngruppen (verteiltes CSCL) und beschreiben im nächsten Abschnitt die Probleme der sozialen und kognitiven Orientierung in CSCL-Systemen. Danach stellen wir die Metapher virtueller Räume vor und skizzieren, wie das Problem der sozialen Orientierung durch eine auf virtuellen Räumen basierende Benutzungsschnittstelle adressiert werden kann. Anschließend gehen wir auf die Methode der Konstruktion von Lernnetzen ein, mit der die kognitive Orientierung unterstützt werden soll. Es wird dann kurz skizziert, wie diese Konzepte in einem aktuellen Prototyp implementiert wurden bzw. implementiert werden sollen.

2 Orientierung in verteilten kooperativen Lernumgebungen

2.1 Das Problem der sozialen Orientierung

Als Problem der *sozialen Orientierung* bezeichnen wir die Schwierigkeit, in kooperativen Lernprozessen notwendige Informationen über die anderen Mitglieder der Lerngruppe zu erlangen. Hierzu müssen etwa Informationen darüber verfügbar sein,

- *wer* aktuell auf welche Weise am Lernprozess teilnimmt (social awareness),
- *welche Rolle* und Funktion die Teilnehmer jeweils innehaben (role identity), und
- *welche Struktur* die Lerngruppe als Ganzes aufweist (group structure).

In traditionellen face-to-face Lernsituationen ist die Anwesenheit einer Person direkt wahrnehmbar, und es werden hinreichend viele soziale und non-verbale Signale vermittelt, um die Rolle anderer Teilnehmer inferieren zu können. Bei computervermittelten Interaktionen, in denen viele dieser Hinweise nicht oder in eingeschränkter Form verfügbar sind (falls z.B. die Interaktion rein textbasiert ist), müssen entsprechende Hilfen zur Verfügung gestellt werden. In unserem Ansatz wird dies durch die Schaffung *virtueller Räume* als Analogon zu Räumen der realen Welt umgesetzt [3].

2.2 Das Problem der kognitiven Orientierung

Ein erfolgreicher kooperativer Lernprozeß erzeugt, ebenso wie ein individueller Lernprozeß, neues bzw. verändertes Wissen. Die Besonderheit kooperativen Lernens liegt darin, daß durch die gemeinsame Bearbeitung von Inhalten nicht nur das eigene Wissen, sondern auch das der anderen Gruppenmitglieder „moderiert" werden muß. Durch den kontinuierlichen Wissensaustausch entsteht „gemeinsames Wissen", welches dann wieder dazu dient, neues individuelles Wissen zu erzeugen [4]. Untersuchungen zeigen, daß unter vielen Bedingungen dieser kooperative Aneignungsprozeß effizienter ist als individuelles Lernen [5] [6].

Allerdings ist es gerade in verteilten Lerngruppen schwierig, hinreichend Metawissen über das Wissen anderer bzw. über das aktuelle gemeinsame Wissen zu bekommen. Bei reduzierten Kommunikationskanälen und in asynchronen Lernsituationen können leicht Defizite in der Konstruktion einer gemeinsamen Wissensbasis entstehen. Wir bezeichnen dies als Problem der *kognitiven Orientierung*.

Das gemeinsame Wissen bezieht sich unter anderem auf folgende Fragen:

- Über welches Wissen besteht in der Lerngruppe *Konsens*, wo gibt es Konflikte?
- Wie sieht die *Schnittmenge* des Wissens aller Lerner aus?
- Welcher Lerngegenstand ist aktuell im *Fokus* des Lernprozesses?

In face-to-face Lernsituationen unterstützen beispielsweise non-verbale Signale die Konstruktion derartiger Informationen. Z.B. kann der Lernende Unverständnis durch Mimik anzeigen und damit die Lerngruppe auffordern, nähere Erläuterungen vorzunehmen. Erlaubt ein CSCL-System nur die zeitversetzte Nutzung durch die Mitglieder der Lerngruppe (asynchrone Kooperation), können ebenfalls im Vergleich zur face-to-face Situation Abstimmungsprobleme auftreten. Wir versuchen, durch eine dynamisch generierte Visualisierung eines sogenannten *Lernnetzes* entsprechende Unterstützung bereitzustellen.

3 Die Metapher virtueller Räume

Sogenannte virtuelle Räume basieren auf der Übertragung abstrakter Eigenschaften realer Räume in eine computergestützte Lernumgebung; wir verfolgen nicht das Ziel, eine virtuelle Realität im Sinne möglichst realitätsnaher 3D-Repräsentationen zu schaffen (vgl. z.B. [7, 8]). Ein virtueller Raum in unserem Sinn wird durch ein unabhängiges Fenster mit Funktionalitäten zur Kommunikation und kooperativen Manipulation von Artefakten repräsentiert. Ein Raum zeichnet sich dadurch aus, daß Benutzer, die sich dort gleichzeitig aufhalten, sowohl Information über die jeweils mitanwesenden Kooperationspartner erhalten, als auch eine identische Sichtweise auf den Inhalt des Raumes (Texte, Grafiken, usw.) besitzen. Virtuelle Räume konstituieren zusammen mit ihren „Bewohnern" eine virtuelle Welt. Virtuelle Räume können betreten und verlassen werden. In virtuellen Räumen können verschiedene Kommunikationsformen etabliert werden: Beispielsweise wird beim Betreten eines Raums automatisch eine Audio-Verbindung mit allen anderen Anwesenden aufgebaut, während Benutzer in verschiedenen Räumen asynchron mittels e-Mail oder Messageboard miteinander kommunizieren. Virtuelle Räume besitzen Grenzen, die eine Isolierung zu anderen Räumen der virtuellen Welt schaffen. Objekte wie Texte und Graphiken können in Räumen sowohl persistent gespeichert als auch durch Personen zwischen verschiedenen Räumen transportiert werden.

3.1 Kognitionspsychologische Begründung

Aus kognitionspsychologischer Perspektive bietet die Strukturierung einer kooperativen Lernwelt durch virtuelle Räume mehrere Vorteile. Zunächst kann dadurch auf direkte Weise an *existierende kognitive Schemata* der sozialen Organisation und sozialer Verhaltensregeln angeknüpft werden [9]: Benutzer wissen, daß Räume mit bestimmten Personen und damit mit bestimmten Ge- und Verboten und Verhaltensregeln assoziiert sind (territoriale Zuordnung). Sie wissen, daß spezifische Räume bestimmten sozialen Zwecken dienen, wodurch wiederum vorhandene Verhaltensskripte für soziale Rollen (z.B. Lehrer oder Schüler) aktiviert werden (funktionale Zuordnung). Beispielsweise wird dadurch definiert, wer einen Dialog eröffnet, oder ob primär symmetrische oder asymmetrische Kommunikation erwartet wird.

Weiterhin konstituieren virtuelle Räume ein - je nach spezifischer Interface-Gestaltung - ein- oder zweidimensionales Bezugssystem. Kognitionspsychologische Untersuchungen haben gezeigt, daß Lern-, Verstehens- und Problemlösungsprozesse als Formen der *Konstruktion mentaler Modelle* konzipiert werden können [10, 11]. Mentale Modelle repräsentieren reale Sachverhalte und Beziehungen mental in Form analoger Bezugssysteme. Mentale Modelle zeichnen sich zum einen durch ihren situativ-dynamischen Charakter aus, d.h. sie werden ständig an den aktuellen Kontext adaptiert, zum anderen sind sie perspektivisch angelegt, d.h. in der Regel um den Protagonisten zentriert. Umgekehrt unterstützt eine Darbietung von Sachverhalten in räumlicher Anordnung die Konstruktion eines nutzbaren mentalen Modells.

Durch die Zuordnung von Personen, Rollen und Funktionen zu abgegrenzten virtuellen Räumen kann die soziale Struktur der Lernwelt als mentales Modell einer räumlichen Organisation repräsentiert werden. Indem den Räumen verschiedene Inhalte zugeordnet werden, kann die soziale Struktur mit der inhaltlichen Gliederung des Lernstoffs übereinstimmen. Die Bewegung durch verschiedene Räume bedeutet dann, daß man sich verschiedenen Lerngruppen anschließen, verschiedene Rollen annehmen und verschiedene Inhalte lernen kann, die sich in jeweils spezifischen Räumen befinden. Durch diese Möglichkeit, in der niedrig-dimensionalen Raumstruktur zu navigieren, wird das in komplexen Informationswelten bekannte „Lost in Hyperspace"-Problem abgemildert.

3.2   Virtuelle Räume im Prototyp CROCODILE

In unserer Konzeption eines kooperativen Lernsystems unterscheiden wir Typen von Räumen hinsichtlich der verfügbaren Kommunikations- und Kooperationsfunktionalitäten. Bislang unterscheiden wir die Raumtypen *private Arbeitsräume* (zur privaten Arbeit, können nicht von anderen Personen betreten werden), *Gruppenräume* (z.B. zur Diskussion zwischen gleichberechtigen Teilnehmern) und *Auditorien* (Vortragsräume mit rollenspezifischer Funktionalität wie unterschiedliche Zugriffsrechte für Trainer und Lerner). Die Konzeption wird z.Zt. im Prototyp CROCODILE (CReative Open COoperative DIstance Learning Environment) realisiert. Hierzu wird die für den Prototyp VITAL [3] entwickelte Benutzungsoberfläche für virtuelle Lernwelten (siehe Abbildung 1) um weitere Strukturierungsmöglichkeiten erweitert.

Abbildung 1 zeigt ein Auditorium mit Trainer und vier Lernern, die sich mit Hilfe von Telepointern auf dem shared whiteboard über Verkehrsregeln informieren. Gleichzeitig ist ein Message Board für einen strukturierten Frage-Antwort-Dialog geöffnet. Zusätzlich ist der sogenannte Room-Locator zu sehen, der eine Übersicht über alle Räume der Lernwelt einschließlich der aktuell dort befindlichen Personen zeigt.

Ein virtueller Raum besteht aus einem gemeinsam genutzten Hypermedia-Whiteboard als zentralem Arbeitsblatt, einer Anzeigeleiste mit Bildern und Namen der aktuell anwesenden Mit-Lerner (je nach Typ des Raumes rollenspezifisch angeordnet z.B. unterschieden in Trainer und Lerner; vgl. Abbildung 1), sowie mit je nach Raumtyp unterschiedlichen Funktionalitäten zur Kommunikation und Dokument-Manipulation. Das Whiteboard selbst basiert auf einem Hypermedia-Dokumentmodell, es können nach Bedarf Seiten neu generiert und beliebig verknüpft werden. Jeder Benutzer kann mit Hilfe eines Telepointers Objekte (Texte, Grafiken, Annotationen, etc.) referenzieren.

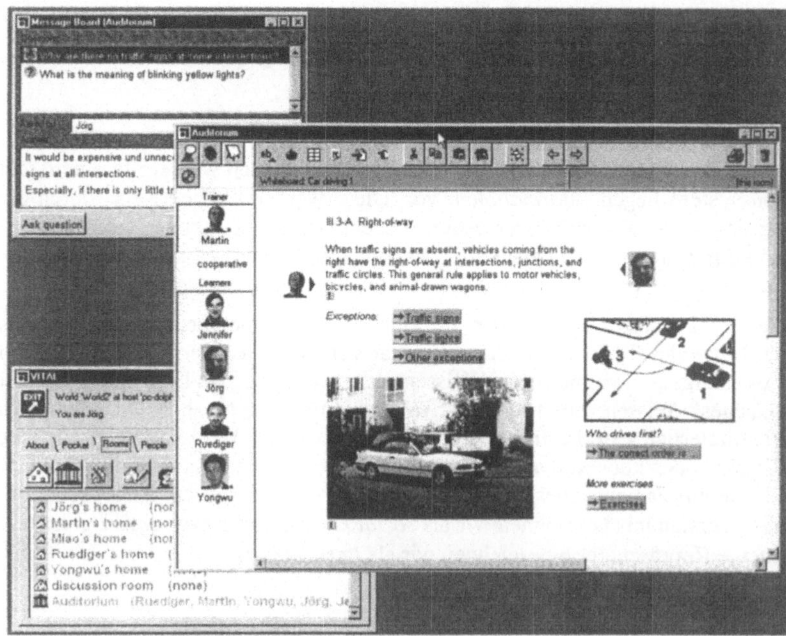

Abb. 1: Auditorium mit Message Board (oben links) and Verzeichnis der Räume (unten links)

Alle vorhandenen Räume, die Teilnehmer der Lernwelt sowie die aktuell im System angemeldeten Lernenden können in separten Fenstern angezeigt werden. Je nach Bedarf können von allen Teilnehmern neue Räume unterschiedlichen Typs angelegt, Objekte zwischen den Räumen transportiert oder kopiert, verschiedene Kommunikationsformen aktiviert (chatboard, Audio), und beliebige Links im Hypermedia-Dokument erzeugt werden.

3.3 Verwandte Ansätze

Die Raum-Metapher ist bereits in mehreren Anwendungen umgesetzt worden, besonders in Informations- und CSCW-Systemen [12, 13, 14]. Eine besonders gelungene Realisation der Raum-Metapher stellt das *TeamWave* System von Greenberg und Roseman dar [15, 16]. In TeamWave wird der Schwerpunkt auf eine reibungslose Transition zwischen synchronen und asynchronen Arbeitsformen gelegt; spezifische kooperative Tätigkeiten werden durch Zusatzwerkzeuge wie Brainstormer, Concept-Mapper u.a. unterstützt. Allerdings besitzt TeamWave kein zugrundeliegendes Hypermedia-Dokumentmodell, so daß die dynamische Generierung komplexer Dokument- und Wissensstrukturen nur eingeschränkt möglich ist.

Die Raum-Metapher wird auch für CVEs (collaborative virtual environments) [7, 8] oder MUDs (multi-user dungeons) verwendet [17]. Während in MUDs Räume meist sehr rudimentär, z.B. als Liste der Benutzer und gemeinsames textuelles Kommunikationsfenster, dargestellt werden, fokussieren CVEs auf eine möglichst realistische Repräsentation von Akteuren mittels Virtual-Reality Techniken. Für beide Bereiche steht die Kommunikation der Benutzer im Vordergrund, das gemeinsame Erzeugen und Manipulieren von Objekten wird nur ansatzweise unterstützt. Gemeinsames Erzeugen von Wissensrepräsentationen ist damit nur eingeschränkt möglich. Ein flexibles Editieren der virtuellen Welt (z.B. Anlegen und Löschen von Räumen durch die Benutzer) ist in aller Regel nicht vorgesehen.

4   Repräsentation sozial konstruierten Wissens

Zunächst muß zwischen Information und Wissen unterschieden werden: In einem Computer-Informationssystem liegen *Informationen* vor. Diese werden üblicherweise durch die Externalisierung des Wissens durch einen Autor erzeugt. Durch die aktive Verarbeitung von Informationen durch den Lerner entsteht in dessen „Kopf" neues Wissen; eine Repräsentation des individuellen Wissens kann medial fixiert bzw. kommuniziert werden.

In kooperativen Lernprozessen werden Informationen von mehreren Lernenden gemeinsam verarbeitet. Entweder fließt dabei Wissen primär von einem Lehrer in Richtung der Lernenden, oder es erfolgt ein symmetrischer Wissensaustausch zwischen Lernenden, oder Lernende erzeugen „neues" Wissen durch kooperative Problemlösungsprozesse. In allen Lernformen können kontinuierlich externe Informationen, z.B. aus dem World Wide Web, einbezogen werden. Ergebnis des kooperativen Lernprozesses ist eine gemeinsam strukturierte Informationsbasis sowie ein gemeinsames Verständnis der Lerner über diesen Wissensbereich. Dieses gemeinsame Verständnis bezeichnen wir als *sozial konstruiertes* bzw. *sozial geteiltes Wissen*. Dessen externe Repräsentation bezeichnen wir als *Lernnetz*.

4.1   Begründung durch konstruktivistische Lerntheorien

Die meisten traditionellen Lerntheorien beschäftigen sich mit individuellem Lernen; dies trifft sowohl auf behavioristische Ansätze [18] als auch auf kognitivistische Ansätze [19] zu. Kooperatives Lernen im engeren Sinn wird erst in konstruktivistischen Lerntheorien [20] und besonders in Theorien situierten Lernens [21, 22] als zentrale Lernform untersucht.

In diesen Ansätzen werden u.a. zwei Faktoren betont, die besonders lernförderlich sind [23]:

- die Möglichkeit der Übernahme *multipler Perspektiven*, und
- das Lernen durch *soziale Konstruktion*.

Ein Sachverhalt wird dann besonders tiefgehend verstanden, besser erinnert und auf andere Kontexte transferiert, wenn er aus unterschiedlichen Perspektiven angeeignet worden ist. Kooperatives Lernen zeichnet sich dadurch aus, daß gerade durch die in der Lernsituation verfügbaren Mit-Lerner multiple Perspektiven erzeugt und zur Reflexion bereitgestellt werden. Es wird so ein multi-perspektivisches Verständnis des Lerngegenstandes aufgebaut. Es ist zu erwarten, daß eine Repräsentation dieses geteilten Wissens in Form eines Lernnetzes sowohl die individuelle Reflexion des Wissens durch den einzelnen Lerner unterstützt, als auch als eine Kommunikationsbasis für weiterführende Lernprozesse der Lerngruppe dient.

4.2   Das Lernnetz im Prototyp Crocodile

Der Zweck des Lernnetzes besteht darin, allen Teilnehmern eine Orientierung darüber zu geben, welche Sachverhalte bislang behandelt wurden und welche Art und welchen Umfang das gemeinsam erarbeitete Wissen aktuell aufweist. Neben dieser statischen Darstellungsfunktion hat das Lernnetz eine dynamische Funktion: die Konstruktion des Netzes selbst ist ein kooperativer Lernprozeß, in dessen Verlauf Wissen ausgehandelt, unterschiedliche Perspektiven integriert und Konflikte explizit gemacht werden. Ein Lernnetz ist daher gleichzeitig Lernmittel und Lernresultat.

Wir gehen davon aus, daß das im Verlauf eines Lernprozesses sozial konstruierte Wissen in Form eines Graphen mit Knoten und Kanten, die Relationen zwischen Knoten bezeichnen, dargestellt werden kann, so wie es seit langem in der Kognitionswissenschaft üblich ist [24, 25]. Dadurch wird eine zweite Ebene über den Informationen der Informationsbasis (den Hypermedia-Dokumenten) aufgespannt, die durch folgende Eigenschaften charakterisiert ist:

- das Lernnetz ist sozial konstruiert, d. h. es bedarf eines *partiellen Konsenses* der Kooperationspartner;
- das Lernnetz ist *summarisch*, d. h. bildet abstrakte Konzepte ab, verweist aber über Links auf konkrete Inhalte in den Hypertext-Dokumenten;
- das Lernnetz ändert sich *dynamisch*, d.h. es wird kontinuierlich modifiziert, erweitert oder reduziert.

Abbildung 2 zeigt als Beispiel den Ausschnitt eines Lernnetzes einer Lerngruppe zum Thema "Projektmanagement".

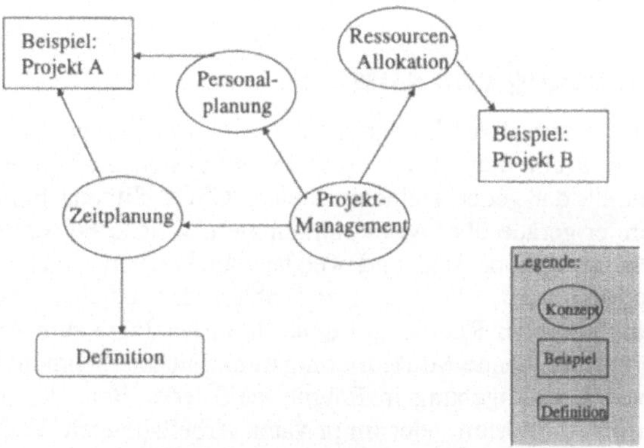

Abb. 2: Ausschnitt aus einem Lernnetz

Das Lernnetz ist jedem Benutzer in einem eigenen Fenster verfügbar. Im Beispiel von Abbildung 2 erhält der Benutzer einen Überblick über die bisher behandelten Themen und wie die Themen zusammenhängen, etwa daß Zeitplanung, Personalplanung und Ressourcen-Allokation Komponenten des Projektmanagements sind. Außerdem kann er erkennen, daß für bestimmte Konzepte konkrete Beispiele existieren. Über die entsprechenden Knoten kann er nun direkt zu den Dokumenten gelangen, in denen die Konzepte bzw. Beispiele beschrieben sind. Wird ein neues Konzept besprochen, kann das Netz modifiziert und nach entsprechender Interaktion ein neuer Knoten eingebaut werden. Lernnetze können auch nach Beendigung einer Lernphase abgelegt und später bei Bedarf wieder konsultiert werden.

Forschungsfragen, die gegenwärtig genauer untersucht werden, sind: Wann genau wird das Lernnetz konstruiert? Wer ist daran beteiligt? Ist die Darstellungsform, d.h. die verfügbaren Konzept- und Relationstypen, dem Lernen förderlich? Da wir als typische Nutzer eine kleine Gruppe von Lernern (ca. 5 – 10 Personen) anzielen, schlagen wir vor, die Konstruktion selbstgesteuert durch die Gruppe zu initiieren. Immer dann, wenn Konzepte als unklar betrachtet werden, kann der Lernmodus gewechselt und von der Dokument- auf die Lernnetz-Ebene gesprungen werden. Dies kann sowohl durch die Gruppe als Ganzes oder durch Teilgruppen initiiert werden.

4.3 Verwandte Ansätze

Im Gegensatz zu Ansätzen der Wissensrepräsentation in wissensbasierten Systemen stellt ein Lernnetz in unserer Konzeption keine systematische Struktur eines Sachverhaltes im Sinne funktionaler und struktureller Beziehungen dar. Das Lernnetz ist vielmehr eine spezifisch für Lernzwecke konstruierte *Überblicks*karte des sozial geteilten Wissens, und ein Mittel, um kooperativ Wissen auszuhandeln. Ähnliche Ansätze finden sich in [26] und [27].

Das Lernnetz weist strukturelle Ähnlichkeiten zu Verfahren des concept mapping [28] und zu kognitionspsychologischen Ansätzen der propositionalen Wissensrepräsentation auf [29]. Es existieren unterschiedlichste Verfahren und Werkzeuge, die das individuelle oder kooperative Anlegen und Explorieren derartiger semantischer Netze unterstützen [30]. Die vorliegende Konzeption greift diese Funktionalität auf und integriert sie mit der in den virtuellen Räumen kooperativ erzeugten Informationsbasis.

5 Zusammenfassung und Ausblick

Computerunterstützes kooperatives Lernen örtlich verteilter Lerner (verteiltes CSCL) erfordert die Lösung zweier Probleme: das der sozialen und das der kognitiven Orientierung. Soziale Orientierung heißt, daß jeder Teilnehmer einer CSCL-Sitzung hinreichend darüber informiert ist, mit wem er gerade über was kommuniziert, welche Personen Mitglieder der aktuell kooperierenden Lerngruppe sind und welche Objekte von welchen Personen wahrgenommen und manipuliert werden können. Das Problem der sozialen Orientierung versuchen wir durch den Einsatz virtueller Räume zu lösen, die in Analogie zum Aufbau der physikalischen Umwelt eine entsprechende Strukturierungsfunktion übernehmen. Durch die Unterteilung der kooperativen Lernumgebung in Räume weiß jeder Benutzer unmittelbar, "wo" er sich befindet - etwa im Auditorium oder im privaten Arbeitsbereich. Weiterhin weiß er, welche Personen sich noch im selben Raum aufhalten, wodurch auf intuitive Weise eine gemeinsame Sicht auf die im Raum vorhandenen Objekte gewährleistet ist.

Das Problem der kognitiven Orientierung besteht darin, einen aktuellen Überblick über das im Verlauf des kooperativen Lernprozesses sozial konstruierte Wissen zu erhalten. Da die Informationsbasis in Form von Hypermedia-Dokumenten durch die beliebigen Verknüpfungsmöglichkeiten schnell an Komplexität zunimmt und so das "Lost-in-Hyperspace"-Phänomen hervorruft, bedarf es einer übergeordneten Abstraktionsebene, die nur die relevanten Konzepte in strukturierter Form abbildet. Als Lösung für das Problem der kognitiven Orientierung haben wir die Konstruktion von Lernnetzen vorgeschlagen. Ein Lernnetz ist eine Visualisierung des sozial geteilten Wissens in Form einer Graphenstruktur, deren Knoten die zentralen Konzepte und deren Relationen die lernrelevanten Beziehungen zwischen den Konzepten bezeichnen. Die im Lernnetz repräsentierten Konzepte verweisen auf die behandelten Themen, visualisieren deren Zusammenhangsstruktur und referenzieren, sofern in der Dokumentenbasis vorhanden, Beispiele und Definitionen. Alle Lerner erhalten so einen einheitlichen Blick auf die Struktur des gemeinsam aufgebauten Wissens. Aus dem Lernnetz selbst kann auf die assoziierten Dokumente gesprungen werden, das Lernnetz dient dabei als eine Art strukturierter Index des Gegenstandsbereichs.

In diesem Beitrag wurden zwei Hauptprobleme örtlich verteilter kooperativer Lernprozesse dargestellt. Es wurden kognitionspsychologisch bzw. lerntheoretisch fundierte Ansätze zur Überwindung dieser Probleme in einer computergestützten kooperativen Lernumgebung präsentiert und deren Umsetzung in ein CSCL-System skizziert. Die Kombination der beiden Konzepte wird von uns in einem Prototyp CROCODILE auf der Basis von COAST [31], einem Framework zur Entwicklung synchroner Groupware, und der virtuellen Lernwelt VITAL [3] am GMD-IPSI umgesetzt. Wir rechnen mit einem lauffähigen Prototyp gegen Ende 1998. Eine systematische Evaluation der technischen und didaktischen Konzepte im Prototypen CROCODILE Anfang 1999 an Universitäten in Deutschland und Österreich befindet sich in Planung

Weiteren Forschungen vorbehalten ist die Frage der Skalierbarkeit: Zwar ist CROCODILE primär für kleinere Gruppen von 5 bis 10 Personen konzipiert, jedoch tauchen bereits in diesem Bereich mit zunehmender Gruppengröße Schwierigkeiten bei der koordinierten Kommunikation auf. Für welche Arten kooperativen Lernens in größeren Lerngruppen die Metapher virtueller Räume tauglich ist, kann bislang nicht beantwortet werden. Auch die Konstruktion und Wartung des Lernnetzes muß besser unterstützt werden: Bislang sind Modifikationen selbständig zu initiieren. Wünschenswert wäre eine Semi-Automatisierung, so daß die kognitive Belastung, die durch die Notwendigkeit der parallelen Aktualisierung der Informationsbasis und des Lernnetzes entsteht, entfällt.

6 Literatur

[1] T. Koschmann (Hg.): CSCL: Theory and Practice of an Emerging Paradigm. Mahwah, 1996: Erlbaum.

[2] F.W. Hesse, B. Garsoffky, A. Hron: Interface-Design für computerunterstütztes kooperatives Lernen. In: L.J. Issing und P. Klimsa (Hg.): Information und Lernen mit Multimedia. Weinheim, 1997 (2. Auflage): Beltz. 254-267.

[3] H.-R. Pfister, C. Schuckmann, J. Beck-Wilson, M. Wessner: The metaphor of virtual rooms in the cooperative learning environment CLear. In: N. Streitz et al. (Hg.): Cooperative Buildings. Integrating Information, Organization and Architecture. Berlin, 1998: Springer. 107-113.

[4] J. Roschelle: Learning by collaborating: Convergent conceptual change. In: T. Koschmann (Hg.): CSCL: Theory and Practice of an Emerging Paradigm. Mahwah, 1996: Erlbaum. 209-248.

[5] A.L. Brown, A.S. Palincsar: Guided, cooperative learning and individual knowledge acquisition. In: L.R. Resnick (Hg.): Knowing, learning and instruction. Essays in Honor of Robert Glaser. Hillsdale, 1989: Erlbaum. 393-451.

[6] D.W. Johnson, R. Johnson, M.B. Stanne, A. Garibaldi: Impact of group processing on achievement in cooperative groups. In: Journal of Social Psychology 130 (1990), 507-516.

[7] Scalable Platform for Large Interactive Networked Environments (SPLINE): www.merl.com/projects/spline

[8] The Distributed Interactive Virtual Environement (DIVE): www.sics.de/dive

[9] R.C. Schank, R.P. Abelson: Scripts, plans, goals, and understanding. Hillsdale, 1977: Erlbaum.

[10] D. Gentner, A.L. Stevens: Mental Models. Hillsdale, 1983: Erlbaum.

[11] S. Dutke: Mentale Modelle: Konstrukte des Wissens und Verstehens. Stuttgart, 1994: Verlag für angewandte Psychologie.

[12] S.K. Card, A.H. Henderson: A multiple, virtual-workspace interface to support user task switching. In: Proceedings of the CHI+GI. Toronto, Canada, 1997.

[13] J.A. Waterworth, G. Singh: Information islands: Private views of public places. In: Proc. of East-West International Conference on Multimedia, Hypermedia and Virtual Reality (MHVR'94). Moskau, 1994.

[14] J.A. Waterworth: Personal Spaces: 3D Spatial Worlds for Information Exploration, Organisation and Communication. In: R. Earnshaw und J. Vince (Hg.): The Internet in 3D: Information, Images, and Interaction. New York, 1997: Academic Press.

[15] S. Greenberg, M. Roseman: Using a room metaphor to ease transitions in groupware. Research Report 98/11/02, Department of Computer Science, University of Calgary, Canada, 1998.

[16] M. Roseman, S. Greenberg: TeamRooms: Network places for collaboration. In: Proceedings of the ACM 1996 Conference on Computer Supported Cooperative Work (CSCW'96). Boston, 1996: ACM. 325-333.

[17] Collaborative Virtual Environments 1998 (CVE'98). URL: http://www.crg.cs.nott.ac.uk/events/CVE98/

[18] B.F. Skinner: Teaching machines. In: Science 128 (1958). 969-977.

[19] J.R. Anderson: Rules of the Mind. Hillsdale, 1993: Erlbaum.

[20] T.M. Duffy, D.H. Jonassen: Constructivism and the technology of instruction: A conversation. Hillsdale, 1992: Erlbaum.

[21] J.G. Greeno, D.R. Smith, J.L. Moore: Transfer of situated learning. In: D.K. Dettermann und R.J. Sternberg (Hg.): Transfer on trial: Intelligence, cognition, and instruction. Norwood, 1993: Ablex. 99-167.

[22] H. McLellan: Situated Learning Perspectives. Englewood Cliffs, 1995: Educational Technology Publications.

[23] H. Mandl, H. Gruber, A. Renkl: Situiertes Lernen in multimedialen Lernumgebungen. In: L.J. Issing und P. Klimsa (Hg.): Information und Lernen mit Multimedia. Weinheim, 1997 (2. Auflage): Beltz. 167-178.

[24] P.R. Churcher: A common notation for knowledge representation, cognitive models, learning and hypertext. In: Hypermedia 1 (1989), 235-254.

[25] D.E. Rumelhart, A. Ortony: The representation of knowledge in memory. In: R. C. Anderson, R.J. Spiro und W.E. Montague (Eds.): Schooling and the acquisition of knowledge. Hillsdale, 1977: Erlbaum.

[26] D. Wan, P.M. Johnson: Computer supported collaborative learning using CLARE: The approach and experimental findings. In: Proceedings of the ACM 1994 Conference on Computer-Supported Cooperative Work (CSCW'94). Chapel Hill, 1994: ACM. 187-198.

[27] N. Streitz, J. Hannemann, M. Thüring: From ideas to arguments to hyperdocuments: Traveling through activity spaces. In: Proceedings of the ACM Conference on Hypertext (Hypertext'89). Pittsburgh, 1989: ACM. 343-364.

[28] K.M. Fisher, M. Kibby (Hg.): Knowledge Acquisition, Organization and Use in Biology. Heidelberg, 1996: Springer.

[29] W. Kintsch: The representation of meaning in memory. Hillsdale, 1974: Erlbaum.

[30] D.H. Jonassen, K. Beissner, M. Yacci: Structural knowledge. Techniques for representing, conveying, and acquiring structural knowledge. Hillsdale, 1993: Erlbaum.

[31] C. Schuckmann, L. Kirchner, J. Schümmer, J.M. Haake: Designing object-oriented synchronous groupware with COAST. In: Proceedings of the ACM 1996 Conference on Computer Supported Cooperative Work (CSCW'96). Boston, 1996: ACM. 30-38.

Adressen der Autoren

Dr. Hans-Rüdiger Pfister
Institut für Integrierte Publikations- und Informationssysteme (IPSI)
GMD - Forschungszentrum Informationstechnik GmbH
64293 Darmstadt, Dolivostr. 15
Email: pfister@darmstadt.gmd.de

Dipl.-Inform. Martin Wessner, M.A.
Institut für Integrierte Publikations- und Informationssysteme (IPSI)
GMD - Forschungszentrum Informationstechnik GmbH
64293 Darmstadt, Dolivostr. 15
Email: wessner@darmstadt.gmd.de

Jennifer Beck-Wilson, M.Ed.
Institut für Integrierte Publikations- und Informationssysteme (IPSI)
GMD - Forschungszentrum Informationstechnik GmbH
64293 Darmstadt, Dolivostr. 15
Email: wilson@darmstadt.gmd.de

An Evaluation of Interaction Techniques for the Exploration of 3D-Illustrations

Ian Pitt*, Bernhard Preim** & Stefan Schlechtweg**

*Department of Computer Science, University College Cork

**Institut für Simulation und Graphik, Otto-von-Guericke-Universität

Zusammenfassung

Dieser Beitrag beschreibt eine empirische Evaluierung von neuartigen Interaktionstechniken in einem hypermedialen Lehr- und Lernsystem, dem Zoom Illustrator. Die Integration von interaktiver 3d-Graphik mit Hypertextmöglichkeiten, die Nutzung von Fisheye-Techniken zur Navigation in textuellen Informationen und die Generierung von Bildunterschriften sind die Kernmerkmale des evaluierten Systems.

Der Zoom Illustrator ist besonders geeignet für die Anatomieausbildung, kann aber auch in anderen Bereichen zur Erklärung räumlicher Zusammenhänge genutzt werden. Die Interaktionstechniken des Zoom Illustrators wurden mit denen eines einfacheren Systems verglichen, das die selben Informationen benutzt und die gleiche Funktionalität bereitstellt. Das System wurde mit 8 Medizinstudenten bzw. jungen Ärzten getestet. Die Evaluierung zeigt die Nützlichkeit und die Akzeptanz der neuen Interaktionstechniken.

Abstract

We present an empirical evaluation of innovative interaction techniques in an educational hypermedia system. The techniques investigated include the integration of 3d-interaction and hypertext-functionality, the use of fisheye techniques for the purpose of navigating in textual information and the generation of figure captions which automatically describe the images presented.

The system under evaluation is particularly useful for anatomy teaching, although it is not restricted to this domain. We compared a fully-fledged version of this system with a cut-down version which used only traditional interaction techniques which was otherwise identical. Both systems use exactly the same geometric models and textual explanations so that the comparison yields reliable results. Eight medical students and recently-qualified medical doctors took part in the evaluation. The results indicate the usefulness and the acceptance of the newly-developed interaction techniques.

1 Introduction

In many areas, learning involves the study of complex 3d-phenomena. In anatomy, for example, highly complex structures are studied under different aspects and viewing directions. To gain an insight into the spatial structure, and to be able to name parts of the objects, are important goals in the learning process. Similar problems arise in technical documentation where the construction of complex technical models must be communicated.

Traditional teaching materials do not support all aspects of this learning process sufficiently. Textbooks (see e.g. [1], [2]) contain valuable drawings but require a lot of flipping through the pages. Numerous illustrations and the text referring to them have to be mentally integrated. The Zoom Illustrator was developed in the light of these observations. It combines techniques from traditional media with the flexibility of a computer system. The system generates illustrations which are adapted to the parts currently being explained, thus allowing a detailed study of interesting parts while maintaining the context.

In this paper we describe a practical study in which we set out to evaluate the major features of the Zoom Illustrator and to establish the value of each in facilitating access to 3d-information.

2   The Zoom Illustrator: An Overview

The Zoom Illustrator and the techniques it embodies have been described in a number of publications, for example [3,4,5,6]. The following description is designed to give a brief overview, concentrating on those aspects of the design which are evaluated in this study.

The Zoom Illustrator is designed to permit the exploration of a 3d-model, displaying the image and its accompanying text within a single window. The user can then scale and rotate the model freely, whereupon the position and size of the text will be adjusted to preserve the context. A typical Zoom Illustrator display is shown in Figure 1 (next page).

Images and texts are located in different areas and connected to each other via reference lines. When the 3d-model is rotated and scaled, the reference lines are updated. To ensure that the image is not occluded by textual information and vice versa, the window is subdivided into a central part for the image (which occupies 50% of the space), as well as a left and a right part for textual descriptions (which occupy 25% each of the screen space available, see Figure 1).

The system is based on polygonal surface models of moderate complexity (~10,000 polygons) which are commercially available. These models are large enough to be interesting but small enough to be handled at interactive rates.

2.1   Use of Fisheye Techniques

Textual information includes labels and more or less detailed explanations which are connected to each other via hyper-links. The explanations were taken from well-established textbooks. Textual information is embedded in rectangular areas. The placement and size of these rectangles is based on fisheye techniques (as introduced in [7]). Other systems, by contrast, open a separate window to display an explanation, making it difficult for the user to create a mental link between the explanation and the object to which it refers. Fisheye techniques provide a useful way to zoom in and out interactively, presenting several levels of detail simultaneously. This makes it possible to explain objects in which the user is interested, while automatically scaling down and repositioning others to provide the necessary space without overlapping information.

The placement and the size of information is based on the degree of interest (DOI), an application-specific value influenced by user interactions. Continuously enlarging entities at the expense of others, the size of which are reduced accordingly, is accomplished by a variant of the continuous-zoom algorithm as presented in [8]. Whenever a rectangle becomes large enough to accommodate more text, an additional explanation is displayed which is sub-sequently hidden automatically if other rectangles are enlarged.

Objects which are textually explained are emphasized within the 3d-model. This strategy – although useful in many cases – is not adequate to illustrate small details. The illustration of small details is accomplished with the 3d-fisheye zoom [3,9]. It makes it possible to scale up small details while automatically repositioning other objects — the same strategy which has been successfully applied to text navigation. The application of fisheye techniques for both symbolic and graphical exploration provides a coherent interface.

2.2   Enhancing Navigation in Textual Information

One important question when applying zoom techniques concerns the automatic hiding of nodes. While the zoom algorithm generally works in a comprehensible manner, it could be irritating if one node disappeared because another node had grown in size. Even if users understand what has happened they might not know what they are supposed to do to get a node back. This raises two questions:

- How to prevent nodes from being closed?
- How to get nodes back which have been closed?

We developed one solution for each question.

To prevent nodes from being closed, we introduce an additional area, a *pinboard*, as a container for some privileged nodes (up to 4) which are not exposed to the zoom algorithm and therefore cannot be closed.

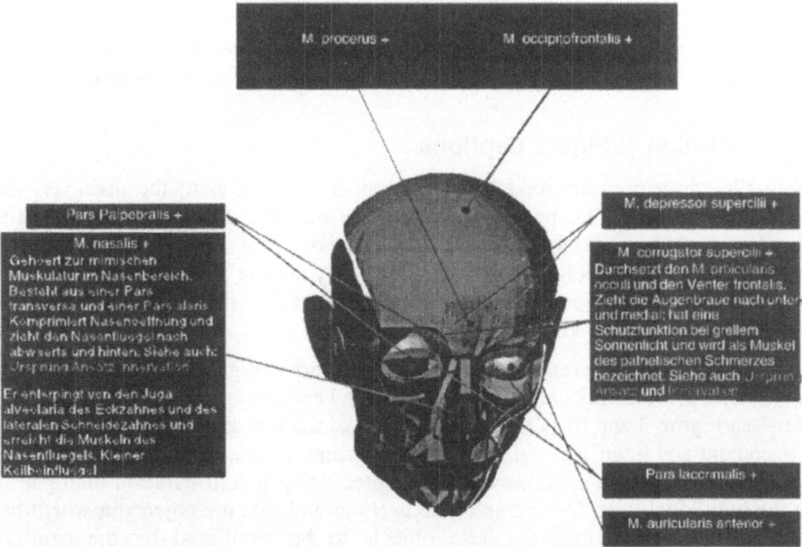

Figure 1: A typical Zoom Illustrator screen display. Text and images are displayed in a single window, with the image in the centre and the text in two side-panels, connected to points on the image by reference lines. More information about an object can be obtained by zooming its label. Surrounding labels will shrink to make space but remain visible. In the upper part is the pinboard, an area controlled by the user. In this example the user has attached two nodes to the pinboard.

The user can initiate an animated movement of a node to the pinboard (above the image) where it remains at a fixed position (see Figure 1). Nodes residing on the pinboard are still connected to the image via reference lines, which are updated if the 3d-model is transformed.

To get a node back once it has disappeared, it should be possible to select it via its label. For this purpose we designed a 3d-widget we call a *roundel* which contains all labels grouped according to their category (see Figure 2). A 3d-widget fits naturally into our 3d-illustration system. The design of this widget is inspired by the work on 3d-interfaces carried out at XEROX Parc (see [10]). The roundel can be rotated by clicking on the disks at the lower and the upper part, one for a rotation to the left and one for the rotation to the right. The colour of the nodes which are closed is a saturated blue (instead of a weak grey for the nodes already presented), to encourage the user to invoke the node. The design and use of the pinboard and the roundel are discussed in [3].

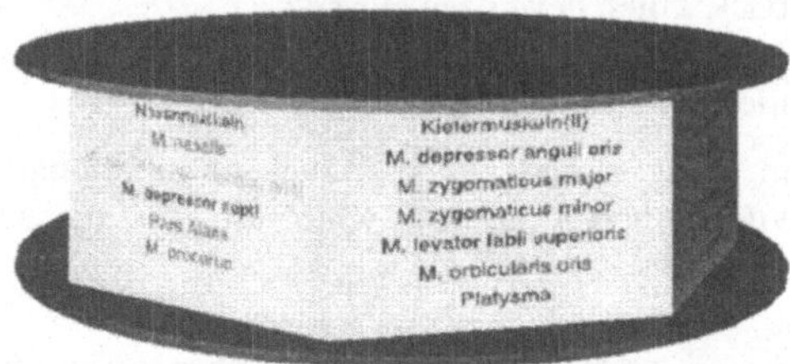

Figure 2: A roundel is used as an overview about all labels available. When a label is selected which does not belong to the illustration it is integrated in the illustration at an appropriate position.

2.3 Incorporation of Figure Captions

The Zoom Illustrator presents models which may be unfamiliar to the user, e.g. anatomic models of inner organs. The presentation of these models is not only a straight-forward rendering but includes an adaptation to the context. Thus objects which are textually explained are emphasized (coloured or enlarged if they are too small). Objects which are not important in a certain dialogue context are deaccentuated by being rendered semi-translucent so that other objects become visible.

All these modifications may remain unnoticed. Therefore, we generate figure captions which automatically describe an image. These captions are kept consistent with the image to which they refer (see Figure 3 and [6,12] for a description of the text-generation methods used). The structure, content and linguistic realisation of these captions are, again, strongly influenced by those in text-books. Figure captions can be adapted using a configuration dialogue to tailor them to the individual user, for example, the user can indicate the objects in which he or she is particularly interested, causing these objects to be monitored by the captions. The incorporation of figure captions is a distinctive concept of the Zoom Illustrator.

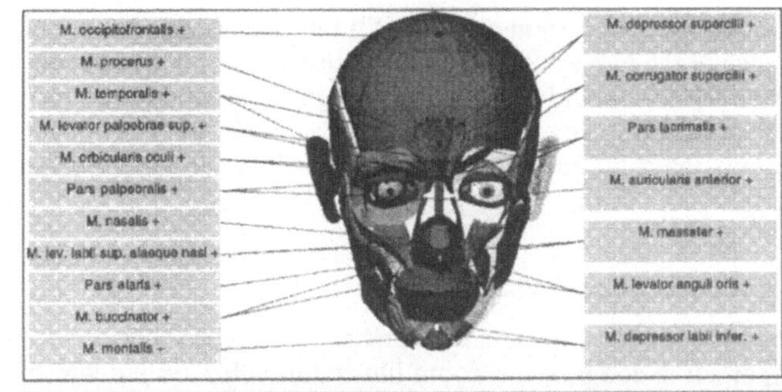

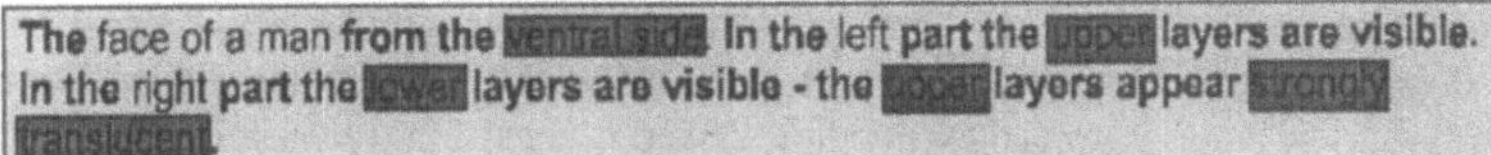

Figure 3: A figure caption describes the model depicted and important aspects of the image generation. In this caption, the difference between the two halves of a symmetric model is described. The dark rectangles correspond to sensitive parts of the caption. The selection of these parts leads to the presentation of pop-up menus with alternatives. Thus, the image generation can be influenced via the caption.

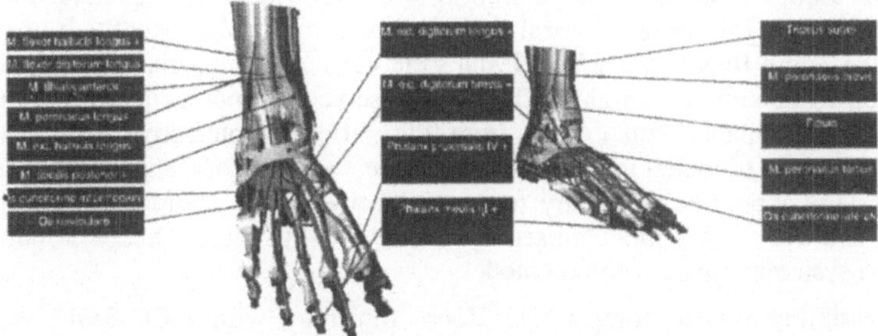

Figure 4: Two instances of a geometric model can be handled independently. Labels which refer to objects visible in both instances are placed between them to support the mental integration of both views.

2.4 Independent Handling of Two Models

A special feature of the Zoom Illustrator is its ability to display two instances of a 3d-model simultaneously. This results in an illustration with two models which can be manipulated independently from each other (see Figure 4). Labels which refer to objects which are visible in both instances are placed between them and connected to both 3d-models.

3 Evaluation

In this evaluation we set out to study the effectiveness of the distinguishing features of the Zoom Illustrator. In order to do this we decided to compare the Zoom Illustrator with another system which is similar in all important respects but uses only conventional interaction techniques. Since we wished to compare a number of features in single study we decided to use a questionnaire. A number of subjects were asked to perform a task using both the Zoom

Illustrator and the comparison system and then fill out a questionnaire in which they should rate the performance of the two systems on a number of scales. A within-subjects design was used because it was felt this would be sensitive enough to reveal differences between the two systems even with a relatively small sample size. In addition, we video-taped a few subjects to get a better impression about the interaction styles they prefer and the frequency of the use of certain features.

3.1 Choice of Comparison System

The choice of comparison system posed a number of problems. In order to obtain useful results it was important to compare the Zoom Illustrator with a system which is similar in all major respects save for those under test. In particular, the comparison system should provide the same level of detail in its models and graphical presentation and have the same basic interactive features. If the comparison system differed significantly in any of these respects it would be impossible to determine reliably the cause of any differences observed in the study.

A number of possible comparison systems were considered and rejected. One obvious approach was to compare the Zoom Illustrator with the VoxelMan (see [12]), a leading, commercially-available interactive anatomy trainer developed at the University of Hamburg. This would have offered a number of advantages, not least that it would have allowed us to compare the Zoom Illustrator with a well-known and widely-used system. However, the VoxelMan offers highly-detailed voxel models which differ significantly from the models used with the Zoom Illustrator. If VoxelMan were used as a comparison system, it would be very difficult to be certain that any differences observed did not stem from the disparity in level of detail available rather then from the different interactive facilities provided. Moreover, the types of model used in the two systems are not compatible, and a considerable amount of work would be necessary to convert between the volume models used by the VoxelMan and the surface models used by the Zoom Illustrator. Thus it is not feasible to compare the systems using a common model.

Another possibility was to compare the Zoom Illustrator with a CD-ROM version of an anatomy textbook (several of which are available). This would allow us to compare the Zoom Illustrator against a known system which has a fairly standard set of interactive features. However, CD-ROMs typically present highly-detailed (static) illustrations which contain far more detail than is available from the rendered images of the Zoom Illustrator. Thus, as with the VoxelMan, it would be very difficult to be certain that any differences observed were not caused by disparities in the level of detail available rather then from the different interactive facilities provided.

Having considered these and other options, it was decided that the Zoom Illustrator should be compared with a program written specifically for the purpose of the evaluation. This avoids the problem of differences in the level of detail available in the illustrations since the comparison system uses the same models as the Zoom Illustrator. The program mimics the operation and look of the Zoom Illustrator in all respects, except that the special interactive features under examination here are excluded. It is implemented with the same toolkit, namely Open Inventor, and the models are presented within the same 3d viewer (see Figure 5). The only differences are that the comparison system allows only one illustration to be viewed at a time, it does not generate figure captions, the explanations are presented in a separate window because no fisheye views are employed, and the user has to place labels manually.

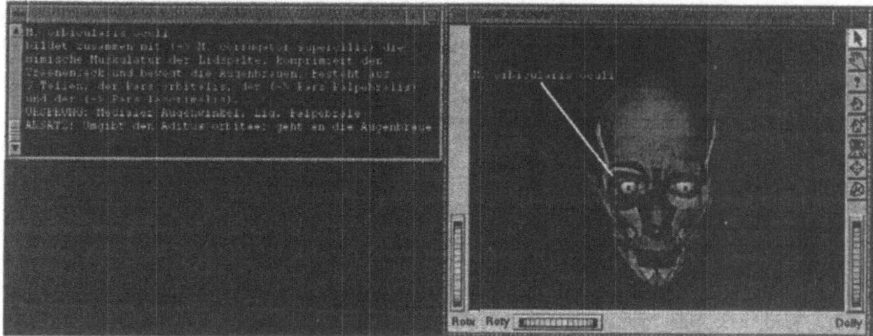

Figure 5: A typical situation working with the comparison system: The user has manually placed a label and asks for an explanation which is subsequently placed in a separate window.

3.2 Design of the Questionnaire

Having chosen a comparison system, we also had to design the questionnaire and decide which criteria to use when comparing the two systems.

Ideally we would have liked to have asked subjects to complete two questionnaires, one for the comparison system and one for the Zoom Illustrator. Subjects would have filled in the relevant questionnaire immediately after working with each system, and we would have obtained two sets of scores which could then be compared statistically.

However, this approach presented a major problem: it was very difficult to devise questions concerning the usefulness of features in the Zoom Illustrator which could be answered sensibly by a subject who was currently using the comparison system but who had not yet used the Zoom Illustrator.

For example, it would make little sense to ask a subject to rate the usefulness of, say, the pinboard, when that subject had not yet seen and used the pinboard. One possible solution would have been to ask more general questions concerning usability, maintenance of context, etc., but this does not fit well with the aim of this study which was to test the specific features of the Zoom Illustrator. Another possibility would have been to test all subjects on the Zoom Illustrator first and then on the comparison system, thus ensuring that when they came to fill-in the questionnaires they had already used all the features mentioned. However, this might have biased the results since subjects often show a preference for the first system with which they are presented, and in consequence it is considered good practice to vary the order of presentation.

For these reasons, it was decided that subjects would be asked to work with both systems (half using the Zoom Illustrator first and half using the comparison system first) and then fill-in a single questionnaire. For each feature, they would be asked to enter a value from 0 to 10, where 10 indicated that the feature was regarded as essential and should definitely be included (as in the Zoom Illustrator), and 0 indicated that the feature was of no value and should be excluded (as in the comparison system). A score of 5 indicated no preference in either direction. Using a single questionnaire in this way makes it more difficult to draw watertight conclusions, but avoids many of the problems inherent in testing features of a user interface.

The scales used in the questionnaire were chosen to reflect what we see to be the important features of the Zoom Illustrator compared with existing systems for viewing 3d-models, namely:

- The ability to view two instances of a model simultaneously
- The use of figure captions
- The use of fisheye techniques
- Automatic placement of labels

The first of these features, the ability to view two instances of a model simultaneously, can be tested on a single scale, as can the use of figure captions. Evaluating the use of fisheye techniques and the placement of labels is more complex, because a number of separate features have been incorporated into the Zoom Illustrator in order to implement fisheye techniques as effectively as possible. For example, the ability to shrink labels is part of the fisheye mechanism, but a further refinement in the Zoom Illustrator is that labels can be shrunk beyond the point of legibility rather than simply vanishing, thus providing a visual reminder of their position and content. Since these are separate features it was felt that they should be investigated separately in the evaluation.

The desire to implement fisheye techniques effectively is also reflected in a number of features which address some of the known problems associated with fisheye techniques. These include the problem of disorientation, which is addressed in the Zoom Illustrator through the use of fixed reference points and user-controlled information spaces (the pinboard and the roundel), and the problem of maintaining a clear link between images and text, which is addressed by ensuring that images and related text always change in tandem and by allowing users to position text within images. It was felt that the usefulness of each of these features should be examined separately.

We were also keen to discover the subjects' views on the use of hypertext links to move between views. These provide an obvious mechanism for linking visible objects to related objects in other parts of the model, but they may also disorientate the user.

The list of features included in the questionnaire was as shown below.

- Ability to view two illustrations at once
- Ability to change the size of the labels
- Ability to shrink labels without hiding/removing them
- The pinboard
- The overview offered by the roundel
- The integration of labels in helping users to understand the illustrations
- The ability to place labels inside an illustration
- Figure captions
- Hypertext functionality

In addition to the nine scales listed above, we introduced two further scales. One was designed to find out whether subjects felt the use of 3d-models was justified, or whether they felt that the spatial relationships could have been conveyed adequately using static images. The other scale concerned the quality of the models. The use of surface models of moderate complexity allows the Zoom Illustrator to produce images in real time, but some might argue that this advantage is outweighed by the resultant loss of image quality. We wanted to find out whether the subjects felt the models had been of adequate quality for the task.

These last two scales differed from the others in that the subjects were asked to indicate their level of agreement with a statement. The possible responses ranged from 0 ('disagree') to 10 ('agree') with a response of 5 indicating no strong feelings in either direction. The two statements were:

- The 3d-presentation aids understanding more than would a 2d presentation.
- The quality and faithfulness of the models is appropriate for teaching topological relations.

3.3 Procedure

Eight subjects took part in the study. All had a basic knowledge of anatomy, being either medical students or probationary (i.e, recently-qualified) doctors. The subjects ranged in age from 22 to 31 and comprised 4 women and 4 men.

Each subject was first given a demonstration of the two systems. The demonstrations lasted for around 5-10 minutes (depending upon the number of questions asked by the subject, etc.) and the order of the demonstrations was varied so that half the subjects were shown the Zoom Illustrator first and half were shown the comparison system first. When the demonstrations were complete and any resulting questions had been answered, the subjects were asked to begin the main part of the experiment. They were asked to perform a simple task using first one system and then the other. The task involved exploring the M. extensor digitorum longus, a long and branching muscle, and correctly locating and tracing all its branches. The time required for completion of the task was typically around fifteen minutes and the order was varied so that half the students performed the task using the Zoom Illustrator first and half performed the task using the comparison system first. A few subjects agreed to be video-taped whilst carrying out the tasks. When they had finished, each subject was asked to complete the questionnaire.

3.4 Results

The results are shown in Table 1. It can be seen that in almost every case the average score is between 5 and 10, indicating that all the Zoom Illustrator's features were regarded as useful. In some cases (e.g., the ability to view two illustrations at once) the score is only slightly greater than 5, but in most cases it is considerably greater than 5.

The only exception to this pattern is the final scale, representing the subjects views on the quality of the models. The average score on this scale was below 5, reflecting a dissatisfaction with the models which was also expressed in informal comments by the subjects.

	S1	S2	S3	S4	S5	S6	S7	S8	Averages	Standard deviation
Ability to view two illustrations at once	10	8	1	3	5	10	4	8	6.13	3.36
Ability to change the size of the labels	7	10	8	10	7	10	10	7	8.63	1.51
Labels shrink without vanishing	7	10	8	8	10	5	10	9	8.38	1.77
The Pin-Wall	8	7	8	5	10	3	10	8	7.38	2.39
The overview offered by the roundel	8	10	8	8	8	5	10	9	8.25	1.58
Integration of labels	10	10	10	10	5	10	2	5	7.75	3.24
Ability to place labels inside illustrations	6	10	8	8	7	9	9	8	8.13	1.25
The figure captions	5	10	8	10	10	7	8	10	8.50	1.85
Hypertext functionality	8	10	10	10	10	9	10	10	9.63	0.74
3d versus 2d presentation	10	10	10	7	10	10	10	10	9.63	1.06
The quality of the models	3	4	5	3	3	5	4	3	3.75	0.89

Table 1: The figures given by the subjects in response to the 11 statements on the questionnaire.

The standard deviation figures (see Table 1) are relatively low in most cases, suggesting that the average values are a good reflection of the subjects' views. This includes the figure for the subjects' views on the quality of the models, which shows a very small standard deviation (0.89). The principal exception is the first scale (ability to view two illustrations at once) which shows a high standard deviation as well as the already-noted low average value. The result obtained in response to the statement regarding the integration of labels also shows a relatively high standard deviation, but the average figure itself is also high.

Since only one set of figures was obtained, it is not strictly appropriate to analyse these results using a standard test of significance. However, it is possible to gain some further insight into the nature of the results by comparing each score with the theoretical result which would have been obtained had there been no preference for either system (i.e., a score of five on each scale, the centre value). This was done using a Wilcoxon matched-pairs signed-ranks test. It was hypothesised that the scores would show a significant preference for or against each feature, but the direction of the preference was not predicted. Thus the results were analysed as for a two-tailed test. It was found that the subjects showed a significant preference ($p <= 0.02$) in favour of all the features tested except for two: the ability to view two models simultaneously, which showed no significant preference for or against; and the quality of the models, which showed a significant preference against.

The analysis of two users who agreed to be video-taped reveals that the newly developed interaction techniques can be used fluently and without help. It was interesting to see that medical students tend to enlarge the geometric model strongly to recognize even the smallest details. These scaling operations provide further evidence that there is not enough detail available.

4   Conclusions

The results show that the subjects regarded all the features evaluated here – with the exception of the simultaneous view facility and the quality of the models used – as useful to them in understanding the spatial relationships encoded in the geometric model. These features include the fisheye zoom to place labels and explanations in an illustration and the various ancillary features introduced in order to counter known or anticipated problems resulting from the use of fisheye techniques. The results therefore support the view that fisheye techniques can be used successfully, provided appropriate mechanisms are introduced to deal with the problem that nodes may disappear inadvertently. Moreover, the generation of figure captions and the automatic placement of labels, both distinguishing features of the Zoom Illustrator, were found to be useful in supporting the learning process.

The empirical evaluation helped us not only to assess the value of different interaction techniques but also to identify some problems and to assess them. The subjects, with their expert knowledge of anatomy, complained about the low level of detail available in the 3d-models used and wished for more information, e.g. about blood vessels and nerves, and more precise information about muscles and bones. More detailed geometric models are available, but these are an order of magnitude more complex than the ones used thus far and require very fast hardware for interactive handling.

A number of other problems were identified from subjects' comments. One is the lack of contrast between adjacent objects, in particular if both belong to the same organ-system and are therefore equally coloured. This is due to the rendering-methods used. To solve this problem, the edges of objects should be enhanced either by image-processing methods or by generating lines which indicate object boundaries. Concerning the labelling, it was sometimes said that it is not clear to which object a label refers or where an object referred to by a label actually starts and ends. The 1:1-connection between a label and an object is not always sufficient. Topographically complex objects, like muscles should be labelled with dedicated strategies to indicate their shape and branching structures. In [5] an approach to generate more appropriate annotations is described.

Moreover, it is interesting to note that students want to be able to underline parts of the texts and to make notes (e.g. hand-written remarks with a pen). They also wish more freedom to "play" with the geometric model, to remove parts and assemble objects themselves. We are thinking about designing a new mode for our learning system in which it can be used as a 3d-puzzle. Users can assemble objects themselves while the system supports them with docking points and snapping mechanisms to ease the 3d-interaction tasks involved.

Some of the features of the Zoom Illustrator deserve more thorough evaluation. In particular, the use of figure captions should be evaluated in detail with regard to the content and structure of the captions and the interaction facilities offered to influence the captions.

5   Acknowledgements

The authors wish to thank Marcel Goetze and Felix Ritter who conducted the tests and analysed the results. Furthermore, our thanks go to Thomas Strothotte for helpful comments on the design and evaluation of the Zoom Illustrator. Many thanks go to Christoph Thomas from the GMD, Sankt Augustin for his detailed comments on the experiment design and the statistical evaluation.

6   References

[1] J. Sobotta (1988) Atlas der Anatomie des Menschen, 19. Edition, J. Staubesand (Ed.), Urban & Schwarzenberg, Munich-Vienna-Baltimore

[2] A. Waldeyer & A. Mayet (1987) Anatomie des Menschen, Walter de Gruyter Verlag, Berlin

[3] B. Preim, A. Raab & T. Strothotte (1997) "Coherent Zooming of Illustrations with 3D-Graphics and Textual Labels", Proc. of Graphics Interface, 19.-23. Mai 1997, pp. 105-113,

[4] B. Preim (1998) Interaktive Illustrationen und Animationen zur Erklärung räumlicher Zusammenhänge, Dissertation, Otto-von-Guericke Universität Magdeburg, Fakultät für Informatik, VDI-Verlag, Reihe 10, Band 532

[5] B. Preim & A. Raab (1998) "Annotation topographisch komplizierter 3D-Modelle", Proceedings of Simulation und Visualisierung, Magdeburg, 5-6th March 1998, SCS-Verlag, Delft, pp. 128-140

[6] B. Preim, R. Michel, K. Hartmann & T. Strothotte (1998) "Figure Captions in Visual Interfaces", Proceedings of ACM Workshop on Advanced Visual Interfaces, L'Aquila, Italy, 24.-27. May), ACM Press, New York, pp. 235-246

[7] G.W. Furnas (1986) "Generalized Fisheye Views", Proc. ACM SIGCHI'86 Conference on Human Factors in Computing Systems, Boston, April, pp. 16-23

[8] J. Dill, L. Bartram, A. Ho & F. Henigmann (1994) "A Continuously Variable Zoom for Navigating Large Hierarchical Networks", Proc. of IEEE Conference on Systems, Man and Cybernetics, pp. 386-390

[9] A. Raab & M. Rüger (1996) "3D-Zoom: Interactive Visualization of Structures and Relations in Complex Graphics", Proc. of Bildanalyse und -synthese, Erlangen, November, pp. 123-132

[10] J.D. Mackinlay, G.G. Robertson & S.K. Card (1991) "The Perspective Wall: Detail and context smoothly integrated", Proc. of CHI '91, New Orleans, April, pp. 173-179

[11] K. Hartmann, B. Preim & T. Strothotte (1998) "Describing Abstraction in Rendered Images through Figure Captions", Workshop on "Combining AI and Graphics for Intelligent User Interfaces of the Future" , European Conference on Artificial Intelligence (ECAI '98), August, 1998, Brighton, UK (to appear)

[12] K.-H. Höhne, B. Pflesser, A. Pommert, T. Schiemann, R. Schubert & U. Tiede (1996) "A Virtual Body Model for Surgical Education and Rehearsal", Computer – Innovative Technology for Professionals, January 1996, pp. 25-31.

Authors addresses

Dr. Ian Pitt
Department of Computer Science,
University College Cork,
Cork,
Ireland
Email: i.pitt@cs.ucc.ie

Dr. Bernhard Preim & Dipl.-Inf. Stefan Schlechtweg
Otto-von-Guericke Universität Magdeburg
Fakultät für Informatik
Institut für Simulation und Graphik
39016 Magdeburg
Email: {bernhard, stefans}@isg.cs.uni-magdeburg.de

Kontinuierliche Prozeßverbesserung durch Integration von Workflow und Intranet

Roland Rolles und Yven Schmidt

Institut für Wirtschaftsinformatik (IWi), Universität des Saarlandes

Zusammenfassung

Kontinuierliche Prozeßverbesserung, organisationales Lernen und Workflow Management sind aktuell diskutierte, innovative Konzepte. Diese in einem integrierten Ansatz miteinander in Einklang zu bringen, ist das Ziel des Prototypen KIWI (Kontinuierliche Prozeßverbesserung durch Integration von Workflow und Intranet). Als hypermediale Lernumgebung im Intranet richtet sich dieser an Mitarbeiter von Unternehmen, die sich mit der Einführung und Nutzung eines Workflow-Management-Systems (WMS) befassen. Über KIWI werden dem Anwender neben dem Verständnis für eine prozeßorientierte Sichtweise und dem Umgang mit WMS auch Methoden und Werkzeuge eines kontinuierlichen Verbesserungsprozesses (KVP) vermittelt. Zum Zwecke einer ständigen Verbesserung von den der Workflow-Anwendung zugrundeliegenden Prozessen erlaubt KIWI zusätzlich die Durchführung dezentraler, modellbasierter Verbesserungsaktivitäten. Dadurch kann im Unternehmen ein Kreislauf von Lernen, Arbeiten, Vorschlagen und Verbessern etabliert werden. In diesem Beitrag werden Konzeption, Aufbau und Inhalte des Systems sowie Einsatzerfahrungen aus der Unternehmenspraxis dargestellt.

Abstract

Continuous process improvement, organizational learning and workflow management are innovative concepts frequently discussed at present. The target of the prototype KIWI (Continuous process improvement by Integration of Workflow and Intranet) is to integrate them in a holistic approach. As hypermedial learning environment in the intranet it addresses employees of enterprises, which deal with the introduction and use of a workflow management system (wms). Apart from the motivation for a process orientated perspective and the training for the wms-handling, KIWI facilitates methods and tools for a continuous improvement process for the user. For the purpose of a constant improvement of the underlying processes KIWI additionally permits the execution of decentralized, model-based improvement activities. Thus a cycle of learning, working, suggesting and improving can be established in the enterprise. In this paper conception, structure and contents of the system as well as practical experiences with this approach are presented.

1 Unterschiedliche Entwicklungen fließen zusammen

Die rasche Weiterentwicklung der Informationstechnik bringt für Unternehmen vielfältige neue Anwendungsfelder mit sich. Vor diesem Hintergrund werden in Forschung und Praxis neue Konzepte erarbeitet, eingesetzt und evaluiert. Das vom BMBF geförderte Forschungsprojekt „Verbesserung von Geschäftsprozessen mit flexiblen Workflow-Management-Systemen (MOVE)"[1] befaßt sich mit der Einführung und Nutzung von Workflow-Management-Systemen (WMS) [1], [2]. Neben Konzepten zur flexiblen Einsetzbarkeit von WMS und Vorgehensweisen zur Beteiligung der Mitarbeiter in den einzelnen Projektphasen steht dabei die Etablierung eines kontinuierlichen Verbesserungsprozesses (KVP) in den beteiligten Unternehmen im Vordergrund. Ein entscheidender Faktor für einen erfolgreichen Workflow-Einsatz und den KVP ist das organisationale Lernen: Mitarbeiter müssen einerseits geschult werden, um das notwendige Prozeßverständnis zu erlangen, mit der neuartigen Vor-

---

1 Fördernummer 01 HB 9606/1. Informationen zu diesem Projekt können im WWW unter http://www.do.isst.fhg.de/move oder http://www.iwi.uni-sb.de/move abgerufen werden.

gangsbearbeitungssoftware umgehen zu können und Wissen über Ziele und Methoden eines Verbesserungsprozesses zu erwerben, und sollen andererseits wieder Wissen an die Organisation in Form von Verbesserungsideen und -vorschlägen zurückgeben. Am Institut für Wirtschaftsinformatik (IWi) der Universität des Saarlandes wurde im Rahmen von MOVE daher der Prototyp KIWI (**K**ontinuierliche Proze**ß**verbesserung durch **I**ntegration von **W**orkflow und **I**ntranet) entwickelt. Dieser erlaubt sowohl eine erste Qualifizierung der Mitarbeiter für den Umgang mit einer Workflow-Anwendung als auch eine ständige Weiterbildung im Sinne eines Training-on-the-job und ist somit als Baustein des Total Quality Learning [3] einsetzbar. Zusätzlich bietet KIWI auch die Möglichkeit zur technischen und methodischen Begleitung dezentraler Prozeßverbesserungsschritte. Der Möglichkeitenraum spannt sich damit zwischen Workflow-Nutzung, kontinuierlichem Verbesserungsprozeß und organisationalem Lernen auf. Als Plattform für den Einsatz im Unternehmen wurde das Intranet gewählt, da es eine leichte Zugänglichkeit mit intuitiven Navigations- und Visualisierungstechniken verknüpft (vgl. Abbildung 1).

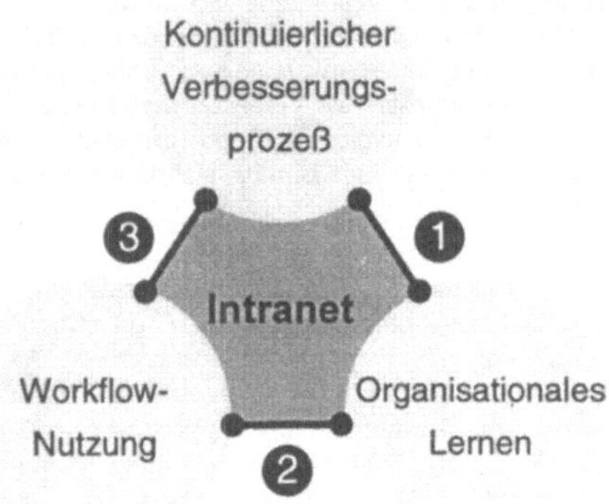

Abbildung 1: Bausteine des Gesamtkonzepts

Es ergeben sich somit 3 Anwendungsfelder, für die KIWI Lösungen anbietet: Das Erlernen der Methoden und Werkzeuge eines kontinuierlichen Verbesserungsprozesses (Abschnitt 2.1), intensive Vorbereitung auf die Workflow-Nutzung (Abschnitt 2.2) sowie das Anwenden der erlernten Methoden im workflow-gestützten, kontinuierlichen Verbesserungsprozeß (Abschnitt 2.3). Abschnitt 2.4 liefert eine Gesamtbetrachtung zu KIWI. Obige Abbildung wird daher in den jeweiligen Unterkapiteln zu Beginn nochmals aufgegriffen, um die Einordnung in den Gesamtzusammenhang zu visualisieren. Die beim praktischen Einsatz des Systems im Unternehmen gemachten Erfahrungen gibt Kapitel 3 wieder. Kapitel 4 faßt zusammen, aus welchen Gründen das Intranet als Integrationsplattform in diesem Anwendungszusammenhang besonders geeignet ist.

2 Bausteine für den erfolgreichen Einsatz

Zentraler Baustein des Ansatzes ist eine computerbasierte, integrierte Lernumgebung, die sowohl den Grundstein für die Nutzung der Workflow-Anwendung legt, als auch für die Beteiligung am kontinuierlichen Verbesserungsprozeß die Voraussetzungen schafft. Diese muß so gestaltet sein, daß die Motivation zur Nutzung des Systems hoch ist. Mit KIWI wurde eine

Lernumgebung geschaffen, in der sich der Benutzer frei bewegen kann und spielerisch ohne das Gefühl von Zwang und Kontrolle den Umgang mit seiner veränderten Arbeitsumgebung und die Partizipation am KVP erlernt. Da für den Mensch die Aufnahme (audio-) visueller Informationen im Vergleich zu rein textuellen Präsentationsformen erheblich leichter ist [4], wurden für die Vermittlung der Inhalte vielfältige Visualisierungstechniken verwendet:

- *Hypertext*: Über Hypertextlinks wurden logisch zusammengehörende Schulungsteile durch Verweise miteinander verknüpft. So können didaktisch gut aufgebaute Lerneinheiten angeboten werden.

- *Bilder*: Zur besseren Visualisierung von Inhalten wird der Text durch Bilder ergänzt, die zur Illustration von Sachverhalten geeignet sind.

- *Animationen:* Bewegte Bilder bringen dem Benutzer nicht-statische Sachverhalte näher, wie etwa das Wandern von Arbeitsmappen über Bearbeiter.

- *Modelle*: Geschäftsprozeß- bzw. Workflow-Modelle spielen in KIWI eine zentrale Rolle. Beim Lernen des Basiswissens zu WMS sind sie Betrachtungsgegenstand und im Verlauf des KVPs werden sie als Objekt von Verbesserungsvorschlägen herangezogen.

- *Screen-Shots*: Durch die Wiedergabe einer Momentaufnahme einer Benutzungsoberfläche wird im Gegensatz zu einer subjektiven verbalen Beschreibung von Bildschirminhalten eine photogenaue, objektive Wiedergabe ebendieser ermöglicht.

- *Screen-Cams*: Screen-Cams sind audio-visuell und ermöglichen die visuelle Wiedergabe von Bildschirmaktivitäten, die in einer Audiospur auch auskommentiert werden können. Die Einzelbilder (frames) einer Screen-Cam-Präsentation stellen wiederum Sceen-Shots dar.

- *Show-Cases* [5]: Videoclips, die einen Ausschnitt aus einem interessierenden Umfeld visualisieren und in einen bestimmten Gesamtzusammenhang einordnen. Diese können einem Benutzer beispielsweise Ausschnitte aus aktuellen oder zukünftigen Tätigkeiten präsentieren.

- *Interaktive Fallstudien:* Fallstudien, die dem Benutzer die Anwendung von gelerntem Wissen verdeutlichen und Interaktion des Benutzers dahingehend erfordern, daß diesem vom System Aufgaben gestellt werden und das System je nach dessen Aufgabenlösung auf unterschiedliche Weise reagiert. Im einfachsten Fall wird hier vom Benutzer beispielweise die Auswahl der richtigen Lösungsalternative bei einer Multiple-Choice-Frage gefordert.

Der Einsatz dieser Visualisierungtechniken zum erfolgreichen Lernen wird auch in den Abschnitten 2.1 und 2.2 noch einmal näher beleuchtet.

Zur ergonomischen Gestaltung derartiger, auf Internettechnologie basierender, Lernumgebungen gibt es Faktoren, die grundsätzlich zu beachten sind und auch bei KIWI Berücksichtigung fanden. Diese sollen hier jedoch nicht Fokus der Betrachtung sein, so daß an dieser Stelle auf die einschlägige Literatur verwiesen sei [6].

2.1   Organisationales Lernen und kontinuierlicher Verbesserungsprozeß

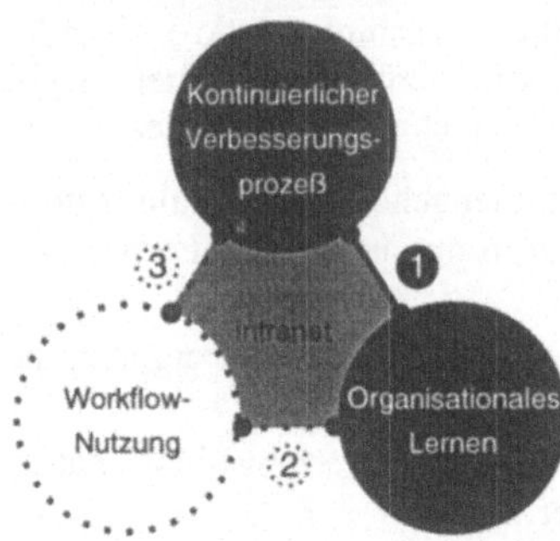

Durch Verbesserungsvorschläge können Mitarbeiter am Arbeitsplatz an der Verbesserung ihrer betrieblichen Situation mitwirken. Mit jeder Umsetzung eines Verbesserungsvorschlages wird quasi ein neuer Standard definiert [7, S. 37f.], der nach Umsetzung eines neuen Verbesserungsvorschlages durch einen wieder neuen Standard ersetzt wird. Da sich dieser Zyklus immer wiederholt, spricht man auch von einem kontinuierlichen Verbesserungsprozeß. Die zugrundeliegende Verbesserungsmentalität stützt sich also in erster Linie auf das Wissen und die Kreativität der Mitarbeiter [8, S. 9].

Als Basisbausteine, von denen das erfolgreiche Gelingen kontinuierlicher Verbesserungsprozesse abhängt, können Motivation, Schulung, Wissen und Kreativität aufgefaßt werden. In der Literatur werden im Zusammenhang mit der Wirksamkeit des Verbesserungswesens sog. Barrieren diskutiert, deren Überwindung die Grundvoraussetzung für das Ausschöpfen von Optimierungspotentialen darstellt [9, S. 43ff]. Eine dieser Barrieren ist die sogenannte Fähigkeitsbarriere, die ausschlaggebende Ursache dafür sein kann, daß ein einzelner Mitarbeiter nicht am Verbesserungswesen teilnehmen kann (für eine detailliertere Untersuchung von Fähigkeits- und Willensbarrieren vgl. [10]). Das Problem des Nicht-Könnens ist mit einer entsprechenden Qualifizierung der Mitarbeiter zu beheben. Sie müssen sich das Wissen aneignen, das bei der Durchführung kontinuierlicher Verbesserungsprozesse von Relevanz ist. Insbesondere methodisches Wissen, konzeptionelles Wissen sowie Prozeßwissen ist hierbei von Bedeutung (für eine Kategorisierung von Wissensarten vgl. [11]).

Die Vermittlung des relevanten Wissens für den kontinuierlichen Verbesserungsprozeß erfolgt in KIWI in der Lerneinheit *KVP*. Dort wird dem Mitarbeiter ein Grundverständnis für die Notwendigkeit zur Durchführung eines kontinuierlichen Verbesserungsprozesses sowie entsprechendes Methodenwissen vermittelt. Zunächst wird jedoch motiviert, welche Idee hinter dem Begriff KVP steht. Weitere Lerneinheiten setzen sich mit dem Ablauf eines KVP, der Schwachstellenanalyse von Prozessen, Sollprozeßentwicklung sowie Werkzeugen für den KVP auseinander.

Um nicht nur Wissen darzustellen und zu reproduzieren soll dem Mitarbeiter auch die Anwendung dieses Wissens demonstriert werden und ihm die Möglichkeit gegeben werden, das gelernte Wissen beispielhaft selbst in einer Fallstudie anzuwenden.

Abbildung 2 zeigt einen Ausschnitt der Demonstration des Ablaufes einer KVP-Sitzung in KIWI. Hier wird die Anwendung der vorher gelernten Methoden eingehend veranschaulicht. Diese beispielhafte KVP-Sitzung wurde didaktisch in Untereinheiten zerlegt. Zu jeder Phase dieser Sitzung werden textuell Hintergrundinformationen geliefert, ein möglicher Ablauf dieser Phase wird in Form einer kurzen Videoanimation visualisiert. So gibt es Show-Cases, die den Ablauf von Workshops zur Modellierung von Sollkonzepten visualisieren und die Bedeutung von Kommunikation und Kooperation im Rahmen des KVP hervorheben. Anhand von Videos wird der Umgang mit Werkzeugen zur Kreativität wie z.B. Ishikawa-Diagramm [12] verdeutlicht.

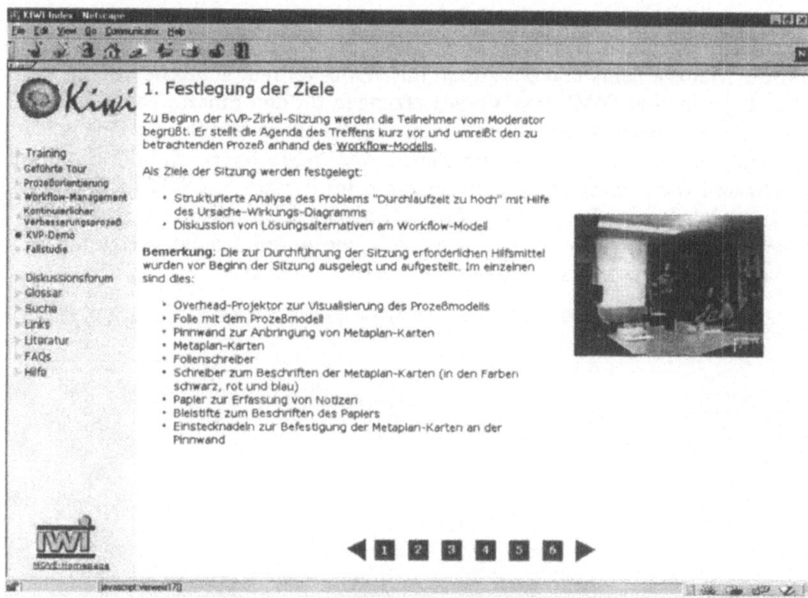

Abbildung 2: Demonstration einer KVP-Sitzung in KIWI

In einer Fallstudie wird der Benutzer durch den kompletten Verbesserungszyklus von der Analyse eines bestehenden Prozesses über die Sollkonzeption und Implementierung bis hin zur Anwendung des verbesserten Prozesses geführt. An geeigneten Stellen hat er Entscheidungen zu treffen, die er auch so bei der tatsächlichen Nutzung der Workflow-Anwendung zu treffen hat. Er wird auf Fehler hingewiesen und bekommt Hinweise und Ratschläge. Somit kann er spielerisch das Verständnis für die neue Technologie und die mit ihr verbundenen Vorgehensweisen erlangen, so daß hiermit gleichzeitig eine Verbindung zu Lernen und Workflow gegeben ist.

2.2 Organisationales Lernen und Workflow-Nutzung

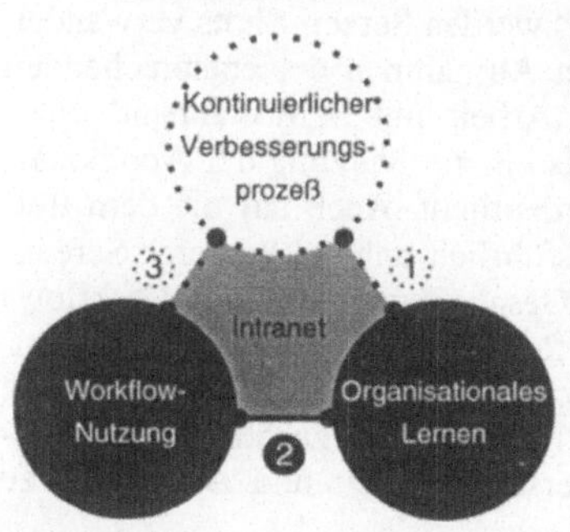

Die unter 2.1 genannten Barrieren lassen sich natürlich auch auf diesen Ausschnitt der Betrachtung übertragen. So gilt es, Fähigkeitsbarrieren durch gezielte Schulung und Vermittlung des relevanten Basiswissens abzubauen sowie Ängste und Vorbehalte gegenüber der neuen Technologie durch eine spielerische Vermittlung der Lerninhalte auszuräumen.

In KIWI werden daher in der Lerneinheit *Prozeßorientierung* die Mitarbeiter zunächst mit dem Prozeßgedanken vertraut gemacht und der Übergang von einer funktionsorientierten zu einer prozeßorientierten Sichtweise motiviert. Dies ist für einen erfolgreichen Workflow-Einsatz unentbehrlich, da die Neuordnung von Kompetenzen, Verantwortung und Aufgaben von den Mitarbeitern getragen werden muß [13, S. 2ff].

Erst in der Folgelerneinheit *Workflow Management* wird der Benutzer mit der Technologie des Workflow Managements und der Arbeit mit WMS vertraut gemacht. Hierzu werden zunächst Grundwissen über WMS und Voraussetzungen für den Einsatz von WMS vermittelt. Weitere Unterlerneinheiten fokussieren darauf, welche Prozesse für eine Unterstützung mit WMS geeignet sind, welche Vorteile der Einsatz von WMS bietet, wie die Einführung von WMS abläuft und wie sich die Arbeitsumgebung beim Einsatz von WMS ändert. Abbildung 3 gibt die Lerneinheit zum Thema Workflow Management sowie die Unterteilung in Unterlerneinheiten wieder. Zur Auflockerung wird auch eine Animation eingesetzt, um zu visualisieren, wie ein Dokument über verschiedene Bearbeiter wandert.

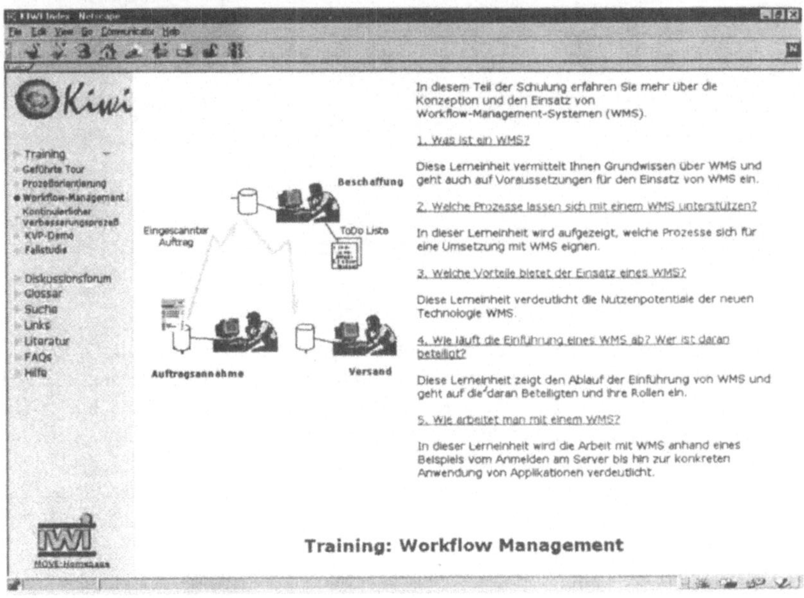

Abbildung 3: Lerneinheit Workflow Management

In der Unterlerneinheit „Wie arbeitet man mit einem WMS" werden Screen-Shots verwendet, um die Arbeit mit der Workflow-Anwendung anhand von Aufnahmen der entsprechenden Bildschirmmasken zu visualisieren. Zusätzlich wird die Arbeit mit WMS anhand eines durchgängigen Beispiels von der Anmeldung am System bis hin zur Nutzung der Applikationen anhand von Screen-Cam-Präsentationen eingängig verdeutlicht. Auch auf die dem Beispiel zugrundeliegenden Geschäftsprozeßmodelle wird ausführlich anhand weiterer Screen-Cam-Präsentationen eingegangen. Bei der Nutzung von Geschäftsprozeß- und Workflow-Modellen ist dabei besonders behutsam vorzugehen, da i.d.R. bei einem Großteil der Anwender keine Methodenkenntnis bzgl. der verwendeten Beschreibungssprachen vorliegt [14]. Daher ist es ratsam, die Modelle jeweils in Kombination mit weiteren, zusätzlich beschreibenden Visualisierungsformen zu verwenden, um die Verständlichkeit und Akzeptanz zu steigern.

2.3 Workflow-Nutzung und kontinuierlicher Verbesserungsprozeß

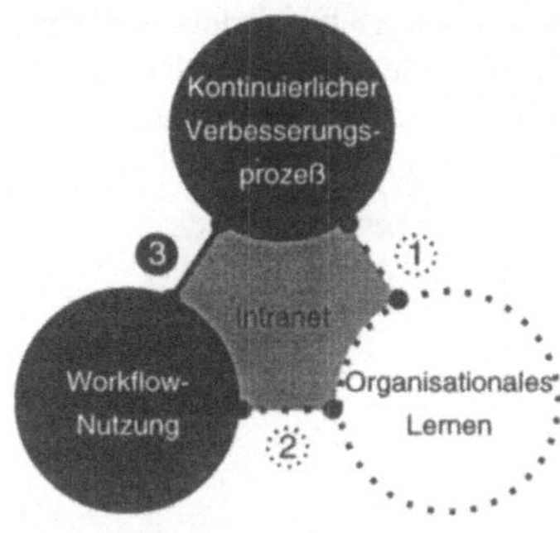

Durch das kontinuierliche Lernen wird der Anwender in die Lage versetzt, sich kritisch mit Organisationsstruktur und Workflow-Anwendung auseinanderzusetzen. Er wird für Schwachstellen sensibilisiert und besitzt Wissen über den Einsatz von KVP-Werkzeugen und Kreativitätstechniken wie etwa Mind Mapping [15], die Problemlösungswerkzeuge von Kaizen [16] oder Methoden der systematischen Strukturierung [17]. Diese in Verbindung mit einem dezentralen Verbesserungsvorschlagswesen einsetzbar zu machen, ist Ziel von KIWI. Zu diesem Zweck werden unterschiedliche Hilfsmittel angeboten und ein solcher KVP methodisch unterstützt.

Die zugrundegelegte Vorgehensweise inklusive der eingesetzten Werkzeuge stellt Abbildung 4 dar. KIWI unterstützt die Anwender dabei, Ansprechpartner für ihre Verbesserungsvorschläge zu ermitteln, ihre Vorschläge per E-Mail einzureichen und mit Kollegen zu diskutieren. Durch die Integration mit dem eingesetzten WMS, die einen automatischen Export der aktuellen Modelle in die Intranet-Umgebung (Erstellung entsprechender Dateien im gif bzw. jpg-Format) zu festgelegten Zeitpunkten vorsieht, kann der Anwender im Intranet auf die Workflow-Modelle zugreifen. Verbesserungsvorschläge können an diesen vermerkt und bei Bedarf über gemeinsame Diskussion am Modell (via Shared Browsing, d.h. das gemeinsame Navigieren durch Webseiten, und Videoconferencing) verfeinert werden. Der Prozeßverantwortliche als Empfänger von Verbesserungsvorschlägen prüft diese und setzt sie gegebenenfalls unmittelbar im Workflow-Modell um. Das betroffene Modell wird hierbei durch Übergabe von Parametern und Kontextinformationen aus der Intranet-Umgebung heraus aufgerufen. Auf einem Schwarzen Brett werden Statistiken zu eingegangenen und umgesetzten Verbesserungsvorschlägen veröffentlicht.

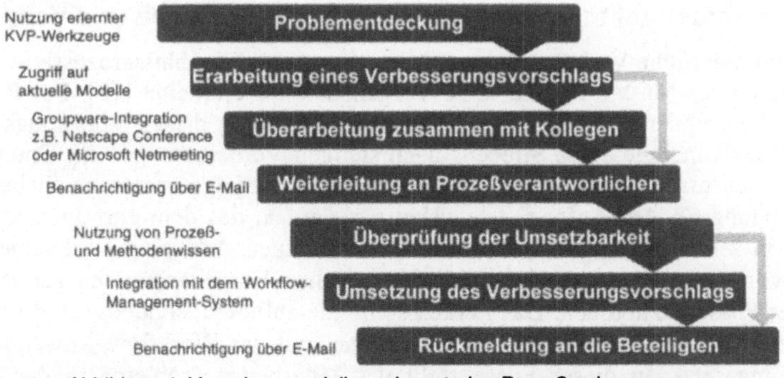

Abbildung 4: Vorgehensmodell zur dezentralen Prozeßverbesserung

Abbildung 5 visualisiert, auf welche Art und Weise die Nutzung dieses KVP-Bausteins erfolgt. Es handelt sich um eine modellbasierte Vorgehensweise, bei der über einen entsprechenden Navigationsbaum (linker Teil der Benutzungsoberfläche) das gewünschte Modell anhand des Geschäftsprozeßkontextes vom Vorschlagseinreicher ausgewählt werden kann. Im vorliegenden Fall ist das Prozeßmodell der Vorkalkulation Gegenstand eines Verbesserungsvorschlags. Bei Auswahl des Modells wird der zuständige Prozeßverantwortliche direkt mitgeliefert. Der weitere Ablauf nach Einreichung erfolgt wie oben kurz beschrieben. Neben der Visualisierung durch Modelle ist im weiteren Projektverlauf auch daran gedacht, z.B. Screen-Shots der Bildschirmmasken der operativen Anwendungen in das System einzustellen, um diese gleichermaßen zum Objekt für Verbesserungsaktivitäten zu machen.

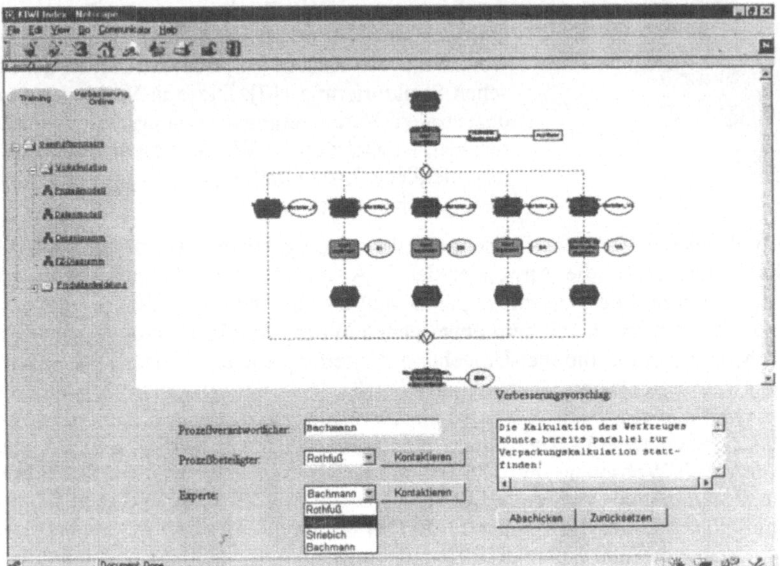

Abbildung 5: Modellbasiertes Verbesserungsvorschlagswesen

2.4   Gesamtbetrachtung: Lernen, Arbeiten, Vorschlagen und Verbessern

Für die kontinuierliche Verbesserung workflow-gestützter Geschäftsprozesse wurde bereits in frühen Phasen des MOVE-Projekts eine Vorgehensweise erarbeitet [18]. Der Prototyp KIWI stellt eine konsequente Weiterentwicklung und Umsetzung dieses Vorschlags dar. Er geht von der Grundannahme eines Strebens nach stetigen Verbesserungen im Unternehmen aus. Dieses Streben manifestiert sich in einer bestimmten Abfolge unterschiedlicher Tätigkeiten (vgl. Abbildung 6). Am Anfang steht dabei das *Lernen*, das dem einzelnen Mitarbeiter das Bewegen in seiner durch Workflow Management neuen Arbeitswelt erleichtert. Auch das Erlernen von Kreativitätstechniken und methodischen Vorgehensweisen zur Problemlösung sind dieser Phase zuzuordnen. Das Lernen sollte nie aufhören, sondern ständiger Wegbegleiter bei den Arbeitsprozessen sein. Beim *Arbeiten* mit der Workflow-Anwendung wird das Erlernte umgesetzt. In dieser Phase stellt der Mitarbeiter Probleme in der Organisation, Schwachstellen in der Anwendung etc. fest. Diese dienen als Grundlage zur Formulierung von Verbesserungsideen, dem *Vorschlagen*. Die Umsetzung der Vorschläge führt schließlich zu einer *Verbesserung* von Prozessen und Anwendung und der beschriebene Zyklus beginnt von vorne.

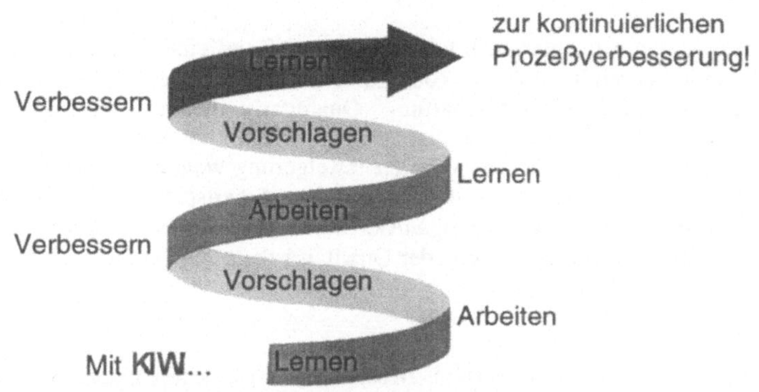

Abbildung 6: Der Weg zur kontinuierlichen Prozeßverbesserung

3 Erfahrungen mit dem praktischen Einsatz

Das Schulungs- und Verbesserungssystem KIWI dient im Betriebsprojekt fischer[2] als Werkzeug zur Unterstützung der Einführungsphase, als Nachschlagewerk bei Fragen zur Nutzung der Workflow-Anwendung, als Grundlage eines kontinuierlichen Lernprozesses und als Wissens- und Methodenspeicher zur Ausschöpfung von Optimierungspotentialen. Mit der Gestaltung des Schulungsbausteins wurden im Rahmen der Qualifizierung von Mitarbeitern bereits positive Erfahrungen gesammelt. Die Hypermedialität und die damit verbundene intuitive Nutzbarkeit des Systems stießen bei den Beteiligten auf ein äußerst positives Echo. Insbesondere die Nutzbarkeit von KIWI zu allgemeinen Schulungszwecken in den Themenbereichen Workflow Management und KVP wie auch die Eignung zur konkreten Anwender- und Benutzungsdokumentation für spezielle Workflows wurde sehr positiv aufgenommen. So können Mitarbeiter bei konkreten Fragen zur Bearbeitung eines bestimmten Falls das System konsultieren, um z.B. auf die richtige Bedienung von Masken, mögliche Reaktionen auf Ausnahmesituationen oder mögliche Ansprechpartner bei Problemen aufmerksam zu werden.

Da kontinuierlicher Lern- und Verbesserungsprozeß sich bei fischer derzeit erst im Aufbau befinden, liegen hierzu noch keine definitiven Praxiserfahrungen vor. Die Rückmeldungen aus den durchgeführten Schulungen lassen eine hohe Mitarbeiterakzeptanz jedoch erwarten. Aus dem bisherigen Feedback der Anwender geht hervor, daß die in KIWI enthaltene Benutzungsdokumentation zur Workflow-Anwendung in Form von auskommentierten Screen-Shots als sehr positiv empfunden wird. Die Multimedialität wurde als motivationsfördernd für die Nutzung des Systems bewertet. Eine detaillierte Befragung der Nutzer und Auswertung der Anwendungserfahrungen wird zu einem späteren Zeitpunkt erfolgen.

4 Fazit: Intranet als Integrationsplattform

Kontinuierliche Lern- und Verbesserungsprozesse lassen sich im Unternehmens-Intranet in idealer Weise miteinander verknüpfen. Dies liegt an der besonderen Eignung der Intranet-Technologie, mit deren Einsatz Nutzenpotentiale sowohl quantitativer als auch qualitativer Art verbunden sind [19]. Hierunter fallen auf qualitativer Seite u.a. die erhöhte Mitarbeiter-

---

2 Die fischer Holding GmbH & Co. KG aus Tumlingen/Waldachtal (Hersteller von Befestigungssystemen) ist eines der Anwenderunternehmungen aus dem MOVE-Projekt.

motivation durch einen erleichterten Informationszugriff und eine verbesserte Informationsverfügbarkeit, eine Vereinheitlichung von Abläufen sowie eine Reduktion telefonischer Rückfragen zwecks Informationsbeschaffung. Quantitativ meßbarer Nutzen entsteht beispielsweise durch die Plattformunabhängigkeit der Infrastruktur, Einsparungen aufgrund reduzierter Papierverwaltung sowie durch Effizienzsteigerung wegen verbesserter Lerneffekte. Auf diese Weise kann ein KVP ideal in den täglichen Arbeitsprozeß integriert werden und muß nicht weiter losgelöst von diesem organisiert werden, wie es bei traditionellen Verfahren wie dem betrieblichen Vorschlagswesen oder Qualitätszirkeln momentan noch der Fall ist.

5 Literatur

[1] Th. Herrmann; A.-W. Scheer; H. Weber (Hg.): Verbesserung von Geschäftsprozessen mit flexiblen Workflow-Management-Systemen: Von der Erhebung zum Sollkonzept. Heidelberg, 1998: Physica.

[2] Th. Herrmann; A.-W. Scheer; H. Weber (Hg.): Verbesserung von Geschäftsprozessen mit flexiblen Workflow-Management-Systemen: Von der Sollkonzeptentwicklung zur Implementierung von Workflow-Management-Anwendungen. Heidelberg, 1998: Physica.

[3] H. Schnauber et al.: Total Quality Learning: Ein Leitfaden für lernende Unternehmen. Berlin u.a., 1997: Springer.

[4] R. Jain: Visual Information Management. Communications of the ACM 40 (1997), 12, S. 31-32.

[5] T. Walter: Visualisierungsmethoden bei Workflow-Management - Prototyping und Showcases. In: Proceedings der DCSCW '98, in Veröffentlichung.

[6] K. Moysich: Empfehlungen aus dem WWW für die WWW-Entwicklung. http://iundg.informatik.uni-dortmund.de/demes/webdsign/webdsign.html, 24. Februar 1998.

[7] M. Imai: Kaizen - Der Schlüssel zum Erfolg der Japaner. 12. Aufl., München, 1994: Wirtschaftsverlag Langen Müller Herbig.

[8] J.-M. Jacobi: Kontinuierlich verbessern: Jeder kann kreativ sein : Das neue BVW. 2. Aufl., Stuttgart, 1997: Dt. Sparkassenverlag.

[9] N. Thom: Betriebliches Vorschlagswesen - Ein Instrument der Betriebsführung: Empirische Erkenntnisse und Gestaltungsempfehlungen. 4. Aufl., Bern u.a., 1991: Lang.

[10] J. Hagemeyer; R. Rolles; Y. Schmidt; J. Bachmann; A. Haas: Dynamische Prozesse durch workflowzentrierte Geschäftsprozeßoptimierung: Herausforderungen in der Praxis. In: A.-W. Scheer (Hg.): Neue Märkte, neue Medien, neue Methoden - Roadmap zur agilen Organisation. 19. Saarbrücker Arbeitstagung, Heidelberg, 1998: Physica, in Veröffentlichung.

[11] G. von Krogh; M. Venzin: Anhaltende Wettbewerbsvorteile durch Wissensmanagement. Die Unternehmung 49 (1995), 6. S. 417-436.

[12] K. Ishikawa: What is quality control? The Japanese way. Englewood Cliffs, N.J., 1985: Prentice Hall.

[13] M. Gaitanides; R. Scholz; A. Vrohlings: Prozeßmanagement - Grundlagen und Zielsetzungen. In: M. Gaitanides et al. (Hg.): Prozeßmanagement: Konzepte, Umsetzungen und Erfahrungen des Reengineering. München u.a., 1994: Hanser. S. 2-19.

[14] J. Hagemeyer; R. Rolles: Erhebung von Prozeßwissen für das Wissensmanagement. IM Information Management & Consulting 13 (1998), 1, S. 46-50.

[15] M. Eipper: Sehen - Erkennen - Wissen: Arbeitstechniken rund um Mind Mapping. Renningen-Malmsheim, 1998: Expert-Verlag.

[16] M. Imai: Kaizen: a. a. O.

[17] E. Mehrmann: Schnell am Ziel: Kreativitäts- und Problemlösungstechniken. Düsseldorf u.a., 1994: ECON Taschenbuch Verlag.

[18] R. Rolles: Kontinuierliche Verbesserung von workflow-gestützten Geschäftsprozessen. In: Th. Herrmann, A.-W. Scheer; H. Weber (Hg.): Verbesserung von Geschäftsprozessen mit flexiblen Workflow-Management-Systemen: Von der Erhebung zum Sollkonzept. Heidelberg, 1998: Physica. S. 109-133.

[19] M. Döge: Intranet: Einsatzmöglichkeiten, Planung, Fallstudien. 1. Aufl., Köln, 1997: O'Reilly. S. 44-45.

Adressen der Autoren

Dipl.-Kfm. Roland Rolles
Institut für Wirtschaftsinformatik (IWi)
Universität des Saarlandes
Postfach 151150
66041 Saarbrücken
Email: rolles@iwi.uni-sb.de

Dipl.-Kfm. Yven Schmidt
Institut für Wirtschaftsinformatik (IWi)
Universität des Saarlandes
Postfach 151150
66041 Saarbrücken
Email: schmidt@iwi.uni-sb.de

Virtuelle taktile Karten –
digitale Stadtpläne für Blinde

Jochen Schneider und Thomas Strothotte

Fakultät für Informatik, Otto-von-Guericke-Universität Magdeburg

Zusammenfassung

Taktile Karten bieten Blinden guten Zugang zu geographischen Informationen. Ihre Herstellung ist jedoch aufwendig und ihr Informationsgehalt im Vergleich zu Karten für Sehende gering. Die Autoren haben daher prototypische Techniken entwickelt und implementiert, durch die Blinde und stark Sehgeschädigte ein geographisches Gebiet flexibel mit Computerhilfe erkunden können. Beim hier vorgestellten Ansatz der virtuellen taktilen Karten werden multimediale Interaktionstechniken zur Unterstützung der geographischen Orientierung von Angehörigen des genannten Benutzendenkreises eingesetzt. Es wird der Entwurf eines Systems vorgestellt, welches statische Handgesten von einer Videokamera übernimmt und akustische Ausgaben erzeugt. Die Informationen in den benutzten digitalen Kartendaten werden auf die Bedürfnisse der Benutzenden und die jeweilige Interaktion angepaßt. In diesem Beitrag werden die Entwicklung des Prototyps und erste Erfahrungen mit ihm geschildert.

Abstract

Tactile maps offer good access to geographic information for blind people. However, compared to maps for sighted people, they are inconvenient to make and contain less information. Therefore, the authors developed and implemented prototypical techniques, by which visually impaired people can explore a geographical area flexibly with the help of a computer. With the concept of virtual tactile maps presented here multimedia techniques of interaction are used to support the geographical orientation of members of the aforementioned group of users. The design of a system is presented which captures static hand gestures through a video camera and produces acoustical output. The information in the digital map data used is adapted to the requirements of the users and the respective interaction. In this paper, the development of the prototype and first experience with it are presented.

1 Eine Vision als Herausforderung

An einer Straßenkreuzung einer beliebigen Stadt steht ein Fußgänger, den man am Langstock als Blinden erkennt. Er hält seine Hände mit den Handflächen nach unten vor sich und bewegt sie konzentriert. Er ist gerade dabei, anhand einer virtuellen taktilen Karte seinen weiteren Weg zu planen. Er hatte sich bereits zu Hause mit der Strecke vertraut gemacht. Nun ist er schon ein ganzes Stück gegangen und möchte sich vergewissern, daß er den restlichen Weg noch beherrscht. Bei näherem Hinsehen bemerkt man, daß sich im Schirm seiner Mütze eine kleine Kamera befindet, die senkrecht nach unten auf seine Hände zeigt. Außerdem, daß der Fußgänger kleine Ohrhörer trägt.

Das beschriebene Szenario stellt eine Zukunftsvision dar. Die Autoren entwickeln seit kurzem ein System zur Erkundung von virtuellen taktilen Karten als am Körper zu tragenden Computer (Wearable Computer [1]). In diesem Bericht wird der Entwurf und der erste Prototyp des Systems vorgestellt.

Unter *virtuellen taktilen Karten* wollen wir digitale Karten verstehen, die nicht grafisch sondern akustisch dargestellt werden, wenn man sie über tastbewegungsähnliche Handgesten bedient. Virtuelle taktile Karten ermöglichen es blinden Benutzenden unterwegs und zu Hause, ein unbekanntes geographisches Gebiet zu erkunden und Strecken darin zu erlernen, so daß sie es danach ähnlich gut begehen können wie nach Benutzung einer taktilen Karte. Ein besonderes Merkmal virtueller taktiler Karten ist, daß sie keine 1:1-Abbildung von taktilen Karten darstellen, sondern auf spezielle Art bedienbare digitale Karten. Die Interaktion mit ihnen erfolgt über mit einer Videokamera aufgenommene Handgesten. Die geographischen Informationen werden akustisch während der Interaktion mit dem entsprechenden System dargestellt, in Form von Sprache und Klängen. Bei der Benutzung auf einem Tisch dient ein taktiles Gitter als Unterlage, welches die Orientierung der bedienenden Hände erleichtert.

So wie es für Blinde und stark Sehgeschädigte in Braille geschriebene Bücher gibt, deren Buchstaben ertastet werden können, gibt es taktile Stadtpläne, deren Straßen und andere Informationszeichen ein Relief darstellen. Derartige Karten können gefaltet oder aufgerollt und so auf eine Reise mitgenommen werden. Die Herstellung taktiler Karten ist hingegen aufwendig, so daß sie relativ teuer und nur in geringer Vielfalt verfügbar sind. Außerdem benötigen die taktilen kartographischen Zeichen mehr Platz als ihre visuellen Entsprechungen, so daß die Karten (bei gleichem Inhalt) größer sind. Das verstärkt die eben geschilderten Probleme, weil für das gleiche Gebiet mehr Karten benötigt werden.

Im MoBIC-Projekt [2], welches am Institut für Simulation und Grafik geleitet wurde, wurde eine auf dem Satellitennavigationssystem GPS basierende elektronische Reisehilfe für blinde und stark sehgeschädigte Fußgänger entwickelt. Dabei ist die Vision eines Computersystems entstanden, welches es blinden Fußgängern ermöglicht, sich mit einem ihnen unbekannten Gebiet vertraut zu machen.

Blinde Computerbenutzende werden durch neuere Interaktionstechniken wie grafische Benutzungsoberflächen eher benachteiligt als unterstützt. Es wird viel Aufwand betrieben, um grafische Darstellungen in akustische zu übersetzen (s. z.B. [3]). Wenn Mustererkennung ein integraler Bestandteil künftiger interaktiver Systeme ist (in [4] wird sie in diesem Zusammenhang gar als "Key Technology" bezeichnet), dann sollte diese Technik auch für Sehbehinderte nutzbar sein. Die hier vorgestellte Anwendung optischer Gesteneingabe ist auf blinde Benutzende zugeschnitten, stellt dabei aber keine Übersetzung eines Systems für Sehende dar. Das neue System ist portabel und soll außerdem erschwinglich sein, indem es größtenteils aus PC-Standardkomponenten besteht.

Die größte Herausforderung beim hier vorgestellten Ansatz liegt in der Gestaltung der Interaktion. Dabei geht es zunächst um die Frage, wie das System Handbewegungen interpretiert und sie als Bewegungen auf einer virtuellen taktilen Karte abbildet. Weiterhin geht es darum, welche Informationen aus der Karte ausgewählt und wie diese akustisch dargestellt werden. Die nächste Herausforderung besteht darin, die digitalen Kartendaten aufzubereiten. Dies ist nötig, weil die Handeingabe relativ grob ist und daher je nach Kartenmaßstab der Abstand zwischen den Details der Rohkarte zu gering wird. Die genannten Aspekte müssen schließlich implementiert werden, indem bekannte Techniken der Gestenerkennung und der akustischen Ausgabe angepaßt und verbunden werden.

Im folgenden Kapitel 2 werden bisherige Ansätze zur Unterstützung von Blinden bei ihrer Mobilität beschrieben, zunächst taktile Karten und ihre Erstellung, anschließend elektronische Reisehilfen. In Kapitel 3 der Entwurf eines Systems mit virtuellen taktilen Karten vorgestellt. Danach wird in Kapitel 4 der erste Prototyp des Systems und Erfahrungen mit diesem geschildert. Zuletzt wird in Kapitel 5 beschrieben, wie die Ansätze und das System in Zukunft weiterentwickelt werden.

2 Verwandte Arbeiten

Der hier beschriebene Ansatz zu virtuellen taktilen Karten stützt sich auf Forschungen aus hauptsächlich drei Gebieten: Studien und praktische Erfahrungen mit taktilen Karten, Arbeiten zu elektronischen Reisehilfen und Methoden der optischen Gestenerkennung. Sie werden in diesem Abschnitt nacheinander beschrieben. Auf den vierten Bereich der Arbeit, die akustische Darstellung, wird im nächsten Abschnitt eingegangen.

2.1 Taktile Karten

Taktile (oder Relief-)Karten bieten Blinden einen geeigneten Zugang zu geographischen Informationen [5]. Die auf einer solchen Karte befindlichen Darstellungen für Punkte, Linien und Flächen werden "Symbole" genannt. Deren Auswahl, Vereinfachung und Umsetzung in taktile Form stellen die wichtigsten Arbeiten beim Erstellen einer taktilen Karte dar [6, 206ff]. Neben taktilen Karten für den Geographieunterricht wird zwischen Orientierungs-, Mobilitäts- und topologischen Karten unterschieden. Orientierungskarten bieten ihren Benutzenden eine allgemeine Übersicht eines bestimmten Gebietes. Mobilitätskarten sind auf die Bedürfnisse von Reisenden zugeschnitten und enthalten Orientierungspunkte. Topologische Karten schließlich enthalten eine bestimmte Strecke, die für Sehbehinderte aufbereitet ist. Auf andere Details wird verzichtet, die Darstellung ist vereinfacht und verzerrt [ebd., 194f].

Taktile Karten stellen ein ausgereiftes Orientierungsmittel dar und besitzen den Vorteil, daß sie portabel sind und daher auf Reisen mitgenommen werden können. Andererseits sind sie nicht einheitlich gestaltet. Ein weiterer Nachteil ist die schwere Verständlichkeit für viele Angehörige der Zielgruppe, u.a. wegen der Notwendigkeit von Braillekenntnissen zum Lesen der Beschriftungen [7]. Taktile Karten sind außerdem nicht flächendeckend verfügbar, weil sie aufwendig manuell hergestellt werden müssen, was sowohl spezielle technische Fertigkeiten als auch Hilfsmittel erfordert [8]. Aus diesen Gründen gibt es schon seit einiger Zeit Bestrebungen, sie durch flexible elektronische Reisehilfen zu ergänzen.

2.2 Elektronische Reisehilfen

Die Reisehilfe für den Nahbereich, die unter Blinden am weitesten verbreitet ist, ist der Blindenstock. Er erlaubt das Erkennen von Hindernissen, die direkt vor einem Fußgänger liegen. Die ersten elektronischen Reisehilfen sollten den Stock ersetzen oder ergänzen, indem sie Hindernisse in größerer Entfernung als er melden, können aber die mittelbare Umgebung ebensowenig wie er berücksichtigen. Ein System mit virtuellen taktile Karten stellt eine Elektronische Orientierungshilfe für den weiteren Raum ("Electronic Orientation Aid (EOA) for larger space", s. [9]) dar. Es dient also einerseits nicht unmittelbar der Mobilität wie die genannten klassischen Reisehilfen für den Nahbereich, andererseits auch nicht der ständigen Orientierungshilfen wie die EOAs, die auf Satellitennavigation aufbauen (ebd.).

Die in [2] beschriebene Elektronische Reisehilfe besteht aus der Reisevorbereitungskomponente MoPS und der Reisehilfe MoODS, die das Satellitennavigationssystems GPS nutzt. Das MoPS verwaltet digitale Karten, die um spezielle Informationen (z.B. Wegbeschaffenheit) erweitert sind. Benutzende können sich in diesen Karten mit Hilfe der Tastatur bewegen und Informationen über ihre aktuelle Position abrufen. Das MoPS gibt Informationen mittels synthetischer Sprache und Braille aus.

In [7] wurde untersucht, inwieweit taktile Karten durch einen Computer mit einem Tasttablett ergänzt oder sogar ersetzt werden können. Dazu haben sehbehinderte Versuchspersonen eine digitale Karte durch ein Tasttablett bedient. Auf das Tasttablett wurde im ersten Versuch eine taktile Karte, im zweiten ein taktiles Gitter gelegt, bzw. wurde im dritten das Tablett ganz leer gelassen. Das Drücken einer Position auf dem Tablett führte zur sprachlichen Ausgabe von Informationen von der wirklich bzw. gedacht vorhandenen Karte.

Die Untersuchungen haben ergeben, daß das Wiederfinden einer Position auf dem leeren Tasttablett für die Versuchspersonen sehr schwierig war. Beim Einsatz von Explorationskarten mit taktiler Karte bzw. reinem taktilen Raster als Auflage für das Tasttablett konnten sie jedoch einen Weg erfolgreich erkunden. Dies wurde durch Tests der mentalen Repräsentation der Route und der Fähigkeit zu ihrem tatsächlichen Ablaufen gemessen. Beim Lernen mit der taktilen Karte wurde allerdings deutlich weniger Zeit benötigt, bis sich die Versuchspersonen sicher genug fühlten, die Route abzugehen. Überraschend war jedoch, daß die erlangte Kenntnis der Route beim taktilen Gitter fast so gut war wie bei den taktilen Karten.

Daß die Versuchspersonen auch ohne taktile Karten eine Route erlernen konnten, läßt sich anhand der Forschung der britischen Psychologin Millar [10] erklären. Sie hat bei ihrer Arbeit mit Kindern herausgefunden, daß Blinde Bilder und Karten beim Ertasten nicht nur über taktile Reize, sondern auch über die dabei ausgeübten Bewegungen an sich erlernen können. Weder Psychologen noch Geographen wissen allerdings sicher, wie Sehgeschädigte oder Blinde eine neue Umgebung erlernen und dieses Wissen auf dem laufenden halten [11].

2.3 Optische Gestenerkennung

Die optische Aufnahme von Handgesten ist zwar inzwischen weit erforscht (s. [12] für eine Übersicht) und sogar in kommerzielle Produkte umgesetzt (z.B. im SIVIT von Siemens), hat jedoch nur in wenigen praktischen Anwendungen Einsatz gefunden. Der grundsätzliche Ansatz der optischen Gestenerkennung existiert schon länger. Ihr Pionier Krueger arbeitet seit Ende der sechziger Jahre an Kunstinstallationen, mit denen Besucher interagieren können. Dabei werden die Besucher über Videokameras aufgenommen und ihre grafischen Abbilder in die Installation zurückprojiziert. Eine Installation namens "Videodesk" hat die Form eines Tisches [13].

Mit dem System "KnowWhere" stellen Krueger u. Gilden eine Anwendung des Videodesk vor, die Blinden geographische Informationen vermittelt [14]. Geographische Daten werden dabei dadurch dargestellt, daß ein Geräusch ertönt, wenn eine Hand ihr imaginäres Abbild berührt. Die Finger agieren auf einer virtuellen, also nicht optisch dargestellten Karte auf einer Tischfläche mit taktilem Gitter. Ihre Bewegungen führen zu akustischen Ausgaben in Abhängigkeit von der Art des geographischen Elementes, welches sie überdeckt haben, entweder sprachlich in Form von Elementnamen oder als Töne. Die virtuelle Karte kann so vergrößert werden, daß ein interessierendes Gebiet den ganzen Tisch einnimmt.

In einem Test mit blinden Versuchspersonen wurden großflächige geographische Informationen dargestellt. Das System wurde mit fünf geburtsblinden Personen mittleren Alters getestet. Sie konnten absolute Positionen leicht finden und danach Puzzlestücke der gelernten geographischen Formen erkennen. Zwei konnten außerdem "annehmbare" Zeichnungen der Formen anfertigen.

Die Arbeiten von Krueger sind generell Vorbilder für den hier beschriebenen Entwurf, letzterer unterscheidet sich in einigen Punkten jedoch erheblich von "KnowWhere". Die Tischfläche des Videodesk besteht aus einem Lichttisch, wodurch die Handerkennung erleichtert wird, allerdings auch die Portabilität des Systems erschwert. Auch wurde die "KnowWhere"-Anwendung nicht eingesetzt, um einen Weg durch ein kleines geographisches Gebiet zu wählen und zu erkunden. Bis auf lineares Zooming kann in dem System keine Auswahl getroffen werden. Schließlich verhindert der Einsatz spezieller Hardware, daß das System je für einen breiten Benutzendenkreis erschwinglich sein könnte.

Starner, Weaver u. Pentland beschreiben einen am Körper tragbaren (wearable) Computer, der über Gebärdensprache gesteuert wird [15]. Das System soll in der Endversion in der Lage sein, Gebärden als gesprochene Sprache auszugeben. Die Gebärden werden optisch über eine Kamera aufgenommen, die sich im Schirm der Mütze des Benutzers befindet und senkrecht nach unten auf die Hände zeigt. Die beschriebenen Forschungsergebnisse zeigen, daß die Vision eines am Körper tragbaren Systems mit virtuellen taktilen Karten bald Wirklichkeit werden kann und deuten Schritte zu ihrer Umsetzung an.

3   Ein System zum Erkunden von Karten

In diesem Abschnitt werden zunächst Anforderungen an ein System zur Erkundung virtueller taktiler Karten formuliert. Dann wird gezeigt, wie sie in einem Entwurf umgesetzt wurden. Schließlich wird ein erster Prototyp des Systems vorgestellt.

3.1   Anforderungen

Um Anforderungen an das System zu definieren, stellt sich zunächst sich die Frage, wie sich blinde Fußgänger auf das Gehen in einem ihnen noch nicht vollständig bekannten Gebiet vorbereiten können. Nach Golledge u.a. müssen sich Blinde, die selbständig navigieren möchten, die entsprechende Umgebung einprägen, Pfadsegmente und Winkel erlernen und beim Gehen wiederfinden [16]. Die genannten Autoren empfehlen als Strategien zum Erlernen einer fremden Umgebung durch Blinde:

1. systematische Erkundung des Gebietes
2. Wahl eines festgelegten Ausgangspunktes (Landmarke)
3. Angabe der Struktur der Umgebung (Regelmäßigkeit von Straßen, etc.)
4. Festlegung von Hinweisen auf Landmarken
5. Entwickeln einer kognitiven Repräsentation des Gebietes
6. Festlegung von Kriterien für die Streckenauswahl
7. Streckenauswahl

Damit ein Computersystem bei diesem Prozeß behilflich sein kann, muß es die nötigen Informationen zu einem Gebiet liefern. In [2] wird beschrieben, welche Art von Informationen eine Reisehilfe per Sprache ausgeben sollte. Dazu wurden sehgeschädigte Fußgänger und Mobilitätstrainer befragt. Als essentielle Informationen auf Karten haben die Befragten Namen und Art einer Straße, Entfernungen und Hinweise auf Gefahrenpunkte genannt. Als wünschenswert bewerteten die Befragten Angaben über Geschäfte, öffentliche Gebäude, Straßenkreuzungen und öffentliche Verkehrsmittel, außerdem die Darstellung eines Umgebungsplans und die Auswahl einer Strecke nach Sicherheit oder Kürze. Als ideale Anforderungen wurden schließlich die Einbeziehung von Informationen über zeitlich begrenzte Hindernisse, z.B. Baustellen, Gerüste und Umleitungen genannt und die Möglichkeit der Ergänzung des Reiseplans um eigene Kommentare.

3.2 Entwurf

Auf Basis der oben geschilderten Erfahrungen und Überlegungen wurden Methoden und Werkzeuge für die Erkundung virtueller taktiler Karten entwickelt. Der Entwurf wurden prototypisch implementiert, und es wurde ein Benutzertest durchgeführt. Die wichtigsten Entwurfsaspekte waren die Dialogführung in Verbindung mit der Aufbereitung digitaler Karten. Diese sollen nun beschrieben werden.

Für die Umsetzung der Vision einer am Körper tragbaren virtuellen taktilen Karte scheiden die zunächst naheliegenden taktilen Ausgabegeräte aus, weil es davon (außer als Forschungsprototypen, s. [17]) nur kleinflächige gibt und sie sperrig sind. Wenn man auf die Benutzung taktiler Karten aus den genannten Gründen ganz verzichtet, kann die Ausgabe also nur akustisch erfolgen. Damit geographische Positionen möglichst einfach ausgewählt werden können, soll dies mit den Händen auf einer Fläche mit taktilem Raster geschehen.

Obwohl Blinde Computer eingeschränkt durch das relative Eingabegerät Maus bedienen können, kommen für sie v.a. absolute grafische Eingabegeräte in Frage, weil ihnen die optische Rückkopplung über die Position des Bildschirmzeigers verwehrt ist. Für die Bedienung des hier beschriebenen Systems bietet sich daher zunächst ein Tasttablett an. Der Einsatz eines solchen Tabletts hat den Vorteil, daß Positionierung absolut und die Positionserkennung sicher ist. Allerdings ist es groß, kann also schlecht transportiert werden. Es bietet weiterhin nur einen Abtastpunkt, kann also nicht mehrere Finger oder ein anderes Objekt erkennen.

Bei dem hier vorgestellten Ansatz wird daher auf den Einsatz eines Tasttabletts verzichtet. Damit die Hände trotzdem frei zur Bewegung und zum Ertasten der schon erwähnten taktilen Orientierung bleiben, werden ihre Positionen berührungsfrei aufgenommen. Daher wurde als Eingabegerät eine Videokamera gewählt, die selbst klein ist und auch mit dem nötigen Stativ zusammen leicht transportiert werden kann. Um den Händen eine Orientierung unter der Kamera zu bieten, wird eine Gitterunterlage aus einem flexiblen Material gewählt. Marken in Form von Spielsteinen dienen der Auswahl von Strecken und Gebieten.

Das im Folgenden beschriebene System unterstützt Blinde beim Erkunden eines Gebietes, indem es sprachlich Straßen und Punkte beschreibt, die sich auf einer durch das Sichtfeld der Kamera begrenzten gedachten Karte unter der Hand befinden. Benutzende können weiterhin eine Strecke erlernen, indem sie sie wiederholt mit ihren Händen verfolgen. Nachdem sie Start- und Zielpunkt gewählt haben, findet das System außerdem eine Strecke als kürzeste Verbindung und vermittelt sie dann in Abhängigkeit der Position der Hände akustisch. Dies geschieht dadurch, daß die laterale Entfernung von der Strecke zur Hand und der relative Fortschritt auf ihr auf zwei Dimensionen der akustischen Darstellung abgebildet werden.

Die Änderung der Balance eines Tones wird im allgemeinen als Seitwärtsbewegung wahrgenommen [18]. Die Balance wird daher im hier vorgestellten System benutzt, um die Entfernung senkrecht zur Strecke darzustellen. Als laterale Entfernung wird die Länge der Normalen vom Cursor zum zu ihm nächsten Segment genommen, positiv nach der einen, negativ nach der anderen Seite der Strecke. Die Kodierung der Entfernung von der Strecke über die Balance hat den Vorteil, daß neben dem Betrag der Entfernung auch ihre Richtung abgebildet wird. Für die Ausgabe dieser Kodierung ist der Einsatz eines Kopfhörers nötig. Weitere einfache tonale Dimensionen sind die Lautstärke und die Höhe eines Tons. Da eine Balanceän

derung als Verschiebung der Lautstärke zwischen zwei Stereolautsprechern wahrgenommen wird, ist eine weitere Verwendung der Lautstärke nicht sinnvoll. Daher wird im hier be-

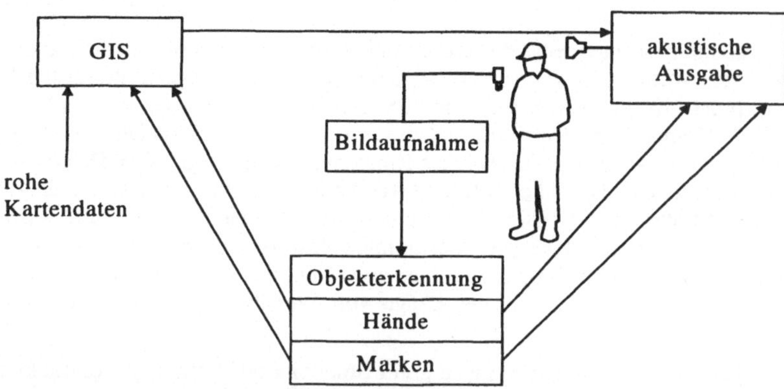

Abb. 1: Blockdiagramm des Systems

schriebenen System die Entfernung eines Fingers auf der Strecke durch die Tonhöhe dargestellt. Die relative Position zwischen Start und Ziel wird die über Tonhöhe kodiert. (Der Ton
ist um so höher, je näher der Finger dem Ziel ist.)

Nach Auswahl einer Route kann eine Beschreibung von Teilstrecken, Wahlpunkten und Rich-
tungen zwischen den Teilstrecken auch über einen Brailledrucker ausgegeben werden, um Benutzenden beim eigentlichen Reisen zur Orientierung zu dienen. Das System soll perspektivisch auf gängigen Heimcomputern eingesetzt werden soll, also MS-Windows-PCs. Neben
den für blinde Benutzende üblichen Ein- und Ausgabegeräten wie Brailletastatur und Sound-
karte wird nur noch eine digitale Kamera und eine Videokarte benötigt.

Im Folgenden wird der Entwurf des System zur Bedienung virtueller taktiler Karten mit Marken und einer speziellen Unterlage zu ihrer Befestigung vorgestellt (s. Abb. 2). Der Entwurf
enthält je eine Komponente zur Bildaufnahme, zur Objekterkennung, zur Generalisierung und
zur akustischen Darstellung.

Die Bildaufnahme liefert das aktuelle Kamerabild über eine Framegrabberkarte an den Computer. In der Objekterkennung werden zunächst Hände und Marken anhand von Schwellwer-
ten segmentiert. Dabei entsteht eine Umrißbeschreibung der Objekte als Polygonzüge. Aus
den Umrissen werden Merkmale errechnet und die Handhaltung nach (vorher erlernten) statischen Gesten (ähnlich dem Verfahren in [19]) und bekannten Marken klassifiziert. Alle Hand-
und Markenmerkmale (Stellung, Position und Richtung) werden an den GIS-Teil des Systems
weitergeleitet. Dort werden die geographischen Daten entsprechend angepaßt, u.a. verzerrt.
Weiterhin werden die Merkmale an den Systemteil zur akustischen Darstellung weiter-
gegeben, wo sie zu der Ausgabe von gesprochenen Informationen und Klängen führen.

3.3 Der erste Prototyp

Es wurde ein Prototyp des eben beschriebenen Entwurfs implementiert. Mit ihm kann man eine virtuelle taktile Karte explorieren, indem man eine Hand mit ausgestrecktem Zeigefinger
unter einer Kamera auf einem taktilen Gitter bewegt. Das Programm setzt Handbewegungen
auf der Interaktionsfläche in Bewegungen auf der Karte um. Wenn die Zeigefingerspitze ein
geographisches Element "berührt", so führt dies zur Ausgabe seines Namens in gesprochener
Sprache.

Die Handerkennungskomponente basiert auf einem in C++ geschriebenen System [20], welches vom ersten Autor nach MS-Windows portiert wurde. Die Oberfläche für die Erkennung und die restlichen Komponenten (u.a. die GIS- Komponente) wurden in Smalltalk geschrieben. Die GIS- Komponente liest Karten in einem am Institut für Simulation und Graphik entwickelten Zwischenformat ein, für welches Konverter von gängigen GIS-Datenformaten vorliegen. Die Komponente verwaltet Attribute wie Name, Art, Position usw. zu geographischen Objekten wie Straßen und Häusern. Sie stellt weiterhin eine Funktion zum Auffinden von Objekten bereit, die in der Nähe einer bestimmten Position liegen. Außerdem ermöglicht sie das Auffinden des kürzesten Weges zwischen zwei Punkten. (Diese Funktion bestimmt einen Weg derzeit noch nach geometrischer Länge von Wegabschnitten, nicht nach Wegeigenschaften, z.B. Sicherheit.)

Für die Ausgabe von Tönen und Sprache greift Smalltalk auf Betriebssystemfunktionen bzw. ein externes Gerät zu. Die Verarbeitungsgeschwindigkeit liegt zwischen 12 Hz, wenn keine Hand vorliegt und 4 Hz, wenn sich geographische Informationen "unter" der Hand befinden. Sie kann sowohl auf Seiten der Erkennung der Hand als auch der der akustischen Darstellung der entsprechenden geographischen Informationen erhöht werden.

Als Herausforderung hat sich die generelle visuelle Orientierung der Gestenerkennung, des Betriebssystems und der Entwicklungswerkzeuge erwiesen. So meldet die für den Zugriff auf die Videokarte eingesetzte Bibliothek eine fehlerhafte Installation beim Laden mit einem grafischen Dialog, der nicht abgestellt werden kann. Gestenerkennungssysteme sind häufig so implementiert, daß sie Fehler und generell ihren Zustand nicht schriftlich ausgeben, sondern grafisch in Fenstern anzeigen. In das genannte Gestenerkennungssystem wurden daher zusätzliche Diagnosemöglichkeiten eingefügt. Daher kann der Zustand des Systems nun auch mit gesprochener Sprache ausgegeben werden. Im Gegensatz zu ständigen grafischen Darstellung erfolgt diese Ausgabe jedoch nur, wenn sich der Zustand des Systems ändert.

4 Test des Prototyps

Zur ersten Einschätzung des oben beschriebenen Prototyps wurde ein Test mit einer blinden Versuchsperson im Alter von etwa 40 Jahren durchgeführt, die Erfahrung mit dem Langstock und mit taktilen Karten hat. Mit dem System wurde ihr die Karte eines Teils der Innenstadt von Magdeburg präsentiert, der ihr bekannt ist. Dazu wurde die Kamera im Labor so an einem Ständer befestigt, daß sie Handbewegungen auf einem Tisch aufnehmen konnte. Auf dem Tisch lag eine Unterlage aus einer Plastikfolie der Größe 43 mal 45 cm mit einem erhabenen Netz mit einem Linienabstand von etwa 9 mm und einer Liniengröße von etwa 1 mm. Die Versuchsperson konnte im Gegensatz zum Tasten einer taktilen Karte nur eine Hand zum Tasten benutzen, weil das System derzeit noch keine zwei Hände unterstützt.

Zunächst wurde die Erkundung des durch die digitalen Daten dargestellten Gebietes getestet. Dazu hat die Versuchsperson die Hand auf der taktilen Unterlage bewegt. Bei Berührung einer gedanklich auf der Unterlage vorhandenen Straße wurde deren Name vom System ausgesprochen. Es wurde anschließend die Kodierung der Entfernung zu einer Strecke getestet. Dazu wurde zunächst eine Strecke ausgewählt, die näherungsweise diagonal durch die Karte führte. Die Versuchsperson hat die Kodierung der Entfernung zum Ziel in Streckenrichtung als annehmbar empfunden, die Kodierung der lateralen Entfernung über die Balance hingegen nicht, was u.a. auf die eingesetzten Lautsprecher zurückzuführen ist.

Wie sich herausgestellt hat, arbeitet der Systemteil zur Gestenerkennung nicht optimal. Beim derzeitigen Aufbau mit Schwarzweißkamera führen Schatten um die Hand zu einem störenden Springen der auf die digitale Karte abgebildeten Handposition. Dieses Springen wird in Zukunft durch Einsatz einer Farbkamera verringert werden. Weiterhin gelingt das Auffinden der Zeigefingerspitze nach Berechnung des Handschwerpunktes und der Lage der Fingerkuppen noch nicht zuverlässig. Schließlich hat sich gezeigt, daß stärker zwischen der Unterlage als taktile Referenz und als Hintergrund für die Gestenerkennung unterschieden werden muß, indem auf der dem Benutzenden zugewandten Seite der Unterlage ein Streifen in Hintergrundfarbe angebracht wird. Wenn Benutzende nämlich ihre Hand auf der ganzen Unterlage bewegen, kann auf dem beschriebenen Streifen die Hand ganz oder teilweise aus dem Sichtfeld der Kamera gelangen.

Zur Kodierung der Entfernung zum Ziel in Streckenrichtung erfolgt eine lineare Abbildung auf reine Töne. Die Richtung wurde damit nur grob vermittelt. Daher soll in Zukunft mit einer exponentiellen Abbildung experimentiert werden. Im darauffolgenden Schritt ist geplant, einen Ton konstanter Höhe gepulst auszugeben, ähnlich der Ausgabe eines Geigerzählers, dessen Pulsfrequenz sich umgekehrt proportional zur Zielentfernung verhält.

5 Ausblick

Nachdem nun ein erster Prototyp erstellt wurde, soll dieser dazu eingesetzt werden, mit Angehörigen der Zielgruppe den beschriebenen Ansatz weiterzuentwickeln. Eine Herausforderung stellt die Kalibrierung des Systems durch blinde Benutzende dar. So hat das System keinen Einfluß auf den Zustand der Kamera, z.B. ihre Lage und Blende, oder auf das Umgebungslicht. Für die Gestaltung einer Benutzungsoberfläche für die Kalibrierung soll auf die Erfahrung von blinden Fotografen zurückgegriffen werden. So gibt es beispielsweise Belichtungsmesser mit Sprachausgabe. Das System soll so erweitert werden, daß es sich selbst diagnostizieren und Benutzenden Hilfestellungen geben kann.

Die Interaktion zur Auswahl von Kartenausschnitten und Routen soll über Bildverarbeitung gesteuert werden. Neben der Route können die Ausmaße der Eingabefläche und damit der virtuellen Karte über weitere Marken, z.B. Scheiben festgelegt werden. Das Verschieben dieser Marken (und das der Marken zur Festlegung von Start- und Zielpunkt) kann das System akustisch melden und die Rücknahme ermöglichen, indem es die Hand akustisch zu dem alten Ort zurückführt. Dieser Ansatz erscheint jedoch wenig praktikabel. Untergrund und Marken werden daher so gewählt, daß die Marken mechanisch nur bewußt verschoben werden können, indem z.B. Unterlage und Marken aus Klettschlaufen und -häkchen oder die Unterlage aus Gummi und die Marken aus einem schweren Material bestehen.

6 Literatur

[1] S. Mann: Wearable Computing: A First Step Toward Personal Imaging. In: IEEE Computer 30 (1997), 2. 25-32.

[2] Th. Strothotte et al.: Development of Dialogue Systems for a Mobility Aid for Blind People: Initial design and Usability Testing. In: Proc. ASSETS '96. New York, NY, 1996: ACM. 139-144.

[3] E.D. Mynatt, G. Weber: Nonvisual Presentation of Graphical User Interfaces: Contrasting Two Approaches. In: B. Adelson, S. Dumais, J. Olson (Hg.): Proc. CHI '94. New York, NY, 1994: ACM. 166-172.

[4] M. Rauterberg, P. Steiger: Pattern Recognition as a Key Technology for the Next Generation of User Interfaces. In: Proc. SMC '96. Piscataway: IEEE, 1996. 2805-2810.

[5] M. Brambring, C. Weber: Taktile, verbale und motorische Informationen zur geographischen Orientierung Blinder. Zeitschrift für experimentelle und angewandte Psychologie 28 (1981), 1. 23-37.

[6] P.K. Edman: Tactile Graphics: New York, 1992: American Foundation for the Blind.

[7] E. Holmes, R. Michel, A. Raab: Computerunterstützte Erkundung digitaler Karten durch Sehbehinderte. In: W. Laufenberg, J. Lötzsch (Hg.): Taktile Medien: Kolloquium über tastbare Abbildungen für Blinde. Freital bei Dresden: Deutsche Blindenstudienanstalt, Blinden- und Sehbehinderten-Verband Sachsen, 1995. 81-87.

[8] R. Michel, Th. Strothotte: Visualisierungstechniken zur computerunterstützten Erzeugung taktiler Karten: das System "Map Wizard". it+ti – Informationstechnik und technische Informatik 39 (1997) 2. 13-18.

[9] G. Jansson. "Spatial Orientation and Mobility for the Visually Impaired". In: B. Silverstone, M.A. Lang, B. Rosenthal, E.E. Fraye, Hg. 1999. The Lighthouse Handbook on Visual Impairment and Rehabilitation. New York: The Lighthouse and Oxford University Press. Forthcoming.

[10] S. Millar: Understanding and Representing Space. Theory and Evidence from Studies with Blind and Sighted Children. Oxford, 1994: Clarendon.

[11] R.M. Kitchin, M. Blades, R.G. Golledge: Understanding Spatial Concepts at the Geographic Scale without the Use of Vision. In: Progress in Human Geography 21 (1997), 2. 225-242.

[12] V.I. Pavlovic, R. Sharma, T.S. Huang: Visual Interpretation of Hand Gestures for Human-Computer Interaction: A Review. In: IEEE Trans. on Pattern Analysis and Machine Intelligence 19 (1997), 7. 677-695.

[13] M.W. Krueger: Artificial Reality II. Reading, MA, 1991: Addison-Wesley.

[14] M.W. Krueger, D. Gilden: KnowWhere™: an Audio/Spatial Interface for Blind People. In: Proc. ICAD `97. Xerox PARC, USA, 1997: Xerox.

[15] Th. Starner, J. Weaver, A. Pentland: A Wearable Computer Based American Sign Language Recognizer. In: Proc. Internat. Symposium on Wearable Computers (ISWC). Los Alamitos, CA u.a., 1997: IEEE Computer Soc. 130-137.

[16] R.G. Golledge, R.L. Klatzky, J.M. Loomis: Cognitive Mapping and Wayfinding by Adults Without Vision. In: J. Portugali (Hg.): The Construction of Cognitive Maps. Dordrecht u.a., 1996: Kluwer. 215-246.

[17] W. Schweikhardt: Interaktives Erkunden von Graphiken durch Blinde. Proc. Software-Ergonomie `85. H.-J. Bullinger (Hg.): Stuttgart, 1985: Teubner. 366-375.

[18] S. W. Mereu: Improving Depth Perception in 3D Interfaces with Sound. Technical Report No. CS95-35. Waterloo, Canada, 1995: University of Waterloo.

[19] U. Bröckl-Fox: Untersuchung neuer, gestenbasierter Methoden für die 3D-Interaktion. Diss. Karlsruhe. Aachen, 1995: Shaker.

[20] G. Schaber: Entwurf und Implementierung eines portierbaren videobasierten 3D-Handgestenerkenners und Evaluierung im Vergleich zu existierenden Eingabemedien, Diplomarbeit, Linz, 1997.

Adressen der Autoren

Dipl.-Inform. Jochen Schneider
Otto-von-Guericke-Universität Magdeburg
Fakultät für Informatik
Institut für Simulation und Graphik
Postfach 4120
39016 Magdeburg
E-Mail: josch@isg.cs.uni-magdeburg.de

Prof. Dr. Thomas Strothotte
Otto-von-Guericke-Universität Magdeburg
Fakultät für Informatik
Institut für Simulation und Graphik
Postfach 4120
39016 Magdeburg
E-Mail: tstr@isg.cs.uni-magdeburg.de

Methodengesicherte Validierung von EU-CON II

Alex Totter, Chris Stary, Thomas Riesenecker-Caba*

Communications Engineering, Institut für Wirtschaftsinformatik, Universität Linz

*Forba - Forschungs- und Beratungsstelle Arbeitswelt, Wien

Zusammenfassung

Im Rahmen dieser empirischen Studie wurde das prozeßgeleitete, Software-ergonomische Instrument EU-CON II validiert. EU-CON II unterstützt die Integration der Bewertung und Gestaltung von Benutzungsschnittstellen. Das Instrument ist, dementsprechend, mehrstufig und -teilig aufgebaut. Es besteht aus einem Leitfaden zur Bewertung und einem Handbuch zur Evaluierung und Gestaltung. Bei der Validierung der Güte wurde abgestuft vorgegangen: Im ersten Schritt wurde die Verständlichkeit des Leitfadens, welcher die schriftliche Befragung von Benutzern vorsieht und sich aus einem Merkblatt, Fragebogen und Informationspaket zusammensetzt, überprüft. Durch eine teilnehmende Beobachtung der Benutzer beim Ausfüllen der Fragebögen wurden zunächst Verständnisprobleme und Unsicherheiten protokolliert und systematisch ausgewertet. In einem zweiten Schritt wurde untersucht, in welchem Ausmaß das Informationspaket Benutzer unterstützt, auftretende Unsicherheiten beim selbst ständigen Ausfüllen des Fragebogens zu klären. Aufbauend auf den Ergebnissen dieser empirischen Validierungsstudie erfolgen nun methodische Verbesserungen des Leitfadens. In einer weiterführenden Untersuchung erfolgt die Validierung des Handbuch für Evaluateure und Designer (ebenfalls mehrstufig).

Abstract

EU-CON II (EU-CONform evaluation and engineering of VDU-work) is a technique that supports the evaluation and design of user interfaces with respect to user needs and task requirements. Since it addresses design and evaluation, it consists of two major parts that are utilized along the process (or cycle) of evaluation and (re)design: a guide for evaluation, and a handbook for evaluation and (re)design. In this paper the results from the validation of the guide for evaluation are presented: In the initial step of validation the guide has been checked for understandability by end users, since it provides the inputs for user-centered evaluation and design. In particular, the questionnaire and briefing sheet have been studied in this step of validation. Users were monitored in the course of reading and filling in information. Success factors and problems were recorded and analyzed systematically. In the second step of the study the usability of the information package in terms of its end user support for defining interactive tasks and judging computer support for their accomplishment has been checked. Based on these empirical results the questionnaire is now improved. In a further step, the handbook for evaluation and design can be validated.

1 Einleitung

Die Software-Ergonomie gewinnt in den letzten Jahren vermehrt Stellenwert im Bereich der Gestaltung und Bewertung sozio-technischer Systeme. Dieser Trend spiegelt sich in der Entwicklung von Standards, z.B. bei ISO in Richtung benutzer-orientierte Software-Entwicklung (ISO 13407, 1997), in der Verabschiedung und Implementierung föderativer Richtlinien, z.B. EU-Richtlinie 90/270/ EWG, und nicht zuletzt, in der Entwicklung von Evaluierungsmethoden, z.B. 9241 Evaluator (Oppermann et al., 1997), wider. So benutzerdienlich und qualitätssichernd diese Entwicklungen einzustufen sind, geben diese Standards, Verfahren und Richtlinien kaum methodisch oder empirisch gesicherte Hinweise zur Operationalisierung

Software-ergonomischer Kriterien, weder im Bereich der Gestaltung, noch im Bereich der Bewertung sozio-technischer Systeme. Weiters zeigen viele Evaluierungsverfahren methodologische Mängel (Stary et al., 1998). Mit wenigen Ausnahmen, z.B. IsoMetrics (Willumeit et al., 1996), wurden auch bislang kaum methodisch gesicherte Validierungsstudien dieser Verfahren veröffentlicht. Folglich ist wenig über die test-theoretische Güte (Validität, Reliabilität und Objektivität) bestehender Bewertungs- und Gestaltungsverfahren bekannt.

Darüber hinaus besteht eine methodologische Kluft zwischen der Gestaltung und Bewertung interaktiver Systeme. So geben zwar viele Evaluierungsverfahren Hinweise auf Software-ergonomische Mängel, unterstützen aber kaum die zielgerichtete Behebung dieser Mängel. Zur Überbrückung dieser Kluft sowie zur Implementierung der EU-Richtlinie 90/270/EWG wurde EU-CON (Stary et al., 1997) entwickelt und in weiterer Folge verfeinert: EU-CON II stellt ein prozeßgeleitetes, integratives Instrument dar, welches die Behebung von Mängel bzw. Probleme aus technischer, arbeitsorganisatorischer, sozialer und kognitiver Sicht bei Bildschirmarbeit unterstützen soll (Stary et al., 1999).

In der Folge werden erste Ergebnisse einer empirischer Validierungsstudie zu EU-CON II prä-sentiert, die wesentliche Erkenntnisse zur Benutzbarkeit des Instrumentes bringt. Kapitel 2 führt das Bewertungs- und Gestaltungsinstrument EU-CON II ein. In Kapitel 3 wird die empi-rische Validierungsstudie des Leitfadens von EU-CON II (Problemstellung, Untersuchungsde-sign, verwendete Methoden, Stichprobe, Durchführung) vorgestellt. Die Präsentation der Er-gebnisse erfolgt in Kapitel 4. Den Abschluß bildet Kapitel 5, das, basierend auf den Ergebnis-sen der Validierungsstudie, Verbesserungsvorschläge für den Leitfaden ableitet.

2 Das Bewertungs- & Gestaltungsinstrument EU-CON II

EU-CON II (Akronym für *EU-CONform Evaluation and Engineering of VDU-Work*) ist ein EU-richtlinienkonformes, prozeßgeleitetes, Software-ergonomisches, integratives Instrument zur Gestaltung und Bewertung von Bildschirmarbeitsplätzen. Es richtet sich zum einen an die von Bildschirmarbeit betroffenen Beschäftigten eines Unternehmens, d.s. die Benutzer interaktiver Computersysteme, sowie an die in einem bzw. für ein Unternehmen inhaltlich Verantwortlichen für die Umsetzung der Richtlinie, d.s. Sicherheitsbeauftragte, Betriebstechniker, Betriebsärzte oder ähnlich ausgebildete Berufsgruppen. Die Unterstützung dieser unterschiedlichen Benutzergruppen sowie der Brückenschlag zwischen Bewertung und Gestaltung erfordern mehrere Phasen des Vorgehens sowie unterschiedliche Komponenten zur Unterstützung der Phasen. In der Folge werden diese Komponenten (Abschnitt 2.1) sowie das Vorgehensmodell (Abschnitt 2.2) vorgestellt.

2.1 Aufbau

EU-CON II setzt sich aus zwei Hauptkomponenten, dem Leitfaden einerseits und dem Hand-buch für Evaluateure und Gestalter andererseits zusammen. Die Handhabung dieser Kompo-nenten wird durch ein spezifisches Vorgehensmodell (siehe Abschnitt 2.2), dem dritten Teil von EU-CON II, unterstützt. EU-CON II soll mit dieser Struktur und den damit verbundenen Inhalten den unternehmensgerechten Umgang mit Software-ergonomischen Verfahren umset-zen (Stary et al., 1998). Dieser praxisnahe Umgang dokumentiert sich vor allem durch die zielgerichtete Unterstützung der unterschiedlichen Benutzergruppen, welche in die Bewertung und Gestaltung von interaktiven Systemen involviert sind: Gestalter, Evaluateure, Beschäftigte. So wird in EU-CON II phasengerecht (Vorbereitung, Durchführung, Analyse, Um/Neugestaltung) Information erhoben und weitergeleitet: In den ersten beiden Phasen wird Wissen nur indikativ durch die Beschäftigten erhoben. Dieses Wissen dient der vertieften Analyse in Phase 3 sowie dem zielgerichteten Zugriff auf Handlungsanleitungen in Phase 4 durch den Evaluateur bzw. Gestalter.

Der *Leitfaden* wird in der Vorbereitungs- und Durchführungsphase benutzt, um den Kontext einer Benutzungsschnittstelle eines Bildschirmarbeitsplatzes zu erfassen und die Beschäftigten zur Bewertung zu befähigen. Er setzt sich aus 3 Teilen zusammen:

1. **Merkblatt**: Das Merkblatt gibt anhand einer kurzen Darstellung den befragten Personen einen Überblick über die derzeitige rechtlichen Bestimmungen und den Hintergrund für die auszufüllenden Fragen, da üblicherweise die Ziele und Bestimmungen des Gesundheitsschutzes am Arbeitsplatz den wenigsten Beschäftigten bekannt bzw. zugänglich sind. Darüber hinaus werden die Inhalte der Bewertung erklärt, um Vorbehalte gegenüber den Zielen der Erhebung auszuräumen, und den Beschäftigten die mit der Bewertung verbundene Möglichkeit der Verbesserung der individuellen Arbeitssituation klar darzustellen.

2. **Fragebogen**: Der Fragebogen setzt die Aussagen der Richtlinie zur Bewertung der Eigenschaften von Bildschirmarbeitsplätzen um. Der Fragebogen ist der Arbeitsgegenstand der Bewertung aus der Sicht der Beschäftigten. Die erste wesentliche Aufgabe bei der Bewertung stellt die Identifikation individueller Aufgaben und Arbeitsschritte dar, auf denen die Messung und danach die Bewertung der Aufgabenangemessenheit basiert. Durch die Erfassung der subjektiven Wahrnehmung von Aufgaben und des subjektiven Zugangs zur Aufgabenbewältigung fällt die Diskussion der Gesamtorganisation von Arbeitsschritten, welche eine objektive, zusätzliche Erhebung, etwa im Sinne einer Workflow-Modellierung, erfordert, weg.
   Die zweite wesentliche Aufgabe bei der Bewertung stellt die Beantwortung der aufgabenunabhängigen Fragen dar, welche im Anschluß an die Fragen zur Aufgabenangemessenheit zu bearbeiten sind. Dabei werden vor allem kognitive Faktoren und technische Aspekte der Adaptierbarkeit interaktiver Computersysteme angesprochen.

3. **Informationspaket**: Um eine mögliche Beeinflussung der Befragten bei der Beantwortung der Fragen zu vermeiden, aber trotzdem Anhaltspunkte und Unterstützung zur Beantwortung der Fragen zur Verfügung zu stellen, wurden sämtliche Beispiele und Erläuterungen getrennt vom Fragebogen in einem sogenannten Informationspaket zusammengefaßt. Das Informationspaket enthält folglich Erklärungen der Fragen, Musterantworten und Beispielantworten aus den Bereichen Produktion und Dienstleistung, damit Beschäftigte etwa vor Ausfüllen des Fragebogens die Bearbeitung nachvollziehen können und die Bedeutung der Fragen für sich klarstellen.

Das in der Folge angeführte Beispiel aus dem Leitfaden (aufgeteilt in Fragebogen und Informationspaket) zeigt anhand einer Frage zur Aufgabenangemessenheit, wie die Benutzer zur individuellen und gleichzeitig aktiven Teilnahme am Bewertungs- und Gestaltungsprozeß gewonnen werden sollen:

Der Fragebogen:

1.1 Gibt es bei der Durchführung der Aufgabe <u>Hindernisse, Erschwernisse oder Unsicherheiten</u>, die Sie der von Ihnen verwendeten Software zuweisen?

☐   nein, ich fühle mich in der Durchführung der Aufgabe durch die EDV insgesamt gut unterstützt.

☐   ja, es gibt Hindernisse, Erschwernisse oder Unsicherheiten.

Geben Sie bitte die Situationen oder Arbeitsschritte an, bei denen Hindernisse, Erschwernisse oder Unsicherheiten auftreten. Beschreiben Sie dabei den Arbeitsschritt und das auftretende Problem. Führen Sie auch diejenigen Arbeitsschritte aus, bei denen Sie das Gefühl haben, daß die Probleme auf fehlende Schulung zurückzuführen sind.

Arbeitsschritte der Aufgabe	Hindernisse, Erschwernisse, Unsicherheiten

Bitte machen Sie Vorschläge, wie die Probleme Ihrer Ansicht nach zu lösen wären:

Arbeitsschritte der Aufgabe	Vorschläge zur Verbesserung von Hindernissen, Erschwernissen, Unsicherheiten

☐   weiß nicht

b) Das Informationspaket: Im Informationspaket gibt es für die Benutzer Erläuterungen zu jeder Frage, unterteilt in Erklärungen und Beispiele (siehe Erklärung der Symbole im Anschluß)

Erklärungen und Beispielantworten zur Frage 1.1:

📄 steht für ERKLÄRUNG   ✍ steht für BEISPIEL

Frage 1.1 Gibt es bei der Durchführung der Aufgabe <u>Hindernisse, Erschwernisse oder Unsicherheiten</u>, die Sie der von Ihnen verwendeten Software zuweisen?

📄**Erklärung:** Diese Frage dient der Beschreibung allgemeiner Behinderungen bei der Durchführung einer Aufgabe entlang Ihres Arbeitsablaufs. Die Ursachen der Behinderungen sollten Ihrer Einschätzung nach bei der verwendeten Software liegen. Stellen Sie fest, daß es Hindernisse, Erschwernisse oder Unsicherheiten gibt, beschreiben Sie diese bitte. Typische Behinderungen können dabei sein: Langes Warten auf erforderliche Daten, nicht einsichtige Abfolgen von Bildschirmanzeigen, unklare Dialogfelder, umständliches Bedienen der Maus.

Beispielantworten für die Auswahl

☒   ja

✍   Arbeitsschritt der Aufgabe   Hindernisse, Erschwernisse, Unsicherheiten
     < Briefwechsel mit ausländischen Kunden >

1. Eingabe des Datums	*Ich muß in Briefen an ausländische Kunden das Datum immer in der Form Tag/Monat angeben. Wenn ich zum Beispiel für den 1.Februar das Datum 1/2 eingebe, ändert sich die Anzeige automatisch in ½.*
2. Manchmal Aussendung von Serienbriefe	*Die kann ich nicht erstellen, da ich mich mit dem Hilfeprogramm nicht auskenne, da alles in Englisch ist.*

✍   Arbeitsschritt der Aufgabe   Hindernisse, Erschwernisse, Unsicherheiten
     < Eingabe von Lagerdaten >

Ablesen der Daten von einem Lieferschein und Eingabe in die Maske.	*Die Daten des Lieferscheins muß ich in unterschiedliche Masken eingeben, geht das nicht auch in einer einzigen?*

Die ausgefüllten Fragebögen dienen schließlich gemeinsam mit dem Handbuch für Evaluateure und Gestalter der Auswertung der Antworten und Bestimmung von Verbesserungsmaßnahmen im Mangelfall.

Das *Handbuch für Evaluateure und Gestalter* dient der Auswertung der Erhebungsergebnisse sowie der Um/Neugestaltung von Bildschirmarbeitsplätzen. Im Mangelfall werden Handlungsanleitungen zur Verfügung gestellt, um den Ursachen nachzugehen bzw. die Mängel zu beheben. Es besteht aus 4 Teilen, kann aber im Rahmen dieses Beitrages nicht ausführlich vorgestellt werden. Die Teile (i) vermitteln *Hintergrundinformation* zur Bewertung von Bildschirmarbeitsplätzen gemäß der EU-Richtlinie, (ii) geben einen *Überblick* über die Anwendung des Leitfadens und den Einsatz des Fragebogens, (iii) führen die grundlegenden *Software-ergonomischen Kenngrößen* ein, welche der Entwicklung des Fragebogens zugrunde gelegt wurden, und (iv) enthalten den Inhalt des *Fragebogens*, welcher mit Hintergrundinformation und Handlungsanleitungen zur Mängelbehebung *erweitert* wurde. Somit wird nicht nur die retrospektive, sondern auch die prospektive Gestaltung von Benutzungsschnittstellen unterstützt.

2.2 Vorgehensmodell

Das *EU-CON II Vorgehensmodell* unterstützt nicht nur die Phasen der Bewertung (Vorbereitung, Durchführung, Analyse), sondern auch die daran anschließende mögliche Umgestaltung des Bildschirmarbeitsplatzes:

- *Vorbereitung.* Die Vorbereitungsphase der Bewertung nach EU-CON II sollte die folgenden beiden betrieblichen Aktivitäten umfassen:
  1. *Information und Anweisung (Briefing) der Beschäftigten* durch die Evaluateure (gegebenenfalls nach Durchsicht des Handbuchs zur Bewertung und Gestaltung). Die Inhalte sollten jenen entsprechen, welche die Beschäftigten auch auf dem Merk- und Informationspaket finden können. Wichtig ist dabei, daß die Beschäftigten die Chance erkennen, durch die Erhebung an *ihrem* Arbeitsplatz erforderliche organisatorische, technische, soziale und individuelle Verbesserungen transparent darstellen zu können.
  2. *Ausgabe der Fragebögen an die Beschäftigten* mit dem Merk- sowie dem Informationspaket, das Erklärungen und Beispielantworten zu den einzelnen Fragen enthält.
  3. *Erstellen der Aufgabenliste:* Diese Aktivität ist die wesentlichste in dieser Phase, da das Verständnis des eigenen Aufgabenbereichs sowie die individuelle Wahrnehmung der Aufgaben die Grundlage für die Erarbeitung von (oft individuell erforderlichen) Verbesserungsvorschlägen und die Ableitung von Verbesserungsmaßnahmen darstellt.
- *Durchführung der Erhebung.* Nach erfolgter Vorbereitung der Inhalte und des weiteren Vorgehens wird zunächst die Datenerhebung durchgeführt:
  4. *Beantwortung der Fragen* mit Hilfe des Merk- und Informationspakets im Fragebogen durch alle Beschäftigte. Bei Bedarf sollte der/die Evaluateur/in für Fragen als Auskunftsperson zur Verfügung stehen.
  5. *Abgabe der Fragebögen* an den/die Evaluateur/in zur weiteren Auswertung.
- *Auswertung der Ergebnisse.* In diesem Schritt wird sowohl einzeln als auch kumulativ ausgewertet:
  6. *Einzelauswertung der Fragen* jedes Fragebogens durch den/die Evaluateur/in, gegebenenfalls unter Zuhilfenahme des Handbuchs zur Bewertung und Gestaltung.

7. *Sammelauswertung und Interpretation* jedes Fragebogens durch den/die Evaluateur/in, gegebenenfalls unter Zuhilfenahme des Handbuchs zur Bewertung und Gestaltung.

8. *Im Mangelfall Erarbeitung von Lösungsvorschlägen* durch den/die Evaluateur/in, gemeinsam mit den Betroffenen, gegebenenfalls unter Zuhilfenahme des Handbuchs zur Bewertung und Gestaltung.

- *Um/Neugestaltung des Bildschirmarbeitsplatzes.* In diesem Schritt werden Verbesserungsmaßnahmen vorgeschlagen, bewertet und gegebenenfalls vorgenommen.

9. *Umsetzung von Verbesserungsmaßnahmen* nach Ursachenidentifikation (gegebenenfalls unter Anleitung des Handbuchs) durch einschlägige Experten (Management, Techniker, Ergonomen, etc.).

10. *Überprüfung der gesetzten Maßnahmen durch den/die Evaluateur/in,* gemeinsam mit der Betroffenen, gegebenenfalls unter Zuhilfenahme des Handbuchs zur Bewertung und Gestaltung - im Mangelfall erneutes Durchlaufen der Bewertungsaktivitäten.

Vor allem durch das zuletzt erwähnte Bündel an Maßnahmen unterscheidet sich EU-CON II von traditionellen Bewertungs- oder Gestaltungsverfahren, wie z.B. IsoMetrics (Willumeit et al., 1996). Es unterstützt nämlich nicht nur die Ursachenidentifikation problematischer Situationen der Bildschirmarbeit, sondern auch die Behebung von Mängeln.

3 Die Validierungsstudie

Da EU-CON II einen integrativen Ansatz zur Bewertung und Gestaltung darstellt, ist das Instrument nicht nur mehrstufig und -teilig aufgebaut, sondern ist auch bei der Validierung der test-theoretischen Güte abgestuft vorzugehen. Als eines der obersten Prinzipien bei der Entwicklung von Test- oder Meßinstrumenten wird in der empirischen Sozialforschung mehrfach die Verständlichkeit der Fragen genannt (vgl. Mummendey, 1995; Lienert et al., 1994). Deshalb wird auch im Rahmen dieser Validierungsstudie im ersten Schritt besonderes Augenmerk auf die Prüfung der Verständlichkeit des Leitfadens, welcher den benutzerspezifischen Teil der Bewertung abdeckt, gelegt. Entsprechend dieser zentralen Fragestellung der Untersuchung und der Struktur des Leitfadens wurde mehrstufig vorgegangen.

Bei der *Prüfung der Verständlichkeit des Fragebogens* wird untersucht, ob die einzelnen Fragen (insgesamt 40 Fragen) für jeden Benutzer eindeutig verständlich und beantwortbar sind. Dies bedeutet, daß jeder Benutzer ohne wesentliche Spezialkenntnisse die ihn/sie bei seiner/ihrer Tätigkeit unterstützende Software hinsichtlich konkreter Fragen bewerten soll. Als einzige Unterstützung bei der Beantwortung der Fragen soll das Informationspaket dienen. Zusätzlich dazu soll die *Prüfung der Verständlichkeit des Informationspaketes* Information darüber liefern, ob die gegebenen Erklärungen und Beispiele die Beschäftigten ausreichend beim Ausfüllen des Fragebogens unterstützen.

Aufgrund dieser Problemstellung wurde ein mehrstufiges Untersuchungsdesign entwickelt:

- Selbst ständiges Ausfüllen des Fragebogens durch Benutzer;
- Bei Beantwortungsschwierigkeiten erfolgt die Verbalisierung des Problems durch die Benutzer;
- Protokollieren des Problems durch Untersuchungsleiter;
- Untersuchungsleiter verweist auf das Informationspaket für die jeweilige Frage;
- Benutzer liest die entsprechenden Erklärungen und Beispiele des Informationspaketes;

- Bei Beantwortungsschwierigkeiten Verbalisieren des Problems durch Benutzer;
- Protokollieren des Problems durch Untersuchungsleiter;
- Verbale Unterstützung durch den Untersuchungsleiter;
- Protokollieren der verbalen Unterstützung des Untersuchungsleiters.

Folgende Instrumente zur Datenerhebung wurden verwendet:
- Sozio-demographischer Fragebogen zur Beschreibung der Stichprobe;
- Leitfaden EU-CON II;
- Protokoll zur Erfassung von Verständnisproblemen.

Insgesamt nahmen 20 Benutzer einer österreichischen Firma an dieser Validierungsstudie teil.
Bei der Auswahl der Untersuchungsteilnehmer wurde darauf geachtet, daß es sich um Arbeit-
nehmer (und keine Studierende) handelt, die an ihren Arbeitsplätzen von Individual-Software
unterstützt werden. Weiters wurde darauf geachtet, daß diese Individual-Software typenkon-
stant gehalten wurde (ein betriebsinternes PPS-System). Zwei Untersuchungsleiter erhoben
die Daten durch teilnehmende Beobachtung und standen bei Fragen zur Verfügung, die nicht
durch den Einsatz des Informationspakets geklärt werden konnten.

4 Ergebnisse

Diese Validierungsstudie liefert erste Ergebnisse zur Verwendung des prozeßgeleiteten, Software-ergonomischen Bewertungs- & Gestaltungsinstrumentes EU-CON II. Im Mittelpunkt
dieser Untersuchung stand die Prüfung der Verständlichkeit des Leitfadens. Nach einer kurzen
Beschreibung der Stichprobe anhand sozio-demographischer Daten werden die Ergebnisse der
Prüfung der Verständlichkeit des Leitfadens präsentiert.

4.1 Beschreibung der Stichprobe

Die an dieser Untersuchung teilnehmende Stichprobe umfaßt 20 Personen (11 Frauen, 9 Männer), wobei sich das Durchschnittsalter auf 32, 9 Jahre beläuft (s=8,7, Minimum 19,
Maximum 53 Jahre). Die durchschnittliche Betriebszugehörigkeit beträgt 8,95 Jahre (s=4,59),
die durchschnittliche Arbeitsplatzzugehörigkeit liegt bei 6,88 Jahren (s=4,68). Im
Durchschnitt arbeiten die Untersuchungsteilnehmer 24,7 Stunden/Woche an einem Bildschirmarbeitsplatz (s= 9,17). Trotz der Verwendung von Individual-Software lassen sich
mehrere Tätigkeitsbereiche unterscheiden (siehe Tabelle 1, die Angaben aus der Tabelle
stammen aus den individuellen Beschreibungen der Untersuchungsteilnehmer und wurden
nicht in Meta-Kategorien zusammengefaßt). Die Gruppe der kaufmännisch Angestellten
(40%) stellt den größten Anteil an der Untersuchung dar.

		Frequency	Percent
Valid	Arbeitstechniker	1	5,0
	Leitende Position	2	10,0
	Kaufmännischer Angestellter	8	40,0
	Sachbearbeiter	2	10,0
	Industriekauffrau	4	20,0
	Technischer Angestellter	2	10,0
	Total	19	95,0
Missing	System Missing	1	5,0
Total		20	100,0

Tabelle 1: Tätigkeitsbereiche der Untersuchungsteilnehmer

4.2   Verständlichkeit des Leitfadens

Verständlichkeit des Fragebogens. Der erste Teil des Fragebogens (Frage 1.1-1.19) beschäftigt sich mit der Messung von Aufgabenangemessenheit der Software. Die Fragen 2.1-2.16 stellen die Beantwortung der aufgabenunabhängigen Fragen dar, welche im Anschluß an die Fragen zur Aufgabenangemessenheit zu bearbeiten sind. Dabei werden vor allem kognitive Faktoren und technische Aspekte der Adaptierbarkeit angesprochen. Die Antworten aus dem Abschlußteil (Fragen 3.1-3.4) sind für die Kategorisierung und Einbeziehung der Benutzer und bei allfälligen arbeitsplatzbezogenen Besonderheiten zu berücksichtigen. Jede der insgesamt 40 Fragen wurde von 20 Personen hinsichtlich ihrer Verständlichkeit geprüft. Weiters wurde das Informationspaket (Erklärung der Frage + Beispiel zu jeder Frage) von jeder dieser Personen parallel dazu bewertet. Betrachten wir den ersten Teil des Fragebogens (Fragen zur Aufgabenangemessenheit 1.1-1.19, Tabelle 2), so traten bei insgesamt 9 von 19 Fragen Verständnisschwierigkeiten beim Fragebogen auf. Bei Frage 1.3 (*Bei der Arbeit mit der EDV arbeiten Sie mit Menus oder geben Befehle ein oder klicken Symbole an. Werden diese Möglichkeiten so dargestellt oder beschrieben, daß Ihnen die Bedeutung durch die Anzeige unmittelbar klar ist?*) hatten insgesamt vier von 20 Personen Probleme. Aus der Analyse der verbalisierten Probleme durch die Benutzer geht hervor, daß die Verständnisprobleme durch die Verwendung der Begriffe „Menus" oder „Symbole" hervorgerufen wurden. In allen vier Fällen wurde das Informationspaket als zusätzliche Unterstützung verwendet (in drei Fällen wurde die Erklärung durchgelesen, eine Person nahm auch die Beispiele zur Beantwortung der Frage zu Hilfe). In einem Fall mußte ein Zusatzkommentar durch die Untersuchungsleiter gegeben werden. Bei Frage 1.15 („*Erhalten Sie bei der Erledigung der Aufgabe unmittelbar ein für Sie verständliches (Zwischen- oder) Arbeitsergebnis, an dem Sie leicht überprüfen können, ob Sie die Aufgabe richtig erledigt haben?*") hatten drei Benutzer Verständnisschwierigkeiten. Bei einer Analyse der Protokolle zeigt sich allerdings, daß die Untersuchungsteilnehmer keine Verständnisschwierigkeiten hinsichtlich der Frage hatten, sondern die von ihnen verwendete Software keine fehlerhafte Eingabe signalisiert und die Untersuchungsteilnehmer nicht wußten, wie sie diese Frage beantworten sollten. Bei den verbleibenden sechs Fragen hatten nur ein oder zwei Personen Verständnisschwierigkeiten, die wiederum in vier Fällen durch die Zuhilfenahme des Informationspaketes beseitigt werden konnten. Nur bei der Frage 1.8 („*Müssen Sie Eingaben (Daten, Befehle) zur Erfüllung Ihrer Aufgabe durchführen, die immer gleich sind und die möglicherweise (zumindest zum Teil) von der Software voreingestellt werden oder automa-*

tisch erfolgen könnten?") mußte in beiden Fällen ein Zusatzkommentar durch die Untersuchungsleiter gegeben werden.

Im zweiten Teil des Fragebogens (Tabelle 3, aufgabenunabhängige Fragen, Frage 2.1-2.16) traten bei 12 von 17 Fragen Verständnisschwierigkeiten auf. Allerdings konnten durch Zuhilfenahme des Informationspaketes alle bis auf eine Frage beantwortet werden. Bei Frage 2.1 *„Die Software soll so gestaltet sein, daß ähnliche Arbeitsergebnisse auf möglichst gleichem Weg erreicht werden (z.B. Kopieren, Ausdrucken usw. immer mit der gleichen Abfolge von Anweisungen). Kommt es an Ihrem Bildschirmarbeitsplatz vor, daß Sie unterschiedlich vorgehen müssen, um im Grunde gleiche Vorgänge auszulösen?"* hatten drei von 20 Untersuchungsteilnehmer Verständnisprobleme. Aus der Analyse der Protokolle ergab sich allerdings, daß einige Benutzer nicht bereit waren, die Frage mit „ja" zu beantworten, da bei einer Beantwortung mit „ja", zwei weitere Fragen qualitativ zu beantworten gewesen wären.

Frage	Problem	Info-paket	Kommentar
1.1	2	2	1
1.2	2	1	
1.3	4	4	1
1.4			
1.5			
1.6	2	2	
1.7	1	1	1
1.8	2	2	2
1.9			
1.10			

Frage	Problem	Info-paket	Kommentar
1.11	1	1	
1.12			
1.13			
1.14			
1.15	3	3	
1.16			
1.17			
1.18	2	2	
1.19			
Gesamt	19	18	5

Tabelle 2: Häufigkeit der Verständnisschwierigkeiten der Fragen 1.1-1.19

Bei Frage 2.8 (*„Können Sie jederzeit feststellen, in welchem Zustand (Warten auf Eingabe, Bearbeitung von Daten, Ausgabe eines Ergebnisses) sich die Software·befindet und wie lange dieser Vorgang andauern wird?"*) konnte aus der Analyse der Protokolle keine eigentlichen Verständnisschwierigkeiten gefunden werden. In diesem Fall wurden die Beispiele des Informationspaketes gelesen, um Anregungen zur Beantwortung der Frage zu gewinnen.

Frage	Problem	Info-paket	Kommentar
2.1	3	3	
2.2	2	2	
2.3	2	2	
2.4			
2.5	2	2	
2.6	2	1	1
2.7	2	2	
2.8	3	3	
2.9			

Frage	Problem	Info-paket	Kommentar
2.10	2	2	
2.11			
2.12	2	2	
2.13	2	2	
2.14	2	2	
2.15	1	1	
2.16			
2.17			
Gesamt	25	24	1

Tabelle 3: Häufigkeit der Verständnisschwierigkeiten der Fragen 2.1-2.17

Im letzten Teil des Fragebogens (3.1-3.4, Tabelle 4) zur Benutzerkategorisierung ergaben sich kaum Verständnisschwierigkeiten.

Frage	Problem	Infoblatt	Kommentar
3.1			
3.2			1
3.3		1	
3.4		1	
Gesamt	0	2	1

Tabelle 4: Häufigkeit der Verständnisschwierigkeiten der Fragen 3.1-3.4

Insgesamt traten bei der Beantwortung der 40 Fragen durch 20 Untersuchungsteilnehmer in 44 Fällen Verständnisschwierigkeiten auf. Durch die Zuhilfenahme des Informationspaketes konnten diese Probleme soweit beseitigt werden, daß in nur sieben Fällen, über alle Fragen hinweg, Kommentare von den Untersuchungsleitern zur Unterstützung der Beantwortung der Fragen durch Benutzer gegeben werden mußten.

Verständlichkeit des Informationspakets. Im ersten Teil des Fragebogens, der Fragen zur Aufgabenangemessenheit der Software zum Inhalt hat, traten weniger Verständnisschwierigkeiten auf als im zweiten, aufgabenunabhängigen Teil des Fragebogens. Allerdings konnte im ersten Teil das Informationspaket weniger zur Klärung von Verständnisproblemen beitragen. Insgesamt mußte in fünf Fällen ein Kommentar von seiten der Untersuchungsleiter abgegeben werden. Im zweiten, aufgabenunabhängigen Teil des Fragebogens konnte in 24 von 25 Fällen von Verständnisschwierigkeiten das Informationspaket die Benutzer soweit unterstützen, daß in nur einem Fall der Untersuchungsleiter Kommentare geben mußte.

5 Verbesserungen und Ausblick

Im Rahmen dieser Untersuchung konnten erste Erfahrungen beim Einsatz des prozeßgeleiteten, Software ergonomischen, integrierten Gestaltungs- & Bewertungsinstruments EU-CON II gesammelt werden. Zentrales Anliegen dieser empirischen Untersuchung war die Prüfung der Verständlichkeit des Leitfadens (Merkblatt, Fragebogen, Informationspaket) durch Beschäftigte.

Wie die Ergebnisse bei Frage 1.3 zeigen, müssen Evaluateure bei der Formulierung der Fraugen, sobald das Instrument zur Bewertung durch Endbenutzer eingesetzt werden soll, beschäftigtengerecht mit der Verwendung von Fachbegriffen umgehen. Dieses Ergebnis deckt sich auch mit der Literatur zur Frageformulierung aus der empirischen Sozialforschung (vgl. Schnell et al.,1992). Die Verständnisschwierigkeiten von Frage 2.1 geben Hinweis auf ein Problem der Antworttendenz. Offene Fragen sind aufwendiger zu beantworten als das Ankreuzen von Antwortalternativen. Dabei spielt die Motivation der Beschäftigen eine entscheidende Rolle. Für den Einsatz des Instruments bedeutet das weiters, daß im Rahmen des „Briefing" die Endbenutzer zu einer aktiven Mitarbeit beim Ausfüllen des Fragebogens motiviert werden sollten. Das kann wiederum nur durch ein geschultes Projektteam gewährleistet werden.

Das Informationspaket ist durch seine Gliederung in „Erklärung der Frage" und „Beispiele zur Frage" für Benutzer sehr gut verständlich. Es wird einerseits als Hilfestellung bei Verständnisschwierigkeiten verwendet, andererseits dienen die Beispiele als Anregung zur Beantwortung der offenen Fragen. Insgesamt mußte nach Einsichtnahme in das Informationspaket in nur sieben Fällen ein Kommentar von seiten der Untersuchungsleiter gegeben werden. Aus der Benutzercharakteristik geht allerdings hervor, daß in diesen Fällen die Beschäftigten noch unerfahren im Umgang mit der Software waren, bzw. sich mit der Untersuchung nicht sehr identifizieren konnten. Allgemein läßt sich aus diesen Ergebnissen schließen, daß sich die Teilung der Verfahrens in die Komponenten Merkblatt, Fragebogen und Informationspaket zur Bewertung bewährt. In einem weiteren Schritt ist nicht nur die Reliabilität des Leitfadens, sondern auch die Güte des Handbuchs zu validieren.

6 Literatur

EG-Richtlinie 90/270/EWG: Mensch-Maschine Schnittstelle. In: Richtlinie des Rates vom 29. Mai 1990 über die Sicherheitsvorschriften bezüglich der Sicherheit und des Gesundheitschutzes bei der Arbeit an Bildschirmgeräten (Fünfte Einzelrichtlinie im Sinne von Artikel 16 Absatz 1 der Richtlinie 89/391/EWG). In: Amtsblatt der Europäischen Gemeinschaften , Vol. 33, L 156, Mindestvorschriften (Artikel 4 und 5), Absatz 3, S. 18 , 21.06.1990.

ISO 13407: Human-Centred Design for Interactive Systems, Commitee Draft, 1997

Lienert, G.; Raatz, U.: Testaufbau und Testanalyse, Beltz Psychologie Verlagsunion, Weinheim, 5. Auflage, 1994.

Mummendey, H.D.: Die Fragebogen-Methode, Hogrefe-Verlag, Göttingen, 2. Auflage, 1995.

Oppermann, R.; Reiterer, H.: Software Evaluation using the 9241 Evaluator, in: Behavior and Information Technology, Vol. 16, No. 4/5, pp. 232-245, 1997.

Schnell, R.; Hill, P.B.; Esser, E.: Methoden der empirischen Sozialforschung, Oldenbourg, München, 3. Auflage, 1992.

Stary, Ch., Riesenecker-Caba, Th., Flecker, J.; Kalkhofer, M.: EU-CON - Ein Verfahren zur EU-konformen Software-ergonomischen Bewertung und Gestaltung von Bildschirmarbeit, vdf, Zürich, 1997.

Stary, Ch., Riesenecker-Caba, Th., Flecker, J.: Implementing the Directive for VDU Work – The EU-State of the Art, in: Behavior and Information Technology, Vol. 18, No. 1/2, 1998.

Stary, Ch., Riesenecker-Caba: EU-CON II - Software-ergonomische Bewertung und Gestaltung von Bildschirmarbeit, Deutsche Bundesanstalt für Arbeit und Gesundheit, Dortmund, 1999.

Willumeit, G. Gediga, K. Hamborg: IsoMetrics: Ein Verfahren zur formativen Evaluation von Software nach ISO 9241/10. In: Ergonomie und Informatik, März (1996), 5-12.

Adressen der Autoren

Alex Totter	Chris Stary	Thomas Riesenecker-Caba
Communications Engineering,	Communications Engineering,	FORBA - Forschungs- und
Institut für Wirtschaftsinformatik,	Institut für Wirtschaftsinformatik,	Beratungsstelle Arbeitswelt Wien
Universität Linz	Universität Linz	
Freistädterstraße 315	Freistädterstraße 315	Aspernbrückengasse 4/5
4040 Linz, Austria	4040 Linz, Austria	1020 Wien, Austria
Totter@ce.uni-linz.ac.at	stary@ce.uni-linz.ac.at	riesenecker@forba.at

Anforderungsanalyse zur Einführung eines Unterstützungssystems bei Software-Entwicklern

Hartmut Wandke, Andreas Dubrowsky und Jens Hüttner

Institut für Psychologie, Humboldt-Universität zu Berlin

Zusammenfassung

Durch die Implementierung eines elektronischen Informations- und Beratungssystems (Hyperbase) sollen Software-Entwickler im Hinblick auf Kooperation und Kommunikation innerhalb der Abteilung unterstützt werden. Vordergründig zur Unterstützung in bezug auf software-ergonomisches Wissen, aber auch anderen damit verbundenen Aktivitäten zum Wissensmanagement, wird das Vorgehen zur Analyse des Ist-Zustandes dargestellt. Diese Anforderungsanalyse diente zur Identifikation des vorhandenen Wissens, seiner Repräsentation sowie der individuellen Verteilung des „Organisationswissens" und sollte darüber hinaus klären, auf welche Art und Weise die Entwickler dieses Wissen erwerben (z. B. mit welchen Hilfsmitteln). Für weitere Entscheidungen wurden diese Ergebnisse ergänzt durch eine prospektive Analyse von Wünschen und Erwartungen der potentiellen Nutzer an ein solches Unterstützungssystem (Inhalte, Darbietung, Motivation zur Nutzung). Die mit Hilfe von Fragebögen, Testaufgaben, Interviews und Gruppendiskussionen erbrachten Ergebnisse zeigen, daß CSCW-Systeme ein geeignetes Medium zur Unterstützung von wissensbasierten Kooperationsprozessen sein können, ihre Effizienz allerdings davon abhängt, inwieweit die Erwartungen und Bedürfnisse der Konsumenten befriedigt werden. Aktive Benutzerbeteiligung ist in dieser Hinsicht unabdingbar.

1 Einleitung

Während in den Frühzeiten der Software-Ergonomie intuitive und empirische Vorgehensweisen beim Entwurf von Oberflächen und Dialogen dominierten, stehen den Software-Entwicklern heute zahlreiche Standards, Regelwerke, Gestaltungsrichtlinien und Gestaltungsempfehlungen zur Verfügung. Neben den immer noch dominierenden papierbasierten Materialien (z. B. [1, 2, 3]) gibt es mittlerweile einige Informations- und Beratungssysteme, die als Hypertexte oder Hypermedia-Anwendungen vorliegen (z. B. [4, 5, 6, 7]). Einige verbinden Werkzeuge zur Oberflächen- und Dialoggestaltung oder zur Evaluation mit Richtlinien und Empfehlungen (z. B. [8, 9, 10]). Alle uns bekannten Systeme dieser Art sind jedoch ausschließlich für die Benutzung durch einzelne Entwickler konzipiert.

Software-Entwicklung ist dagegen meist ein hochgradig arbeitsteiliger Prozeß, der ein erhebliches Ausmaß an Kooperation zwischen allen daran Beteiligten erfordert. Für den Fall der Kooperation zwischen Entwicklern und Benutzern - oft als Benutzerpartizipation bezeichnet - ist dies auch auf den Veranstaltungen dieser Konferenzserie thematisiert worden. Dagegen spielt die Kooperation bei der Bearbeitung software-ergonomischer Fragen so gut wie keine Rolle. Dies zeigt sich auch im Fehlen von kooperativ zu benutzenden Unterstützungssystemen für die software-ergonomische Gestaltung.

Das verwundert um so mehr, da Konzepte wie „Lernende Organisation" oder „experience factory" (z. B.[11]) in der Diskussion um die Wettbewerbsfähigkeit eines Unternehmens eine zunehmende Rolle spielen. Ein strategischer Faktor ist in diesem Zusammenhang das im

Unternehmen vorhandene Wissen, dessen Nutzung und Management im Fokus der fachübergreifenden Diskussion steht. Da dieses Wissen in der Regel verteilt ist, kommt der Kooperation und Kommunikation ein hoher Stellenwert zu. Für den Fall der Software-Entwicklung konnte dies auch empirisch sehr gut nachgewiesen werden: Im IPAS-Projekt [12, 13, 14] wurden durch Mitarbeiterbefragungen in 29 verschiedenen Software-Entwicklungsprojekten 1679 Tätigkeiten analysiert. Davon waren mehr als 30 % direkt oder indirekt kommunikativen Tätigkeiten zuzuordnen (21% Beratungen, Gespräche; 12% Organisation). Weitere 12% waren dem selbständigen Wissenserwerb gewidmet und solchen kommunikativen Tätigkeiten, wie der Konzipierung und Durchführung von Mitarbeiter-schulungen sowie Reisetätigkeiten und Teilnahme an Weiterbildungen. Nicht nur die Häufigkeit kooperativer und kommunikativer Tätigkeiten ist beeindruckend, sondern auch ihre Relevanz: Besonders erfolgreiche Software-Entwickler unterscheiden sich von ihren Kollegen neben anderem durch Kommunikations- und Kooperationskompetenz [14, 15]. Trennt man Software-Projekte danach, ob zwischen den Projektmitarbeitern über- oder unterdurchschnittlich viel kommuniziert wird (unter Fortlassung der Projekte mit mittlerem Kommunikationsaufwand), dann ergibt sich, daß in Projekten, in denen mehr kommuniziert wird, die Termin- und Kostenpläne besser eingehalten werden, ein größerer Projekterfolg erreicht wird, die entstandene Software später leichter änderbar ist und insgesamt eine bessere Teameffizienz vorliegt.

Wenn die Kommunikations- und Kooperationsprozesse sich nicht nur spontan entfalten, sondern gezielt durch organisatorische und/oder technische Maßnahmen gefördert werden, sprechen wir von Wissensmanagement.

Die Modellvorstellungen zum Wissensmanagement sind zahlreich; einen interessanten Ansatz bildet dabei das modulare Konzept von Probst et al. [16], da es insbesondere für angewandte Fragestellungen einen brauchbaren Zugang liefert. Die (miteinander vernetzten) Kern-prozesse dieses Konzeptes stellt die folgende Tabelle dar:

Baustein	spezifische Fragestellungen
Wissensidentifikation	Wie schaffe ich Transparenz über das im Unternehmen vorhandene Wissen? Wer weiß was?
Wissenserwerb	Welche Wissensquellen lassen sich außerhalb des Unternehmens erschließen? Welche sind relevant?
Wissensentwicklung	Wie bilde ich (auf individueller und kollektiver Ebene) neue Fertigkeiten / Fähigkeiten aus?
Wissensverteilung/ Wissensaustausch	Wie bringe ich das notwendige Wissen an den richtigen Ort?
Wissensnutzung	Wie stelle ich die Nutzung des zur Verfügung stehenden Wissens sicher? Wie beseitige ich bestehende Barrieren?
Wissensbewahrung	Wie bewahrt / pflegt das Unternehmen sein Wissen und schützt sich vor Wissensverlust?

Tab. 2: Kernprozesse des Wissenmanagements nach Probst, Raub & Romhardt [16]

In bezug auf **Wissensverteilung / Wissensaustausch** bieten sich neue Medien geradezu an. Durch die Verwendung geeigneter Informations- und Kommunikationssysteme der im Rahmen der CSCW-Forschung entwickelten Groupware (zur Klassifikation siehe [17]) läßt sich breit verteiltes Erfahrungswissen der Organisationsmitglieder bündeln und bereitstellen, wovon insbesondere Teamarbeitsprozesse profitieren [18, 19].

Die Grundidee des hier vorzustellenden Ansatzes besteht darin, Kommunikation und Kooperation zwischen Entwicklern zu fördern, indem ihnen ein kooperativ zu benutzendes hypermediales Informations- und Beratungssystem (im folgenden kurz Hyperbase genannt) im Rahmen eines Intranets zur Verfügung gestellt wird. Diese Hyperbase soll ihnen zunächst eine Menge von Basis-Informationen zur Software-Ergonomie anbieten, so wie es auch die oben genannten, für die individuelle Nutzung konzipierten Informations- und Beratungssysteme tun.

Zugleich bietet die Hyperbase aber auch die Möglichkeit, die Erfahrungen, die im Prozeß der Gestaltung von Benutzungsschnittstellen von einzelnen Entwicklern oder kleinen Teams von 2-3 Personen gemacht wurden, aufzubereiten, abzuspeichern und sie so allen Mitgliedern einer größeren Organisationseinheit zur Verfügung zu stellen. Dabei können und sollen Erfahrungen, die sich auf Prozesse beziehen, die nicht zur Gestaltung von Benutzungs-schnittstellen gehören, ausdrücklich miteinbezogen werden.

Mit der Einführung einer solcher Hyperbase ist die Erwartung verbunden, daß sich der Anteil von Kommunikation- und Kooperationsprozessen bei der software-ergonomischen Ge-staltung von Systemen erhöht und daß Entwicklerteams letztlich mit weniger Aufwand bessere Benutzungsschnittstellen produzieren. Natürlich ist auch bei der Entwicklung und Einführung der Hyperbase in allen Phasen eine partizipative Beteiligung aller Mitarbeiter / Gruppen sicherzustellen [20, 21].

2 Fragestellung

Wie bei allen Vorhaben dieser Art empfiehlt es sich, vor der Entwicklung eines solchen Systems durch eine Anforderungsanalyse den Unterstützungsbedarf und die Erwartungen der zukünftigen Benutzer an solch eine Hyperbase zu erfassen. Im einzelnen fragten wir:

4. Benötigen die von uns untersuchten Software-Entwickler überhaupt Unterstützung auf dem Gebiet der Software-Ergonomie? Wie ausgeprägt ist ihr Wissen auf diesem Gebiet? Wie gut können sie es anwenden?

5. Wie ist das Wissen auf verschiedene Personen verteilt? Lohnt sich ein Austausch von Wissen?

6. Spielen Kenntnisse zur Software-Ergonomie aus der Sicht der Entwickler überhaupt eine wichtige Rolle in ihrer Arbeit?

7. Welche Rolle wird dem Erfahrungswissen auf dem Gebiet der Software-Ergonomie zugeschrieben?

8. Wird Erfahrungswissen jetzt schon dokumentiert? Welche Techniken werden dabei ein-gesetzt?

9. Auf welchen Wegen wird jetzt Wissen erworben?

10. Welche Unterstützungsformen wünschen sich die Software-Entwickler?

11. Welche Inhalte und Funktionen sollte eine Hyperbase zum Thema Software-Ergonomie und Wissensaustausch zur Verfügung stellen?

Software-Entwickler befinden sich generell in einer Doppelrolle. Einerseits sind sie Entwickler von Software, die von anderen benutzt wird, andererseits sind sie selbst Benutzer

z. B. von Betriebssystemen, Entwicklungswerkzeugen und Standardsoftware. In unserer Untersuchung ging es allerdings nur um ihre Rolle als zukünftige Benutzer der konzipierten Hyperbase zur Software-Ergonomie.

3 Methodik und Durchführung

3.1 Stichprobe

Alle befragten Personen sind diplomierte Informatiker und arbeiten als Entwickler/Programmierer in einem Unternehmen, das komplexe elektronische Anlagen mit einem hohen Anteil von Bedien-Software erstellt. Die Arbeitsaufgabe erfordert im wesentlichen die Spezifikation und Entwicklung datentechnischer Systemkonzepte unter Berücksichtigung ihrer konkreten Nutzungsbedingungen, die Integration der Anwendungen in vorhandene Umgebungen sowie die Pflege der Produkte. Die Abteilung ist dabei in mehrere Teams aufgeteilt, die jeweils eigenständige Projekte bearbeiten.

In die Untersuchung einbezogen wurden insgesamt 24 Mitarbeiter, davon waren 6 Teamleiter, 18 Entwickler.

Testaufgabe, Fragebogen & Interview	18 Entwickler	6 Teamleiter
Durchschnittsalter	34	41
Beschäftigungsdauer im Rahmen der Software-Entwicklung allgemein	11	17
Beschäftigungsdauer im Rahmen der Entwicklung unternehmensspezifischer Software	7	12

Tab 2: Angaben zu befragten Personen (Jahre; gerundete Mittelwerte)

Die Teamleiter und Entwickler unterscheiden sich hinsichtlich aller drei Merkmale signifikant (a=0,05) voneinander; sie sind älter und beruflich länger mit der Entwicklung von Software befaßt.

3.2 Methoden und Ablauf der Untersuchung

Die Untersuchung vollzog sich in mehreren Schritten:

Zu Beginn gab es eine ca. zweistündige Zusammenkunft aller Beteiligten (Software-Entwickler, Management, Mitglieder des Untersuchungsteams). In dieser Zusammenkunft wurde das Anliegen der Untersuchung vorgestellt und mit den Entwicklern diskutiert. Ferner wurde der weitere Ablauf erläutert und die verschiedenen Methoden erklärt. Dann wurden Fragebögen ausgeteilt, die von den Entwicklern individuell am Arbeitsplatz oder zu Hause ausgefüllt werden sollten. Die Fragebögen enthielten Fragen zu biographischen Daten, zur Bedeutung der Software-Ergonomie in den Projekten der Abteilung, zu den Kenntnissen auf dem Gebiet der Software-Ergonomie und zur Kenntnis und Relevanz von Informations-quellen. Bei der Auswahl und Formulierung der Fragen orientierten wir uns Jen früheren Studien [22, 23, 24, 25, 26] und berücksichtigten neuere Entwicklungen am Lehrstuhl. Zum Ende der Zusammenkunft bearbeiteten die Entwickler eine heuristische Evaluationsaufgabe, die von Nielsen [27] unter der Bezeichnung *Travel Weather* entwickelt und als Übungs-aufgabe publiziert wurde. Wir wählten diese Aufgabe, weil sie vom Inhalt her recht ähnlich zu den realen Projekten der Untersuchungsteilnehmer war. Die Aufgabe war überschaubar und in gut 45 Minuten zu bearbeiten. Nielsen bietet Lösungen als Ergebnis einer Evaluation durch vier Usability-Experten zum Vergleich an.

Am Nachmittag nach der Startsitzung begannen die Einzelinterviews mit den Entwicklern. Die Teilnehmer brachten zu den Interviews die ausgefüllten Fragebögen mit. Die Interviews bezogen sich zum Teil auf die Antworten im Fragebogen (insbesondere wurden Begründungen zu einzelnen Antworten erfragt), zum Teil waren es auch freie Fragen, die den Entwicklern Gelegenheit gaben, über ihre Arbeitsverfahren, Erfahrungen, Wünsche und Erwartungen zu berichten. Die Interviews wurden auf Tonband aufgezeichnet.

Nach den Interviews trat eine längere Auswertungsphase ein, während der die Lösungs-vorschläge für die Testaufgabe und die Antworten in den Fragebögen ausgewertet wurden. Außerdem wurden die Interview-Daten transkribiert und inhaltsanalytisch ausgewertet.

Alle Ergebnisse wurden ca. zwei Monate später in einem gemeinsamen Workshop mit den Entwicklern vorgestellt. Sie dienten dazu, den Entwicklern eine Rückmeldung über das software-ergonomische Wissen in der Abteilung, über die Erfahrungen und Bedarfsanmeld-ungen der Untersuchungsteilnehmer zu geben und sie zur Mitarbeit zu motivieren. Zugleich legten die Ergebnisse die Grundlage für die dreistündige Diskussion in zwei Teilgruppen, die sich zum einen mit den Inhalten und Funktionen einer zukünftigen Hyperbase und zum anderen mit Organisationsfragen rund um die Entwicklung, Nutzung und Pflege der Hyperbase beschäftigten.

4 Ergebnisse

4.1 Aktueller Wissensstand zur Software-Ergonomie

Von den 31 Usability-Problemen, die im Original von Nielsen [27] beschrieben wurden, sind von den Probanden insgesamt 24 erkannt worden. Dabei wurden im Mittel von jedem Teilnehmer sieben Probleme benannt. Die Spanne reicht von vier bis zu zehn Problemen. Die Probanden haben zusätzlich insgesamt 26 neue Probleme aufgelistet, die die bei Nielsen (1993) einbezogenen Experten nicht angesprochen hatten. Im Mittel hat jeder Teilnehmer 3 neue Gestaltungsmängel oder Verbesserungsmöglichkeiten genannt (Spanne von eins bis vier). Die nebenstehende Abbildung faßt die Ergebnisse zusammen.

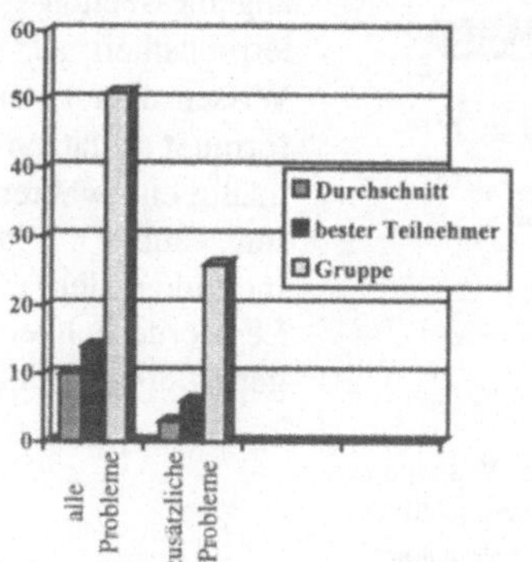

Abb. 1: Die linke Säulengruppe zeigt die Häufigkeiten für alle 50 software-ergonom-ischen Mängel bzw. Verbesserungsvorschläge, die von den Teilnehmern insgesamt benannt wurden. Die rechte Säulengruppe stellt diese Häufigkeiten für die Gestaltungsaspekte dar, die in der Beschreibung von Nielsen nicht erwähnt wurden. Der Vorteil der Nominalgruppe wird in beiden Kategorien sehr deutlich.

Zur Zeit existieren keine Vergleichsdaten, so daß keine definitive Aussage über das Ausmaß des individuell verfügbaren und anwendbaren Wissens getroffen werden, obwohl der Mittelwert von 22% individuell erkannter Probleme durchaus akzeptabel erscheint. Wichtiger aber als der Durchschnittswert ist die Relation zwischen individueller Leistung und Leistung der Gruppe.

Unterschiede zwischen Teamleitern und Entwicklern sind hinsichtlich genannter Mängel nicht ausweisbar. Dieser Befund wird durch den Fragebogen (Selbsteinschätzung zum software-ergonomischen Wissen) bestätigt, da Teamleiter und Entwickler hier gleichermaßen über ihren eigenen Wissensstand urteilen.

4.2 Subjektive Selbsteinschätzung des Wissensstandes

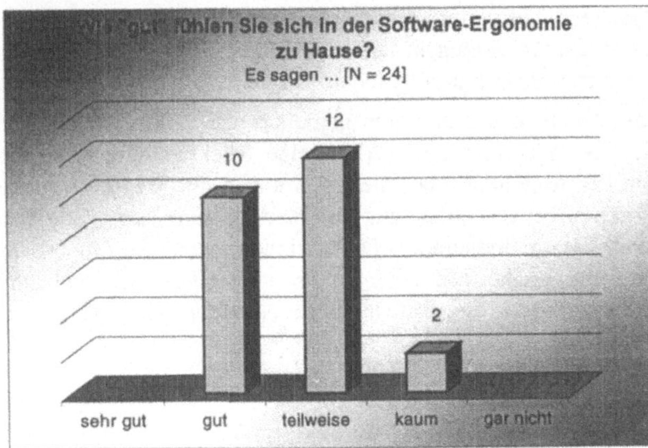

Abb. 2 : Selbstbild in bezug auf software-ergonomische Kenntnisse

Wie spiegelt sich die Situation aus der Sicht der Entwickler wider? In welchem Maße fühlen sich die befragten Entwickler auf dem Gebiet der Software-Ergonomie zu Hause?

Selbstbild und Leistung in den Testaufgaben korrelieren allerdings nicht signifikant, was auch an der geringen Streuung der Fragebogendaten liegt.

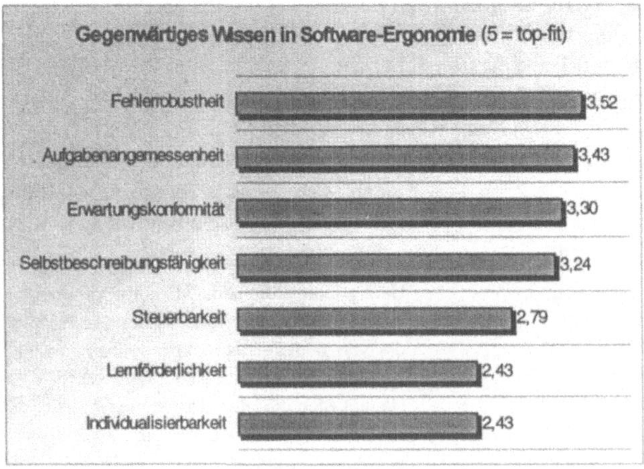

Fragt man bei der Selbsteinschätzung nach dem Wissen um die Kriterien der ISO 9341/Teil 10 so zeigt sich, daß die Entwickler meinen, relativ viel über Aufgabenangemessenheit und Fehlerrobustheit zu wissen.. Ihr Wissen über Erwartungskonformität schätzen sie mittelmäßig ein, während das Wissen über Steuerbarkeit, Individualisierbarkeit und Lernförderlichkeit eher als gering eingeschätzt wird.

Abb. 3: Selbsteinschätzung der Entwickler hinsichtlich ihres Wissens zu den verschiedenen Kriterien der Software-Ergonomie (Rating 1 - 5). Die Unterschiede zwischen den einzelnen Kriterien sind nicht signifikant.

Das schlägt sich sowohl in den Ergebnissen der Testaufgabe als auch in den Antworten im Rahmen der Interviews nieder.

In bezug auf die Selbsteinschätzung zum software-ergonomischen Wissen sind keine Unterschiede zwischen Teamleitern und Entwicklern feststellbar. Dieser Befund wird durch die Ergebnisse der Testaufgabe bestätigt, da Teamleiter und Entwicklern hier gleichermaßen gut abschneiden.

4.3   Wie ist das software-ergonomische Wissen verteilt?

Das ist die zentrale Frage, wenn es um den Austausch von Wissen geht. Ein Austausch macht z. B. wenig Sinn, wenn alle Mitarbeiter in etwa über das gleiche Wissen verfügen. Das dies nicht so ist, läßt sich am Beispiel der Lösungen der Testaufgabe zeigen.

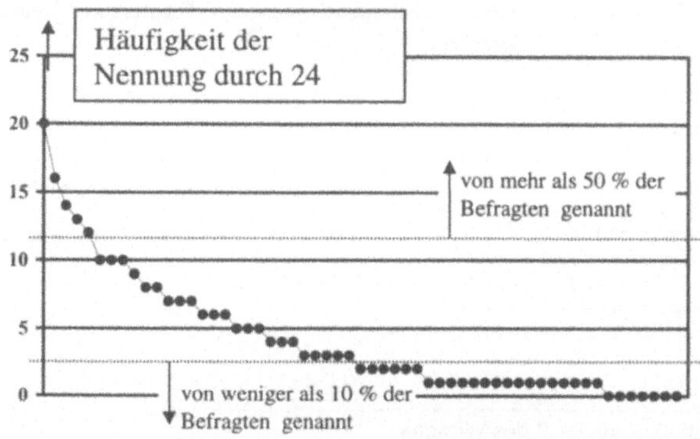

Abb. 4: Absolute Häufigkeit, mit der die einzelnen Probleme (insgesamt 57) genannt wurden. Jeder Punkt steht für die Häufigkeit, mit der die einzelnen Probleme jeweils individuell entdeckt wurden. Das Bild zeigt, daß nur fünf Probleme übereinstimmend von mehr als 50% der Entwickler genannt wurden, während 29 von weniger als 10% entdeckt wurden.

Wenn es um Wissensaustausch geht, spielt nicht nur die Verteiltheit des Wissens eine Rolle, sondern auch die Symmetrie oder Asymmetrie der Verteilung. Einige Entwickler wissen weniger als andere. Sie könnten sicher von einem Wissensaustausch profitieren. Lohnt es sich jedoch für diejenigen, die jetzt schon über mehr Wissen verfügen, andere daran partizipieren zu lassen? Um diese Frage zu beantworten haben wir gegenübergestellt, welche Probleme bei der Testaufgabe von den leistungsstarken Testpersonen (dadurch definiert, daß sie mehr als elf Probleme benannt haben) und den leistungsschwachen Testpersonen (dadurch definiert, daß sie weniger als neun Probleme benannt haben) benannt worden. Folgendes Bild ergibt sich:

1 •••	2 •••	3 ••••	4 ••••	5	6	7 ••	8
9 ••••	10 ••••	11 ••••	12 ••	13 •••	14 ••	15 ••	16 ••
17 ••	18 ••	19 ••	20	21 •••••	22	23	24 •
25 ••	26 ••	27	28 ••	29 •••••	30 ••	31 ••	32 ••
33 ••	34 ••••	35 •••	36 ••	37 ••••	38 •••	39 ••	40 ••••
41 ••••	42 ••••	43 ••••	44 •••	45 ••••	46 ••••	47 ••••	48 •
49 •	50 •	51 •	52 •	53 •	54 ••••	55 ••••	56 ••••
57 ••••							

Abb. 5: Die Nummern kennzeichnen die software-ergonomischen Probleme. Nr. 1 bis 31 entsprechen den Angaben von Nielsen (1993). Die Nummern 32 bis 57 kennzeichnen zusätzlich gefundene Probleme.

Je mehr Punkte auf ein Problem entfallen, desto weniger verteilt sich das Wissen auf verschiedene Personen. Felder ohne Punkte kennzeichnen dabei die Probleme, die überhaupt nicht angesprochen wurden (siehe auch Tab. 3).

Die Probleme 21 und 29 werden von jeweils nur einer leistungsschwachen Testperson angesprochen.

nur von einer leistungsschwachen Testperson entdeckt	•••••
nur von einer leistungsstarken Testperson entdeckt	••••
Von mehreren leistungsstarken Testpersonen entdeckt	•••
Sowohl von leistungsstarken als auch von leistungsschwachen Testpersonen entdeckt	••
Weder von leistungsstarken noch von leistungsschwachen Testpersonen entdeckt	•
Von keiner der 24 Testpersonen entdeckt	

Tab. 3: Verteiltheit des Wissens

4.4 Wo kommen Wissen und Erfahrung her?

In den Interviews wurde sehr deutlich, daß viele Entwickler auf individuell archivierte Problem- oder Fallsammlungen ihrer Karriere zurückgreifen. Von den im Interview befragten 18 Entwicklern dokumentieren insgesamt 15 auftretende Probleme, etwaige Fehler oder Lösungen einzelner Projektabschnitte individuell für sich. Etwa 1/3 der Entwickler dokumentiert neben allgemeinen Problemen auch solche, die direkt im Zusammenhang mit software-ergonomischen Fragen stehen.

Die Art und Weise der individuellen Dokumentation von Erfahrung ist dabei sehr verschieden: die Spanne reicht von einfachen papierbasierten Sammlungen wie Notizen (sog. „schlaue Bücher") und Papierskizzen (mock-ups) bis hin zu elektronisch archivierten Beispieloberflächen, „digitalen" Tagebüchern (die Vorgehensweisen, Probleme und Fehler für einzelne Schritte enthalten) und eingescannter Literatur.

Darüber hinaus interessierte uns, wovon die individuelle Erfahrung der Mitarbeiter in bezug auf software-ergonomisches Wissen besonders geprägt ist und aus welchen anderen Quellen sie ihr Wissen schöpfen.

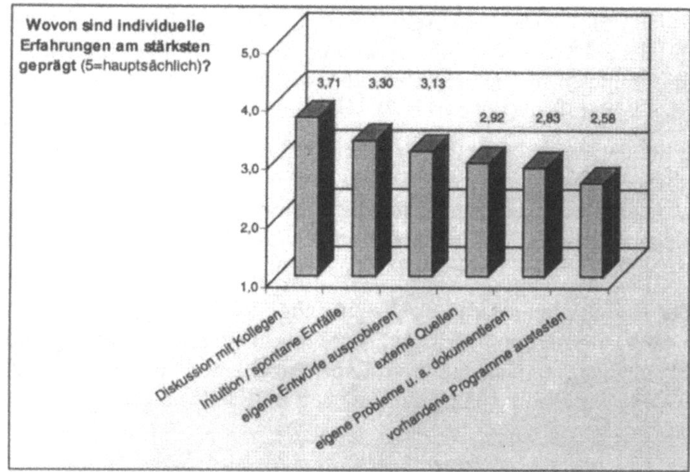

Abb. 6 : Wovon individuelle Erfahrung geprägt ist

Im Fragebogen geben die Entwickler an, daß „individuelle Erfahrung" neben externen Quellen am stärksten durch die Diskussion mit Kollegen geprägt sei. Diese Wertung schlägt sich sowohl in den Fragebogendaten (siehe Abb. 6) als auch in den Interviews und den Diskussionen im Rahmen der Workshops nieder.

Deutlich wurde sehr schnell, daß elektronische Unterstützungssysteme den direkten Kontakt der Mitarbeiter nicht ersetzen können, sondern immer eine Ergänzung sein sollen. Entwickler wünschen sich bspw. in einer Hyperbase Informationen darüber, wer von den Kollegen was macht, woran einzelne Teams arbeiten, wofür es Lösungen gibt; sie möchten vorab Ansprechpartner und Anregungen finden, dann zielorientiert mit diesen Kollegen diskutieren und so die Rückkopplung zu eigenen Erfahrungen herstellen.

Ebenfalls im Fragebogen erfaßt wurde, welche externen Quellen wie häufig von den Entwicklern benutzt werden, um das Wissen über die Benutzerfreundlichkeit von Software zu erweitern (siehe Abb. 7). Dabei dominieren Software-Werkzeuge (z.B. Starview), Fachliteratur, sog. Vorbildlösungen der Abteilung, dokumentierte Probleme / Fehler sowie der Austausch mit Ergonomie-Spezialisten (bspw. im Rahmen von Seminaren). Standards bzw. Guidelines verlieren demgegenüber an Bedeutung. Gleiches gilt für selbst archivierte Sammlungen von Positiv- bzw. Negativbeispielen, die jedoch im bisher vorliegenden Umfang auch nicht sehr zahlreich sind (lediglich zwei Kollegen archivieren Positiv-/Negativbeispiele).

Netzbasierte elektronische Informationssysteme (WWW, Newsgroup) werden kaum benutzt; dies liegt daran, daß an einigen Arbeitsplätzen kein Internetanschluß verfügbar oder die Netzkapazität unzureichend ist. Klassische Einzelplatzsysteme zur Unterstützung (Beratung und Information, Expertensysteme, User Interface Management Systeme) stehen nur eingeschränkt zur Verfügung und werden so gut wie nicht benutzt. Die Möglichkeit zur Teilnahme an Kongressen besteht kaum, diese Form der Wissensgewinnung schneidet am schlechtesten ab.

4.5 Anforderungen der Entwickler an ein Unterstützungssystem

Für die Erfassung der Anforderungen und Wünsche, die Entwickler an ein Unterstützungssystem haben, wurden im wesentlichen drei Quellen herangezogen: Fragebogen, Interview und Workshop-Diskussion.

Im Fragebogen wurden die Entwickler einerseits befragt, wie häufig sie welche Informationsquellen nutzen (siehe 4.4). Darüber hinaus wollten wir wissen, wie hilfreich sie die vorgegebenen Informationsquellen für sich einschätzen. Zwischen beiden Fragen zeigen sich klare Unterschiede in den Rangfolgen (siehe Abb. 7).

Es zeigt sich deutlich, daß die derzeitige Rezeption externer Quellen nicht den tatsächlichen Anforderungen der Entwickler entspricht. Sie schreiben den einzelnen Quellen generell mehr Potential zu, als ihnen die gegenwärtige Nutzung ermöglicht. Sie wünschen sich durchweg eine Intensivierung der Informations- und Weiterbildungsangebote; dies sowohl zu Themen der Software-Ergonomie als auch zu anderen Themen (siehe unten).

In der Rangreihe (hinter der sich die favorisierte Quellennutzung niederschlägt) dominieren der direkte Austausch mit Spezialisten, der Zugriff auf Vorbildlösungen der Abteilung und dokumentierte Fehler / Probleme, die Nutzung von Expertensystemen, Beratungs- und Informationssystemen sowie die Verwendung von Standards und Beispiellösungen.

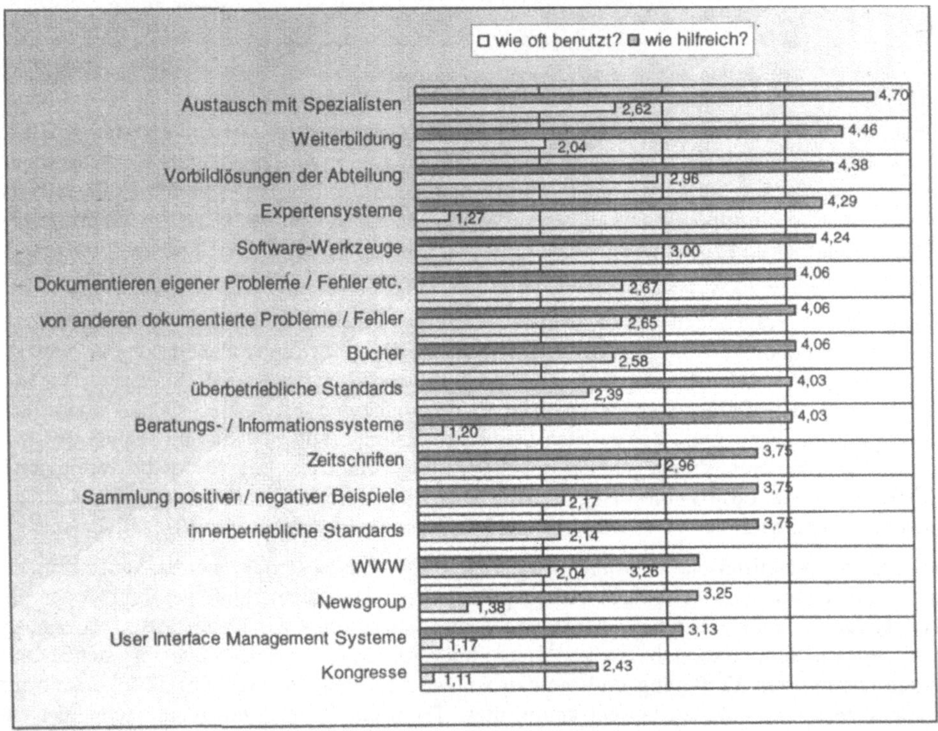

Abb. 7: momentane Häufigkeit der Nutzung bestimmter Quellen und deren Bewertung (5= sehr oft/sehr hilfreich)

Bei den Interviews mit den Entwicklern konnten die individuellen Ideen und Wünsche der Teilnehmer zu den Inhalten und der Realisierung der Hyperbase erfragt werden. Was sollte in einer Unterstützungsumgebung enthalten sein? Wie stellen sich die Entwickler die Gestaltung des Systems vor? Die Antworten lassen sich in drei große Bereiche gliedern, die alle von dem Gedanken der intensiven Kommunikation zusammengehalten werden.

1. Die meisten Ideen beschäftigen sich mit dem Thema Software-Ergonomie.
2. Es gibt Wünsche nach Informationen auf Projektebene - vor allem soll der Austausch zwischen den Projekten durch ein netzbasiertes Informationssystem gefördert werden.
3. Den dritten Bereich bilden die "sonstigen" Anforderungen an eine solche Hyperbase.

Die Schwerpunkte innerhalb dieser drei Bereiche lassen sich wie folgt komprimieren:

Unter der Rubrik **Softwareergonomie** werden Tips/Tricks und Erfahrungen zum Software-Entwurf gefordert. Dies sollte in einer sehr praktischen Sprache und ganz konkret erfolgen. In einer komprimierten Form soll das an guten und schlechten Beispielen orientierte Lehrbuchwissen (Farben, Schriften etc.) aufbereitet vorliegen und durch das Wissen der Kollegen laufend ergänzt werden (Verknüpfungen der Inhaltsteile untereinander). Hierzu gehört der Wunsch nach Verweisen auf entsprechende Verordnungen und Normen (innerbetriebliche und externe). Vorhandene (teilweise unbekannte) Verwendungsvorschriften des Hauses sollen aufbereitet in einen Hypertext o. ä. eingebunden sein. Eine Verbindung der alltäglichen Aufgaben mit den Maßgaben der Software-Ergonomie soll eine Sammlung aller in der Abteilung entwickelten Oberflächen herstellen, die durch 2-3 Personen nach SE-Kriterien beurteilt worden sind.

Um die Konstruktion der Hyperbase aus inhaltlicher und zeitlicher Perspektive zu optimieren, mußten die aus Sicht der Entwickler zunächst wichtigsten Inhalte zu Themen der Software-Ergonomie identifiziert werden. Dazu wurde im Rahmen des Workshops die jeweils individuelle Gewichtung potentieller Inhalte und Funktionen einer Hyperbase ermittelt; 7 Teilnehmer konnten in der von uns vorgegebenen Matrix individuell pro Zelle einen Punktwert von 0 bis 4 vergeben. Die folgende Abbildung stellt die über alle Teilnehmer gebildete Matrix dar. Graue Zellen kennzeichnen die zunächst wichtigsten Inhalte und Funktionen:

Inhalte / Funktionen	Generische Schnittstellenobjekte	Generische Funktionen und Interaktionstechniken	Aufgabenbezogene Gestaltung der Funktionalität	Aufgabenbezogene Gestaltung der Semantik (Objekte und Funktionen)	Strukturierung der Oberfläche (Gruppierung etc.)	Kodierung der Interaktionselemente (Form, Farbe, Text, Icon)	Geräte & Medien für die Interaktion (bspw. Spezialgeräte, VR, Spracheingabe etc.)	Punktzahl
Planen			12	12	4		4	32
Beraten	20	20	22	28	23	15	12	140
Vorschlagen	12	13	16	18	17	10	13	99
Entscheiden	4	4	5	5	8	5	8	39
Informieren	17	14	25	25	22	20	13	136
Ausführen	5	3		5		3	1	17
Überprüfen	1	5					1	7
Σ	59	59	80	93	74	53	52	470

Tab. 4: Häufigkeitsverteilung gewünschter Inhalte und Funktionen

Die **projektbezogenen übergreifenden Informationen** sollen eine Kommunikationslücke schließen. Wie in vielen vergleichbaren Unternehmen verbleiben Informationen in kleinen Gruppen, obwohl sie durchaus für alle Mitarbeiter von Interesse wären. Als Basisanforderung wurde mehrmals eine Endauswertung zu jedem einzelnen Projekt genannt, die mit einer Personenliste gekoppelt ist, auf der vermerkt ist: Wer hat was gemacht? Arbeitet jemand z. Z. an ähnlichen Problemen wie ich? Welche Klassenbibliotheken gibt es, welche sind von wem entwickelt worden? usw.

Weiterhin sollten Infos zu allen aktuell laufenden Projekten (wer macht aktuell was, Prozeßverlauf, welche Tools werden verwendet, wie sehen die Oberflächen aus, ist eine Datenbankanbindung realisiert usw.) bereitstehen und eine mit Beispielen unterlegte Sammlung von Problemen oder Fehlern, die in den verschiedenen Projekten aufgetreten sind (veranschaulicht an Graphiken, Quellcodes, Texten, Screendumps etc.).

Unter der Rubrik **sonstige Anforderungen** wurden allgemeine Informationen summiert, die sich auf Schwierigkeiten mit bestimmten Betriebssystemen, Eigenheiten einzelner Programmiersprachen und Administratorproblemen beziehen. Gewünscht werden aber auch Informationen zu aktuellen Entwicklungstools, neuer Hardware und Eigenschaften der technischen Geräte, die von der Software angesteuert werden. Einen eigenständigen Schwerpunkt nehmen Informationen zu Datenbanken (Schnittstellen, Anlegen von DB, Anwendungsfälle) ein.

Auch außerhalb einer elektronischen Unterstützungsumgebung - oder parallel zu dieser - wurden immer wieder eher traditionelle Möglichkeiten der abteilungsinternen Kommunikation und Weiterbildung eingefordert. Genannt wurde die stärkere Förderung des Austauschs von und der Diskussion über Zeitschriftenartikel, eine regelmäßige interne Schulung durch vorbereitete Vorträge von Mitarbeitern aus der Abteilung mit hin und wieder auch externen Referenten. Ganz wichtig scheint eine institutionalisierte, regelmäßige Diskussionsrunde über aktuelle Probleme / Méthoden / Entwicklungsfragen über die Grenzen der Projektteams hinweg. Dazu gehören auch Feedbackrunden mit mehreren Kollegen zu aktuellen Gestaltungslösungen bei denen dann aufgetretene Fehler/Irrwege gemeinsam diskutiert und Lösungen gefunden werden können.

5 Diskussion und Ausblick

Wissensmanagement erfordert zunächst die Identifikation des vorhandenen Wissens und die Analyse der Kommunikations- und Kooperationsprozesse innerhalb einer definierten Organisationseinheit. Mit dem beschriebenen Vorgehen war es im konkreten Fall möglich zu entscheiden, ob und in welcher Form Entwickler in bezug auf software-ergonomische Kenntnisse Unterstützung benötigen. Es wurde eine Methodenkombination aus Testaufgaben, Fragebögen, Interview- und Moderationstechniken zusammengestellt, mit der neben Aussagen zum aktuellen Wissensstand (Wer weiß was, entspricht die Selbsteinschätzung der tatsächlichen Leistung, wovon hängt die jeweils individuelle Erfahrung ab, welche Quellen werden als hilfreich beurteilt?) erneut nachgewiesen werden konnte, daß die einzelnen Gruppenmitglieder über jeweils anderes Wissen verfügen. Dieser Beleg der sog. *Verteiltheit des Wissens* war ein ausschlaggebendes Argument zur Differenzierung weiterer Maßnahmen.

Da Benutzerbeteiligung ein notwendiges Merkmal zur erfolgreichen Gestaltung von Unterstützungsmitteln ist, wurden im Rahmen von Interviews und Workshops Bedürfnisse und Erwartungen der potentiellen Nutzer an ein Unterstützungssystem ermittelt sowie Ideen zur möglichen Umsetzung generiert und auf Machbarkeit geprüft. Die anschließenden

Schritte sind nur bedingt parallel möglich; deshalb kam es darauf an, die Implementierung notwendiger Inhalte und Funktionen des elektronischen Unterstützungssystems auch zeitlich zu strukturieren.

Die Einführung der Hyperbase soll ebenfalls begleitet und evaluiert werden. Hier ist wichtig, die Interaktion der Benutzer mit dem System bzw. seine Nutzung (Ort, Zeit, Motive, Erfolge usw.) zu analysieren. Darüber hinaus ist zu prüfen, inwieweit sich durch die Nutzung der Hyperbase Kooperation und Kommunikation in der Abteilung verändern und sich die Qualität (prozeß- und produktbezogen) der Software-Entwicklung verbessert.

Nicht unerwähnt bleiben soll die Möglichkeit, daß die hier vorgestellte Methodenkombination leicht auf ähnliche Fragestellungen in anderen Unternehmen adaptiert werden kann. Es bietet sich darüber hinaus an, die geschilderte Vorgehensweise auf andere Probleme und Aufgaben des Wissensmanagements zu übertragen bzw. zu verallgemeinern.

6 Literatur

Smith, L.S. & J. N. Mosier (1986): Guidelines for designing user interface software. Bedford: The Mitre Corporation.

IBM (1991): System Application Architecture: Common User Access. Cary: IBM.

ISO 9241 (1992): Visual Display Terminals (VDT´s) used for office tasks, Ergonomic Requirements, Part 10: Dialogue Principles.

Pearl, G. & Moorhead, A.J. (1989): NaviText™ SAM. Westford: Northern Lights Software.

Iannella, R. (1991): BRUITΔSAM™. Gold Coast: Bond University.

Vanderdonckt, J. & F. Bodart: (1996): The "Corpus Ergonomicus": a Comprehensive and Unique Source for Human-Machine Interface Guidelines, In: Ozog, A. F. & G. Salvendy (Hrsg): Advances in Applied Ergonomics. Proceedings of 1st International Conference on Applied Ergonomics ICAE'96 (Istanbul, 21-24 May 1996). Istanbul: West Lafayette, (S. 162-169).

Wandke, H. & J. Hüttner (1995): Unterstützung bei der Gestaltung von Benutzungsschnittstellen durch die Bereitstellung von Software-Ergonomie-Wissen in einem Informations- und Beratungssystem. In: H.-D. Böcker (Hrsg.): Software-Ergonomie 95, Mensch-Computer-Interaktion: Anwendungsbereiche lernen voneinander. Stuttgart: Teubner, (S.383-393).

Forbrig, P.; Gorny, P. & A. Viereck (1993): Unterstützung des Software-Design-Prozesses durch EXPOSE. In: W. Coy, P. Gorny, I. Kopp und C. Skarpelis (Hrsg.): Menschengerechte Software als Wettbewerbsfaktor. Stuttgart: B. G. Teubner, (S. 463-479).

Balzert, H., (1994): Das JANUS-System: Automatisierte, wissensbasierte Generierung von Mensch-Computer-Schnittstellen. Heidelberg: Springer-Verlag, Issue 9, (S. 22-35).

Reiterer, H. (1995): IDA - A Design Environment for Ergonomic User. In: K. Nordby et al. (1995): Human-Computer Interaction, Interact ´95, IFIP Conference. London: Chapman & Hall, (S. 305-310).

Basili, V. R.; Caldiera, G. & H. D. Rombach (1994): The Experience Factory. In: J. J. Marciniak (Hrsg.): Encyclopedia of Software Engineering. Vol. 1. New York: John Wiley & Sons.

Frese, M. & Brodbeck, F. C. (1992): Psychologische Aspekte der Software-Entwicklung (Ergebnisse des IPAS-Projekts). In: IBM Nachrichten (42) 1992, H. 309, (S. 14-19).

Brodbeck, F. C. & M. Frese (1994): Produktivität und Qualität in Software-Projekten - Psychologische Analyse und Optimierung von Arbeitsprozessen in der Software-Entwicklung. München: R. Oldenburg Verlag.

Brodbeck, F. C. (1996): Kommunikation und Leistung in Projektarbeitsgruppen. Eine empirische Untersuchung an Software-Entwicklungsprojekten. Aachen: Shaker Verlag.

Sonnentag, S. (1995): Excellent software professionals: Experience, work activities, and perceptions by peers. Behaviour & Information Technology, 14, (S. 289-299).

Probst, G.; Raub, S. & K. Romhardt (1998): Wissen managen: Wie Unternehmen ihre wertvollste Ressource optimal nutzen. Wiesbaden: Gabler (2. Auflage).

Warnecke, G.; Stammwitz, G. & F. Hallfell (1998): Intranets als Plattform für Groupware-Anwendungen. In: Industrie Management 1/98, (S. 24-28).

Oberquelle, H. (1991 a): CSCW- und Groupware-Kritik. In: Oberquelle, H. (Hrsg.): Kooperative Arbeit und Computerunterstützung. Göttingen: Verlag für Angewandte Psychologie.

Oberquelle, H. (1991 b): Kooperative Arbeit und menschengerechte Groupware als Herausforderung für die Software-Ergonomie. In: Oberquelle, H. (Hrsg.): Kooperative Arbeit und Computerunterstützung. Göttingen: Verlag für Angewandte Psychologie.

Herrmann, T. & T. Walter (1997): Participatory Design and Cyclic Improvement of Business Processes with Workflow Management Systems. In: Proceedings of the Association Information Systems 97 (AIS Indianapolis), (S. 925 - 928).

Hoffmann, M. & K.-U. Loser (1997): Mitarbeiter-orientierte Modellierung und Planung von Geschäftsprozessen bei der Einführung von Workflow-Management. In: Ortner, E. (Hrsg.): Proceedings des EMISA-Fachgruppentreffens 1997: Workflow-Management-Systeme im Spannungsfeld einer Organisation (Darmstadt).

Beimel, J., Hüttner, J. & H. Wandke (1993): Kenntnisse von Programmierern auf dem Gebiet der Software-Ergonomie: Stand und Möglichkeiten zur Verbesserung. In: Arbeits-, Betriebs- und Organisationspsychologie vor Ort. Bonn: Deutscher Psychologen Verlag, (S. 72-82).

Hüttner, J. & H. Wandke (1993): What do system designers know about software ergonmics and how to improve their knowledge? In: H. Luczak, A. Cakir and G. Cakir (Hrsg.): Work with display units 92. Amsterdam: Elsevier, (S. 304-308).

Wandke, H. (1995): Individualisierung als zentrales Problem der Unterstützung und wie man sie erreichen kann. In: Hacker,W., Rothe, H.-J., Wandke, H. und Ziegler, J. (Hrsg.): Entwicklung und Einsatz wissensorientierter Unterstützungsysteme. Bremerhaven : Wissenschaftsverlag NW Verlag für neue Wissenschaft, (S.109-132).

Wandke. H. (1995): Struktur begrifflichen Wissens im menschlichen Gedächtnis - Anregungen für den Software-Entwurf ? In: Dzida, W. und Konradt, U. (Hrsg.): Psychologie des Software-Entwurfs. Göttingen und Stuttgart : Verlag für Angewandte Psychologie, (S. 60-83).

Hüttner, J. (1997): Software-Ergonomie in mittelständischen Softwarehäusern - eine Befragung auf der CeBIT96. Poster in: Lischkowsky, R.; Velichkowsky, B. M. & Wünschmann, W. (Hrsg.): Software-Ergonomie ' 97. Berichte des German Chapter of ACM Nr. 49. Stuttgart : B. G. Teubner, (S. 362 ff.).

Nielsen, J. (1993): Usability Engineering. Cambridge, MA: Academic Press, (S. 273 ff.).

Adresse der Autoren

Prof. Dr. Hartmut Wandke
Humboldt-Universität zu Berlin
Institut für Psychologie
Oranienburger Str. 18
D – 10178 Berlin

Hartmut.Wandke@rz.hu-berlin.de

Dipl.-Psych. Jens Hüttner
Humboldt-Universität zu Berlin
Institut für Psychologie
Oranienburger Str. 18
D – 10178 Berlin

Huettner@rz.hu-berlin.de

Dipl.-Psych. Andreas Dubrowsky
Humboldt-Universität zu Berlin
Institut für Psychologie
Oranienburger Str. 18
D – 10178 Berlin

Dubrowsky@rz.hu-berlin.de

Aufgabenorientierte Visualisierung eines komplexen verfahrenstechnischen Prozesses unter Verwendung dreidimensionaler Computergrafik

Carsten Wittenberg

Fachgebiet für Systemtechnik und Mensch-Maschine-Systeme,
Universität Gesamthochschule Kassel

Zusammenfassung

In diesem Beitrag wird eine Methodik zur Gestaltung von Mensch-Maschine-Schnittstellen vorgestellt, die Visualisierungstechniken der dreidimensionalen Computergrafik verwendet, um Prozeßelemente, -größen und Zusammenhänge zum Führen und Überwachen von technischen Systemen darzustellen. Prozeßgrößen und Zusammenhänge zwischen Prozeßelementen werden durch die Verwendung von prägnanten Farb- und Formkodierungen sichtbar gemacht. Grundlage dieser Visualisierungsmethodik ist der Bildüberlegenheitseffekt, der das menschliche Vermögen beschreibt, bildliche und damit nichtsprachliche Informationen leichter und schneller aufzunehmen und zu verarbeiten als sprachliche Informationen. Zur transparenteren Darstellung wird das zu führende System hierarchisch aufgabenorientiert strukturiert. Als Anwendungsgebiet ist die Prozeßführung in dem Bereich der thermischen Verfahrenstechnik gewählt worden.

Summary

In this contribution a methodology is presented for the design of human-machine interfaces, which uses visualisation techniques of the three-dimensional computer graphics, in order to visualise elements, variables and relations for controlling and monitoring of technical systems. Process variables and relations between elements are made visible by the use of well-known colour and form coding. Basis of this visualisation methodology is the picture superiority effect, which describes the human characteristic, to process figurative and not-linguistic information more easily and faster than linguistic information. For a more transparent representation the technical system is structured task-oriented. As area of application the process controlling of a technical system within the area of thermal process engineering was selected.

1 Einleitung und Problemstellung

Steigende Anforderungen hinsichtlich Qualität, Ökonomie und Ökologie als auch die immer leistungsfähiger und preiswerter werdenden Automatisierungseinrichtungen führen zu einem steigenden Automatisierungsgrad in verfahrenstechnischen Produktionsanlagen. Derartige äußerst komplexe und dynamische Industrieanlagen bedürfen aber nach wie vor der Überwachung und Führung durch den Menschen in einer Leitwarte. In Folge der fortschreitenden Automatisierung nimmt jedoch die Anzahl der beteiligten Menschen ab, während die Zahl der zu überwachenden Prozeßeinheiten und -elemente steigt (z.B. [1][2]) Gleichzeitig sinkt aber auch die Zahl der notwendigen Eingriffe in den technischen Prozeß seitens des Menschen. In der Position des reinen Beobachters ist der Mensch aber nicht mehr in das Prozeßgeschehen eingebunden, er verliert immer stärker den Kontakt zu dem Prozeß. Dieser hohe Automatisierungsgrad führt zum Verlust der manuellen Fertigkeiten in der Prozeßführung, während die Verantwortungsbereiche des einzelnen Menschen immer größer werden [3][4].

Die früher üblichen lokalen Leitwarten sind aufgrund der damaligen technischen Gegebenheiten von großen Tafeln mit elektromechanischen Anzeigegeräten nach der Single – Sensor

– Single – Indicator – Philosophie (SSSI) geprägt worden [5]. Die damit verbundene Darstellungsart setzt beim Menschen eine ausgeprägte Fähigkeit zum schlußfolgernden Denken voraus, welche aber durch die sehr hohe Anzahl von Informationen über den Prozeß sehr erschwert wird. Stetig weiterführende Entwicklungen auf dem Gebiet der Computertechnologie haben zur Verdrängung der elektromechanischen Anzeigegeräte durch die Einführung von rechnergestützten Leitsystemen mit Tisch- und Wandmonitore geführt. Die Gestaltung der neuen grafischen Benutzerschnittstellen richtet sich aber immer noch nach dem Bild der früheren Leitwarten. Die starke räumliche Konzentrierung der Informationen auf den Bildschirmen im Gegensatz zu der räumlich sehr verteilten Anordnung von Anzeigegeräten in den früheren Leitwarten erschwert es aber dem Menschen, die für die aktuelle Situation relevanten Informationen zu erkennen und aus der Flut an Informationen herauszufiltern.

Trotz rasant steigender grafischer Leistung bei den verwendeten Systemen werden menschliche Faktoren sowohl hinsichtlich der Informationsaufnahme als auch des Problemlösens und Planens nicht ausreichend bei dem Design von leittechnischen Benutzerschnittstellen berücksichtigt. Doch gerade die Wahrnehmung von relevanten Prozeßinformationen inmitten der immensen Anzahl von dargestellten Informationen ist für das schnelle und sichere Identifizieren von Störfällen von entscheidender Bedeutung. Ist dann schließlich ein Störfall eingetreten, muß der Operateur unter Verwendung von Problemlöse- und Planungsvorgänge den Prozeß wieder in einen sicheren und zufriedenstellenden Zustand überführen. Derartige kritische Situationen haben die Eigenschaft, daß sie in der Regel neuartig und ungewohnt für die beteiligten Menschen sind. Da in der Regel Routineprozeduren in solchen abnormalen Situationen nicht mehr wirksam sind, muß der Mensch eine geeignete Anwort basierend auf seinem Wissen über das funktionale Systemverhalten selber generieren [6].

Eine den menschlichen Eigenschaften angepaßte Gestaltung der Schnittstelle zu dem technischen Prozeß kann den Mensch in dieser Gesamtsituation unterstützen und die Sicherheit und Leistungsfähigkeit des technischen Prozesses erhöhen. Dagegen kann eine dem Menschen nicht angepaßte Gestaltung (z.B. durch Informationsüberlastung) zu Bedienfehlern führen [7], welche schwerwiegende Folgen nach sich ziehen können.

2 Exemplarische Anwendung: Rektifikationsprozeß zur thermischen Trennung eines Zweistoffgemisches

Als exemplarische Applikation wird ein Rektifikationsprozeß zur thermischen Trennung (Destillation) eines Zweistoffgemisches in seine Bestandteile (Benzol und Toluol) zugrunde gelegt. Der Prozeß besteht aus einer Kolonnensäule mit 10 Glockenböden, einem Vorlagebehälter, zwei Sammelbehältern für die Endprodukte Benzol und Toluol, zwei Wärmetauschern zum Erhitzen des Zulauf- und Sumpfstromes sowie aus drei Kondensatoren zum Abkühlen des Kopfstromes und der Endprodukte (siehe Abb. 1).

Der mathematische Kern des zugehörigen Prozeßsimulators wurde am Institut für Systemdynamik und Regelungstechnik der Universität Stuttgart als Trainingssimulator entwickelt [8]. Daraus folgt, daß der simulierte Prozeß hinsichtlich des Automatisierungsgrads und der vorhandenen Prozeßinformationen nicht dem derzeit möglichen Stand der Technik entspricht, da mit der Simulation bestimmte manuelle Fertigkeiten der Anlagenfahrer trainiert werden sollen [9]. Dieser simulierte verfahrenstechnische Prozeß besitzt aber gerade deshalb eine ausreichend hohe Komplexität und hohe produktionstechnische Anforderungen zur Erprobung von neuartigen Visualisierungsmethoden zur Prozeßführung. Einige bereits abgeschlossene Forschungsprojekte zur Gestaltung von Bedienoberflächen [10][11] sowie zur systemtechnischen

Entwicklung von grafischen Nutzerschnittstellen (Graphical User Interface GUI) [12][13]
konnten mit dieser Simulation erfolgreich durchgeführt werden.

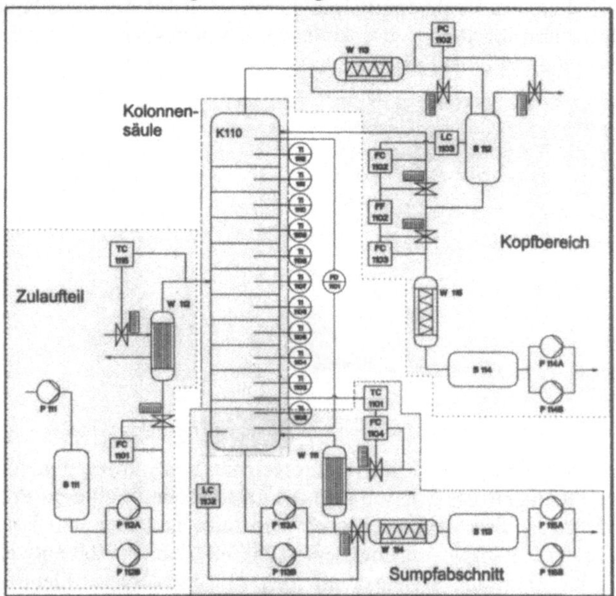

Abb. 1: Fließbild der Rektifikationsanlage [10]

3 Vermittlung von Prozeßwissen

Durch die steigende Komplexität technischer Prozesse und der damit einher gehenden Ver-
mehrung der relevanten Prozeßgrößen und Abhängigkeiten ist es extrem wichtig, dem Menschen das Wissen über den Prozeß hinsichtlich seiner funktionalen Eigenschaften und Zusammenhänge zu vermitteln. Die grafische Benutzeroberfläche, die sowohl die Kommunika-
tionsschnittstelle zwischen dem Menschen und dem technischen Prozeß darstellt als auch ein
Abbild des technischen Prozesses wiedergibt, bestimmt maßgeblich die Vermittlung von
Wissen über das System und damit die Effektivität der Eingriffe des Menschen in den Prozeß.
Zur Steigerung dieser Effektivität muß eine Prozeßvisualisierung die notwendigen Informationen in Form einer dem Menschen angepaßten Strukturierung und Darstellung anbieten.

3.1 Mentales Modell

Auf Basis des Wissens über das technische System bildet sich der Mensch ein mentales Modell. Bei einem mentalen Modell handelt es sich um die Vorstellungen, die sich der Mensch
hinsichtlich der Funktionalität von dem Prozeß gebildet hat (siehe Abb. 2). Die Bedeutung
von mentalen Modellen besteht darin, daß Korrektheit und Differenziertheit des mentalen
Modells die Qualität der an ihnen orientierten kognitiven Prozessen und Handlungstätigkeiten
bestimmen [14]. Prozeßeingriffe durch den Menschen sind also um so effektiver, je angemessener das der Handlungsregulation zugrunde liegende mentale Modell ist [15]. Die Denkvor-
gänge bei der Planung von Handlungsschritten können dabei als eine Art dynamischer ko-

gnitiver Simulation des mentalen Modells beschrieben werden, die analog zu Simulationen mathematischer Modelle auf Rechnersystemen iterativ solange mit jeweils veränderten Parametern – den möglichen Eingriffen in das technische System - wiederholt wird, bis ein zufriedenstellendes Resultat erreicht ist [16].

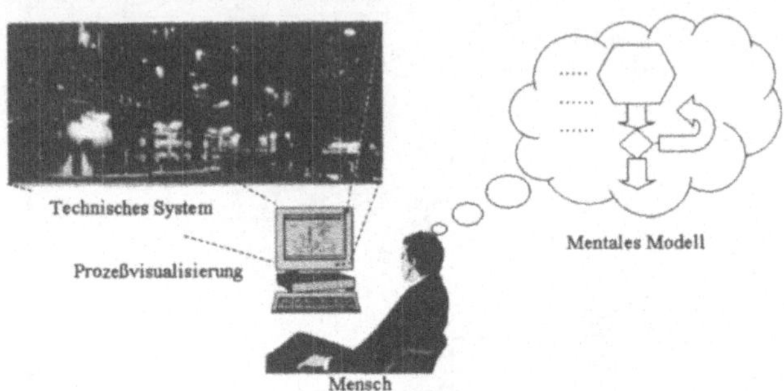

Abb. 2: Prozeß, Prozeßvisualisierung und mentales Modell

Auf der Basis der über die Prozeßvisualisierung vermittelten Informationen und funktionalen Zusammenhänge aktualisiert der Mensch sein mentales Modell von dem Prozeß fortwährend durch seine Erfahrungen. Das mentale Modell wird also nicht explizit konstruiert, sondern graduell in der aktiven täglichen Auseinandersetzung mit dem Prozeß entwickelt. Bei diesem kontinuierlichen Identifikationsvorgang kann man zwischen Systemidentifikation (Feststellung einer abgegrenzten Anordnung von aufeinander einwirkenden Gebilden) und Prozeßidentifikationen (Analyse des Transports bzw. der Umformung von Materie, Energie und/oder Information) unterscheiden [17][18]. Der Mensch identifiziert bei diesem Vorgang zuerst die beteiligten Elemente. Anschließend werden die Relationen zwischen den Elementen identifiziert, wobei zwischen Relationswissen (Erkennen eines Zusammenhanges), Vorzeichenwissen (Richtung des Zusammenhanges) und Wirkstärkenwissen (Gewichtung der Relation) unterschieden wird [17].

Mentale Modelle sind trotz der grundsätzlich funktionalen Struktur durch einen starken bildhaft-anschaulichen Charakter geprägt und können durch bildliche Mittel gefördert werden. Generell wird der bildhaften Visualisierung objektiver Gegebenheiten eine große Bedeutung für die Erzeugung mentaler Modelle zugesprochen [16].

3.2 Aufgabenorientierte Strukturierung des Prozesses

Um komplexe technische Systeme in allen Aufgabensituationen für den Menschen beherrschbar zu machen, ist der Einsatz einer funktional-orientierten Strukturierung der Prozeßinformationen notwendig, welche eine transparente Darstellung des Prozesses ermöglicht und damit das notwendige Prozeßwissen zur Bildung des mentalen Modells vermitteln kann.

Mit der Führung eines jeden technischen Systems sind für den Menschen Aufgaben und Ziele verbunden, die sein Handeln bestimmen. Mit der Erfüllung dieser Aufgaben sind wiederum bestimmte Zustände von Prozeßeinheiten und –elemente verknüpft. Diese Zustände sind abhängig von weiteren Prozeßeinheiten und –elemente, mit denen der Mensch bei der Ausführung der jeweiligen Aufgabe interagiert. Durch Analysen ([19], z.B. [9]) können diese Zusammenhänge ermittelt und als Basis einer strukturierten Visualisierung des technischen Systems verwendet werden. Das Beispiel der Rektifikationsanlage soll dies erläutern.

Die übergeordnete Aufgabe beim Führen dieses Prozesses lautet:

- *Produziere Benzol und Toluol in geforderter Qualität.*

Diese Aufgabe läßt sich in verschiedene Unteraufgaben gliedern:

- *Führe ausreichend Gemisch zu.*
- *Trenne Gemisch in der Kolonne.*
- *Erzeuge Temperaturprofil.*
- *Halte das Rücklaufverhältnis optimal.*
- *Führe das Endprodukt Benzol ab.*
- *Führe das Endprodukt Toluol ab.*
- ...

Beispielsweise müssen sich die Füllstände im Sumpf und im Vorlagenbehälter zur Erfüllung
der Unteraufgabe *Führe ausreichend Gemisch zu* innerhalb eines bestimmten Wertebereiches
befinden. Der Mensch reguliert diese Füllstände durch das Betätigen von Pumpen und Venti-
len. Bei der Visualisierung der Aufgabe ist es also nur notwendig, die beteiligten Elemente
und Zustandsgrößen darzustellen. Weitere Prozeßeinheiten und –elemente wie beispielsweise
Wärmetauscher, die sich zwar auch in dem beteiligten Untersystem befinden, aber nicht zur
Aufgabenerfüllung beitragen, müssen nicht dargestellt werden. Die aufgabenorientierte Prozeßvisualisierung entlastet so den Menschen bei der Identifizierung und vermittelt gleichzeitig die möglichen Prozeßeingriffe zum Erfüllen der jeweiligen Aufgabe.

Werden zusätzlich die Erfüllungsgrade der jeweiligen Aufgaben visualisiert, so kann der
Mensch die Dringlichkeit des Eingreifens in den Prozeß erkennen und sein Handeln entsprechend planen.

4 Darstellung von Prozeßinformationen

Die Überwachung von Prozeßvariablen ist eine zentrale Tätigkeit in der Prozeßführung. Dabei müssen etliche Informationen gleichzeitig von dem Menschen aufgenommen und verarbeitet werden. Diese menschliche Informationsaufnahme und –verarbeitung ist seit Ende des
letzten Jahrhunderts ein Forschungsschwerpunkt in der Psychologie. Ein Ergebnis dieser Untersuchungen ist vor allem die Unterscheidung zwischen sprachlicher und nichtsprachlicher
(z.B. bildlich dargestellter) Information. Die Vorteile von bildlicher Information - der sogenannte *"Bildüberlegenheitseffekt"* - sind dabei ermittelt worden. In [20] wird in diesen Zusammenhang das sogenannte multimodale Modell eingeführt. Dieses Modell beschreibt verschiedene Gedächtnisbereiche, die jeweils von der Modalität des auslösenden Reizes abhängig sind und sehr unterschiedliche Leistungsmerkmale besitzen. In Abb. 3 wird der für die
Wahrnehmung von Prozeßinformationen wichtige Bereich der visuellen Information in dem
multimodalen Modell hervorgehoben.

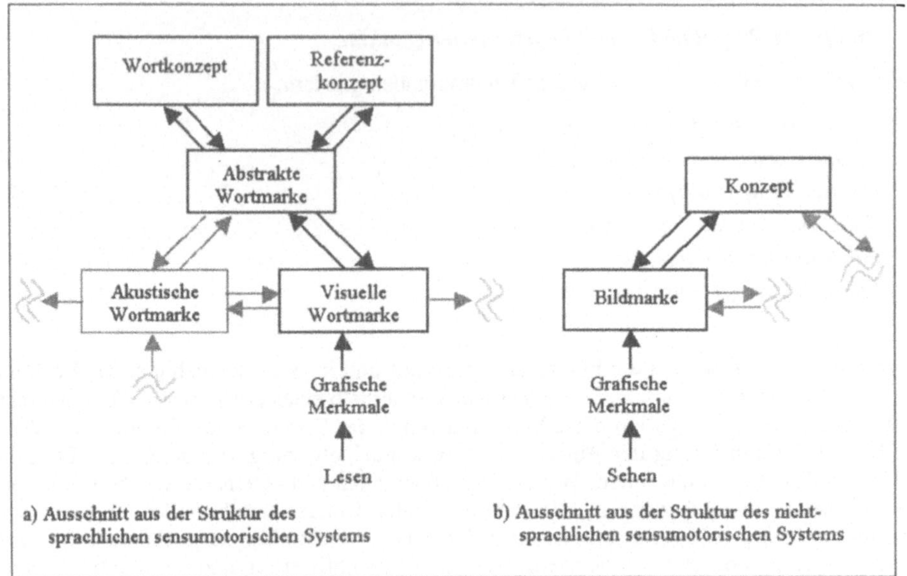

Abb. 3: Multimodales Modell des sensumotorischen Systems hinsichtlich der
visuellen Informationsaufnahme (abgeleitet nach [20], siehe auch [21])

Bei der Struktur des sprachlichen sensumotorischen Systems wird durch die unterschiedlichen
Reize (visuell oder akustisch), mit denen die informationsverarbeitenden Prozesse ausgelöst
werden, zwischen visuellen und akustischen Wortmarken unterschieden. Eine übergeordnete
abstrakte Wortmarke integriert diese modalitätsspezifischen Wortmarken. Bedingt durch die
bedeutungsmäßige Unbestimmtheit von Sprache sind der abstrakten Wortmarke sogenannte
Wort- und Referenzkonzepte zugeordnet. In dem Wortkonzept wird das linguistisches Konzept repräsentiert. Das Referenzkonzept ist die Repräsentation der nonverbalen Konzepte
(hier: die Prozeßinformationen), auf die mit der sprachlichen Information verwiesen wird.

Einfacher und direkter ist die Struktur des nichtsprachlichen sensumotorischen Systems, da
hier die modalitätsspezifische Marke (hier: die Bildmarke) direkt mit dem zugehörigem Konzept verbunden ist. Die bildliche Information muß aber einer entsprechenden Bildmarke eindeutig zuzuordnen sein, um eine schnelle und korrekte Informationswahrnehmung und Zuordnung zu dem entsprechenden Konzept zu gewährleisten.

Durch die rasante Leistungssteigerung in der Computertechnik bietet sich die Verwendung
von neuen Technologien wie die dreidimensionale Computergrafik in der Prozeßleittechnik
an. Mit Hilfe der dreidimensionalen Computergrafik können komplexe Informationswelten
grafisch bildlich-anschaulich visualisiert und manipuliert werden. Der Mensch kann diese
Informationen je nach Ausprägung der Visualisierung interaktiv erleben. Forderungen nach
bildlicher –Darstellung komplexer Systeme zur Ausnutzung des Bildüberlegenheitseffektes
und zur Förderung der Bildung des mentalen Modells durch bildliche Darstellungen (s.o.)
können erfüllt werden.

5   Visualisierung von Prozeßelementen und Prozeßgrößen

Die sogenannten virtuellen Prozeßelemente sind zur bildlich-anschaulichen Prozeßvisualisierung eingeführt worden [21]. Es werden dabei die Prozeßelemente basierend auf der Abbildung von verbreiteten und damit bekannten Musterelementen aus dem Einsatzgebiet unter Verwendung der dreidimensionalen Computergrafik modelliert. Dabei ist relevant, daß visuelle Charakteristika dieser Elemente dargestellt werden, um bereits vorhandene Bildmarken zur Erkennung des Elements seitens des Menschen zu aktivieren und dadurch notwendige Lernprozesse zu minimieren. So wird beispielsweise eine Pumpe basierend auf der weit verbreiteten Chemie-Normpumpe in Blockbauweise [22] modelliert (siehe Abb. 4).

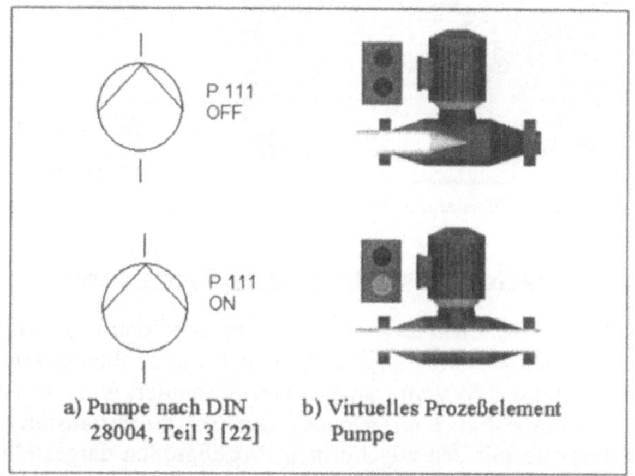

Abb. 4: Vergleich der Darstellung einer Pumpe

Zum Führen eines Prozesses ist die Kenntnis der für die aktuelle Situation bedeutsamen Prozeßgrößen unabdingbar [23]. Allerdings ist dabei nicht immer der absolute numerische Wert der Prozeßgröße gefordert. Häufig reicht auch ein Vergleich mit Sollwerten oder eine qualitative Darstellung der Größe. So ist in vielen Situationen nicht der absolute Füllstand eines Behälters entscheidend, sondern die Frage, ob der Inhalt noch zur Erreichung des Produktionsziels innerhalb eines Zeitabschnitts ausreicht. Entsprechend der Vorgehensweise bei der grafischen Modellierung der virtuellen Prozeßelemente werden in diesem Konzept Interaktionsformen entwickelt, die den Bildüberlegenheitseffekt ausnutzen.

Visuelle Größen - z.B. Füllstände oder Durchflüsse - werden bei dieser Visualisierungstechnik realitätsnah dargestellt. Nicht-visuelle Größen wie Temperaturen oder Drücke werden mit Mitteln der Farb- bzw. Formkodierung dargestellt. So assoziiert der Mensch mit Farben bzw. Farbverläufen bestimmte Eigenschaften wie die Temperatur [24][25]. Formänderungen werden mit Druckänderungen assoziiert [26] und eignen sich ebenfalls sehr gut zur Visualisierung von Prozeßinformationen. Soll- und Grenzwerte, die nur bei Abweichungen vom Sollzustand angezeigt werden, erleichtern die Identifizierung des aktuellen Prozeßzustandes und vermindern die Informationsflut. Ein Beispiel ist in Abb. 5 dargestellt. Der Füllstand in dem Tank ist unter dem Grenzwert gesunken. Der Grenzwert wird somit sichtbar, gleichzeitig wird ein verbreitetes Alarmsymbol angezeigt.

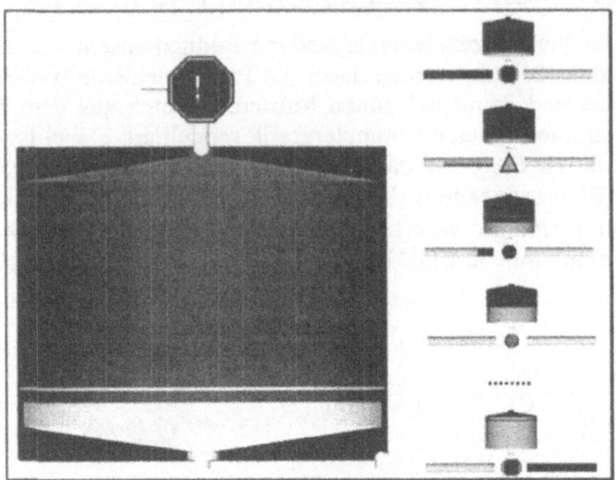

Abb. 5: Darstellung von Füllstand und Grenzwert in einem Tank sowie mögliche Zustände

6 Aufgabenorientierte Darstellung des Prozesses

Das gesamte System *Rektifikationsanlage* wird entsprechend den in [9] durchgeführten Analysen aufgabenorientiert strukturiert dargestellt, so daß der Problemlöseraum für den Menschen verkleinert wird und das System transparenter präsentiert wird. In Abb. 6 wird die bereits oben erwähnte Aufgabe *Führe ausreichend Gemisch zu* visualisiert. Im linken Fenster werden die Prozeßelemente mit den zugehörigen Prozeßgrößen dargestellt. Im rechten Fenster, dem sogenannten Zielmonitor, wird die Erfüllung der Ziele mit den zugehörigen Prozeßelementen visualisiert.

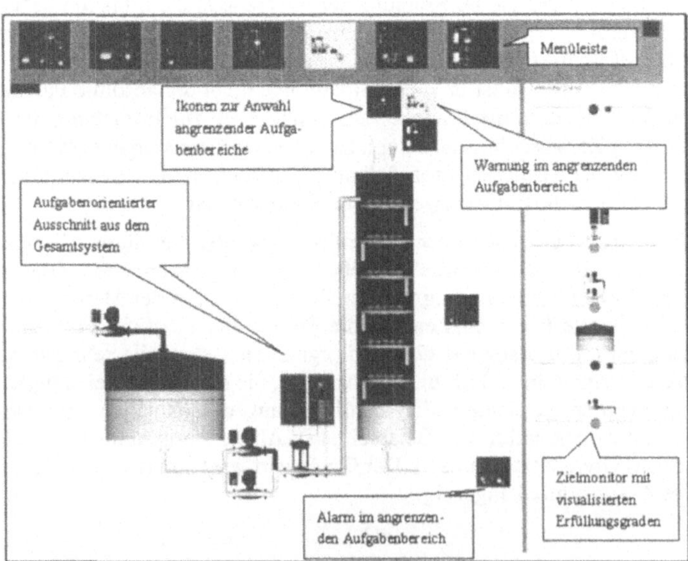

Abb. 6: Aufgabenvisualisierung *Führe ausreichend Gemisch zu*

In diesem Zielmonitor wird nicht nur dargestellt, ob der Zustand des Prozeßelementes oder des Teilsystems dem Zielzustand entspricht, also ob das Ziel erfüllt wird, sondern es wird auch der Grad der Abweichung von dem Ziel in Form von farbigen Säulen dargestellt. Die Farbe und die Länge der Säulen zeigen an, ob ein Ziel erfüllt ist bzw. sich noch im Toleranzbereich befindet (grüner Ausschlag) oder ob eine Warnung (gelber Ausschlag) beziehungsweise ein Alarm (roter Ausschlag) vorliegt. Zusätzlich zu den Säulen werden Ikonen verwendet, um den Zustand der Prozeßelemente hinsichtlich ihrer Zielerfüllung zu visualisieren. Auch hier werden weit verbreitete Symbole (Ziel erfüllt: grüner Kreis; Warnung: gelbes Dreieck; Alarm: rotes Achteck) verwendet. Ist ein Ziel entsprechend den vorgegebenen Sollwerten erfüllt, so wird das Kreissymbol in der Mitte in einem leuchtendem Grün dargestellt. Bei Abweichungen, die aber noch in dem Toleranzbereich liegen, wird die Säule und das Kreissymbol in einem abgeschwächten Grün gezeichnet. Anhand der Säulenlänge und der Richtung kann der Mensch erkennen, wie weit sich das Ziel in dem jeweiligen Bereich befindet, d.h. bei welcher Prozeßeinheit sie/er als erstes eingreifen muß. Liegt eine Warnung oder ein Alarm vor, so werden auch die Elemente hervorgehoben, mit denen Einfluß auf die Prozeßgrößen genommen werden kann, für die eine Zielverletzung vorliegt. Die Interaktion zwischen dem Menschen und Prozeßelementen geschieht direkt per Maus. Die Navigation zwischen den einzelnen Fenstern ist sowohl über die Menüleiste am oberen Bildrand als auch in den einzelnen Unterfenstern möglich. Dazu sind an den Nahtstellen zu benachbarten Aufgabenbereichen Ikonen dargestellt. Werden diese Ikonen angewählt, so wird der gewählte Aufgabenbereich dargestellt. Massen- und Wärmeflüsse werden an den Nahtstellen durch Pfeile dargestellt, deren Durchmesser abhängig von dem jeweiligen Fluß dargestellt werden. Wärmeflüsse werden zusätzlich entsprechend ihrer Temperatur farbig gekennzeichnet. Der Mensch kann so leicht Zusammenhänge zwischen den Unteraufgaben erkennen und diese seinem mentalen Modell hinzufügen.

7 Diskussion

Das in diesem Beitrag vorgestellte Konzepte hat zum Ziel, komplexe technische Systeme strukturiert unter Verwendung der dreidimensionalen Computergrafik bildlich-anschaulich darzustellen. Die Verwendung von bildlicher Informationskodierung· soll die Informationsaufnahme und Informationsverarbeitung des Menschen erleichtern. Durch die Aufgabenorientierung wird das gesamte System übersichtlicher und transparenter dargestellt, der Mensch erhält einen besseren Zugang zu der Informationswelt des Systems. Gleichzeitig wird die Fülle an Informationen auf die situationsabhängige notwendige Anzahl beschränkt. Die Leistungsfähigkeit dieser Visualisierungstechnik wird im Rahmen von Experimenten untersucht.

8 Literatur

[1] Thierfelder, H. G.: Mit neuester Technik bestehende Kraftwerke modernisieren. ABB Technik 2/1997,
     pp. 15-24.

[2] Herbst, L., Rieger, W.: Modernste Prozeßvisualisierung im Kraftwerk Schkopau. ABB Technik 1/1997,
     pp. 13-18.

[3] Reinig, G., Winter, P., Linge, V. und Nägler, K.: Trainingssimulatoren: Engineering und Einsatz.
     Chemie Ingenieur Technik (69) 12/97, pp. 1759-1764.

[4] Bainbridge, L.: Ironies of Automation. Automatica, Vol. 19 (1983), No. 6, pp. 775-779.

[5] Raichle, P.: Erfahrungen mit unterschiedlich gestalteten Leitwarten in kontinuierlich arbeitenden Prozeßanlagen. In: Kolloquium Leitwarten. Köln, TÜV Rheinland e.V. 1984, pp. 595-601.

[6] Vicente, K. J., Rasmussen, J.: The Ecology of Human-Machine Systems II: Mediating "Direct Perception"
     in Complex Work Domains. ECOLOGICAL PSYCHOLOGY, 2(3) 1990, pp. 207-249.

[7] Grams, T.: Bedienfehler und ihre Ursachen (Teil 2). atp – Automatisierungstechnische Praxis, 40 (1998) 4,
     pp. 55-60.

[8] Gilles, E.D., Holl, P., Marquardt, W., Schneider, H., Mahler, R. , Brinkmann, K. & Will, K.-H.: Ein Trai-
     ningssimulator zur Ausbildung von Betriebspersonal in der Chemischen Industrie. Automatisierungs-
     technische Praxis 32 (1990) Nr. 7, pp. 343-350.

[9] Wiese, F., Settgast, D.: Analyse von Prozeßführungsproblemen bei einer Destillationskolonne. Unveröffent-
     lichte Studienarbeit. Fachgebiet für Systemtechnik und Mensch-Maschine-Systeme, Universität Ge-
     samthochschule Kassel, 1993.

[10] Ali, S., Heuer, J., Hollender, M.: Partizipative Erstellung von Bedienoberflächen für Prozeßleitsysteme
     durch interaktive Gestaltung und Bewertung. Endbericht zum DFG-Forschungsvorhaben Kz. JO 139/6-1
     Partizipative Bediengestaltung. Berichtszeitraum: 1.7.92 - 30.6.94 . IMAT - Bericht MMS-14, 1995;
     ISSN 0940-094 X Nr. 14, 1995.

[11] Ali, S., Heuer, J., Hollender, M.: Partizipative Erstellung von Bedienoberflächen für Prozeßleitsysteme
     durch interaktive Gestaltung und Bewertung. Endbericht zum DFG-Forschungsvorhaben Kz. JO 139/6-2
     Partizipative Bediengestaltung. Berichtszeitraum: 1.7.94 - 30.6.96. IMAT - Bericht MMS-17, 1995;
     ISSN 0940-094 X Nr. 17, 1997.

[12] Tiemann, M.: Evaluation of a methodology for human-machine interface development.
     In: P. A. Wieringa and H.G. Stassen (Eds.): Proc.15th European Annual Conference on Human Decision
     Making and Manual Control, Soesterberg, The Netherlands, 1996, pp. 2.3-1 - 2.3-12.

[13] Borys, B.-B. and M. Tiemann: The DIADEM software development methodology extended to multimedia
     interfaces. In: Proc. Annual Conference 1997: Human Factors and Ergonomics Society Europe Chapter,
     Advances in Multimedia and Simulation, Bochum, 1997, pp. 117-125.

[14] Heuer, J., Ali, S., Hollender, M., Rauh, J.: Vermittlung von Mentalen Modellen in einer chemischen
     Destillationskolonne mit Hilfe einer Hypermedia-Bedienoberfläche. In: E. Schoop, R. Witt & U. Glowalla
     (Eds.): Hypermedia in der Aus- und Weiterbildung, Schriften zur Informationswissenschaft, Bd. 17,
     Universitätsverlag Konstanz, 1995, pp. 27-42.

[15] Hacker, W.: Arbeitspsychologie. Bern: Verlag Hans Huber, , 1986.

[16] Dutke, S.: Mentale Modelle: Konstrukte des Wissen und Verstehens – Kognitionspsychologische Grundla-
     gen für die Software-Ergonomie. Göttingen: Verlag für Angewandte Psychologie 1994.

[17] Funke, J.: Wissen über dynamische Systeme: Erwerb, Repräsentation und Anwendung.
     Berlin: Springer 1987.

[18] Isermann, R.: Identifikation dynamischer System 1 - Grundlegende Methoden. 2. neubearbeitete und
     erweiterte Auflage Berlin: Springer 1992.

[19] Johannsen, G.: Mensch-Maschine-Systeme. Berlin: Springer 1993.

[20] Engelkamp, J.: Das menschliche Gedächtnis. Göttingen: Verlag für Psychologie 1990.

[21] Wittenberg, C.: Unterstützung der menschlichen Informationsaufnahme durch Prozeßvisualisierung mittels
     virtueller Prozeßelemente. In: K.-P. Gärtner (Hrsg.): Menschliche Zuverlässigkeit, Beanspruchung und be-
     nutzerorientierte Automatisierung, pp. 173-182. Bonn: DGLR 1997.

[22] Ignatowitz , E.: Chemietechnik. Haan-Gruiten: Verlag Europa – Lehrmittel, 1992.

[23] Färber, G., Polke, M., Steusloff, H.: Mensch-Prozeß-Kommunikation. Chemie-Ingenieur-Technik 57 (1985)
     Nr. 4, pp. 307-317.

[24] Frieling, H.: Licht und Farbe am Arbeitsplatz. Bad Wörishofen: Verlagsgemeinschaft für Wirtschaftspublizi-
     stik, 1982.

[25] Heller, E.: Wie Farben wirken. Reinbek: Rowohlt, 1989.

[26] Houssidas, L.: An exploratory study of the perception of causality. The British Journal of Psychology.
     Monograph Supplements. Cambridge University Press 1964.

Adresse des Autors

Dipl.-Ing. Carsten Wittenberg
Fachgebiet für Systemtechnik und Mensch-Maschine-Systeme
Institut für Meß- und Automatisierungstechnik, Fachbereich 15 Maschinenbau
Universität Gesamthochschule Kassel
34109 Kassel
Email: carsten@imat.maschinenbau.uni-kassel.de

Der automatisierte Wissenserwerb im Kontext der Kommunikation. Ein Vorschlag zur Entwicklung von Expertensystemen durch die Experten[1]

Susanne Ziegler

Arbeitsbereich Technikbewertung und Technikgestaltung

TuTech Technologie GmbH der Technischen Universität Hamburg-Harburg

Zusammenfassung

Der Text diskutiert zwei Konzepte der Wissensakquisition, die sich im Hinblick auf die Bedeutung der Kommunikation zwischen Wissensingenieuren und Experten unterscheiden. Bei der Selbstakquisition wird der Erwerb des Erfahrungswissens durch die Wissenseingabeschnittstelle des Systems automatisiert und den Experten übergeben. Beim kommunikativen Wissenserwerb steht die Kommunikation zwischen Wissensingenieur und Experte im Mittelpunkt. Auf der Basis empirischer Befunde zur Umsetzung der Selbstakquisition bei der Entwicklung eines Diagnose- und Informationssystems wird gezeigt, daß auch die Selbstakquisition Kommunikationsprozesse zwischen Wissenschaft und Erfahrungswissen erzeugt und voraussetzt. Zur Umsetzung der Selbstakquisition in der betrieblichen Praxis wird deshalb ein Modell ihrer kommunikativen Einbettung vorgeschlagen.

Abstract

The process of knowledge acquisition for the development of expert systems can be organized in two different ways. The concept of direct knowledge acquisition supports the expert with knowledge acquisition tools which automize parts of the knowledge transformation process. The concept of indirect knowledge acquisition is caracterized by communication processes between the knowledge engineer and the expert. This contribution shows that direct knowledge acquisition creates communication processes of its own, which support the implantation of direct knowledge acquisition in industrial domains where scientists and workers communicate. The contribution concludes with an example for the integration of direct knowledge acquisition and communication.

1 Zwei Methoden der Entwicklung von Expertensystemen: Selbstakquisition und kommunikativer Wissenserwerb

Um den Entwicklungsprozeß für Expertsysteme zu rationalisieren und sie für industriebetriebliche Anwendungen attraktiver zu machen, entwickelt die Informatik spezielle Wissenseingabewerkzeuge, die den Wissenserwerb unterstützen und automatisieren. Diese Werkzeuge haben jedoch nicht den Wissensingenieur zum Adressaten, sondern die Fachexperten selbst. Beim "direkten Wissenserwerb" [1] oder der "Selbstakquisition" [2, 3] akquiriert die Wissenseingabeschnittstelle einer Expertensystemshell das Erfahrungswissen der Fachexperten und generiert aus deren Eingaben das eigentliche Expertensystem. Nicht der Wissensingenieur, sondern der Fachexperte ist Nutzer der Wissenseingabeschnittstelle. Aber auch die Vermittlungs- und Explikationsleistungen des Wissensingenieurs sollen entfallen.

---

1 Dieser Text basiert auf Erfahrungen des Forschungsprojekts "Kooperierende Diagnostik-Expertensysteme bei der Komplexitätsreduktion sehr großer Wissensbasen" (Projektkennziffer: 01 HP 8440). Es wird vom Programm "Arbeit und Technik" im BMBF gefördert. Laufzeit: 7/95-10/98.

Dieses Konzept wird in Industriebetrieben und der wissenschaftlichen Community erprobt und kontrovers diskutiert. Seine Attraktivität für industrielle Anwendungen wird mit dem Verzicht auf den Wissensingenieur begründet. Die Kommunikation von Wissensingenieur und Experte lasse nämlich zahlreiche ungeklärte Mißverständnisse und eine hohe Fehlerrate des Systems bei beträchtlichen Entwicklungskosten erwarten [1]. Beim direkten Wissenserwerb sollen sich mit dem Wissensingenieur auch die kommunikationsbedingten Mißverständnisse aus dem Wissenserwerb herauskürzen, so daß sich die Treffsicherheit des Expertensystems bei geringeren Entwicklungskosten erhöht.

Kommunikationsprozesse zwischen Fachexperte und Wissensingenieur können in der Tat von Mißverständnissen, Übersetzungsproblemen und Fehldeutungen geprägt sein, die zu fehlerhaften Systemen führen. Im Wissenserwerb sind Akteure mit differenten Sozialisationsmustern, Berufsrationalitäten und Arbeitsstilen im Rahmen einer gemeinsamen Systementwicklung aufeinander angewiesen. Ausgangspunkt des Wissenserwerbs sind nämlich Wissensdefizite auf Seiten beider Akteure. Der Wissensingenieur möchte das Erfahrungswissen des Experten, das er selbst nicht kennt, für die Steuerung und Unterstützung von Arbeitsprozessen im Unternehmen nutzbar machen. Dazu stützt er sich auf die Algorithmen des Computers. Der Experte kann sein Erfahrungswissen im Dialog mit dem Wissensingenieur ohne die Einbindung in seinen Arbeitskontext häufig nur bruchstückhaft explizieren. Aber auch frühere Erfahrungen oder Vorstellungen der Akteure über die Intentionen und Orientierungen des Dialogpartners können zu Explikationsschranken werden. Ein auf Facharbeiterniveau qualifizierter Experte z.B. geht die Explikation seines Erfahrungswissens mit Befangenheit an, wenn er vom akademisch qualifizierten Ingenieur eine Kritik seiner Arbeitsrationalität befürchtet.

Aber auch eine von Mißverständnissen weitgehend freie Kommunikation stellt keine Garantie für valide Expertensysteme dar. Eine "asymmetrische" Dialoggestaltung kann ein "Modellmonopol" [4, S. 92 f.] desjenigen erzeugen, der das Gespräch führt. Ein Modellmonopol wird erreicht, wenn ein Dialogpartner Widersprüche in den Aussagen des anderen benutzt, um dessen Perspektive unbegründet erscheinen zu lassen und die Systemperspektive als die "logisch konsistentere und unumstößlich Geltende" [4, S. 93] ins Spiel zu bringen. Diese Dialogführung kann zur Folge haben, daß Teile des für die Systemaufgaben relevanten Erfahrungswissens nicht expliziert werden und für die Systemimplementation nicht genutzt werden können. Kann die Selbstakquisition da nicht zu besseren Systemen führen als asymmetrische Dialoge unter der Kontrolle des Wissensingenieurs?

Ein weiteres Argument für die Selbstakquisition ist die Bedeutung des Fachexperten für die Systementwicklung und Systemwartung. Der Experte wird aufgrund der Fortentwicklung der Bezugsdomäne des Systems und der damit verbundenen Erneuerung seines Erfahrungswissens ohnehin zum "unersetzlichen Systembetreuer", und die mit dem Systemeinsatz intendierte "Ent-Subjektivierung" des Erfahrungswissens kann letztendlich gar nicht gelingen [5, S. 217 f.]. Ist es angesichts der geringen Chance des "egoless programming" in der Expertensystemtechnik nicht konsequent, die Position des Experten zu stärken und ihm eine höhere Autonomie bei der Systemgestaltung einzuräumen?

Ein Autonomiezuwachs der Experten läßt sich jedoch auch auf anderem Wege erreichen. Die Selbstakquisition zielt auf die Beseitigung von kommunikationsbedingten Mißverständnissen zwischen Wissensingenieur und Fachexperten ab. Der Verzicht auf den Wissensingenieur ist jedoch nicht die einzig denkbare Alternative. Geht man davon aus, daß Mißverständnisse die Systementwicklung beeinflussen, wird ein Wissenserwerbsmodell denkbar, das die Koopera-

tion zwischen Wissensingenieur und Experten beibehält und die Unabweisbarkeit von
Mißverständnissen zum Ausgangspunkt des Fortgangs der Kommunikation macht. Die
Akteure sensibilisieren sich für die unvermeidlichen Unterbrechungen, Irritationen,
Deutungsdifferenzen und Fehlschlüsse, um sie gezielt für die Fortsetzung des
Kommunikationsprozesses und die Vertiefung der Wissensexplikation zu nutzen [2]. Die
Kommunikation wird nicht zur Etablierung eines Modellmonopols genutzt, sondern für einen
wechselseitigen Verständigungsprozeß instrumentalisiert [6]. Widersprüche bei der
Explikation des Erfahrungswissens dienen nicht dessen Entlarvung als einer unterlegenen
Wissensform, sondern werden zum Zweck ihrer Aufklärung identifiziert und thematisiert.
Auf diese Weise überwinden Wissensingenieur und Experte die Problemblindheit des
Experten im Hinblick auf sein Erfahrungswissen, aber auch die Problemblindheit des
Wissensingenieurs im Hinblick auf die algorithmische Überdeckung des Erfahrungswissens.
Das Bild des unternehmensrelevanten Erfahrungswissens kann sich auf diese Weise
abrunden, und der Experte avanciert vom systemnaiven zum aufgeklärten Experten, der dazu
übergeht, selbst Teile seines Wissens systemgerecht zu modellieren [3].

2 Die Selbstakquisition und der Fortgang der Kommunikation

Projekterfahrungen mit der Umsetzung der Selbstakquisition zeigen, daß eine Synthese aus
beiden Ansätzen des Wissenserwerbs einen gangbaren Weg für die Praxis darstellt. Die Nutzung eines Wissenseingabewerkzeugs durch die Experten macht Kommunikationsprozesse
nämlich nicht überflüssig.[2] Vielmehr drängen sich der Kommunikation auch hier Differenzen
zwischen der Systemrationalität und dem Erfahrungswissen als Thema auf. Auch die Wissenseingabeschnittstelle kann die Experten mit einer fremden Rationalität und einem
vorgegebenen Wissenserwerbsmodell konfrontieren. Paradoxerweise liegt darin kein
Widerspruch zu einem nutzerorientierten Vorgehen der Informatik bei der Gestaltung der
Wissenseingabeschnittstelle. Gerade wenn Erfahrungen mit verschiedenen Nutzergruppen in
die Gestaltung einfließen, kann sich im Hinblick auf eine spezifische Expertengruppe ein
kontextfremdes Design ergeben [8, 9]. Deshalb können bei der Nutzung der
Wissenseingabeschnittstelle durch die Experten Verständnisblockaden, Mißverständnisse und
Verunsicherungen auftreten, die kommunikativ identifiziert und geklärt werden müssen, wenn
sich die Selbstakquisition bewähren soll. Auf diese Weise generiert auch die Automatisierung
des Wissenserwerbs Kommunikationsprozesse.

Diese These soll anhand von empirischen Befunden aus einem Forschungsprojekt illustriert
werden, das die Umsetzung der Selbstakquisition in einem Industriebetrieb erprobt hat. Dabei
ging es um die Implementation eines wissensbasierten Diagnosesystems zur Unterstützung
der Fehlersuche an komplexen technischen Anlagen. Die Rolle der Fachexperten nahmen
Facharbeiter ein, welche die im Kundenauftrag gefertigten Maschinen bei den Kunden
montieren und im Fehlerfall entstören. Als Implementationssoftware wurde der Diagnostik-
Shell-Baukasten D3 [10] genutzt.

Im Rahmen des Projekts ließ sich beobachten, daß die Nutzung des Wissenseingabewer-
zeugs durch die Facharbeiter einen Bedarf nach Kommunikation auf mehreren Ebenen

---

2 Kritische Beobachter des Software-Engineering räumen der Technisierung der Systementwicklung durch
Werkzeuge nur einen geringen Stellenwert im Hinblick auf die Projekteffektivität ein, sehen jedoch einen
starken Zusammenhang zwischen Effektivität und Kommunikation. Werkzeuge und Effektivität tragen nur mit
4% zur Effektivität von Softwareprojekten bei, Teamarbeit dagegen mit 45 %, Projektorganisation mit 27 %,
und Qualifikation mit 21 % [7, S. 241]

erzeugte, und zwar auf der Ebene der Explikation und Bearbeitung des Erfahrungswissen mit dem Wissenseingabewerkzeug (2.1), auf der Ebene der Deutungen des Potentials der Expertensystem-Shell und des Umgangs mit ihr (2.2), auf der Ebene des Wissensaustauschs unter den Facharbeitern (2.3) und auf der Ebene der Kommunikation zwischen Wissenschaft und Facharbeit (2.4).

2.1 Der Kommunikationsbedarf bei der Bearbeitung des Erfahrungswissens

Die Wisseneingabeschnittstelle setzt eine Zerlegung des Erfahrungswissens in isolierte Bestandteile, die Subsumtion dieser Teile unter Systemkategorien wie "Diagnose" und "Symptom" und die Erstellung einer flexiblen Reihenfolge der Bearbeitung durch den Computer voraus. Dieser Zergliederungs- und Subsumtionsprozeß ist jedem erfahrenen Wissensingenieur geläufig. Bei den systemnaiven Experten unter den Facharbeitern war das jedoch nicht der Fall. Diese Leistung wurde erst im Rahmen einer gemeinsamen Wissensmodellierung mit verschiedenen Vermittlungsschritten möglich. Dazu explizierten die Facharbeiter zunächst einen kleinen Bereich ihres störfallbezogenen Erfahrungswissens im Dialog mit einer Wissenschaftlerin. Dabei wurde unterschieden, welche Störungen die Facharbeiter an den Maschinen wahrnehmen, und welche Handlungsanforderungen sich aus ihrer Sicht daraus ergeben. Die Störungen wurden den Symptomen, und die Handlungsanforderungen den Diagnosen zugeordnet. Auf diese Weise verstanden die Facharbeiter das Prinzip der Wissenszerlegung und waren besser in der Lage, eine Zuordnung zu "Symptomen" und "Diagnosen" vorzunehmen. Im Rahmen dieses gemeinsamen Arbeitsprozesses konnten sie eigene Vorstellungen von den Leistungen und der Arbeitsweise des Systems entwickeln. Sie erklärten sich z.B. die Funktion einer Diagnose als "Tip, den das System dem Nutzer geben soll".[3]

2.2 Deutungsdifferenzen des Systempotentials

Unterschiedliche Deutungen des Systempotentials und des Umgangs mit dem System wurden zum Gegenstand kommunikativer Aushandlungsprozessezwischen Wissenschaft und Facharbeit. Das Nutzungspotential des Diagnostik-Shell-Baukasten D3 ist nicht auf eine spezifische Nutzergruppe, wie z.B. Mediziner, Botaniker, oder eben Facharbeiter beschränkt. D3 folgt vielmehr dem Grundsatz eines flexibel-generalisierten Designs, um Diagnoseprozesse in höchst unterschiedlichen Bereichen zu modellieren. Als Shell-Baukasten bietet es verschiedene Wissensmodelle und Problemlöser an, über deren Auswahl und Kombination der Anwender entscheiden soll [12]. Ein systemnaiver Experte kann weder diese Entscheidung treffen, noch kann er aus seinem Nutzungskontext heraus die konfigurative Systemgestaltung verstehen. Aus der Perspektive der Informatik erweitert die Konfigurativität der Expertensystem-Shell das Anwendungspotential, und die verschiedenen Problemlöser dienen der systeminternen Komplexitätsreduktion. Den Facharbeitern erscheint dieses Design jedoch hochkomplex und wie eine Vernachlässigung des Gebots der Nutzerorientierung. Ihre Arbeit mit dem Shell-Baukasten war deshalb zunächst vom Bedürfnis nach Komplexitätsreduktion geprägt. Sie bildeten kurzschrittige Symptom-Diagnose-Ketten und nutzten zunächst lediglich einen Problemlöser. Um ihre Zielvorstellungen zu realisieren, entwickelten sie einen eigenwilligen Umgang mit den Systemmöglichkeiten. Dieser Umgang

---

3 Diese Erfahrungen wurden im Rahmen der Entwicklung und Evaluation eines Tutorials gemacht, daß den Facharbeitern eine strukturierte Anleitung für den Wissenswerb mit dem Diagnostik-Shell-Baukasten D3 bietet [11].

schöpfte jedoch das Potential des Shell-Baukastens nicht aus und blieb hinter dem Anspruch
der Verwissenschaftlichung des Erfahrungswissens zurück. Deshalb wurde den Facharbeitern
ein anderer Problemlöser vorgeschlagen, der das Systempotential für die Umsetzung ihrer
Zielvorstellungen besser nutzte. Um die Attraktivät dieses Problemlösers für die Facharbeiter
zu erhöhen, wurde er in ein zentrales Eingabeformular integriert und eine einfache
Umgangsform mit ihm ermöglicht, die den Eingabeaufwand im Vergleich zur früheren
Arbeitsweise verringerte.

2.3 Der Wissensaustausch der Experten

Die Selbstakquisition erzeugt auf Seiten der Experten den Bedarf nach der Kommunikation
mit anderen Experten, aber auch mit den Facharbeiterkollegen im gewohnten Arbeitskontext
der Montage. Die Isolation vor dem Computer wurde als kontraproduktive Vereinzelung
empfunden, weil sie keinen direkten Austausch über maschinentechnische oder
systemtechnische Fragen zuließ. In den Diagnostik-Shell-Baukasten D3 sind Methoden der
Verteilten Künstlichen Intelligenz eingebunden [13], um die Komplexität einzelner
Wissensbasen zu reduzieren und die Systementwicklung und Systemwartung auf mehrere
Köpfe zu verteilen. Diese Erweiterung von D3 ermöglicht nicht nur die getrennte Bearbeitung
der Wissensbasen durch mehrere Experten, sondern auch ein dezentrales, räumlich verteiltes
Arbeiten der Experten an unterschiedlichen Standorten des Unternehmens. Die Einzelarbeit
sollte jedoch nicht zur ausschließlichen Sozialform der Selbstakquisition werden.

Die Facharbeiter-Experten verstehen unter Selbstakquisition nämlich nicht nur die Arbeit am
Computer, sondern auch das Sammeln von Erfahrungsdaten im angestammten Arbeitsumfeld.
Die Präsenz auf Montagen oder Serviceeinsätzen beugt der Überalterung des Erfahrungswis-
sens vor. Die Gelegenheit, maschinenbezogene Problemlösungen im Austausch mit den
Kollegen zu erarbeiten und zu erproben, vermeidet jene kontraproduktive Problemblindheit,
die sich einstellen kann, wenn die Selbstakquisiton als "Robinsonade" des Experten vor dem
Computer stattfindet. Da Unklarheiten bezüglich des Status der Experten Mißtrauen erzeugen
können, setzt der Austausch mit den Kollegen nicht nur Sozialkompetenz auf Seiten der
Experten voraus, sondern auch eine offizielle Statuszuweisung von Seiten der Leitungsebene
[14].

2.4 Die Thematisierung des Verhältnisses von Wissenschaft und Facharbeit

Die Selbstakquisition macht schießlich auch das Verhältnis von Wissenschaft und Facharbeit
in der Sozialdimension zum Kommunikationsthema. Facharbeiter registrieren das Interesse
der Wissenschaft an ihrem Erfahrungswissen. Gleichwohl sehen sie Wissenschaftler nicht
ausschließlich als Personen, die ihrem Erfahrungswissen Anerkennung zollen, sondern auch
als potentielle Kritiker ihrer Arbeitsrationalität und ihrer vermeintlichen oder tatsächlichen
Wissensdefizite. Wie reagieren Wissenschaftler von der Universität oder Akademiker in der
eigenen Firma, wenn sie sich die Wissenbasen der Facharbeiter ansehen und bemerken, daß
die Orthographie nicht immer korrekt ist und die Formulierungen nicht dem akademischem
Standard entsprechen? Die Explikation des Erfahrungswissens im Expertensystem macht
soziale Grenzziehungen deutlich. Die Selbstakquisition gibt den Facharbeitern zwar die
Chance, eigene und für die künftigen Systemnutzer verständliche Formulierungen zu wählen;
dieser Autonomiezuwachs hat jedoch auch zur Folge, daß Korrekturen durch einen
Wissensingenieur entfallen. Solche Eingriffe werden nicht zwangsläufig als Attitüde der
"Besserwisserei" aufgefaßt. Eine gemeinsame Wissenserwerbsphase, bei der Wissenschaft

und Facharbeit kooperieren, bietet den Facharbeitern die Gelegenheit, das Thema der "Bloßstellung" anzuschneiden, sich der Solidarität der Wissenschaftler zu versichern und sich von diesen hin und wieder die verräterische "Schreibarbeit" am Computer abnehmen zu lassen. Sie schafft auch einen Rahmen, um vor den Akademikern ein wenig die Überlegenheit des eigenen Maschinenwissens zu inszenieren. Was macht es dann noch aus, wenn auffällt, daß Wissenschaftler nicht nur schneller tippen als Facharbeiter, sondern (vorläufig) auch noch die besseren Systembediener sind?

3   Die kommunikative Einbettung der Selbstakquisition

Die Automation des Wissenserwerbs über die Wissenseingabeschnittstelle generiert einen Kommunikationsprozeß zwischen Wissenschaft und Erfahrungswissen. Der Grund liegt darin, daß auch die Arbeit der Experten mit der Wissenseingabeschnittstelle eine Konfrontation differenter Rationalitäten bedeutet, wenn die Wissenseingabeschnittstelle eine für die Nutzergruppe fremde Rationalität objektiviert. Die Selbstakquisition des Erfahrungswissens durch die Wissenseingabeschnittstelle ersetzt zwar den Wissensingenieur, nicht aber die vollständige Transformationsarbeit. Diese fällt den in der Regel systemnaiven Experten anheim. Deshalb muß auch die Selbstakquisition durch einen Vermittlungsprozeß gestützt und angeleitet werden.

Die Experten kennen weder das Potential einer Expertensystemshell, noch haben sie Erfahrung in der systemkonformen Zerlegung und Formalisierung ihres Erfahrungswissens. Deshalb ist für den Einstieg in die Selbstakquisition eine enge Kooperation zwischen den Fachexperten und einem system- und programmierkundigen Akteur mit hoher Beobachtungskompetenz sinnvoll, der als Vermittler zwischen den Entwicklern der Expertensystemshell im Wissenschaftssystem auf der einen Seite und den betrieblichen Anwendern auf der anderen Seite agiert. Die Kooperation beschränkt sich jedoch nicht auf einen dialogischen Gedankenaustausch, sondern treibt die Verständigung durch die gemeinsame Modellierung prototypischer Wissensbasen voran. Im Mittelpunkt steht die Explikation des Erfahrungswissens, aber auch die zielgruppenbezogene Explikation der kontextfremden Elemente, mit denen die Wissenseingabeschnittstelle und das Systemverhalten die Experten konfrontieren. Die Zusammenarbeit ermöglicht darüberhinaus die Evaluation der Wissenseingabeschnittstelle im Hinblick auf ihr Adaptionspotential an den Bedarf der Expertengruppe und regt inkrementelle Innovationen der Systemoberfläche an. Die gemeinsame Wissensmodellierung kann die Akzeptanz einer Expertensystemshell aus der Wissenschaft besser unterstützen als eine reine Zuliefer-Abnehmer-Beziehung zwischen den Shell-Entwicklern und den Experten. Die Phase der engen Kooperation und gemeinsamen Wissensmodellierung leitet den Übergang der Experten zu größerer Selbständigkeit im Umgang mit dem System ein. Der Vermittler kann sich schrittweise aus dem Modellierungsprozeß zurückziehen.

4   Literatur

[1] F. Puppe: Problemlösungsmethoden in Expertensystemen. Berlin u.a., 1990: Springer Verlag.

[2] T. Malsch: Vom schwierigen Umgang der Realität mit ihren Modellen. Künstliche Intelligenz zwischen Validität und Viabilität. In: T. Malsch, U. Mill (Hg.): ArBYTE: Modernisierung der Industriesoziologie? Berlin, 1992: Edition Sigma Rainer Bohn Verlag. S. 157-184.

[3] T. Malsch, R. Bachmann, M. Jonas et. al.: Expertensysteme in der Abseitsfalle? Fallstudien aus der industriellen Praxis. Berlin, 1993: Edition sigma Rainer Bohn Verlag.

[4] J. Pasch: Softwareentwicklung im Team. Mehr Qualität durch das dialogische Prinzip bei der Projektarbeit. Berlin u.a., 1994: Springer Verlag.

[5] C. Kehrwald: Die Genese von Expertensystemen als Rationalisierungsprojekte der Gesellschaft. Eine empirische Analyse der Erfahrungen betrieblicher und wissenschaftlicher Akteure in der Entwicklung und Anwendung von Expertensystemen. In: C. Bender, M. Luig (Hg.): Neue Produktionskonzepte und industrieller Wandel. Industriesoziologische Analysen innovativer Organisationsmodelle. Opladen, 1995: Westdeutscher Verlag. S. 168-224.

[6] T. Malsch: Die Informatisierung des betrieblichen Erfahrungswissens und der "Imperialismus der instrumentellen Vernunft": Kritische Bemerkungen zur neotayloristischen Instrumentalismuskritik und ein Interpretationsvorschlag aus arbeitssoziologischer Sicht. Zeitschrift für Soziologie 16 (1987), 2, S.77- 91.

[7] F. C. Brodbeck: Warum es sinnvoll ist, Kommunikation und Kooperation in Software-Entwicklungsprojekten verstärkt zu kultivieren: Ergebnisse aus einer empirischen Untersuchung. In: K. H. Rödiger: Software Ergonomie '93. Stuttgart, 1993: Teubner Verlag. S. 237-248.

[8] S. Ziegler, S. Schwingeler: Rekontextualisierung als Konzept einer Systemschulung: Ein Tutorial für die Selbstakquisition mit dem Expertensystem-Shell-Baukasten D3. In: J.-P. Pahl (Hg.): Lern- und Arbeitsumgebungen zur Instandhaltungsausbildung. Seelze-Velber, 1997: Kallmeyer'sche Verlagsbuchhandlung. S. 171-186.

[9] S. Schwingeler , S. Ziegler, T. Malsch: Kontextualität als Orientierungsgröße für die Implementation von Expertensystemen. Erscheint in: J.-P. Pahl (Hg.): Instandhaltung. Arbeit - Technik - Bildung.

[10] F. Puppe, U. Gappa, K. Poeck, S. Bamberger: Wissensbasierte Diagnose- und Informationssysteme. Berlin u.a., 1996: Springer Verlag.

[11] S. Ziegler, S. Schwingeler: Tutorial für die Entwicklung von Wissensbasen mit dem Expertensystem-Shell-Baukasten D3. Technologie GmbH der Technischen Universität Hamburg Harburg, Arbeitsbereich Technikbewertung und Technikgestaltung, 1998: Manuskript.

[12] F. Puppe, K. Poeck, U. Gappa, S. Bamberger, K. Goos: Wiederverwendbare Bausteine für eine konfigurierbare Diagnostik-Shell. KI (1994) 2, S. 13-18.

[13] S. Bamberger: Cooperating Diagnostic Expert Systems to Solve complex Diagnosis Tasks. Proc. der deutschen Konferenz für KI (KI '97), Berlin u.a., 1997: Springer Verlag.

[14] S. Schwingeler, S. Ziegler: Leitfaden für die betriebliche Einführung des wissensbasierten Diagnose- und Informationssystems D3. Technologie GmbH der Technischen Universität Hamburg Harburg, Arbeitsbereich Technikbewertung und Technikgestaltung, 1998: Manuskript.

Adresse der Autorin

Susanne Ziegler
TuTech Technologie GmbH der Technischen Universität Hamburg-Harburg
Arbeitsbereich Technikbewertung und Technikgestaltung
FSP 1-11
21071 Hamburg
Email: tg-suzi@wiso.wiso.uni-dortmund.de

Workshop "Gestaltung von virtuellen und beGreifbaren Mensch-Computer-Schnittstellen"

F. Wilhelm Bruns, Bernd Robben, Ingrid Rügge

Forschungszentrum Arbeit und Technik (artec), Universität Bremen

Zusammenfassung

Speziell für die Gestaltung von Arbeitsplätzen in der Produktion hat es in den letzten Jahren eine Reihe von Ansätzen gegeben, neuartige Mensch-Computer-Interaktionen zu unterstützen. Als Ein- und Ausgabemedien werden nicht mehr Bildschirm, Tastatur und Maus benutzt, sondern Techniken der Virtual Reality sowie zunehmend auch der Kopplung von greifbaren Gegenständen mit Computerwelten (Graspable User Interfaces). Für die Gestaltung solcher Arbeitsumgebungen spielen Fragen der Erfahrung und Wahrnehmung auf eine andere, grundsätzlichere Art und Weise eine Rolle, da die BenutzerInnen in diesem Anwendungsgebiet ganz andere Fähigkeiten und Fertigkeiten besitzen und benötigen als ihre KollegInnen im "Schreibtischbereich". Im interdisziplinären Austausch sollen in diesem Workshop spezielle Fragen zur Gestaltung dieser neuartigen Mensch-Computer-Interaktion behandelt werden, zugespitzt auf das Thema multimodaler Wahrnehmung in multisensorischen (auditiv, visuell, haptisch) computergestützten Arbeitsumgebungen.

Abstract

During the last years considerable amount of work has been spend developing new kinds of human-computer-interaction, especially for workplaces in production areas. Instead of monitor, mouse, and keyboard as input/output devices Virtual Reality techniques are being used and – with increasing importance –the coupling of tangible objects with computer worlds (Graspable User Interfaces). Users in production areas have and need other capabilities and skills than their colleagues working in office environments. Therefore issues of experience and perception play a different and fundamental role for the creation of these work environments. During our workshop questions concerning the design of these new ways of human computer interaction will be discussed in interdisciplinary exchange. We will focus on the issue of multimodal perception in multisensoric (auditory, visual, haptic) computerized work environments.

1 Motivation

Die ArbeitspsychologInnen Fritz Böhle und Brigitte Milkau beschreiben in ihrem Buch "Vom Handrad zum Bildschirm" die Veränderung der Arbeit in der Produktion durch das Eindringen der Automatisierung mittels des Computers. Sie warnen, daß das menschliche Handeln durch selektive Wahrnehmung, konzentrierte Aufmerksamkeit sowie rein analytisches und logisches Denken eingeengt wird. Der umfassende Gebrauch aller Sinne, intuitives Handeln und Reagieren, der dialogisch-interaktive Umgang mit der Maschine unter Einbeziehung des ganzen Körpers – das subjektivierende Arbeitshandeln, wie sie es nennen – spielt für ein (auch im ökonomischen Sinne) effektives Handeln in der Produktionsarbeit eine entscheidende Rolle. BenutzerInnenfreundliche Schnittstellen zum Computer zu entwickeln, heißt in diesem Kontext, mehr zu gestalten als eine selbsterklärende, aufgabenangemessene Bildschirmgrafik. Es geht um die Gestaltung einer intuitiven Mensch-Maschine-Kommunikation als menschengerechtes soziotechnisches System.

Ein intensiver Diskurs zu diesem Themenkomplex begann bereits im Herbst 1997 auf einem in Bremen durchgeführten Workshop "Vom Bildschirm zum Handrad – Computer(be)nutzung nach der Desktop-Metapher" Mit großer interdisziplinärer Beteiligung wurde eine Bestandsaufnahme der Forschungsansätze im deutschsprachigen zu neuartigen

Benutzungsschnittstellen für Anwendungen in der Produktion erbracht, die eine Reihe weiterführender Fragestellungen aufwarf: Wie verändert sich das Arbeitsfeld von KonstrukteurInnen, ModellbauerInnen und FertigungstechnikerInnen durch neue Informations- und Kommunikationstechniken? Wie beeinflussen abweichende Habitualisierungen in unterschiedlichen Kulturen die Gestaltungsmöglichkeiten von Mensch-Maschine-Schnittstellen? Wann sind detailgetreue 3D-Darstellungen mit Virtual Reality Techniken sinnvoll, wann behindern sie eine aufgabengerechte Wahrnehmung und Erkenntnis? Welche Abstraktionen im Bereich der Sinneswahrnehmungen sind möglich? Im Rahmen der Software Ergonomie Tagung soll auf diesem Workshop der Diskurs fortgesetzt werden, indem das Problem der veränderten Wahrnehmung in einer solchen komplexen Umgebung in den Blick genommen wird:

- Sind Sehen und Schauen unveränderliche Fähigkeiten der menschlichen Wahrnehmung oder ergibt sich durch die Gewöhnung an andere Bildmedien ein veränderter Blick?

- Welche Rolle spielt das Greifen für das Begreifen? Wie geht der haptische Sinn in Abstraktions- und Sinnbildung ein?

- Wo liegen die Grenzen des Gegenständlichen und des Anschaulichen für das Verständnis komplexer Strukturen?

- Welche Rolle spielt der Wechsel zwischen verschiedenen Abstraktionsebenen für die Durchdringung eines Gebiets? Welche Bedeutung haben dabei die Übergänge zwischen verschiedenen Sinnesmodalitäten?

Besonders für Anwendungen im Produktionsbereich hat es in den letzten Jahren eine Reihe von Entwicklungen gegeben, neuartige Mensch-Computer-Interaktionen zu unterstützen. Als Ein- und Ausgabemedien werden nicht mehr Bildschirm, Tastatur und Maus benutzt, sondern Techniken der Virtual Reality sowie zunehmend auch der Kopplung von greifbaren Gegenständen mit Computerwelten (Graspable User Interfaces). Für solche Umgebungen spielen Fragen der Erfahrung und Wahrnehmung noch auf eine grundsätzlichere Art und Weise eine Rolle als bei den üblich gewordenen graphischen Benutzungsschnittstellen. Im interdisziplinären Austausch sollen Fragen zur Gestaltung dieser neuartigen Mensch-Computer-Interaktion diskutiert werden, und zwar zugespitzt auf das Thema Wahrnehmung in derartigen multisensorischen (auditiv, visuell, haptisch) Umgebungen.

- Welche anderen Wahrnehmungsfähigkeiten erwirbt der Mensch bei der Interaktion in Virtual oder Augmented Reality Umgebungen?

- Wie lassen sich Übergänge zwischen begreifbaren, sinnlich erfahrbaren Gegenständen und vom Computer generierten Welten gestalten?

- Die Bedeutung von Handeln und Erfahrungensammeln für das Erlernen von Zusammenhängen wird immer wieder hervorgehoben. Wie können solche Erkenntnisse bei der Entwicklung von Computerumgebungen berücksichtigt werden?

- Wie erweitert der interaktive Umgang mit der Maschine die Wahrnehmung und die Handlungsfähigkeit? Wie schränkt er diese ein?

- Wie lassen sich technische Systeme realisieren, die Übersetzungen zwischen Abstraktionsebenen, zwischen Sinnesmodalitäten und ihrem gegenseitigen Zusammenspiel unterstützen?

- Was bedeutet die Variabilität der menschlichen Wahrnehmung für die Gestaltung von Computer-generierten Bildern und Grafiken beim Entwurf von Benutzungsschnittstellen?

Welche dieser spannenden Fragen im Workshop diskutiert werden, ergibt sich aus den Interessen der ReferentInnen und TeilnehmerInnen. Angesprochen fühlen sollen sich vor allem

InformatikerInnen, ArbeitswissenschaftlerInnen, WahrnehmungspsychologInnen, IngenieurInnen, PhilosophInnen, PädagogInnen, PraktikerInnen, kurz alle, die sich mit der Gestaltung multisensorischer neuer Computerumgebungen insbesondere für den Einsatz in der Produktion beschäftigen.

2. Impulse

Als Impulse für die Diskussion und zur Motivation legen vier ReferentInnen ihre jeweilige Sichtweise auf den Themenbereich in kurzen Vorträgen dar. Im folgenden sind Auszüge aus den Positionspapieren wiedergegeben. Die Standpunkte aller TeilnehmerInnen sind zu finden unter der URL: http://www.artec.uni-bremen.de/field1/Workshop99

2.1 Subjektivierendes Arbeitshandeln

Auch bei der Arbeit mit hochtechnisierten Systemen ist neben wissenschaftlich fundiertem Fachwissen, analytischem Denken und systematisch-planmäßigem Handeln ein besonderes "Erfahrungswissen" erforderlich. In der Praxis wird hier von "Materialgefühl", "Gespür für Maschinen", von "blitzartigen Entscheidungen ohne langes Nachdenken" sowie Improvisation und Intuition gesprochen. Solche Arbeitspraktiken erweisen sich weder als veraltet noch minderwertig oder unzuverlässig – im Gegenteil: Gerade hierin liegen besondere Leistungen des Menschen im Umgang mit technischen Systemen.

Untersuchungen des Arbeitswissenschaftlers Fritz Böhle zeigen, daß dieses Erfahrungswissen eine eigenständige Form von Wissen ist. Dabei geht es nicht nur um die Anwendung von Erfahrungen, die in der Vergangenheit angesammelt wurden. Wichtig ist vor allem der Aspekt des *Erfahrens* bzw. des *Erfahrungmachens*. Das Erfahrungswissen beruht auf einer besonderen Methode der Auseinandersetzung mit konkreten Gegebenheiten und zwar sowohl was deren Erkenntnis als auch was den praktischen Umgang hiermit betrifft. Mit dem Konzept "subjektivierenden Arbeitshandelns" haben Böhle und Milkau dies systematischer analysiert; hier einige Charakteristika:

- Eine komplexe sinnliche Wahrnehmung, die sich über sämtliche Sinne (Hören, Sehen, Tasten, etc.) sowie über körperliche Bewegung vollzieht und die vom subjektiven Empfinden nicht abgelöst ist. Sie richtet sich nicht nur auf eindeutig definierte oder meßbare, sondern auch auf eher diffuse und vielschichtige Informationsquellen (Geräusche, Farbänderungen etc.).

- Sinnliche Wahrnehmungen solcher Art sind verbunden mit assoziativem und anschaulichem Denken. Denken erfolgt hier nicht nur in Begriffen, sondern vor allem in Form von Bildern, erlebten Bewegungsabläufen oder akustischen Ereignissen. Sinnliche Wahrnehmungen und mentale Prozesse sind verbunden mit praktischen Handlungen, die auch im Umgang mit "Sachen" dialogisch-interaktiv vollzogen werden.

- Gefühle und subjektive Empfindungen sind bei diesen Formen sinnlicher Wahrnehmung, des Denkens und praktischen Handels nicht ausgeschlossen, sondern vielmehr ein wichtiger Bestandteil.

Das subjektivierende Arbeitshandeln ist notwendig insbesondere zur Bewältigung nicht planbarer und nicht exakt beschreibbarer Arbeitsanforderungen. Diese treten gerade auch bei fortschreitender wissenschaftlicher Durchdringung und Technisierung immer wieder in neuer Form auf. Vieles weist darauf hin, daß nicht nur im Umgang mit Technik, sondern auch bei der Technikentwicklung dieses Konzept eine wichtige Rolle spielt.

2.2 FUTURION – Realität durch virtuelle Lernumgebungen begreifen

Dr. Ulrich Karras von der Firma Festo Didactic GmbH & Co stellt die Realisierung eines lernförderlichen integrierten Übergangs von der realen in die virtuelle Welt vor, in dem heutige Anforderungen an Lernumgebungen Eingang gefunden haben, denn Multimedia hat neue Trends in modernen Bildungskonzepten gesetzt:

- Lernen muß in einer offenen Lernumgebung stattfinden können,
- Lernen muß Spaß machen,
- Lernen muß individuell gestaltbar sein und
- Lernen muß an beliebigen Orten stattfinden können.

Da im Bereich technischer Aus- und Weiterbildung das Arbeiten mit realer Technik trotz Multimedia und Simulation von fundamentaler Bedeutung ist, gilt es Konzepte zu entwickeln, den didaktischen Nutzen der beiden Lernwelten miteinander zu verknüpfen. Das DeskTop Training Studio FUTURION bietet einen Lösungsansatz. Es bietet ein Lernboard zum Aufbau einer realen Automatisierungsumgebung, eine Master-Unit bestehend aus technischen Meßeinrichtungen und einem Bilderkennungssystem, das eine interaktive Kopplung zur virtuellen Lernumgebung auf einem Rechner ermöglicht. Die virtuelle Lernumgebung besteht aus verschiedenen grafischen Darstellungen der zugehörigen Hardwareumgebung und einem multimedialen Lernprogramm, das interaktiv den Umgang mit der Automatisierungsanwendung auf dem Lernboard unterstützt.

Im Rahmen von zwei Forschungsprojekten, L3 – Lebenslanges Lernen gefördert durch das BMBF und BREVIE gefördert durch die EU, wird einerseits die pädagogische Eignung einer solchen neuen Lernumgebung evaluiert und analysiert und andererseits die Technologie bzgl. Mensch-Computer Interface und Telelernen weiterentwickelt.

2.3 Sehen in medialen Umgebungen

Bernd Robben, Informatiker:
Bei computerisierten Benutzungsumgebungen erfolgt die Darstellung der durch den Computer berechneten komplexen und abstrakten Modelle hauptsächlich über visuelle Displays. Damit folgen InformatikerInnen der wissenschaftlichen Tradition der Schrift, die sprachliche Gedanken visualisiert. Auf den Displays erscheint computergenerierte Graphik, die sich allerdings von der klassischen Schrift wesentlich unterscheidet und oft bildhafte Elemente enthält.

Um den Charakter des (neuen) Sehens in Computerumgebungen zu verstehen, werfe ich einige Schlaglichter auf die Geschichte des Blicks im Westen, die sich nach Regis Debray in drei Hauptphasen einteilen läßt: Die erste Phase, das Bild als *Idol*, beginnt mit der Erfindung der Schrift. Das Bild hat eine transzendente *Präsenz* und wird selbst als sehend aufgefaßt. Die zweite Phase, das Bild als *Kunstwerk*, beginnt mit dem Buchdruck. Das Bild wird als Schein *repräsentiert* und als Abbild gesehen. Die dritte Phase, das Bild als *Visualisierung,* hat mit den neuen Medien angefangen. Das Bild erscheint durch numerische Simulation und wird als eine *Vision* von vorher nicht Dagewesenem aufgefaßt.

Diese neue Möglichkeit des Sehens in virtuellen Computerwelten kennzeichne ich als "Pictorial Turn", einen Ausdruck W.J.T. Mitchells übernehmend. Ich verstehe diesen Begriff aber eingeschränkter als Mitchell, der damit einen Paradigmenwechsel in der Philosophiegeschichte kennzeichnet. Computer ermöglichen einen interaktiven Umgang mit Darstellungen: In einem Textverarbeitungssystem ist das Wissen von Setzern gespeichert, so daß mit dem Computer der Text maschinell gesetzt werden kann, was am Bildschirm sogleich sichtbar wird. Mit einem Bildbearbeitungsprogramm lassen sich abstrakte Parameter eines Bildes verändern und dadurch experimentierend neue Bilder auf eine Art und Weise erzeugen, wie es weder klassischen MalernInnen noch FotografenInnen möglich wäre. Eine KomponistIn am Computer kann mit der Beschreibung von Musik automatisch durch einen Synthesizer dessen Klänge ertönen lassen. Dabei kann sie an Parametern der Musik "drehen", wie es vorher KomponistInnen nie vermochten.

Der Kern des "Pictorial Turn" liegt darin, daß der Computer sehr unterschiedliche Notationssysteme darstellen kann und eine technische Übersetzung zwischen ihnen ermöglicht. Was das bedeutet, möchte ich an Beispielen aus dem Produktionsbereich erläutern. Hier haben wir als Notationssysteme Entwurfskizzen, technische Zeichnungen, Schaltpläne, Gleichungssysteme, Stücklisten, bildliche Veranschaulichungen, aber auch gegenständliche Modelle, und in gewissem Sinne sind die technischen Anlagen selbst Notationssysteme. Alle diese Notationssysteme sind im Raum sichtbar, aber auf unterschiedliche Art und Weise. Außer der direkten Sicht auf abstrakte Symbole von Modelldarstellungen braucht die IngenieurIn Imaginationsvermögen für die Bedeutung ihres Modells, ein inneres Auge wie es Ferguson ausdrückt. Zumindest teilweise kann im Zeitalter der Visualisierung die Imagination des abstrakten Modells durch eine technische computergenerierte Übersetzung zwischen unterschiedlichen Notationssystemen sichtbar werden. Was das für das Imaginationsvermögen und den verständigen Umgang mit Notationen im Produktionsbereich bedeutet, soll beispielhaft erörtert werden.

2.4 Abstraktion von Tönen

Die Künstlerin Zorah Mari Bauer wird sich auf der Grundlage ihrer eigenen künstlerischen Arbeit, die seit Mitte der 80er Jahre kontinuierlich den Strukturwandel im Einflußbereich der neuen Medien reflektiert, vor allem Fragestellungen der Abstraktionsbildung zuwenden.

"arbeiten in haptisch, materieller Erfahrung, fühlbar, handgreiflich... das ist Kontrapunkt und Vorläufer meiner gegenwärtig exzessiv immateriellen Arbeit am Computer (was bedeutet: Powerknopf OFF = Wunderwelt weg). Jetzt die Hände bloß noch als taubes Mausschiebewerkzeug. 30 x 25 cm, meine Spielwiese... das ist die Größe eines Mauspads. Rechnen Sie sich aus, wieviele Kilometer Sie im Laufe eines Projekts so von Hand abarbeiten – eine sportliche Höchstleistung!" [1]

So hat der kreative Prozeß eine Akzentverschiebung erfahren, weg von der Konzentration auf das künstlerische Endprodukt und hin auf den Produktionsprozeß selbst. Dieser wird definiert durch multifunktionale Partituren, die als für die jeweiligen Produktionserfordernisse entworfene Abstraktionssysteme nicht nur objektiv faßbare Parameter formalisieren, sondern auch diffuse (z.B. emotionale) Qualitäten durch den Entwurf sinnfälliger "Notationsweisen" beschreiben können. Sie plädiert

- *für ein "Workout der Abstraktionsmuskulatur"*
  Für den Produktionsprozeß bedeutet dies: dem ausführenden Personal - von der OperatorIn bis zur ArbeiterIn - wird fortan mehr als nur die Fähigkeit abverlangt, Produktionsnormative kompetent auszuführen. Vor dem Hintergrund der Partiturisierung möglichst vieler Bestandteile einer Produktion bedeutet dies, daß verstärkt auch die Kompetenz der individuellen Auswertung und Abstraktion der Produktionsabläufe eine Rolle spielen wird und der Rückfluß dieses persönlichen Erfahrungswissens.

- *für eine "lustvolle" Ökonomie des Lernens*
  Die Aneignung von Know How krankt häufig an der Vorgehensweise nach der alten Maxime "was Hänschen nicht lernt, lernt Hans nimmermehr". Vermittelt wird meist nur das Wissen um fachspezifische Inhalte, von denen dann während der beruflichen Lauf-

bahn möglichst lange gezehrt werden soll. Auch hier wird sich das Augenmerk weg von der Vermittlung kurzfristig obsoleter Inhalte verstärkt auf das Training der strukturellen Aspekte der Wissensaneignung selbst, auf den Prozeß des Lernens an sich richten müssen. Alternative Spielformen dieser Disziplin (z.B. das unmittelbare, kopierend-imitierende Lernen) werden hier genauso zu erproben sein, wie etwa eine verstärkte "Notierung" subjektiver Parameter (zur Optimierung individueller Stärken und Vorlieben, aber auch zur Reflexion von Ängsten und Hemmnissen im Umgang mit neuen Techniken), im Interesse, den Prozeß des Lernens subjektiv so "lustvoll" und dadurch objektiv so ökonomisch wie möglich zu gestalten.

3 Ergebnisse

Wir erwarten von diesem Workshop präzisierte Fragen über die Erhaltung von sinnlichen Qualitäten in computerisierten Umgebungen und Hinweise auf das Entstehen *neuer* sinnlicher Qualitäten durch die computerisierten Umgebungen.

4 Literatur

[1] Z.M. Bauer: Vortrag "multifunktionale ordnungssysteme" anläßlich der multimediale5 im ZKM 1997.

[2] F. Böhle, B. Milkau: Vom Handrad zum Bildschirm. Frankfurt u.a., 1988: Campus.

[3] W. Bruns et al. (Hg.): Vom Bildschirm zum Handrad - Computer(be)nutzung nach der Desktop-Metapher. Workshop-Dokumentation. artec-paper 59, 1998.

[4]. R. Debray: Vie et mort de l'image. une histoire du regard en Occident. Paris, 1992: Editions Gallimard.

[5] E. Ferguson: Das innere Auge – Von der Kunst des Ingenieurs. Basel u.a., 1993: Birkhäuser.

[6] W.J.T. Mitchell: Der Pictorial Turn. In: Ch. Kravagna (Hg.): Privileg Blick, Kritik der visuellen Kultur. Berlin, 1997: Edition ID-Archiv.

[7] I. Rügge et al. (Hg.): Arbeiten und begreifen: Neue Mensch-Maschine-Schnittstellen. Münster, 1998: LIT Verlag.

Adressen der Autoren

Prof. Dr. F.W. Bruns, Dipl.-Inform. B. Robben, Dipl.-Inform. I. Rügge
Forschungszentrum Arbeit und Technik (artec)
Universität Bremen
Postfach 33 04 40
28334 Bremen
Email: bruns, robben, ingrid@artec.uni-bremen.de

Workshop „Nutzerunterstützung durch Klassifikationssysteme im Unternehmen"

Andreas Gronski

sd&m software design & management

GmbH & Co. KG, München

Paul Königer

Siemens AG, München

Zusammenfassung

Die reiche Erfahrung bestehender Klassifikationssysteme kann entscheidende Beiträge für die Bewältigung der Informationsmengen in Unternehmen leisten. Da sich die Rollen von Informationsproduzenten und Informationsnutzern im elektronischen Medium deutlich verschoben haben, sind Klassifikationssysteme und Anwendungsumfeld neu aufeinander abzustimmen. Die Einführung von Klassifikationen kann nur als Prozeß der Konventionalisierung erfolgen, deren Herausbildung also unterstützt werden muß.

Abstract

There is a wealth of traditional classification systems that has not yet been fully exploited for coping with information floods in the modern enterprise. The roles of information providers and information users have shifted dramatically, therefore classification systems and their application environments need to be reconsidered. The introduction of classification systems requires that the development of new conventions in the use of information is supported.

1 Problem

Mit der steigenden Bedeutung von Information als Produktionsfaktor und dem gleichzeitigen Explodieren der Informationsmenge wachsen die Anforderungen an die Verfügbarkeit von Informationen exponentiell. Dies betrifft in besonderem Maße *unstrukturierte* Informationen wie beispielsweise Projektberichte, Präsentationen, Nachrichten, E-mails.

Mehr Informationen bedeuten natürlich nicht automatisch auch einen erhöhten Wissensstand bei den Informationsempfängern. Das Problem verschärft sich üblicherweise mit zunehmender Verantwortung [1] [2]. Dabei ist nicht allein der quantitative Umfang an Daten das Problem. Wesentliche Ursache für das Empfinden einer Informationsflut ist der unstrukturierte Charakter der erhaltenen Informationen. Wüßten nämlich die Informationsempfänger beim Erhalt bereits, um welche Art von Informationen es sich handelt (also: über was diese informieren), so wäre die erforderliche Selektionsleistung, um aus den vielen angebotenen Einheiten die jeweils relevanten zu identifizieren, einfacher und damit das Problem der Informationsmenge objektiv reduziert. [3]

In seltsamen Kontrast zu der Informationsflut liegt gleichzeitig ein Informationsmangel vor. Informationen fehlen immer häufiger in den Situationen, in denen sie gebraucht werden [2]. Oft sind die benötigten Informationen sogar im System vorhanden, fehlen aber dort, wo sie benötigt werden. Das zentrale Problem ist in diesem Fall der Zugriff. Es hilft wenig, von der Existenz einer Informationseinheit Kenntnis zu haben, wenn es nicht möglich ist, auf diese mit einem dem Nutzen der Informationseinheit angemessenen Aufwand zuzugreifen. Dabei sind hier nicht allein ein schneller Algorithmus und eine flache Suchhierarchie entscheidend, sondern die *inhaltsorientierte Anordnung der Informationseinheiten*.

2 Rahmenbedingungen

2.1 Begriffsklärung

In der Literatur ist keine einheitliche Definition des Begriffs 'Information' zu finden. Deshalb scheint es gerechtfertigt, den Informationsbegriff dem hier behandelten Thema pragmatisch anzupassen. Als unstrukturiert werden hier solche Informationen bezeichnet, die nicht in automatisierten Verfahren in der dort üblichen formularähnlich aufbereiteten Form vorkommen. Genauer betrachtet ist das Attribut 'strukturiert' abhängig von der Perspektive des Betrachters, denn die Strukturierung des einen Systems kann für ein anderes völlig uninterpretierbar sein. Unstrukturiert sind beispielsweise die meisten über *Emails* transportierten Informationen.

Die Abgrenzung von Informationen zu Daten wird wie folgt bestimmt: Daten sind nach Regeln gebildete Zeichenketten, im hier behandelten Fall insbesondere digitale Kodierungen und Zahlen, die innerhalb einer formal definierten Umgebung konvertiert und maschinell verarbeitet werden können. Informationen sind für den Menschen, und nur für diesen, interpretierbar, benötigen aber Daten als Träger. Maschinen tauschen also untereinander Daten aus und transformieren diese anhand von definierten Regeln, während Menschen Informationen austauschen und diese vor ihrem Wissenshintergrund (mehr oder weniger präzise) verstehen. Diese Bestimmung ist hilfreich, da im behandelten Kontext das Zusammenspiel von Mensch und Maschine im Vordergrund steht und dabei dem Bedeutungsaspekt für ersteren großes Gewicht zukommt.

2.2 Der Prozeß des Medienwandels

Mit dem Übergang konventioneller Medien in die elektronische Form (Beispiel: Aus dem Brief wird die E-mail, aus der Werbebroschüre der Website, aus der zentralen Dokumentenablage das Intranet) stellen sich zusätzliche Anforderungen. Die undifferenzierte Übertragung strukturierender Mechanismen von konventionellen auf moderne Medien führt zu zahlreichen Defiziten bei Verfügbarkeit, Orientierung und Etikettierung von Dokumenten.

Jeder Ansatz, der eine Klassifikation von Informationen im Unternehmen anstrebt, muß diesen Übergangsprozeß berücksichtigen. Im Wesentlichen handelt es sich um die Lösung der Information von ihrem Träger, also die Abstraktion des Inhaltes von seiner physischen Form. Sämtliche Klassifikationssysteme, die auf diese physische Form rekurrieren (z.B. Leitz-Ordner für Briefwechsel oder Diskettenschränke für Datenträger) sind nicht geeignet, Informationen vollständig zu erfassen: Im Leitz-Ordner bleiben E-mails unberücksichtigt, im Diskettenschrank können keine Werbebroschüren eingehängt werden. Der Schlüssel zur Lösung dieses Problems liegt in der Identifizierung der die jeweiligen Informationseinheiten beschreibenden Metainformationen und in der angemessener Zuordnung von durch diese Metainformationen geweckter Erwartungshaltung und tatsächlichem Profil der Informationen [1].

2.3   Konventionalisierter Umgang mit Informationen

Ein großer Teil des erfolgreichen Umgangs mit Informationen ist auf das Bestehen von Konventionen zurückzuführen. Man könnte so weit gehen zu postulieren, daß im Sinne eines Schichtenmodells der Kommunikation eine oberste Schicht der Umgangskonventionen erst das umfassende Verstehen ermöglicht. (vgl. Abbildung 1).

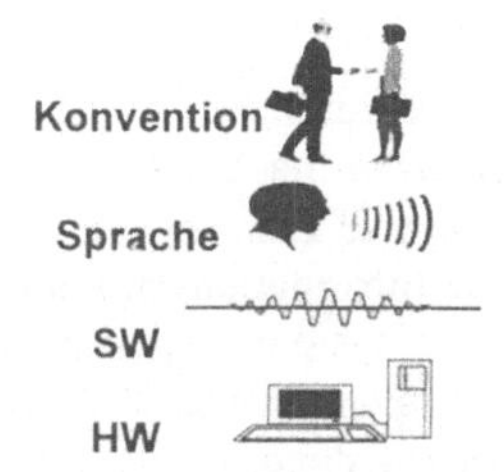

Abbildung 1: Konvention als oberste Kommunikationsschicht

Konventionen bestehen sowohl bezüglich einzelner, mit bekannten Metainformationen gekennzeichneter Informationseinheiten, als auch bei der Nutzung von Informationslagern. In beiden Fällen wird von den regelmäßigen Nutzern die Erwartungshaltung aufgebaut, daß neue Informationseinheiten einer bekannten Klasse oder eines vertrauten Informationslagers größtenteils die Eigenschaften besitzen, die in den bereits bekannten Vertretern der genannten Gruppen als typisch angesehen werden. Eine wesentliche Funktion dieser Erwartungshaltungen liegt in der Reduzierung der Komplexität der informationellen Umwelt. Klassifikationen von Informationen können als Konventionen in diesem Sinne angesehen werden, wenn sie hinreichend in der Nutzergemeinde verankert sind. Dies ist für viele konventionelle Formen von Informationen bereits gegeben, beispielsweise für Sicherheitshinweise oder Werbeschriften. Für die meisten digitalisierten Formen von Informationen fehlen Umgangskonventionen dagegen noch.

3   Lösungsansätze

Entscheidende Beiträge zur Beantwortung dieser Herausforderungen können davon erwartet werden, daß man die Erfahrungen nutzbar macht, die in den vergangenen Hundert Jahren in der Anwendung der großen Klassifikationssystem gemacht wurden [9]. Eine zentrale Unterscheidung dieser Systeme ist die zwischen *Prä-Koordination* und *Post-Koordination*. Präkoordinative Systeme ordnen ihre Informationen vorab in eine stabile Systematik ein – das Paradebeispiel dafür ist die Bibliothek, in der jedes Buch einen festen Platz hat. Postkoordinative Systeme treffen die Zuordnung im Einzelfalle jeweils neu und bauen darauf, daß die Information mit genügend Erkennungsmerkmalen ausgestattet ist, die eine hilfreiche Zuordnung gestatten. Nach den Erfahrungen mit den Suchmaschinen des Internet erscheint dies als eine nicht leicht einzulösende Erwartung. [5]

3.1   Präkoordinative Klassifikation

3.1.1 Monohierarchie

Monohierarchien kommen dadurch zustande, daß man ein Sachgebiet systematisch von oben nach unten immer feiner aufteilt – der klassische Fall einer Baumstruktur. Monohierarchien für unstrukturierte Informationen existieren im Bibliotheks- und Archivbereich bereits seit langem. Dabei werden neue Informationseinheiten in das bestehende Klassifikationssystem gemäß ihrem Inhalt eingeordnet. Beginnend bei der ersten Differenzierungsstufe, die bei universellen Klassifikationssystemen meist aus den wissenschaftlichen Disziplinen besteht, muß in jeder Stufe erneut eine nähere Bestimmung des Inhalts vorgenommen werden. Ist die adäquate Position in der Hierarchie erreicht, erhält die Informationseinheit einen eindeutigen Eintrag und eine Kennung, die diese Position wiedergibt.

3.1.2 Facette

Wesentlich größere Flexibilität bietet die Erweiterung der Monohierarchie durch Facetten. Der Einsatz von Facetten wurde erstmals 1933 in der *Colon-Klassifikation* nach Shiyal R. Ranganathan verwendet, um neben Büchern auch kleinere Dokumente wie Zeitschriftenaufsätze (Mikrodokumente) geeignet zu klassifizieren. Zunächst werden Informationseinheiten dazu wie bei der Internationalen Dezimalklassifikation in ein Hierarchiesystem eingeordnet. Dieses ist bei der Colon-Klassifikation deutlich weniger ausgebildet. Der zweite Teil der Zuordnung findet anhand von Facetten statt. Es gibt dazu die fünf Facettenkategorien *Personality (P), Matter (M), Energie (E), Spare (S)* und *Times (T)* [4]. Für die jeweils spezifischen Anwendungsgebiete werden für die ersten drei Kategorien individuelle Facetten definiert. Mit derart detaillierten Facetten lassen sich Sachverhalte sehr viel genauer beschreiben, als dies mit einer reinen Monohierarchie möglich ist.

3.1.3 Unscharfe Gruppen

Die Vorteile einer Klassifikation werden zu Nachteilen, wenn sich Informationsschwerpunkte verschieben, da dann die ein effizientes Retrieval ermöglichende Starrheit des Systems hinderlich wird und bei jeder Änderung der Definitionen für alle Beteiligten ein Nachholbedarf entsteht. Somit scheint monohierarchische und Facettenklassifikation nur für wenig dynamische Informationen geeignet zu sein – und damit für den größten Teil der unstrukturierten Informationen in Unternehmen ungeeignet. Weitere Schwierigkeiten ergeben sich durch zu geringe Spezifität und Probleme bei der Objektivität der Zuordnung [5].

Eine Konsequenz daraus ist, statt streng universeller Kategorien Beschreibungsdimensionen als unscharfe Gruppen einzusetzen. Bei diesen wird der Anspruch aufgehoben, für jede Information eine vollständige Beschreibung nach global definierten Merkmalen zu erstellen. Das Netz von Informationseinheiten hat dann weder eine eindeutige Wurzel noch einen eindeutigen Weg zwischen zwei Einheiten. Typische Gebilde dieser Art sind Hypertextsysteme, die dezentral gepflegt werden. Wenn ein Anwender in ein solches System eintritt, ordnen sich die Elemente abhängig vom Eintrittspunkt verschieden an.

Die Beschränkung auf 'Inseln' der detaillierten Klassifikation umgeht pragmatisch das verbreitete Problem universeller Klassifikationssysteme, durch die Komplexität und Inkohärenz der menschlichen Begriffswelten zu schwerfällig zu werden, um neue Entwicklungen und Verschiebungen der Wortbedeutungen schnell und mit geringem Aufwand zu integrieren.

3.2 Postkoordinative Klassifikation

Klassifikation ist nicht nur auf Dokumente beschränkt, ein wesentlicher Teil betrifft auch die Definition bestimmter *Rollen* im Unternehmen. Ein Informationsrezipient hat je nach Aufgabe und Vorwissen unterschiedliche Anforderungen an die Dokumente; der Informationsproduzent erzeugt abhängig von seiner Funktion Information unterschiedlicher Kategorien.

Kern der Postkordination ist daher der Abgleich der erzeugten Beschreibung von Informatiosobjekt und Nutzeranforderung. Werden neue Informationen erzeugt, so sollte eine Interessentengruppe über Existenz und Zugriffsmöglichkeiten benachrichtigt werden. Diese Vorgänge lassen sich weitgehend automatisieren, wie Beispiele u.a. des *Information Filtering* belegen. [8]

Besonders interessant ist es, mehrerere Teilklassifikationen auf verschiedenen Ebenen zu verwenden. Dazu bieten sich vielfältige Beschreibungskriterien an. Beispielsweise ließen sich (a) Fachgebiet, (b) Detaillierungsgrad und (c) Gültigkeitszeitraum definieren. Informationseinheiten werden dann durch einen Vektor (a,b,c) beschrieben. Dieser könnte mit den Anforderungen des Suchenden verglichen und die geeignetsten Informationseinheiten ausgewählt werden.

3.3 Umsetzung im Unternehmen

Wesentliche Elemente der Umsetzung eines Kategoriensystems implizieren auch kulturelle Aspekte des Umgangs mit Information. Der Einsatz ist nur möglich, wenn eine breite Akzeptanz bei den Anwendern sichergestellt ist. Zusätzlich kann nennenswerter Aufwand dadurch gespart werden, daß ein Wissenstransfer aus den etablierten Disziplinen eingeleitet wird: Die Fachrichtung *Dokumentation und Information* hat hierzu wichtige Erfahrungen beizusteuern; die Anpassung auf Unternehmensbelange ist jedoch noch zu leisten. Ein Klassifikationssystem für Dokumente im elektronischen Medium hat andere Rahmenbedingungen als bestehende betriebliche Informationssysteme:

- Das lebende System bildet die Unternehmenskultur ab. Es benötigt daher Feedbackzyklen, um Konventionen einzuführen und zu stabilisieren und muß sich beständig um die Akzeptanz aller Beteiligten bemühen.
- Durch Bedeutungsverschiebungen müssen Ungenauigkeit und Fehler zugelassen werden.
- Ein funktionierendes Klassifikationssystem im Unternehmen wird nie einen statischen Charakter haben können. Die Dynamik des Systems muß aber gesteuert werden.

Erfolgreiche Kooperation kann durch ein derartiges Klassifikationssystem erreicht werden, wenn sie von gemeinsamen Konventionen getragen wird. Die heute verwendeten Konventionen sind im Unternehmen auf zwei Fronten Angriffen ausgesetzt:

a) Zunehmend findet Kooperation auch international und bereichsübergreifend statt. Dabei können die Beteiligten zunächst nur auf wenige gemeinsame Konventionen aufbauen. Folglich müssen besonders im Umgang mit unstrukturierten Informationen viele ursprünglich durch Konventionen geregelte Abstimmungen entweder expliziert oder neue Konventionen gezielt aufgebaut werden. Welcher Weg zu wählen ist, hängt im Wesentlichen von der Gültigkeitsdauer der benötigten Konvention ab.

b) In konventionellen Medien etablierte Zuordnungskonventionen von Trägermedium und transportierter Information können nicht direkt ins Medium Computer übertragen werden. Zu diesem Defizit kommen weitere Informationsklassen hinzu, die in konventionellen Medien kaum auftraten und daher auch dort keine Konventionen haben.

Es erscheint sinnvoll, zur Vermittlung zwischen diesen Fronten eine eigene Rolle zu definieren: den *Informationsmanager*. Seine Aufgabe ist es, die Dynamik des Systems zu reflektieren und zu steuern. (vgl. Abbildung 2) Indem er den Prozeß des Entstehens und Verschwindens von Konventionen in der Unternehmenskommunikation steuert, kann das den Konventionen inhärente Effizienzpotential ausgeschöpft werden. Auf diese Weise läßt sich die

im digitalen Medium verlorengegangene Effizienz nicht nur wiedererlangen, sondern so steigern, daß die verfügbaren Möglichkeiten des neuen Mediums tatsächlich gewinnbringend genutzt werden können.

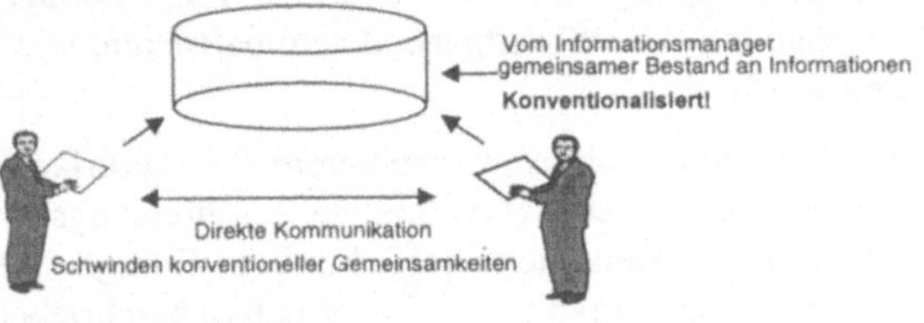

Abbildung 2: Das Arbeitsgebiet des Informationsmanagers

4 Ziele des Workshops

Mit der vorgestellten theoretischen Basis können nun konkrete Klassifikationsschemata gemäß den jeweiligen Anforderungen entwickelt werden. Entscheidende Bedeutung kommt dabei der Bestimmung geeigneter Klassifikationskriterien zu. Die Referenten haben hierzu bereits Vorschläge gemacht [6] [7]. Daran anknüpfend stellt sich die Frage nach der Automatisierbarkeit der Klassifizierung unstrukturierter Informationen. Ein Prototyp wird einige Kriterien berücksichtigen und eine einsatzfähige Benutzeroberfläche bieten. Dieser wird dann Ausgangspunkt für den zweiten Diskussionsschwerpunkt des Workshops sein, die unternehmensweite Anwendbarkeit. Der Workshop wird als moderierte Diskussion durchgeführt. Der Diskussionsverlauf ist grundsätzlich offen, wird sich aber an folgenden Kernfragen orientieren:

- Wo liegen die Möglichkeiten, wo die Grenzen der Klassifikation unstrukturierter Informationen im Unternehmen?

- Welches sind die zentralen Unterschiede zwischen konventionellem und elektronischem Medium und inwiefern muß dies bei der Umsetzung berücksichtigt werden?

- Inwiefern können konkrete Aussagen zur Wahl der einzusetzenden Klassifikationsschemata, Klassifikationskriterien und begleitender organisatorischer Maßnahmen bei der Umsetung im Unternehmen getroffen werden?

5 Literatur

[1] *Königer, Paul / Janowitz, Karl* (1995): Drowning in Information, but Thirsty for Knowledge. In: International Journal of Information Management, Vol. 15, No. 1/95, S. 5-16

[2] *Touche Ross and Co.* (1994): Information Management Survey. Touche Ross: London

[3] *Schenk, David* (1997): Data Smog. Harper Collins: New York

[4] *Ranganathan, Shiyali R.* (1965): The Colon Classification. Rutgers University Press: New Brunswick, N.J.

[5] *Weinberg, Bella H.* (1996): Complexity in Indexing Systems -- Abandonment and Failure: Implications for Organizing the Internet. ASIS 1996 Annual Conference Proc. Oct. 19-24, 1996. *Online* http://www.asis.org/annual-96/ElectronicProceedings/weinberg.html; (22.6.1997)

[6] *Königer, Paul / Reithmayer, Walter* (1998): Management unstrukturierter Informationen. Campus Verlag, Frankfurt / New York

[7] *Gronski, Andreas* (1997): Entwicklung eines Instrumentariums zur Behandlung unstrukturierter Informationsflüsse in Unternehmensnetzen. Diplomarbeit Universität Paderborn, Lehrstuhl Informatik & Gesellschaft.

[8] *Kuhlen, Rainer* (1997): Abstracts - Abstracting - Intellektuelle und maschinelle Verfahren. In: Buder [u.a] (Hrsg.): Grundlagen der praktischen Information und Dokumentation. Band 1. K.G. Saur: München

[9] *Manecke, Hans-Jürgen* (1997): Klassifikation. In: Buder [u.a.] (Hrsg.): Grundlagen der praktischen Information und Dokumentation, Band 1. K.G. Saur: München

Adressen der Autoren

Andreas Gronski
sd&m software design & management GmbH & Co. KG
Thomas-Dehler-Straße 27, 81737 München
Email: gronski@sdm.de

Paul Königer
Siemens AG
Open Enterprise Computing
Otto-Hahn-Ring 6, 81739 München
Email: pako@mch.sni.de

Workshop „Was leisten ergonomische und arbeits-psychologische Verfahren im Software-Entwicklungszyklus?"

Dr. Günther Gediga, Dr. Kai-Christoph Hamborg

Fachbereich Psychologie, Universität Osnabrück

Dr. W. Hampe-Neteler

TÜV Informationstechnik GmbH, Essen

Zusammenfassung

In dem Workshop werden Einsatzmöglichkeiten von Methoden zur Analyse und Gestaltung ergonomischer Software im Rahmen des gesamten Entwicklungszyklus thematisiert. Im Schwerpunkt geht es darum, zu erörtern, wann im Entwicklungszyklus ergonomische Gestaltung effektiv ansetzen sollte bzw. kann, welche Methoden diesbezüglich Verwendung finden können. Die Bearbeitung des Themas soll sich nicht auf die Software-gestaltung im engeren Sinn („reine" Produktgestaltung) beschränken, sondern auch den Entwicklungsprozeß und (psychologische) Aspekte der Arbeitsgestaltung, die in Entwicklungsprojekten zu berücksichtigen sind, einbeziehen. Gegebenenfalls soll für die angesprochenen Fragen Entwicklungsbedarf bestimmt werden.

Abstract

This Workshop shall stress the applicability of techniques for designing ergonomic software in the whole development life cycle. The main aim will be the discussion in which phase of the life cycle an ergonomic design should take place and which techniques should be applied. The Workshop should not be limited to software design in the sense of pure product design but it will also include (psychologically based) aspects of work design which should be considered in development projects.

1 Thema der Workshops

Der Schwerpunkt software-ergonomischer Forschung hat sich in den letzten Jahren von Fragen der Arbeits- und Aufgabengestaltung (s. die Titel der Software-Ergonomie Tagungen 89 und `93: „Aufgabenorientierte Systemgestaltung und Funktionalität" und „Von der Benutzeroberfläche zur Arbeitsgestaltung") zur praktischen Umsetzung verfügbarer software-ergonomischer Erkenntnisse und der Anwendung und Weiterentwicklung von Methoden im Rahmen des „Usability Engineering" verschoben.

Diese Entwicklung ist sicherlich auch darin begründet, daß aktuelle rechtliche Verordnungen Umsetzungskonzepte für software-ergonomisches Handeln in der Praxis fordern und damit einen praktischen Handlungsbedarf erzeugt haben. Der Bedarf an praktikablen arbeits-wissenschaftlichen Instrumenten zur aufgaben- und nutzerbezogenen Softwaregestaltung ist darüber hinaus schon vor längerer Zeit registriert worden [2].

Die Integration von Konzepten und Methoden der Software-Ergonomie und des Software-Engineering wird damit zu einer wichtigen Forschungs- und Praxisfrage, sowohl in Bezug auf das zu gestaltende Produkt als auch auf die Organisation des Entwicklungsprozesses [s. 4, S. 199f].

Wohl bestehen allgemeine Konzepte zur Organisation eines „Usability Engineering Life Cycle" [1, 5], die den Entwicklungszyklus konzeptionell von der Analyse menschlicher Arbeit bis zur Gestaltung und ergonomischen Bewertung von Programmen abdecken [s. 3]. Es fehlen jedoch Konkretisierungen und die Prüfung dieser Konzepte an praktischen Problemstellungen. Dieses war auch eines der Ergebnisse des Workshops „User-Perceived Quality of Interactive Systems" [6, 7] auf der Software-Ergonomie Tagung 1997.

Auch wurde in entsprechenden „Usability Life Cycle"-Modellen die geforderte Integration arbeitspsychologischer und software-ergonomischer Analyse- und Gestaltungsmethoden [s. 3] bisher noch nicht überzeugend realisiert.

Vor dem Hintergrund dieser Bewertung aktueller Entwicklungstendenzen und praktischer Anforderungen im Bereich der Software-Ergonomie und des Software-Engineering sollen in dem Workshop Einsatzmöglichkeiten von Methoden zur Gestaltung ergonomischer Software im Rahmen des gesamten Entwicklungszyklus an Hand von Fallbeispielen dargestellt und diskutiert werden.

Die Konzentration richtet sich dabei auf zwei wesentliche Punkte:

1.) wann bestehen im Entwicklungszyklus geeignete Zeitpunkte, zu denen ergonomische Gestaltung effektiv ansetzen sollte bzw. kann,

2.) welche Evaluations-, Analyse- und Gestaltungsmethoden können oder sollten Verwendung finden und wo besteht Entwicklungsbedarf für neue bzw. Überarbeitungsbedarf für existierende Methoden.

Die Bearbeitung des Themas soll sich nicht alleine auf die Softwaregestaltung im Sinne der Produktgestaltung beschränken, sondern auch den Entwicklungsprozeß und (psychologische) Aspekte der Arbeitsgestaltung mit einbeziehen.

2 Ziele des Workshops

Der Workshop soll dazu beitragen, die genannten praktischen und konzeptionellen Problempunkte bei der Umsetzung software-ergonomischer Intervention im Rahmen des gesamten Entwicklungszyklus zu systematisieren und Lösungsansätze zu diskutieren.

Drei konkrete Ziele werden in dem Workshop verfolgt:

1. Problematisierung software-ergonomischen Handelns in praktischen Beratungs- und Entwicklungsprojekten an Hand von Fallbeispielen,
2. Reflektion der software-ergonomischen „Handlungsspanne" in Entwicklungsprojekten nach methodischen und organisatorischen Aspekten an Hand der Fallbeispiele,
3. Formulierung von Thesen zum effektiven Einsatz software-ergonomischer und arbeitspsychologischer Analyse- und Gestaltungsmethoden, ggf. Bestimmung von Handlungsbedarf.

3   Zeitlicher Ablauf

Der Workshop besteht aus 2 Teilen: Im ersten Teil werden an Hand von Fallbeispielen Möglichkeiten und Probleme der ergonomischen Arbeits- und Softwaregestaltung beschrieben und diskutiert (Dauer: ca 2 Stunden). Die Fallbeispiele werden geschlossen präsentiert (jeweils 10 Minuten) und dann im Plenum diskutiert. Im zweiten Teil des Workshops werden Thesen zur Problemlösung erarbeitet (siehe Abschnitt 2, Punkt 3; Dauer: ca. 1 Stunde).

4   Form der Durchführung

4.1   Präsentation von Fallbeispielen.

Ausgangspunkt für die Bearbeitung des Themas in dem Workshop sollen Fallbeispiele sein. In den Fallbeispielen werden praktische Entwicklungsprojekte dargestellt, in denen in unterschiedlichen Umfang software-ergonomische Interventionen stattgefunden haben.

Die Darstellung der Fallbeispiele wird sich an der folgenden Struktur orientieren:

A)   Kurzchrakterisierung des vorgestellten Fallbeispiels

B)   Charakterisierung des zugrundeliegenden Softwareentwicklungsprozesses /-projektes

C)   Eingesetzte Methoden/Verfahren und Einsatz der Methoden/Verfahren im Beispiel

E)   Was hat die geplante Durchführung der Methoden beeinflusst?

F)   Was hat der Einsatz der Methoden bewirkt?

G)   Welche (Neben-) Effekte der Verfahrensanwendung traten auf?

H)   Katalysatorfunktionen: Wurden Ergebnisse des Verfahrens genutzt, um andere laufende Verfahren zu unterstützen bzw. zu beschleunigen

I)   Abstimmung/Verträglichkeit mit softwaretechnischen Methoden und Entwicklungswerkzeugen

J)   Ergebnisse für die Anwendung oder Weiterentwicklung von Methoden und/oder Verfahren

K)   Zusammenfassung/Gesamteinschätzung

4.1.1 Referenten

Die Fallbeispiele werden von den folgenden Referenten dargestellt:

* Dr. Ahmed Cakir, ERGONIOMIC Institute Berlin. (Thema wird noch festgelegt).
* Prof. Dr. Heiner Dunkel/Bildungswissenschaftliche Hochschule Flensburg, Universität. (Thema wird noch festgelegt).
* Dr. Hampe-Neteler, TÜVIT Essen. Thema: „Entwicklungsbegleitung zur ergonomischen Evaluierung und Gestaltung einer Liegenschaftsverwaltungs-Software"
* Dr. G. Gediga, Dr. K.-C. Hamborg, Universität Osnabrück. Thema: „Formative Evaluation als unterstützende Maßnahme bei dem Redesign von Sparkassen-Software".

4.2 Diskussion

Die TeilnehmerInnen des Workshops sind aufgefordert, sich an der Diskussion der Fallbeispiele und der Formulierung von Thesen zu dem Thema (siehe Abschnitt 2) zu beteiligen.

5 Literatur

[1] Gould, J. D., Boies, S. J. & Ukelson, J. (1997). How To Design Usable Systems. In M. Helander, T.K.
     Landauer & P. Prabhu (eds.). *Handbook of Human-Computer Interaction. Second completely revised edition.*
     Amsterdam: Elsevier Science B.V.

[2] Hamborg, K.-C. & Schweppenhäußer, A. (1993). Zur Bedeutung psychologischer Arbeits- und
     Aufgabenanalyse für die Softwaregestaltung. In: K.H. Rödiger (Hrsg.). *Softwareergonomie '93. Von der
     Benutzungsoberfläche zur Arbeitsgestaltung.* Stuttgart: Teubner.

[3] Hampe-Neteler, W. (1994). *Software-ergonomische Bewertung zwischen Arbeitsgestaltung und Software-
     Entwicklung.* Frankfurt:Lang.

[4] Maaß, S. (1993). Software-Ergonomie. Benutzer- und aufgabenorientierte Systemgestaltung. *Informatik
     Spektrum*, 16, S. 191-205.

[5] Nielsen, J. (1992). The Usability Engineering Life Cycle. *IEEE Computer*, 25, 3, 12-22.

[6] Vossen, P., Gediga, G. & Hamborg, K.-C. (1997). *User-Perceived Quality of Interactive Systems.* Workshop
     auf der Tagung Softwareergonomie '97 in Dresden. [http://www.psycho.uni-osnabrueck.de/se97work/].

[7] Vossen, P. (1997). User Percieved Quality of Interactive Systems. *Ergonomie & Informatik*, 31, S. 32-34.

Adressen der Autoren

Dr. Günther Gediga,
Universität Osnabrück,
FB Psychologie
Fachgebiet Methodenlehre
Seminarstr. 20
49069 Osnabrück

Dr. Kai-Christoph Hamborg
Universität Osnabrück,
FB Psychologie
Fachgebiet Arbeits- und
     Organisationspsychologie
Seminarstr. 20
49069 Osnabrück

Dr. Wolfgang Hampe-Neteler
TÜV Informationstechnik GmbH
Im Teelbruch 122
D-45219 Essen-Kettwig

Praktischer Workshop:
Werkzeuge zur Visualisierung
von Unternehmensinformationen

Hans-Günter Lindner, Christoph G. Thomas

HumanIT
Human Information Technologies GmbH
GMD TechnoPark

GMD - Forschungszentrum
Informationstechnik GmbH
Institut für angewandte Informationstechnik
Forschungsbereich Mensch-Maschine-
Kommunikation (FIT.MMK)

1 Einleitung

In der täglichen Praxis führte der Mangel an ergonomisch gestalteter Software dazu, dass entscheidungsrelevante Informationen häufig nur indirekt von Entscheidern genutzt werden. Dabei erfordert gerade betriebswirtschaftliche Standardsoftware Benutzungsschnittstellen, die an die Anforderungen ihrer Nutzer ausgerichtet sind. Dies lässt sich in der Regel erst dann erreichen, wenn die intensive Auseinandersetzung mit persönlichen Erfahrungen der Nutzer aus der Praxis Grundlage der Entwicklungen ist.

Dieser Workshop präsentiert Führungskräften, Beratern und Projektleitern neueste Entwicklungen ergonomischer Standardsoftware zur betriebswirtschaftlichen Entscheidungsfindung:

- Das SAP Business Information Warehouse
- humanIT InfoZoom - eine neuartige Methode zur Visualisierung von Informationslandschaften und die
- SAP Ad-hoc-Query zur leichten Extraktion von Inhalten.

Die Erfahrungen der Teilnehmer sollen zu einem Leitfaden verdichtet werden, der einen erfolgreichen Einsatz in der Praxis erleichtert. Da der Workshop an den Bedarfen der Teilnehmer angepasst werden soll, sind die konkreten Inhalte erst kurz vor der Durchführung bekannt. Im folgenden werden neben dem Zeitplan und einer Grobbeschreibung der Inhalte die Softwarewerkzeuge präsentiert.

2 Die Motivation

Entscheidungsrelevante Unternehmensdaten werden heutzutage in fast allen Bereichen von betrieblicher Standardsoftware abgebildet. Für diese Abbildung sind häufig umfangreiche Analysen und Modellbildungen notwendig. Die dafür notwendigen Anstrengungen lassen in der betrieblichen Praxis wenig Raum für eine nutzerorientierte Versorgung mit entscheidungsrelevanten Informationen. Insbesondere im Personalmanagement wird es immer wichtiger, schnell und einfach einen aktuellen Gesamtüberblick über die wichtigsten Ressourcen des Unternehmens selbsttätig ermitteln zu können. Dies erfordert eine Anpassung der Informationsdarstellung an die unterschiedlichen Interessen und Sichtweisen der Nutzer.

Einfache Benutzung, Überblick und leichte Auswertungen versprechen viele Verkaufsbroschüren konventioneller Software. Es wäre also alles bereits getan, wenn wir der Werbung glauben schenken könnten. Realität ist jedoch, dass meist nur Spezialisten die Unternehmensdaten analysieren und die Entscheidungsunterstützung vorbereiten können. Die Information kommt daher meist nur mehrfach gefiltert an die Entscheider. Hoher Schulungsaufwand oder die Überfrachtung mit Funktionalität können Gründe für eine eingeschränkte Nutzung sein.

3 Die Ziele

Der Workshop vermittelt Kenntnisse, wie Nutzer schnell einen Überblick über ihre Daten gewinnen können. Das Finden und Analysieren entscheidungsrelevanter Informationen, hier beispielhaft aus dem Bereich Personal, wird mit ergonomischen Softwarewerkzeugen direkt am Computer erarbeitet.

Für das praktische Training wird als Basis SAP R/3 verwendet. Komponenten für die Aufbereitung der Informationen sind die *SAP Ad-hoc-Query*, das *SAP Business Information Warehouse* (BW) und *InfoZoom* von *humanIT*.

Ziel des Workshops soll ein gemeinsam erarbeiteter Leitfaden für den erfolgreichen Einsatz von Reportingsystemen sein. Dabei sollen Aspekte der Personalentwicklung und der Ergonomie Berücksichtigung finden.

4 Die Inhalte

Die intensive Auseinandersetzung mit den Interessen und Zielen der Teilnehmer in der Praxis erfolgt in einem Zielabgleich zu Beginn des Workshops. Anschließend wird das Basiswissen zur Nutzung der Softwaresysteme vermittelt. Hierbei wird für die praktischen Übungen auf praxisrelevante Trainingsdaten aus dem SAP R/3 System zurückgegriffen.

Das Praxisbeispiel wird ausgehend von einfachen Abfragen zur Deckung des Standardinformationsbedarfs stufenweise komplexer gestaltet. Damit soll sowohl die Handhabung von einfach benutzbaren Abfragewerkzeugen wie die Ad-hoc-Query als auch der Umgang mit dem BW und InfoZoom vermittelt werden. Am Beispiel sollen Funktionen des Datawarehousing und Datamining auf einfache Nutzbarkeit und Intuitivität beurteilt werden.

Grundlegende Fragen der täglichen Report- und Analysenerstellung sollen beantwortet werden:

- Wie kann ich Daten einfach extrahieren und Standardreports erzeugen?
- Wie kann ich schnell detaillierte Fragen im Unternehmensalltag beantworten?
- Wie kann ich einfach entscheidungsrelevante Daten analysieren?

In Gruppenarbeit sollen unterschiedliche Aspekte bei der Nutzung und Einführung herausgearbeitet werden. Die Teilergebnisse sollen die Grundlage für weitere Besprechungen und letztendlich für einen gemeinsamen Leitfaden zur Planung und Einführung von Reportingsystemen darstellen.

5 Die Softwarelösungen

Im folgenden werden die Funktionen der Softwarelösungen kurz dargestellt. Tiefergehende Beschreibungen können bei SAP AG unter http://www.sap-ag.de und humanIT Human Information Technologies GmbH unter http://www.humanit.de gefunden werden.

5.1 SAP Ad-hoc-Query

Die Ad-hoc-Query ist eine einfache und schnelle Softwarelösung zur Selektion und Weiterverarbeitung von Personalinformationen aus SAP R/3. Die Ergebnisse stehen sehr schnell zur Verfüfung, da der Zugriff direkt auf die Datenbank erfolgt.

Die Ad-hoc-Query liefert Grunddaten zur umfangreichen Gestaltung von Berichten, ohne dass aufwendige Definitionen erstellt werden müssen. Die Selektions- und Ausgabefelder sind frei wählbar. Die Menge der selektierten Daten wird vor der Ausführung angezeigt und kann mit Hilfe unterschiedlicher Softwarelösungen wie z.B. InfoZoom angezeigt und weiterverarbeitet werden.

Die Ad-hoc-Query kann von mehreren Bereichen aus SAP R/3 heraus genutzt werden. Beispielsweise kann die Ad-hoc-Query direkt aus dem Menü des Moduls HR heraus, über den Manager's Desktop oder über das Personalinformationssystem *HIS* aufgerufen. Dabei kann die Selektion erfolgen wie in Abbildung 1 dargestellt oder über die Organisationsstruktur.

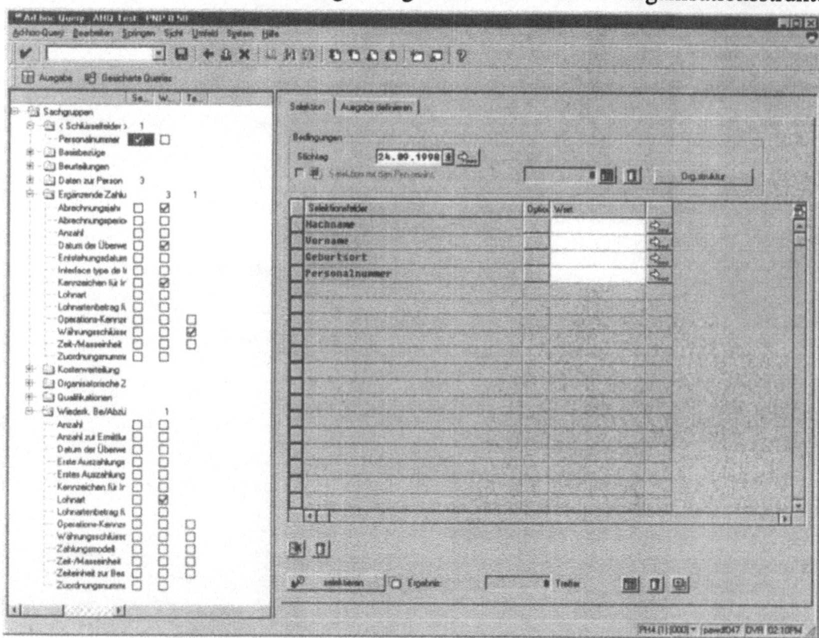

Abbildung 1: Die Benutzungsschnittstelle der Ad-hoc-Query

5.2 SAP Business Information Warehouse

Das SAP Business Information Warehouse (BW) ist eine eigenständige Anwendungsumgebung mit eigener Datenbank, die Informationen aus verschiedenen Datenquellen bezieht und für Abfragen sowie Analysen konzipiert ist. Das BW verbindet modernste Data-Warehouse-Technologien mit dem betriebswirtschaftlichen Know-How der SAP und ist ein eigenständiges R/3 System mit eigenen Release-Phasen.

Das Datenmanagement wird mit Metadaten unterstützt. Dabei können beliebige Daten integriert werden, egal ob diese aus R/3 stammen oder nicht. Intuitive Benutzungsschnittstellen und mächtige OLAP-Funktionalitäten (OLAP = OnLine Analytical Processing) stehen für eine Vielzahl von Nutzertypen zur Verfügung. Die inhaltliche Unterstützung orientiert sich an den Standardprozessen und Modellen von SAP. Kennzahlensystem und Benchmarking muss nicht getrennt implementiert werden, sondern steht sofort zur Verfügung.

Mit Hilfe des Business Explorer können Standardabfragen und –reports anderen Nutzern bereitgestellt werden. Die Ergebnisse können mit dem Business Explorer Browser betrachtet werden, der unterschiedliche Möglichkeiten der Visualisierung zulässt (Abbildung 2).

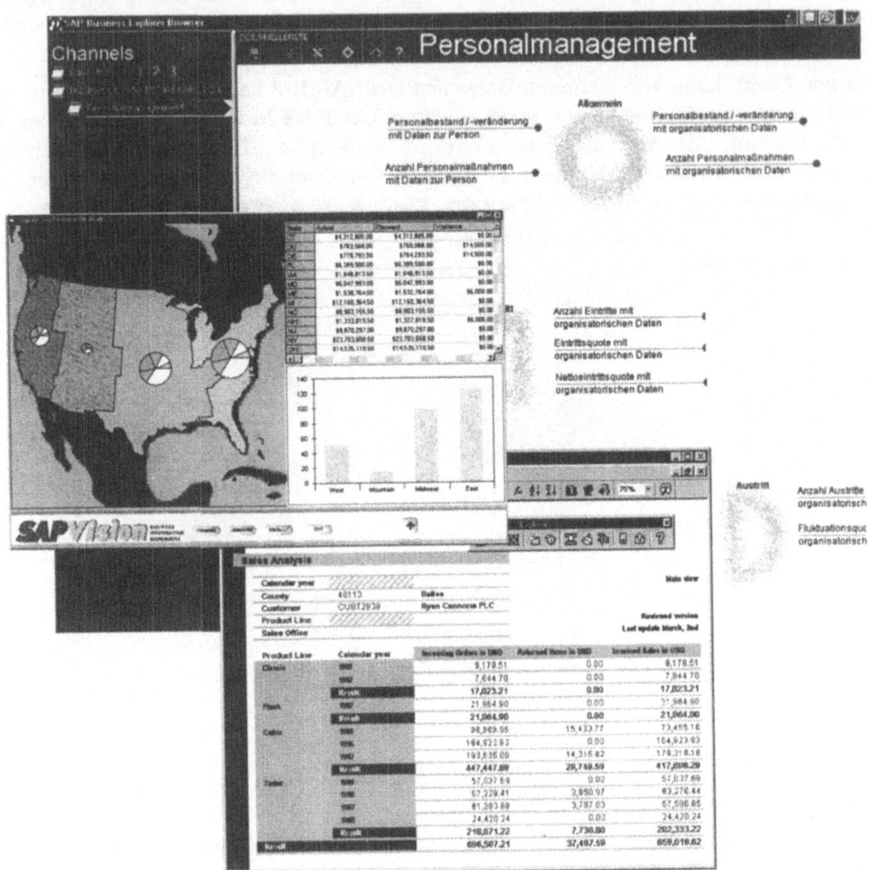

Abbildung 2: Beispiele zur Visualisierung durch den Business Explorer

Die Generierung von Abfragen erfolgt mittels des Business Explorer Analysers (Abbildung 3), der die Zusammenstellung von Standardmessgrössen, Kennzahlen und Dimensionen aus den InfoCubes - die zentralen Datenbehälter für Berichte und Auswertungen - erlaubt.

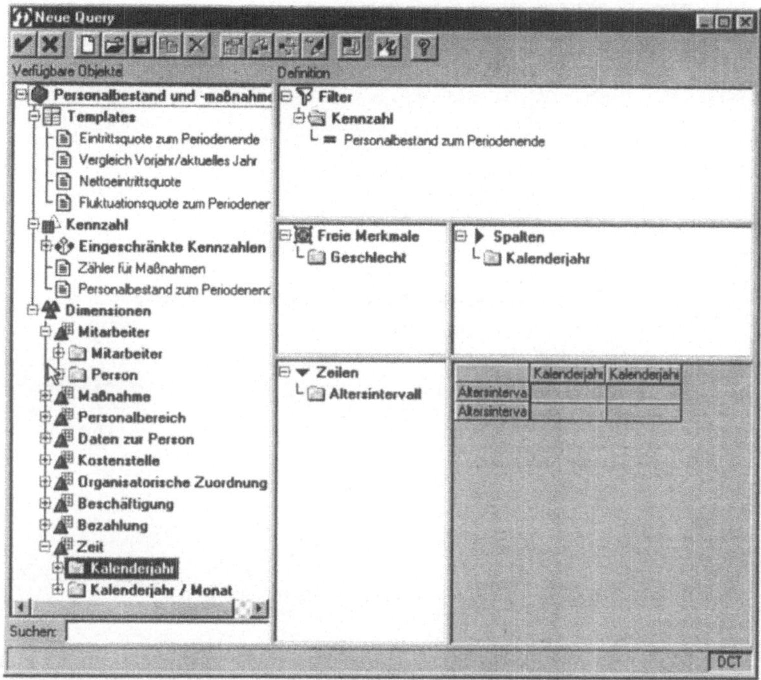

Abbildung 3: Business Explorer Analyser zur Generierung von Abfragen und Reports im BW

5.3 humanIT InfoZoom

InfoZoom ist eine neue Methode, Informationen in großen Datenmengen zu visualisieren und diese schnell auszuwerten. Mit wenigen Mausklicks können entscheidungsrelevante Zusammenhänge über die Ad-hoc-Query und die ABAP-Query aus R/3 gefunden und untersucht werden.

Das Arbeiten mit diesem universellen Werkzeug ist so einfach wie Fotografieren mit einem Zoom-Objektiv: Beliebige Ausschnitte aus der Informationslandschaft können ausgewählt, vergrößert und untersucht werden. Zusammenhänge werden sichtbar und sofort dokumentierbar. Das Ergebnis kann als Geschäftsgrafik oder als interaktiver Bildausschnitt weitergereicht werden - auf Diskette, CD oder direkt im Internet.

Die folgenden Abbildungen zeigen Beispiele der Analyse von Personalinformationen mit mehr als 200 Tochterunternehmen. Abbildung 4 zeigt die Übersicht über mehr als 1000 Mitarbeiter; dabei wird jedes Merkmal getrennt sortiert visualisiert. Links in der Abbildung sind Merkmalsbezeichnungen aus den Stammdaten wie z.B. Personalnummer, Name oder Geburtsdatum dargestellt. Hier lassen sich aber auch andere Merkmale abbilden, die unterschiedliche Dimensionen darstellen können. Damit werden InfoCubes für eine einfache Analyse leicht zugänglich gemacht. Dimensionswechsel müssen vorher nicht durchdacht, sondern können einfach durch Mausklick erstellt und die Auswirkungen der Auswahl sofort beobachtet werden.

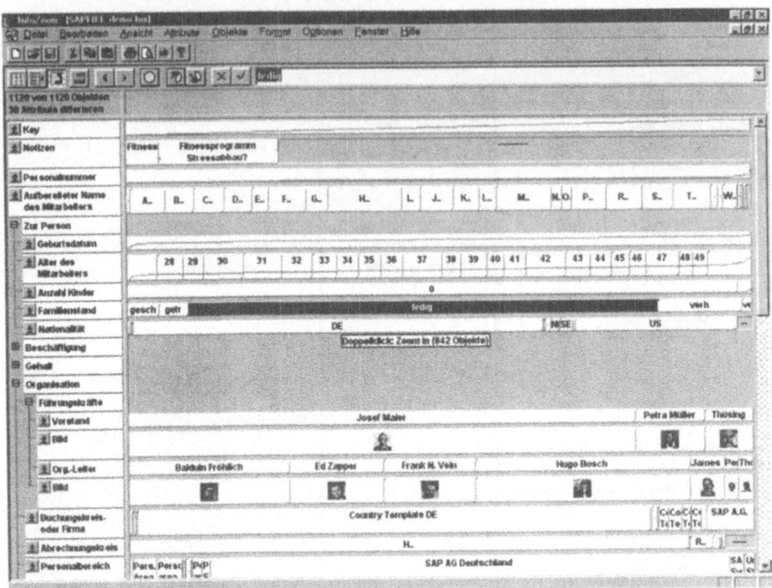

Abbildung 4: Übersicht über mehr als 1000 Personen

Für detaillierte Analysen können alle Datensätze nach einzelnen Merkmalen getrennt sortiert dargestellt werden. Abbildung 5 zeigt eine geschachtelte Sortierung nach Führungshierachien bei gleichzeitiger Visualisierung von Kennzahlen. Diese Kennzahlen werden in InfoZoom in Abhängigkeit der persönlichen Sichtweise angepasst.

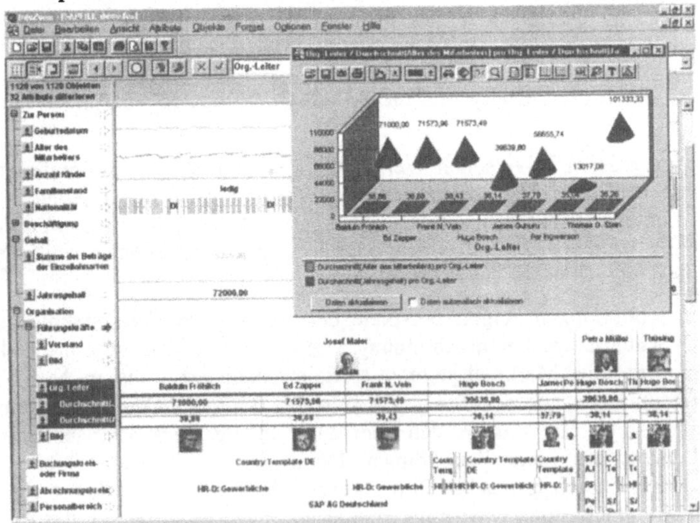

Abbildung 5: Interaktive Kennzahlenanalyse mit InfoZoom

Die Erfahrungen mit InfoZoom zeigen, dass der Schulungsaufwand trotz der Verwendung einer neuen Methode zur Visualisierung sehr gering ist. Schon nach ca. 10 Minuten kann der Nutzer 80% der Funktionalität nutzen.

Adressen der Autoren

Hans-Günter Lindner
HumanIT
Human Information Technologies GmbH
GMD TechnoPark
Rathausalle 10
53754 Sankt Augustin
Lindner@humanit.de

Christoph G. Thomas
GMD - Forschungszentrum
Informationstechnik GmbH
Institut für angewandte Informationstechnik
Forschungsbereich Mensch-Maschine-Kommunikation
(FIT.MMK)
Schloß Birlinghoven, 53754 Sankt Augustin
christoph.thomas@gmd.de

Workshop „Visualisierung von entscheidungsrelevanten Daten"

U. Bleimann[1], H. Reiterer[2], T. M. Mann[2], G. Mußler[2]
Fachhochschule Darmstadt[1], Universität Konstanz[2]

Zusammenfassung

Dieser Beitrag hat zum Ziel, die Inhalte des Workshops „Visualisierung von entscheidungsrelevanten Daten" kurz vorzustellen. Dieser Workshop konzentriert sich auf die Präsentation und Diskussion unterschiedlicher Strategien zur Visualisierung entscheidungsrelevanter Daten, wie man sie heute typischerweise in entscheidungsunterstützenden Informationssystemen (Management Support Systems) findet. Aus dem Themenkomplex der Entscheidungsunterstützung werden drei wichtige Bereiche exemplarisch vorgestellt, um sie im Workshop eingehender zu behandeln: Visualisierung von (Kenn-) Zahlen mit Geschäftsgrafiken, Visualisierung von Daten mit Raumbezug durch GIS-Komponenten, sowie Metaphern zur Visualisierung von unternehmensexternen Informationen.

Abstract

This paper describes the content of the workshop „information visualization for decision support". The workshop is focused on the presentation and discussion of a variety of strategies for information visualisation, as it can be found in todays Decision Support and Management Support Systems. From the diversity of information that can be visualised, the workshop takes three into account: visualisation of business graphics, visualisation of data with a spatial reference by GIS components and metaphors for the visualisation of external information.

1 Ziele und Zielgruppe

Das explosive Wachstum des Internet, die fortschreitende Computerisierung aller Geschäftsbereiche und die Einführung von Data Warehouses in vielen Unternehmen haben zu der Einsicht geführt, daß Visualisierungstechniken als ein wesentliches Werkzeug für viele technische und kommerzielle Bereiche angesehen werden müssen [1]. Die Fachdisziplin Informationsvisualisierung (Information Visualization) als Teildisziplin der Software-Ergonomie (Human Computer Interaction) hat hierzu in den letzten Jahren eine Vielzahl neuer Erkenntnisse und Visualisierungsstrategien entwickelt.

Der Workshop konzentriert sich auf die Präsentation und Diskussion unterschiedlicher Strategien zur Visualisierung *entscheidungsrelevanter Daten*, wie man sie heute typischerweise in entscheidungsunterstützenden Informationssystemen (Management Support Systems) findet. Die zunehmende Bedeutung derartiger Systeme (z.B. in Form von MIS Management Information Systems, DSS Decision Support Systems oder EIS Executive Information Systems) und der damit einhergehende, immer umfänglicher werdende Einsatz hat dazu geführt, daß Fragestellungen hinsichtlich adäquater Formen der Darstellung von entscheidungsrelevanten Daten vermehrt an praktischer Relevanz gewinnen.

Als Zielpublikum spricht der Workshop einerseits Anwender entscheidungsunterstützender Informationssysteme an. Für diese Zielgruppe bietet der Workshop Anregungen hinsichtlich zu beachtender Gestaltungsprinzipien und möglicher Visualisierungsformen.

Andererseits wendet sich der Workshop auch an die Hersteller von MIS-, DSS-, EIS- und GIS-Tools sowie OLAP-Front-Ends. Für diese Zielgruppe bietet der Workshop Anregungen hinsichtlich der Erweiterung ihrer Tools um entsprechende Visualisierungsmöglichkeiten – unter Beachtung entsprechender Gestaltungsprinzipien.

2 Ablauf des Workshops

Der Workshop ist ganztägig. Im Anschluß an eine theoretische Einführung werden in drei thematischen Blöcken die spezifischen Themenstellungen in ähnlicher Art und Weise behandelt. Nach einer wissenschaftlichen Einführung in das jeweilige Thema berichtet ein Hersteller über die Möglichkeiten zur Visualisierung von entscheidungsrelevanten Daten mittels heutiger Tools und anschließend ein Anwender über seine praktischen Erfahrungen beim Einsatz derartiger Tools. Die einzelnen Themenblöcke werden dann durch eine ausführliche Diskussion abgeschlossen. Durch diese Vorgehensweise soll sichergestellt werden, daß der Workshop eine gute Mischung bietet zwischen wesentlichen theoretischen Erkenntnissen aus dem Bereich der Forschung und in der Praxis gemachten Erfahrungen und verfügbaren technischen Möglichkeiten.

3 Visualisierung

Die Bezeichnung Informationsvisualisierung (Information Visualization) hat sich in den letzten Jahren als eigenständiger Fachbegriff für den Einsatz graphischer Mittel bei der Erschließung großer Datenbestände mit Hilfe von Computer herausgebildet. Bereits Ende der achtziger Jahre führte der Einsatz entsprechender Techniken bei der Bearbeitung wissenschaftlicher Datenbestände zu einer immer intensiveren Auseinandersetzung mit Fragen der Visualisierung ·(Scientific Visualization). Hintergrund waren seinerzeit die neuen Möglichkeiten der Rechnertechnik, die den Forschern über entsprechende Darstellungen neue Einblicke in mental nur schwer erschließbare Datenbestände ermöglicht. Anfang der 90er prägten Robertson, Mackinlay und Card den Begriff Information Visualization für die Übertragung des Konzeptes auf nicht wissenschaftliche Informationsbereiche, in denen ebenfalls mit komplexen Datenmengen gearbeitet wird [2]. Insbesondere für den Bereich der Entscheidungsunterstüzung spielt die Suche nach der adäquaten Visualisierung, unter Ausnutzung der Möglichkeiten von Softwaresystemen der neuesten Generation, eine zunehmend wichtigere Rolle.

Unter Visualisierung werden in der Literatur zwei generelle Facetten beschrieben: Datenpräsentation versus Datenexploration. Bei der Datenpräsentation steht die Vermittlung an sich bekannter Sachverhalte durch geeignete Darstellungsformen im Mittelpunkt der Überlegung. Bei der Datenexploration soll eine Erleichterung der Erkennung und Aufdeckung neuer oder unbekannter thematischer Zusammenhänge durch geeignete Visualisierung ermöglicht werden („Visual Data Mining"). Die Übergänge sind dabei als fließend anzusehen. Beiden Facetten ist gemeinsam, daß der Erkenntnisgewinn mit Hilfe visueller Darstellungen unterstützt werden soll. Im Fall der Präsentation steht die Vermittlung im Vordergrund, im Falle der Exploration die Entdeckung. Im Rahmen des Workshops werden beide Thematiken behandelt.

Was der Workshop nicht bieten kann, ist die Antwort auf die Frage: Welche Visualisierung ist die beste? Wie Jung in einem Beitrag zur Visualisierung raumbezogener Daten darlegt, hängt die Effektivität einer Visualisierung mindestens von folgenden Parametern ab [3]:

- Aufgabenstellung der Visualisierung,
- verfügbare Ausgabemedien,
- spezielle Eigenschaften der Zielgruppe.

Die Aufgabenstellung beschränkt sich hierbei nicht nur auf die Gegenpole Präsentation oder Exploration. So vergleicht Jung beispielsweise die Effektivität bestimmter Darstellungen für die Aufgaben: Werte ablesen, Werte finden, Werte vergleichen und Trends erkennen, wobei je nach Aufgabe durchaus unterschiedliche Visualisierungsformen die besten Werte erreichen.

Yang et. al. haben 1993 moderne Visualisierungskomponenten als funktionsreich, aber wissensarm charakterisiert [4]. Ihre Aussage bezog sich auf GIS-Visualisierungen. Sie gilt jedoch sicher auch für andere Bereiche der Informationsvisualisierung. Der Workshop soll dazu beitragen, die bekannten Erkenntnisse für einzelne Bereiche vorzustellen, um den Teilnehmern auf der Suche nach geeigneten Visualisierungsformen zu helfen.

3.1 Geschäftsgrafiken

Als erstes Anwendungsfeld der Visualisierung von entscheidungsrelevanten Daten wird der Bereich der Geschäftsgrafiken vorgestellt. Diese gelangen insbesondere *zur Visualisierung von wesentlichen (Kenn-)Zahlen* zum Einsatz. Dabei steht jedoch nicht im Vordergrund welche Unternehmenskennzahlen darzustellen sind, sondern vielmehr an welchen Richtlinien die jeweiligen Darstellungen sich sinnvollerweise zu orientieren haben. Gerade im Bereich der Unternehmenskennzahlen findet man in heutigen entscheidungsunterstützenden Informationssystemen zwar eine Vielzahl von technischen Möglichkeiten, deren Anwendung erfolgt aber Vielfach unter Vernachlässigung elementarer Gestaltungsprinzipien aus dem Bereich der Visualisierung (z.B. inadäquater Einsatz von 3D, von Farbe, von Grafiktypen). Bei der Untersuchung einer adäquaten Visualisierung von Kennzahlen bewegt man sich dabei in unterschiedlichen Wissensgebieten, wie beispielsweise der Wahrnehmungspsychologie, der Kognitionspsychologie, den Arbeiten aus dem Bereich Human Factors und der Software-Ergonomie (Human-Computer Interaction), aber auch der Visualisierung statistischer Daten. Aus der Agglomeration der genannten Wissensgebiete ergeben sich Forderungen für die Visualisierung von Geschäftsgrafiken. Als Folge der Auseinandersetzung mit den genannten Forschungsgebieten, können Gestaltungsprinzipien für die Darstellung von Geschäftsgrafiken abgeleitet werden:

- Entlastung des Arbeitsgedächtnisses [5]
- automatisierte Informationsverarbeitung [5]
- diagramminterne Auswertungssteuerung
- wahrnehmungsgerechte Datendarstellung [6]
- augenphysiologisch basierte Farbgebung
- kombinierte Darstellung von Diagramm und Tabelle [7]

Die zu beachtenden Gestaltungsprinzipien werden anhand konkreter Praxisbeispiele vorgestellt (z.B.„Briefing Book").

3.2 GIS-Komponenten

Als weiteres Anwendungsfeld der Visualisierung von entscheidungsrelevanten Daten soll die Präsentation von Daten mit Raumbezug diskutiert werden. In der Literatur finden sich umfassende Diskussionen über die spezifischen Charakteristika von Geografischen Informationssystemen (GIS) in Abgrenzung beispielsweise zu reinen Kartendarstellungssystemen. Als Arbeitsdefinition wird im Rahmen des Workshops der Begriff „GIS-Komponente" als Oberbegriff verwendet, für alle Komponenten zur Visualisierung von Daten mit räumlichem Bezug.

GIS-Komponenten in Verbindung mit Management Support Systemen werden heute beispielsweise für Standortplanungen, Markt- und Konkurrenzbeobachtungen, Risikoanalysen oder auch visuelles Data Mining eingesetzt. Die Anbieter von GIS-Komponenten verweisen auf die Vorteile und neuen Möglichkeiten, die durch den Einsatz dieser Visualisierungstools für den Benutzer entstehen können. Im Rahmen des Workshops werden empirische Untersuchungen präsentiert, die diese Vorteile erforschen [8-11]. Anschließend werden verfügbare GIS-Komponenten und deren praktische Anwendung im Rahmen der Entscheidungsunterstützung vorgestellt.

3.3 Externe Informationen

Der Visualisierung unternehmensexterner Informationen, wie sie auch im WWW zu finden sind, kommt im Rahmen von entscheidungsunterstützenden Systemen eine immer wichtigere Rolle zu. Gerade strategische Unternehmensentscheidungen basieren vielfach auf externen Informationen [12]. Dabei werden unternehmensexterne in Abgrenzung zu unternehmensinternen Informationen gesehen. Unternehmensinterne Informationen sind Informationen, die in den einzelnen Unternehmensbereichen anfallen und in einem direkten Zusammenhang mit dem Zweck des Unternehmens stehen [13]. Im Gegensatz dazu liegen die Quellen der unternehmensexternen Informationen außerhalb der jeweiligen unternehmerischen Wertschöpfungskette, haben aber einen direkten Einfluß auf diese. Ein Beispiel soll dies verdeutlichen: Ein Teil der Wertschöpfungskette eines Unternehmens sei der Einkauf von Rohstoffen. Der Einfluß auf die Wertschöpfungskette sei der jeweilige Rohstoffpreis, der beispielsweise durch die politische Lage innerhalb des rohstoffliefernden Landes beeinflußt werden könnte. Dieses Beispiel macht zweierlei Dinge deutlich:

- Externe Informationen (z.B. Rohstoffpreis, politische Lage) können für Unternehmen eine immense Bedeutung haben.

- Externe Informationen müssen im Bezug zu internen stehen, um Handlungsrelevanz zu erlangen.

Bei unternehmensexternen Informationen handelt es sich oft um unstrukturierte Daten, wie beispielsweise verbale Unternehmensanalysen, Pressemitteilungen, Produktankündigungen oder auch Mitteilungen von Regierungen [14]. Die adäquate Visualisierung derartiger Daten setzt das Finden geeigneter Metaphern voraus. Metaphern werden dabei eingesetzt, um den Anwender durch vertraute Zusammenhänge und Abläufe auf neue Situationen einzustimmen [5]. Im Bereich der externen Informationen ist es eine wesentliche Forderung, daß unternehmensexterne im gleichen Kontext wie die korrespondierenden internen Informationen dargestellt werden [15]. Ist dieses nicht der Fall, so ist es wahrscheinlich, daß der Anwender dem „Browsingsyndrom" erliegt, und das eigentliche Ziel, die Erarbeitung, der für ihn entscheidungsrelevante Information, aus den Augen verliert. Als Folgerung ergibt sich, daß die zu wählende Metapher diese Kontextunterstützung bieten muß. Bezieht man die Überlegung mit ein, daß es sich bei der Zielgruppe Entscheidungsträger meist um Personen des mittleren bis Topmanagement handelt, so rückt die Intuitivität der Bedienung mehr und mehr in den Vordergrund und somit auch die Vertrautheit mit der einzusetzenden Metapher. Beispiele für derartige Metaphern sind beispielsweise die Zeitungsmetapher, die Cockpit Metapher oder die Schreibtisch Metapher.

Eine in diesem Kontext nicht zu vernachlässigende Voraussetzung für die adäquate Visualisierung ist die Klärung des Anwendungsgebietes. So sind externe Informationen über Mitbewerber (bspw. zur Marktanalyse) sicherlich anders zu visualisieren, als Informationen über neue Technologien. Dies ist durch die unterschiedlichen Zielgruppen und den damit verbundenen unterschiedlichen Aufgabenstellungen zu begründen. Gerade vor diesem Hintergrund erscheint der Ansatz hybrider Metaphern erfolgversprechend. Welche Metapher in welcher Gestaltung letztendlich zum Einsatz kommt, kann jedoch nur durch die intensive Kooperation mit den Anwendern, festgelegt werden.

Nutzung externer Informationen impliziert, daß diese Information vorab gefunden werden muß. Gerade die Popularisierung des Internet mit seinen beeindruckenden Möglichkeiten hat vielen Anwendern die Grundproblematik des Information Retrieval praktisch vor Augen geführt. Suchprobleme sind iterativer Natur: Der Kenntnisstand verändert sich bereits während der Suche und definiert vielfach das ursprüngliche Problem neu. Visualisierung externer Informationen und die Anwendung von Visualisierungstechniken zur Unterstützung der Recherche nach externen Informationen sind von daher kaum voneinander zu trennen.

Daß die Anwendung von Information Retrieval-Ansätzen nicht nur zur Information führt, sondern selbst Information generiert, wird gegenwärtig unter den Bezeichnungen Text Mining oder Web Mining zum Thema gemacht. Auch hier spielt Visualisierung eine wesentliche und zweifache Rolle: Als Schnittstellentechnik, um die Ergebnisse der Analysen zu vermitteln oder aber um durch die Wahl und Darstellung eines geeigneten Informationsraumes den Benutzer selbst mit seinen Fähigkeiten zur räumlichen Wahrnehmung als "Data Miner" einzusetzen.

4　Ausblick

Als Ausblick werden im Rahmen des Workshops einige ambitionierte Visualisierungsansätze aus dem Forschungsbereich präsentiert (z.B. WebBook, WebForager). Dies soll den Teilnehmern Denkanstöße liefern, wie man herkömmliche Formen der Visualisierung entscheidungsrelevanter Daten erweitern oder ergänzen könnte.

5　Literatur

[1] N. Gershon, S. Eick , S. Card: Information Visualization. Interactions 2 (1998), 2, 9-15.

[2] G.G. Robertson, J.D. Mackinlay, S.K. Card: Cone Trees: Animated 3D Visualizations of Hierarchical Information. In: Proceedings ACM CHI'91 Conference. New York, 1991: ACM. 189-194.

[3] V. Jung: Eine wissensbasierte Umgebung zur Benutzerunterstützung bei der Visualisierung raumbezogener Daten. In: geoinformatik_online, (1997), 3. http://gio.uni-muenster.de/beitraege/ausg3_97/jung/paper.html (11. November 1998)

[4] J. Yang, E.M. Siekierska, D.R.F. Taylor: Concept Design and Knowledge Acquisition for the Thematic Map Design Advisory System (TMDAS). In: The Canadian Conference on GIS '93. Ottawa, 1993: The Canadian Institute of Geomatics. 475-484. Nach [2]

[5] C. Marshall, C. Nelson, M. Gardiner: Design Guidelines. In:. M.M. Gardiner, B. Christie, (Hg.): Applying cognitive psychology to user-interface design. Chichester, 1989: Wiley. 221-278.

[6] W. Cleveland, R. McGill: Graphical perpection: theory, experimentation, and application of the development of graphical methodes. Journal of the American Statistical Association 79 (1984), 531-545.

[7] I. Benbasat, A.S. Dexter, P. Todd: An experimental program investigating color-enhanced and graphical information presentation: An integration of the findings. In: Communications of the ACM, 29 (1986), 11, 1094-1105.

[8] J.B. Smelcer, E. Carmel: Do geographic information systems improve decision making? An experiment comparing maps and tables. Washington DC, 1994: Kogod College of Business Administration. http://lattanze.loyola.edu:80/lattanze/research/wp1094.023.html (11. November 1998)

[9] M.D. Crossland, B.E. Wynne, W.C. Perkins: Spatial decision support systems: An overview of technology and a test of efficacy. In: Decision Support Systems, 14 (1995), 3, 219-235.

[10] V. Jung: Experimentelle Bestimmung der Effektivität kartographischer Visualisierungsverfahren für quantitative GIS-Daten. Forschungs- und Arbeitsbericht GRIS 96-1. TH Darmstadt, Fachbereich Informatik. 1996. Als Manuskript gedruckt.

[11] B.E. Mennecke, M.D. Crossland, B. Killingsworth: An Experimental Examination of Spatial Decision Support Effectiveness: The Roles of Task Complexity and Technology. In: On-Line Conference Papers Association for Information Systems 1997 Americas Conference. Indianapolis, 1997: http://hsb.baylor.edu/ramsower/ais.ac.97/papers/mennecke.htm (11. November 1998)

[12] E. Turban, J. Aronson: Decision Support Systems and Intelligent Systems, 5th Edition. New Jersey, 1998: Prentice-Hall Int.

[13] J. Biethahn, B. Huch: Informationssysteme für das Controlling- Konzepte, Methoden und Instrumente zur Gestaltung von Controlling-Informationssystemen. Berlin u.a., 1994: Springer. 36.

[14] H. Watson, K. Rainer, C. Koh: Executive Information Systems: A Framework for Development and a Survey of Current Practices. In: H. Watson, K. Rainer, G. Houdeshel (Hg.): Executive Information Systems: Emergence, Development, Impact. New York, 1992: John Wiley & Sons, Inc., 93.

[15] R. Sprague, H.Watson: Decision Support for Management.New Jersey, 1995:Prentice-Hall, Inc. 300-318.

[16] H. Mucksch, W. Behme: Die Notwendigkeit einer entscheidungsorientierten Informationsversorgung. In: H. Mucksch, W. Behme (Hg.): Das Data Warehouse Konzept: Architektur – Datenmodelle – Anwendungen 2.Aufl.. Wiesbaden, 1997: Gabler

Adressen der Autoren

Prof. Dr. U. Bleimann
Fachhochschule Darmstadt
Fachbereich Informatik
Schöfferstr. 8b
D-64295 Darmstadt
Email: bleimann@fbi.fh-darmstadt.de

Prof. Dr. H. Reiterer, T. M. Mann, G. Mußler
Universität Konstanz,
Fakultät für Mathematik und Informatik
Fach D73
D-78457 Konstanz
Email: Harald.Reiterer@,uni-konstanz.de

Design und Evaluation von EDV-Lernumgebungen

Maria Bannert

Fachbereich Psychologie, Universität Koblenz-Landau

Vor einigen Jahren noch waren Computerarbeitsplätze typischerweise mit ein paar wenigen und vergleichsweise übersichtlichen Programmen ausgestattet. Heutzutage dominiert eine große Anzahl verschiedener „Office -" und anderer Produkte den Computerarbeitsplatz, die in den meisten Fällen um ein vielfaches komplexer und umfangreicher sind als ihre jeweiligen Vorgänger. Zwar sind die heutigen EDV-Programme häufig benutzerfreundlicher, dennoch ist ein Mindestmaß an Schulung für den Großteil der Nutzerinnen und Nutzer unabdingbare Voraussetzung, um den effektiven Gebrauch der Computerprogramme für die Bewältigung der täglichen Arbeit zu gewährleisten. Wie sollen aber gute EDV-Schulungsmaßnahmen gestaltet sein, und welche Empfehlungen lassen sich diesbezüglich aus der aktuellen instruktionspsychologischen Forschung ableiten?

Ausgehend von dieser allgemeinen Fragestellung wurden in einem Forschungsprojekt Qualifizierungsmaßnahmen für Computernutzerinnen und -nutzer aus einer instruktionspsychologischen Perspektive systematisch analysiert. Die zentrale Zielsetzung bestand darin, relevante pädagogisch-psychologische Aspekte zu identifizieren und deren Bedeutsamkeit für den Erfolg von EDV-Schulungsmaßnahmen zu untersuchen.

Zu diesem Zweck wurde auf der Basis instruktionspsychologischer Befunde exemplarisch ein Kurskonzept mit einer computerunterstützten Lernumgebung zur Einführung in das Textverarbeitungsprogramm Word für Windows 6.0 entwickelt. Dieses wurde in insgesamt 14 Kursen mit Landauer Studierenden durchgeführt und zugleich umfassend evaluiert. Dabei wurde mittels einer quasi-experimentellen Versuchsanordnung die Bedeutung verschiedener Unterstützungsformen hinsichtlich verschiedener Erfolgskriterien untersucht. Experimentell variiert wurden zum einen die Art der Unterrichtsmethode (traditionelles Vorgehen vs. Selbstlernmaterial), zum anderen die Art der Benutzerschnittstelle (Standardschnittstelle vs. Trainingsschnittstelle). Als abhängige Variablen wurden Kursdurchführungszeit, Lernerfolg und Teilnehmerzufriedenheit erfaßt.

In dem Poster-Beitrag werden die quasi-experimentelle Feldstudie und deren zentrale Ergebnisse näher dargestellt sowie ihre Implikationen für die Gestaltung künftiger EDV-Schulungsmaßnahmen erörtert. Damit soll gleichzeitig ihr Stellenwert für diesen Forschungsgegenstand in theoretischer und methodischer Hinsicht diskutiert werden.

Adresse der Autorin

Dr. phil. Maria Bannert, Dipl.-Psych.
Universität Koblenz-Landau
Fachbereich Psychologie
Abteilung Allgemeine und Pädagogische Psychologie
Im Fort 7, 76829 Landau
Email: bannert@rhrk.uni-kl.de

Hyper-Skript –
Entwicklung und Nutzung
von verteilten Multimediaskripten

Andreas Brennecke[1], Friedrich-L. Holl[2], Reinhard Keil-Slawik[1], Jörg Meier[2], Harald Selke[1]

[1]Heinz Nixdorf Institut, Universität-GH Paderborn
[2]Fachbereich Wirtschaft, Fachhochschule Brandenburg
Email: anbr@uni-paderborn.de

Im Projekt Hyper-Skript werden Verfahren und Werkzeuge zum Einsatz und zur Pflege verteilter Skripte entwickelt und erprobt. Dabei stehen die arbeitsteilige Erarbeitung und Pflege von multimedialen Unterrichtsmaterialien, ihre Integration in eine lokale oder persönliche Lehr- und Lernumgebung und die aktive Bearbeitung des Materials im Vordergrund. Neuartig ist dabei vor allem die spezifische Form der Arbeitsteilung bezüglich Entwicklung und Nutzung: statt eines Verbundes, der für andere produziert, wird ein Verbund von Produzenten aufgebaut, die zugleich auch Nutzer sind. Prototypisch wird dabei ein verteiltes Multimediaskript zu den Themenbereichen Software-Ergonomie und Gestaltung von Multimediasystemen erstellt und von den Projektpartnern am jeweiligen Standort in unterschiedlichen Lehrveranstaltungen eingesetzt.

Ein modularer Aufbau der Inhalte und die Evaluation unter alltagspraktischen Bedingungen bei Zugrundelegung unterschiedlicher Ausbildungskontexte und technischer Infrastrukturen sollen die Übertragbarkeit und Erweiterbarkeit des Ansatzes sichern. Klassische Autorensysteme basieren auf der Annahme einer standardisierten Lehreinheit, die durch einen Autor bzw. eine Autorengruppe erstellt und dann durch viele andere genutzt wird. Dies führt in der Regel zu abgeschlossenen Systemkonzeptionen, die sich schwer oder gar nicht in das spezifische lokale Lehr-/Lernarrangement einbetten lassen. Weiterhin fehlt eine explizite Kooperationsunterstützung bei der Erstellung der Materialien; die notwendige Pflege, Aktualisierung und Erweiterung langfristig bestehender Dokumentenbestände wird nicht berücksichtigt bzw. unterstützt.

Demgegenüber basiert der Ansatz der Hyper-Skripte auf der Tatsache, daß es im Universitätsalltag angesichts der Kurzlebigkeit von Wissensbeständen und aufgrund gestiegener Aktualitätserfordernisse zunehmend wichtiger ist, zum einen Materialien verschiedenster Art einfach und schnell in Lehrveranstaltungsunterlagen einzubinden und zum anderen die Bereitstellung, Pflege und Aktualisierung arbeitsteilig zu erledigen. Das Hyper-Skript ist aufgrund seiner Produktionsstruktur individualisierbar angelegt und soll den Nutzern an allen Lernorten als vollständiger elektronischer Benutzerarbeitsplatz zur Verfügung stehen, deshalb finden ausschließlich (WWW-)Standardtechnologien Verwendung.

Das Hyper-Skript enthält alle für die Durchführung der Lehrveranstaltungen notwendigen Materialien. Diese werden verteilt erstellt, indem die Autoren neben abgestimmten Lehrinhalten auch ihre unterschiedlichen Kompetenzen und Forschungsschwerpunkte einbringen. So ist es möglich, die Materialien flexibel in unterschiedlichen Veranstaltungen zu nutzen, in denen aber auch eigene nicht geteilte Ansätze vorgestellt werden. Im Hyper-Skript müssen somit je nach Erfordernis gemeinsame und individuelle Materialien verwaltet werden.

Um diesen unterschiedlichen Anforderungen gerecht zu werden, ist das Hyper-Skript aus mehreren Schichten aufgebaut.

- Die *Datenbasis* oder *Diensteebene* stellt Materialien zur Verfügung, die in sich abgeschlossene Dokumente sind oder aus Sammlungen von Animationen, Bildern o. ä. bestehen. Diese werden auf der Ebene der Basiselemente nicht nach didaktischen Gesichtspunkten aufbereitet und sollen universell eingesetzt werden können. Jedem Bereich in der Datenbasis ist eine verantwortliche Person zugeordnet, die für notwendige Änderungen, Aktualisierungen und Vollständigkeit sorgen muß. Beispielsweise enthält ein Bereich die in der Software-Ergonomie zu berücksichtigende Normen (z. B. die DIN/EN 9241) und Gesetzestexte (z. B. die Bildschirmarbeitsverordnung). Zuständig sind Autoren in den Bereichen, in denen sie ohnehin das aktuelle Wissen aufbereiten müssen. Im Gegenzug können sie selbst von den Bereichen der anderen Autoren profitieren.

- In den *gemeinsamen Verbundobjekten* werden die Materialien von den Autoren kooperativ strukturiert und nach didaktischen Gesichtspunkten aufbereitet. Diese Verbundobjekte stellen *Lernzieleinheiten* dar, deren Inhalt und didaktische Konzeption somit von allen beteiligten Autoren getragen wird. Die Elemente in dieser Ebene sind somit verbindlich, abgestimmt und genehmigt. Die Elemente können Kompositionen von Basiselementen, Dokumente mit neuem Inhalt, die Bezüge zwischen Basiselementen herstellen (Zusammenfassende Texte, Grafische Übersichten) oder auch Foliensätze für die Vorlesungen sein.

- Die Materialien einer *individuellen Lehrveranstaltung* unterliegen letztendlich der Freiheit aber auch der lokalen Verantwortung des jeweiligen Dozenten. Persönliche Forschungsschwerpunkte, eigene Ansätze oder ein anderes didaktisches Vorgehen erfordern teilweise den Einsatz weiterer persönlicher Materialien. Diese sind nicht an die anderen beiden Ebenen des Hyper-Skriptes gebunden, sollten aber nach Möglichkeit die gemeinsamen Materialien nutzen.

Als Vorteile der ergeben sich für die Autoren und gleichzeitigen Nutzer eines Hyper-Skriptes, daß

- durch die *Kooperation* beim Erstellen gemeinsam genutzter Lehreinheiten, zur Vermittlung spezifischer Lernziele unter Berücksichtigung geeigneter didaktischer Orientierungen, die Qualität der Materialien ansteigt. Das Erreichen der definierten Lernziele wird von allen Kooperationspartnern somit in verschiedenen Lehrveranstaltungen überprüft und findet Eingang in die Weiterentwicklung der Materialien.

durch *Arbeitsteilung* der Aufwand für die jeweils Beteiligten sinkt, da bestimmte Teile durch andere gepflegt werden.

Für den Entwicklungs- und Überarbeitungsprozeß sowie die langfristige Pflege und Aktualisierung der Materialien wird eine angemessene technische Unterstützung realisiert. Dafür muß die (WWW-)Umgebung um Werkzeuge erweitert werden, z. B. zur Abstimmung über gemeinsame Lernzieleinheiten oder um Änderungen in der Datenbasis mitzuteilen. Im Rahmen des Projektes soll sowohl das methodische Vorgehen zur Erstellung verteilter Skripten entwickelt werden, als auch ein Hyper-Skript zu den Themenbereichen Software-Ergonomie und Gestaltung von Multimediasystemen entstehen.

Das am 1. Oktober 1997 angelaufene Projekt wird als dreijähriges Modellvorhaben durch die Bund-Länder-Kommission und die Länder Nordrhein-Westfalen und Brandenburg gefördert.

Ein Wirkmodell für eine marktorientierte Konzeption und Produktion von Softwareprodukten

Helmut Degen

Arbeitsbereich Informationswissenschaft, Freie Universität Berlin

Bisherige Ansätze, die sich mit der Konzeption von Softwareprodukten im gestalterischen und technischen Sinne auseinandersetzen, gehen von folgenden Annahmen aus: Bei dem betroffenen Menschen handelt es sich um einen Benutzer (Annahme des Software-Engineering und der Software-Ergonomie), der Mensch hat ein vernunftbasiertes Interesse (psychologisches Paradigma der HCI-Forschung), und bei Softwareprodukten handelt es sich um digitale Werkzeuge bzw. Arbeitsmittel (Software-Ergonomie). Diese Annahmen haben Konsequenzen für das, was von den Softwareprodukten erwartet wird, also für die Produktleistungen. Diese beziehen sich auf eine gute Bedienbarkeit, die häufig durch die Anwendung von Styleguides erreicht werden soll.

Ausgehend von der Beobachtung, daß viele Produkte (z. B. Autos, Möbel, Kleidung) in unterschiedlichen Erscheinungsformen auf dem Markt angeboten werden, werden hier die Annahmen verändert: Es ist nicht mehr der Benutzer, der sich für ein Softwareprodukt interessiert, sondern allgemein ein Marktteilnehmer, ein sogenannter Verwender, dessen Verhaltensweisen auch gegenüber Softwareprodukten durch seine Vernunft (kognitiv) und durch sein Gefühl (emotional) beeinflußt werden. Weiterhin wird die Verhaltensweise des Verwender durch sein soziales Umfeld beeinflußt. Bezugspunkt ist damit nicht mehr allein der Benutzer bzw. die wahrnehmbare Gestaltung von Softwareprodukten, sondern einerseits der Marktteilnehmer, andererseits alle marktrelevanten Produktleistungen, in denen technische, aber auch gestalterische Aspekte (Erscheinungsform) eines Softwareprodukts berücksichtigt werden. Diese Produktleistungen werden in einem sogenannten Wirkmodell zusammengefaßt.

Das Wirkmodell besteht aus einer Mittel- und einer Leistungsseite. Die Mittel enthalten die Konstruktions- und Gestaltungsmittel. Durch sinnvolle, zielgerichtete Mittelkombinationen entsteht ein Softwareprodukt, das Leistungen umfaßt. Zwischen den Mitteln und den Produktleistungen besteht demnach eine Wirkbeziehung. Die Produktleistungen können in kognitiv orientierte Sachleistungen, die aus Bewirkungs-, Bedienungs-, Ökonomie-, Sicherheits-, Service- und Anwendungsleistungen, und emotional orientierten Anmutungsleistungen, die aus Empfindungs- und Antriebsleistungen bestehen, unterteilt werden. Mit dem Wirkmodell kann eine zielgruppengerechte Softwarekonzeption und -produktion erfolgen. In Tab. 1 wird am Beispiel des lebensstilorientierten Marksegmentierungsmodell der SINUS-Milieus zeigt, daß unterschiedliche Zielgruppen - hier ausgewählte Milieus - unterschiedliche Ansprüche an Softwareprodukte haben.

Die Praxistauglichkeit des Wirkmodells wird beispielhaft anhand des Gestaltungsstils für Softwareprodukte aufgezeigt. So tendieren die Stilpräferenzen des modernen bürgerlichen Milieus und des hedonistischen Milieus zu einem emotionalisierenden Gestaltungsstil, d. h. Empfindungsleistungen haben einen überdurchschnittlich hohen Stellenwert im Gegensatz zu den unterdurchschnittlich eingestuften gestalterischen Bewirkungsleistungen und Bedienungsleistungen. Softwareprodukte dieser Art werden beispielsweise von der Fa. Metacreations (SOAP, SuperGOO usw.) angeboten. Das liberal-intellektuelle Milieu und das moderne Arbeitnehmermilieu bevorzugen eine Gestaltung, die sich an einer guten Bedienbarkeit orientieren sollte, also kognitiv orientiert ist, d. h. die gestalterischen Bewirkungsleistungen und die Bedienungsleistungen haben hier einen überdurchschnittlich hohen Stellenwert im Gegensatz zu den unterdurchschnittlich eingestuften Empfindungsleistungen. Von diesen Zielgruppen wird eher ein Microsoft Windows-Stil oder vielleicht sogar ein technizistischer X-Windows-Stil bevorzugt.

Ansprüche an Produkt-leistungen (Index = 100 entspricht dem westdeutschen Durchschnittswert)	Modernes bürgerliches Milieu	Liberal-intellektuelles Milieu	Modernes Arbeitnehmer-milieu	Hedonistisches Milieu
Sachleistungen				
Gestalterische Bewirkungsleistungen	85	129	161	78
Bedienungsleistungen	95	144	146	87
Anmutungsleistungen				
Empfindungsleistungen	110	87	122	114
Bevorzugter Gestaltungsstile	Romantisch emotional	Technizistisch-nüchtern	Sachlich-nüchtern	Verspielt emotional

Tab. 1: Anspruchsprofile ausgewählter westdeutscher SINUS-Milieus (Quelle: [1])

Das Wirkmodell enthält neben Designstilen andere Produktleistungen. Insgesamt ist es mit dem Wirkmodell möglich, zielgruppenspezifische Leistungsprofile zu erstellen, die dann als Vorgabe für eine marktorientierte Softwarekonzeption und -produktion dienen können.

Folgende Konsequenzen ergeben sich für die Software-Ergonomie:

- Auffassung des Menschen nicht mehr nur als Benutzer, sondern in seinem gesamten Lebensumfeld (Mensch als Marktteilnehmer/Verwender)
- Modellierung emotional bedingter Verhaltensweisen des Menschen
- Untersuchung des Einflusses unterschiedlicher Designstile bzw. emotionalisierender Reize auf die Bedienbarkeit von Softwareprodukten

Literatur

[1] Spiegel-Verlag & Manager Magazin (Hrg.): Sonderauswertung der Online-Offline-Studie. Brief vom 2. Juni und 1. September 1998. Hamburg: Spiegel-Verlag.

Adresse des Autors

Dipl.-Inform. Helmut Degen
Freie Universität Berlin
Arbeitsbereich Informationswissenschaft
Malteserstr. 74-100
12249 Berlin
E-Mail: degen@zedat.fu-berlin.de

ProVision3D - Eine Virtual Reality Workbench zur Visualisierung geschäftsprozeßbezogener Organisationsinformationen im virtuellen Raum (Videobeitrag)

Feng Gu, Arno Mitritz, Thomas Bendig und Nicole Henseler

Fachbereich Informatik, Technische Universität Berlin

1 Einleitung

Mit diesem Videobeitrag präsentieren wir einen Prototyp von ProVision^{3D} zur integrierten dreidimensionalen Visualisierung und zur direkten Manipulation geschäftsprozeßbezogener Organisationsinformationen im virtuellen Raum. Die Architektur von ProVision^{3D} wurde in [1] beschrieben. Die adaptiven und adaptierbaren Aspekte wurden in der Workbench-Entwicklung berücksichtigt [2]. Diese Präsentation beinhaltet zwei Szenarien. Während sich das erste Szenario auf Visualisierung von Modellstrukturen und deren übergreifenden Relationen konzentriert, setzt das zweite Szenario die Darstellungsschwerpunkte auf die intuitive Darstellung von Modellelementen durch 3D-graphische Metaphern. In bezug auf das erste Szenario ergibt sich in Bild 1 ein Überblick über die Komponenten. Dargestellt in der Workbench wurden ein Geschäftsprozeß, ein Organigramm und eine Landkarte.

2 Visualisierung und Interaktion

Die Visualisierung von Geschäftsprozessen in ProVision^{3D} basiert auf der „Graph im Graph" Methode, die eine stufenweise detailliertere Darstellung einzelner Unterprozesse im Prozeßmodell ermöglicht [2]. Des weiteren wurden Animationsfunktionen zur Beschreibung dynamischer Informationsflüsse zwischen Prozeßschritten und Änderung von Leistungs-kennzahlen zwecks Simulation entwickelt. Das Organigramm wurde als Cone Tree [3] visualisiert. Die Organisationseinheiten im Organigramm wurden durch intuitive 3D-Metaphern repräsentiert. Als drittes Komponente bietet ProVision^{3D} eine Landkarte zur geographischen Positionierung realer organisatorischer Einheiten.

Zur Interaktion kann der Benutzer die Objekte im virtuellen Raum auswählen, drehen, anordnen, oder abfragen. Die drei Komponenten in der Workbench lassen sich voneinander unabhängig bewegen und skalieren, um unterschiedliche Betrachterblickwinkel zu erzeugen. Die Navigation und Interaktion erfolgt durch 3D-Ein-/Ausgabegeräte.

3 Technische Ausstattung

Der Prototyp von ProVision^{3D} läuft auf einem ONYX2 Graphikrechner. Die Interaktion mit ProVision^{3D} erfolgt durch einen Sensorstift oder eine 3D-Maus als Eingabegerät sowie eine 3D-Brille als Ausgabegerät. Die Benutzerpositionen werden durch ein Tracking System in

Echtzeit erfaßt. Implementiert wurde der Prototyp objektorientiert in C++ und unter Einsatz der Graphikbibliothek OpenGL Optimizer.

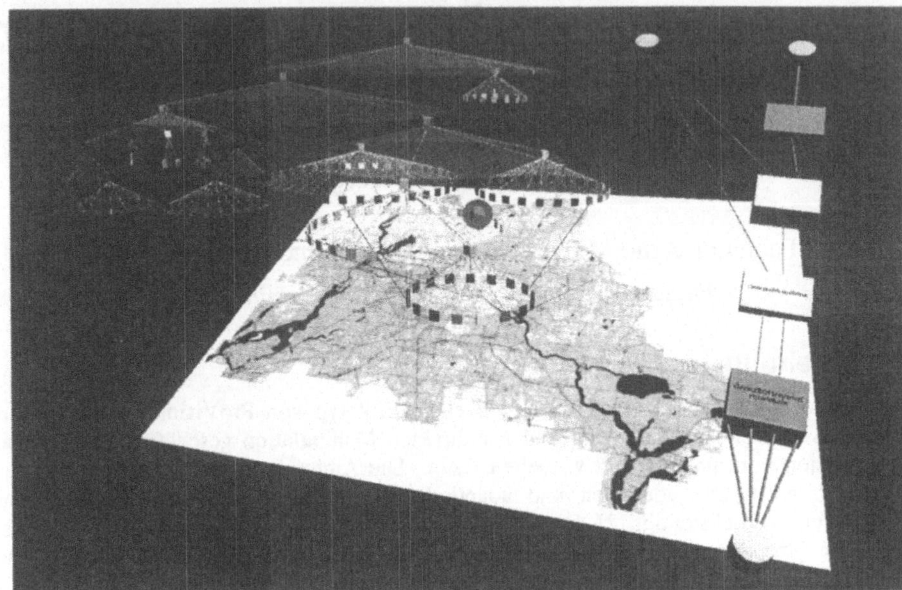

Bild 1   Prototyp von ProVision^{3D} (Ausschnitt)

4   Literatur

[1] H. Krallmann, F. Gu, A. Mitritz: ProVision^{3D} - Eine Virtual Reality Workbench zur Modellierung, Kontrolle und Steuerung von Geschäftsprozessen im virtuellen Raum. In: Zeitschrift für Wirtschaftsinformatik 40 (1998) 5

[2] F. Gu; A. Mitritz, H. Krallmann: Adaptivität und Adaptierbarkeit in einer Virtual Reality-Workbench zur Gestaltung, Kontrolle und Steuerung von betrieblichen Vorgängen. In: R. Schäfer und M. Bauer (Hg.): ABIS-97: 5. Workshop Adaptivität und Benutzermodellierung in interaktiven Softwaresystemen. Saarbrücken, 1997. 13-24

[3] G. Robertson, J. Mackinlay, S. Card: Cone Tree: Animated 3D Visualizations of Hierarchical Information. In: CHI '91 Proceedings: Human Factors in Computing Systems. New York, 1991: ACM Press. 189-194

Adressen der Autoren

Feng Gu, M.Sc.
Dipl.-Inform. Arno Mitritz
Dipl.-Inform. Thomas Bendig
Dipl.-Inform. Nicole Henseler

Technische Universität Berlin
Fachbereich Informatik
Fachgebiet Systemanalyse und EDV
Sekr. FR 6-7
Franklinstr. 28/29
D-10587 Berlin
Email: {gu | am | bendig | henseler}@cs.tu-berlin.de

Anwendbarkeit software-ergonomischer Kriterien auf Einsatz und Gestaltung von Agententechnologie

Patrick Schumacher

GMD - FIT.MMK

Helge Kahler

Institut für Informatik III
Universität Bonn

Die menschengerechte Ausgestaltung informationstechnologischer Systeme ist ein altes Ziel software-ergonomischer Forschung. Teilweise ganz ähnliche Ziele, wenn auch mit anderen Methoden, werden mit Softwareagenten verfolgt, welche in den letzten Jahren einen starken Aufschwung erleben. Es scheint aussichtsreich, beide Disziplinen nicht länger isoliert, sondern im Zusammenhang zu betrachten und anzuwenden, um mögliche Synergien zu nutzen. Auf der einen Seite setzen Agenten an Kernproblemen der Software-Ergonomie an, auf der anderen Seite könnten sie ihr durch neuartige Fragestellungen frische Impulse geben. Denn zunehmend werden Softwareagenten auf eine Weise eingesetzt, die sich mit traditionellen software-ergonomischen Zielen nicht mehr ausreichend beschreiben lassen. Die Idee ist, software-ergonomischer Grundsätze im Agentenkontext durch Auswahl, Konkretisierung und Anreicherung fortzuschreiben, so dass diese Einsatz und Gestaltung von Agentenanwendungen verbessern.

Es lässt sich erkennen, dass Agenten über ein großes Potential verfügen, software-ergonomische Grundsätze zu verwirklichen und fortzusetzen. Daraus leitet sich die Idee ab, die Software-Ergonomie um Agentenkonzepte anzureichern bzw. eine agentenbasierte Sicht auf die Software-Ergonomie zu propagieren. Davon könnte nicht nur die Software-Ergonomie, sondern noch vielmehr die Technik der Agentenentwicklung profitieren, deren Anwendungen bislang selten Benutzerbelange ausreichend berücksichtigen und software-ergonomische Erkenntnisse in der Regel vernachlässigen.

Besonders relevante Grundsätze der Software-Ergonomie und verwandter Gebiete werden im Agentenkontext zu möglichst präzise formulierten Kriterien, angereichert mit Beispielen und Erläuterungen, transformiert. Diese Kriterien bilden die Grundlage für einen „Leitfaden zur agentenbasierter Softwareentwicklung". Der Leitfaden verfolgt als übergeordnete Ziele (1) die Verbesserung der Benutzerfreundlichkeit von Agentensystemen, (2) die Umsetzung des Delegationskonzeptes und (3) den effektiven Einsatz von Softwareagenten.

Die hier vorgestellte Anwendung „bizzyB" ist ein von der GMD entwickeltes prototypisches Brokeringsystem, das an der Mailänder Handelskammer eingesetzt wird. Hier unterstützt es professionelle Informationsbroker dabei, Firmenkontakte innerhalb und außerhalb der Mailänder Region herzustellen. bizzyB dient den Brokern als integrierte Schnittstelle, um parallel in den zahlreichen Firmenadress-Datenbanken zu suchen und eine einheitliche Produktklassifikation von Unternehmen verwenden zu können. Weiterhin können auf Kunden individuell zugeschnittene Suchprofile dauerhaft gespeichert und wiederholt ausgeführt werden.

Um den potentiellen Nutzerkreis auch auf weniger professionelle Nutzer zu erweitern, muss die Komplexität der Benutzerschnittstelle reduziert sowie die Lernförderlichkeit des Systems erhöht werden. Die Realisierung erfolgt weitgehend mit Hilfe von leitfadenorientiert implementierter Agenten. Als wichtige Entwicklung möchten wir sogenannte Wizards vorstellen, die durch Dialoge dem Benutzer anhand typischer Beispielaufgaben die Funktionalität des Systems vermitteln. Somit gewinnt der lernende Benutzer einen ersten Überblick über die Fähigkeiten und Anwendungsbereiche von bizzyB. Wizards ermöglichen durch beispielhaftes Lernen die Entwicklung eines Verständnisses für Information Brokering-Systeme. Sie leiten den Benutzer schrittweise durch seine Aufgabe, bewahren ihn vor falschen oder irrelevanten Aktionen. Teile seiner Aufgaben werden dabei an den Wizard delegiert.

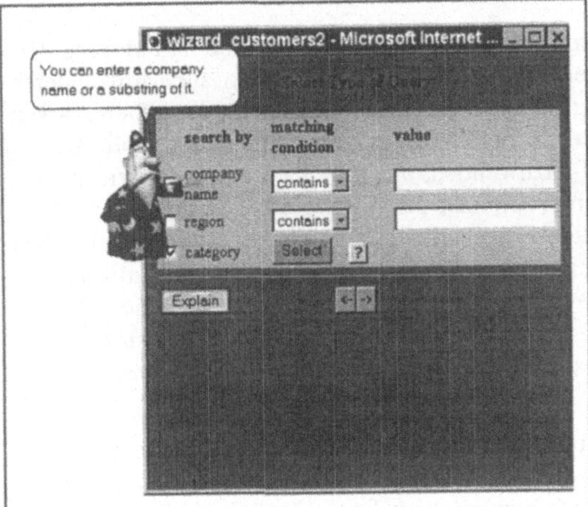

Abbildung: Der Zauberer erklärt die Bedeutung einzelner Formularelemente

Wichtigstes Ziel der Wizardkomponente ist die Erhöhung der Lernförderlichkeit des Gesamtsystems. Nebenstehende Abbildung zeigt eine Momentaufnahme des Wizards, der dem Benutzer beim Anlegen eines bestimmten Firmenprofiltyps hilft. Die Profilerstellung geschieht, wie bei Wizards typisch, auf mehreren kleinen, aufeinander folgenden Formularen. Jede dieser Formularseiten wird kurz von einem als Zauberer visualisierten Agenten erläutert.

Der Agent kann sprechen, sich bewegen und auf Bildschirmelemente deuten. Auf Wunsch des Benutzer erläutert der Agent jederzeit die aktuelle Seite und besonders erklärungsbedürftige Elemente erneut. Die Funktion des Agenten ist nicht auf bloße Erklärungen beschränkt. An verschiedenen Stellen treten Agent und Benutzer vielmehr in einen Dialog, innerhalb dessen es zu einer Arbeitsteilung kommt, sobald der Benutzer zu erledigende Arbeitsanteile an den Agenten delegiert. Weil die Formularseiten vorausgefüllt sind, ergeben sie in Kombination auch ohne Modifikation durch den Benutzer bereits eine sinnvolle Anfrage (Programming by Demonstration).

In Tests mit dem erweiterten bizzyB-System äußern sich mit dem Originalsystem vertraute Versuchspersonen durchweg positiv über die Wizards. Das Konzept, Profilbildungen in mehrere, halbautomatisierte Schritte zu zerlegen, wird - von der konkreten Reihenfolge einmal abgesehen - begrüßt. Die Einbeziehung software-ergonomischen Wissens in die Entwicklung von Agentensystemen hat also Vorteile. Designschwächen werden erkannt, neue und alternative Entwicklungsmöglichkeiten tun sich auf, was letztlich zu einer gesteigerten Gebrauchstauglichkeit der Anwendung führt.

Adresse der Autoren

Patrick Schumacher
GMD – FIT.MMK
Schloss Birlinghoven
53754 St. Augustin
patrick.schumacher@gmd.de

Helge Kahler
Universität Bonn
Institut für Informatik III
Römerstraße 164
53117 Bonn
kahler@informatik.uni-bonn.de

Double-Fading-Support als Trainingskonzept für Softwaresysteme vergleichbarer Komplexität mit unterschiedlich realisierter Benutzerschnittstelle?

Ina Weinberg und Detlev Leutner

Institut für Psychologie, PH Erfurt

Zusammenfassung

Wer als Anfänger vor der Aufgabe steht, sich in eine neue, komplexe Softwareanwendung einzuarbeiten, hat bekanntermaßen oft immense Probleme. Um solche Anfänger im Erkennen und Korrigieren ihrer Benutzerfehler in zweifacher - zunehmend zurückgenommener - Weise zu unterstützen, wurde das Double-Fading-Support-Konzept [1] entwickelt. Zu Trainingsbeginn erhält der Lernende dabei sehr detaillierte Handlungsanleitungen zur Aufgabenbearbeitung sowie ein Softwaresystem, das auf wenige grundlegende Funktionen reduziert ist. Im Trainingsverlauf werden beide Unterstützungsangebote (Support) schrittweise reduziert (Double-Fading). Zu Trainingsende erhält der Lernende schließlich nur noch sehr kurze Handlungsanleitungen und das volle Softwaresystem. Durch beide Unterstützungsangebote und deren schrittweise Zurücknahme wird das Ziel einer lang anhaltenden wechselseitigen Passung zwischen dem Lernenden und dem Softwaresystem angestrebt.

Der Trainingserfolg sollte sich maßgeblich verbessern, wenn man nach kognitionspsychologischen Theorien [z.B. 3] den Detailliertheitsgrad schriftlicher Anleitungen zur Aufgabenbearbeitung von hoch anleitend bis explorierend reduziert. Ähnliches sollte für eine Erweiterung des anfänglich stark reduzierten Funktionsumfanges zu einem vollen gelten. Für ein solches Vorgehen wurde der Training-Wheels-Ansatz von Carroll [2] weiterentwickelt. Die Ergebnisse einer empirischen Untersuchung zu einer komplexen CAD-Software zeigen die Wirksamkeit des Double-Fading-Support-Konzeptes [1]: Positive Auswirkungen hat zum einen eine im Trainingsverlauf sehr spät reduzierte Anleitungsdetailliertheit hinsichtlich des Erfolges zum Trainingsende (statistisch signifikanter Wechselwirkungseffekt) und zum anderen eine Erweiterung des anfänglich stark reduzierten Funktionsumfanges hinsichtlich des gesamten Trainingserfolges (statistisch signifikanter Haupteffekt). Beide Effekte treten zudem verstärkt bei Vpn mit geringen kognitiven Fähigkeiten auf (statistisch signifikanter Aptitude-Treatment-Interaction-Effekt). Bisher ist allerdings noch ungeklärt, ob sich diese Effekte in ähnlicher Weise auch bei Softwaresystemen vergleichbarer Komplexität, jedoch mit unterschiedlich realisierter Benutzerschnittstellen zeigen.

Zur Beantwortung dieser Frage wurden zwei Softwaresystemen mit unterschiedlicher Benutzerschnittstelle verglichen. Bei dem einen Softwaresystem (CADdy) ist die Benutzerschnittstelle gekennzeichnet durch tiefe Menüs <u>ohne</u> *icons* und bei dem anderen (AutoCAD) durch sehr breite Menüs in Kombination <u>mit</u> *icons*. Es liegen Softwaretrainings-Daten von 88 Vpn vor, die anhand folgender *between subjects*-Faktoren kodiert sind: „Benutzerschnittstelle der Software" (tiefes Menü <u>ohne</u> *icons* vs. breites Menü <u>mit</u> *icons*), „Fading des Detailliertheitsgrades der Anleitung" (im Trainingsverlauf sehr spät reduzierte Handlungsanleitung vs. konstante Handlungsanleitung mittleren Detailliertheitsgrades) und „Fading der Reduktion des Funktionsumfanges" (ja vs. nein). Innerhalb einer Woche lernten die Vpn über neun Selbstlernlektionen hinweg in ca. 30 Zeitstunden das jeweilige CAD-System zu benutzen. Zur Kodierung der Meßzeitpunkte dienten die beiden *within-subjects*-Faktoren „Aufgabentyp" (Übungsaufgabe mit Handlungsanleitung vs. Testaufgabe ohne Handlungsanleitung) und „Trainingsabschnitt" (Beginn vs. Mitte vs. Ende), bei denen jeweils drei Lektionen zusammengefaßt wurden.

Die Ergebnisse zeigen, daß erfolgreiches Lernen bei der Aufgabenbearbeitung *sowohl* von der Benutzerschnittstelle des Softwaresystems *als auch* vom Fading des Detailliertheitsgrades der Handlungsanleitung abhängig ist (statistisch signifikanter Wechselwirkungseffekt): Bei einem Softwaresystem mit sehr tiefen Menüs <u>ohne</u> icons (CADdy) verbessert eine im Trainingsverlauf sehr spät reduzierte Handlungsanleitung den Trainingserfolg. Genau umgekehrt verhält es sich bei einem Softwaresystem mit sehr breiten Menüs <u>mit</u> *icons* (AutoCAd). Bei dieser Benutzerschnittstelle verbessert sich der Trainingserfolg bei konstanten Anleitungen mittleren Detailliertheitsgrades. Der Effekt der Funktionsumfangsvariation ist ebenfalls statistisch signifikant: Die Reduktion des Funktionsumfanges wirkt sich grundsätzlich positiv auf den Trainingserfolg aus. Dabei zeigen sich tendenziell größere Vorteile bei einem Softwaresystem mit sehr tiefen Menüs <u>ohne</u> *icons* (CADdy) im Vergleich zu einem Softwaresystem mit sehr breiten Menüs <u>mit</u> *icons* (AutoCAd). Im übrigen rufen die Vpn bei AutoCAD die zu benutzenden Funktionen häufiger per *icons* und seltener per Pull-Down-Menü auf und dies speziell bei konstanter Anleitung.

Die Ergebnisse weisen darauf hin, daß bei Softwaretrainings nach dem Double-Fading-Support-Konzept die Zugriffsmöglichkeit auf *icons* eine wichtige Rolle zu spielen scheint. Dieser Frage wurde in einem weiteren Trainingsexperiment nachgegangen, bei dem das Softwaresystem selbst und die Breite der Menüs konstant gehalten und die Benutzerschnittstelle nur im Hinblick auf die Verfügbarkeit von *icons* variiert wurde (Auto-CAD <u>mit</u> *icons* vs. AutoCAD <u>ohne</u> *icons*). Die Auswertung des durchgeführten Trainingsexperimentes steht jedoch noch aus, so daß die Ergebnisse leider noch nicht berichtet werden können.

Literatur

[1] I. Weinberg: Probleme von Softwaretrainings und Lösungsansätze: Instruktionspsychologische Entwicklung und Evaluation eines Double-Fading-Support-Konzeptes für komplexe Softwaresysteme. Aachen, 1998: Shaker.
[2] J. M. Carroll: The Nurnberg funnel. Designing minimalist instruction for practical computer skill. Cambridge, 1990: MIT Press.
[3] J. R. Anderson: Rules of the mind. Hillsdale, 1993: Erlbaum.

Anmerkung

Die vorliegende Untersuchung wurde aus Mitteln der Deutschen Forschungsgemeinschaft gefördert (Kennzeichen: Le 645/3-1).

Adressen der Autoren

Ina Weinberg
PH Erfurt
Institut für Psychologie
Abt. Instruktionspsychologie
99006 Erfurt
Email: Weinberg@ipsych.ph-erfurt.de

Prof. Dr. Detlev Leutner
PH Erfurt
Institut für Psychologie
Abt. Instruktionspsychologie
99006 Erfurt
Email: Leutner@ipsych.ph-erfurt.de

Zehnder
Informations-systeme und Datenbanken

Von Prof. Dr.
Carl August Zehnder
Eidg. Technische Hochschule
Zürich

6., völlig neubearbeitete und
erweiterte Auflage. 1998.
335 Seiten. 16,2 x 22,9 cm.
(Leitfäden der Informatik)
Kart. DM 59,80
ÖS 437,– / SFr 47,–
ISBN 3-519-32480-6

Wer Informatikmittel praktisch
nutzen will, braucht dazu nicht nur
Geräte und Programme, sondern
auch Daten.

Wie diese Daten erkannt und
nutzungsgerecht in eine computergestützte Datenbank übergeführt
werden können, zeigt dieses Buch

- für die Anwendung:
  mit einer systemunabhängigen
  Entwurfsmethode, mit vielen
  Beispielen und mit Hinweisen
  zur direkten Umsetzung

- für die Ausbildung:
  mit systematischen Einführun-
  gen, mit speziellen Hintergrund-
  abschnitten und mit Übungs-
  möglichkeiten auf einem verbreiteten Datenbankverwaltungssy-
  stem (»MS-Access«)

Studierende und Praktiker finden
hier eine abgerundete und leser-
freundliche Grundlage für den
Einstieg in die Welt der Informationssysteme und Datenbanken.

Aus dem Inhalt
Leben mit Information – Daten-
bank-Grundlagen – Entwurf des
Datenschemas – Datenmanipula-
tion – Physische Datenorganisation
– Datenintegrität – Architektur
einer Datenbank – Datensysteme
auf mehreren Rechnern – Aufbau
und Betrieb von Informations-
dienstleistungen – Literaturver-
zeichnis

Preisänderungen vorbehalten.

B. G. Teubner Stuttgart · Leipzig

Kneuper/Müller-
Luschnat/Oberweis
(Hrsg.)
**Vorgehensmodelle
für die betriebliche
Anwendungs-
entwicklung**

Herausgegeben von
Dr. **Ralf Kneuper**
TLC GmbH Frankfurt/Main
Günther Müller-Luschnat
FAST e.V. München und
Prof. Dr. **Andreas Oberweis**
Johann Wolfgang Goethe-
Universität Frankfurt/Main

1998. 305 Seiten
mit 41 Bildern. 16,2 x 22,9 cm.
(Teubner-Reihe
Wirtschaftsinformatik)
Kart. DM 69,80
ÖS 510,– / SFr 63,–
ISBN 3-8154-2605-7

Vorgehensmodelle für die betriebliche Anwendungsentwicklung beantworten die Fragen: Wie, in welchen Abschnitten, mit welchen
Ergebnissen und mit welchen Personen muß ein Projekt zur Anwendungsentwicklung durchgeführt
werden?
Dieses Buch gibt einen Überblick
über den Stand von Wissenschaft
und Praxis zu dieser Thematik.
Behandelt werden u.a. Begriffe und
Geschichte des Themas, Standards
für Vorgehensmodelle, Vorgehensmodelle für verschiedene Projekttypen und Werkzeugunterstützung.
Das Buch soll sowohl dem Praktiker bei der Diskussion, Auswahl
und Erstellung von Vorgehensmodellen im Unternehmen helfen,
als auch Dozenten und Studenten
im Hauptstudium Informatik und
Wirtschaftsinformatik einen Überblick über das Fachgebiet geben.

Aus dem Inhalt
Grundlagen – Begriffliche Grundlagen für Vorgehensmodelle –
Genealogie von Entwicklungs-
schemata – Vorgehensmodelle und
ihre Formalisierung – Modellie-
rungssprachen für Vorgehensmodelle – Beschreibung von Vorgehensmodellen mit FUNSOFT-
Netzen – Vorgehensmodelle für
spezielle Projekttypen – Vorgehensmodelle für objektorientierte
Systementwicklung – Ein Vorgehensmodell für Workflow-Management-Anwendungen – Ein Vorgehensmodell für das Software
Reengineering – Vorgehensmodelle
für die Entwicklung wissensbasier-
ter Systeme – Iteratives Prozeß-
Prototyping (IPP®) – Modellgetrie-
bene Konfiguration des R/3-
Systems – Praktischer Einsatz von
Vorgehensmodellen – Werkzeugunterstützung beim Einsatz von
Vorgehensmodellen – Organisatorische Gestaltung des Einsatzes
von Vorgehensmodellen – Ein Vorgehen für das Einführen eines Vorgehens (modells) – Erste Stan-
dardaufwandsschätzung für ein
größeres Projekt mit Hilfe des
V-Modells

Preisänderungen vorbehalten.

B.G.Teubner Stuttgart · Leipzig

Haux/Lagemann/ Knaup/Schmücker/ Winter
Management von Informationssystemen

Analyse, Bewertung, Auswahl, Bereitstellung und Einführung von Informationssystemkomponenten am Beispiel von Krankenhausinformationssystemen

Von Prof. Dr. **Reinhold Haux**
Dr. **Anita Lagemann**
Dr. **Petra Knaup**
und **Paul Schmücker**
Universität Heidelberg
Prof. Dr. **Alfred Winter**
Universität Leipzig
Unter Mitarbeit von
Anke Häber
Universität Heidelberg

1998. II, 221 Seiten.
16,2 x 22,9 cm.
(Leitfäden der Informatik)
Kart. DM 48,–
ÖS 350,– / SFr 43,–
ISBN 3-519-02944-8

Eine umfassende und systematische Informationsverarbeitung wird heute in praktisch allen Unternehmen benötigt. Hierfür sind leistungsfähige Informationssysteme notwendig. Diese sind zu planen, zu steuern und zu überwachen, kurz: zu managen. Damit das Informationssystem eines Unternehmens in erster Linie ein Qualitätsfaktor für das Unternehmen ist und nicht primär als unnötiger Kostenfaktor wirkt, muß das Management von Informationssystemen systematisch betrieben werden.

Das vorliegende Buch soll auf möglichst einfache und praxisbezogene, aber dennoch gehaltvolle Weise in die Grundlagen des Managements von Informationssystemen einführen. Schwerpunkt der Darstellung ist dabei das sogenannte taktische Management von Informationssystemen. Dies beinhaltet Projekte für die Systemanalyse, -bewertung, -auswahl, -bereitstellung und -einführung, die adäquat geplant, durchgeführt und abgeschlossen werden müssen.

Aufgrund vielfältiger eigener Erfahrungen der Autoren mit dem Management von Krankenhausinformationssystemen sind die Beispiele aus diesem Anwendungsgebiet gewählt. Die vorgestellten Methoden und Aktivitäten gelten jedoch für Informationssysteme im allgemeinen.

Das Buch richtet sich an Studierende, die sich in ihrem Studium mit Informationssystemen befassen, z.B. an Studierende der Informatik, der Medizinischen Informatik, der Wirtschaftsinformatik. Die Einführung eignet sich jedoch auch für Praktiker und Wissenschaftler, die mit der Analyse, Bewertung, Auswahl, Bereitstellung und Einführung von Informationssytemkomponenten befaßt sind. Nicht zuletzt mag das Buch auch für Entscheidungsträger in Unternehmen von Interesse sein.

Preisänderungen vorbehalten.

B. G. Teubner Stuttgart · Leipzig

Berichte des German Chapter of the ACM

Band 34: Friedrich/Rödiger, Computergestützte Gruppenarbeit (CSCW)
Fachtagung vom 30. 9. bis 2. 10. 1991 in Bremen. 314 Seiten, DM 69,–/ÖS 504,–/SFr. 62,–

Band 35: Hoffmann, Eiffel
Fachtagung am 25./26. 5. 1992 in Darmstadt. 112 Seiten, DM 42,–/ÖS 307,–/SFr. 38,–

Band 36: Schweiggert, Wirtschaftlichkeit von Software-Entwicklung und -Einsatz
Fachtagung am 21./22. 9. 1992 in Ulm. 272 Seiten, DM 62,–/ÖS 453,–/SFr. 56,–

Band 37: Ludewig/Schneider, Software Engineering im Unterricht der Hochschulen SEUH '92
Workshop am 27./28. 2. 1992 in Stuttgart. 132 Seiten, DM 46,–/ÖS 336,–/SFr. 41,–

Band 38: Raasch/Bassler, Software Engineering im Unterricht der Hochschulen SEUH '93
Workshop am 25./26. 2. 1993 in Hamburg. 190 Seiten, DM 49,–/ÖS 358,–/SFr. 44,–

Band 39: Rödiger, Software-Ergonomie '93
Fachtagung vom 15. bis 17. 3. 1993 in Bremen. 330 Seiten, DM 78,–/ÖS 569,–/SFr. 70,–

Band 40: Coy/Gorny/Kopp/Skarpelis, Menschengerechte Software als Wettbewerbsfaktor
Arbeitstagung am 27./28. 1. 1993 in Bonn. 647 Seiten, DM 138,–/ÖS 1007,–/SFr. 124,–

Band 41: Züllighoven/Altmann/Doberkat, Requirements Engineering '93: Prototyping
Fachtagung vom 25. bis 27. 4. 1993 in Bonn. 383 Seiten, DM 88,–/ÖS 642,–/SFr. 79,–

Band 43: Hußmann/Paech, Software Engineering im Unterricht der Hochschulen SEUH '94
Workshop am 24./25. 2. 1994 in München. 178 Seiten, DM 54,–/ÖS 394,–/SFr. 49,–

Band 44: Spillner/Breymann, Software Engineering im Unterricht der Hochschulen SEUH '95
Workshop am 23./24. 2. 1995 in Bremen. 141 Seiten, DM 48,–/ÖS 350,–/SFr. 43,–

Band 45: Böcker, Software-Ergonomie '95
Fachtagung vom 20. bis 23. 2. 1995 in Darmstadt. 424 Seiten, DM 98,–/ÖS 715,–/SFr. 88,–

Band 46: Daldrup, Menschengerechte Softwaregestaltung
252 Seiten, DM 54,–/ÖS 394,–/SFr.49,–

Band 47: Schweiggert/Stickel, Informationstechnik und Organisation
Fachtagung am 28./29. 9.1995 in Ulm. 276 Seiten, DM 64,–/ÖS 467,–/SFr. 58,–

Band 48: Forbrig/Riedewald, Software Engineering im Unterricht der Hochschulen SEUH '97
Workshop am 27./28. 2.1997 in Rostock. 124 Seiten, DM 54,–/ÖS 394,–/SFr. 49,–

Band 49: Liskowsky / Velichkovsky / Wünschmann, Software-Ergonomie '97
Fachtagung vom 3. bis 6. 3. 1997 in Dresden. 370 Seiten, DM 108,–/ÖS 788,–/SFr. 97,–

Band 50: Sommer / Remmele / Klöckner, Interaktion im Web –
Innovative Kommunikationsformen
Fachtagung am 12./13. Mai 1998 in Marburg. 221 Seiten. DM 68,–/ÖS 496,–/SFr. 61,–

Band 51: Herrmann / Just-Hahn, Groupware und organisatorische Innovation
Fachtagung vom 28. bis 30. September 1998 in Dortmund.
370 Seiten, DM 98,–/ÖS 715,–/SFr. 88,–

Band 52: Dreher / Schulz / Weber-Wulff, Software Engineering im Unterricht
der Hochschulen SEUH '99
Workshop am 25./26. Februar 1999 in Wiesbaden. 165 Seiten, DM 64,–/ÖS 467,–/SFr. 58,–

Band 53: Arend / Eberleh / Pitschke, Software-Ergonomie '99
Fachtagung vom 8. bis 11. März 1999 in Walldorf. 397 Seiten, DM 118,–/ÖS 861,–/SFr. 106,–

Preisänderungen vorbehalten

B. G. Teubner Stuttgart · Leipzig